A Professional Guide to Audio Plug-ins and Virtual Instruments

WATFORD LRC

A Professional Guide to Audio Plug-ins and Virtual Instruments

Mike Collins

AMSTERDAM BOSTON HEIDELBERG LONDON NEW YORK OXFORD
PARIS SAN DIEGO SAN FRANCISCO SINGAPORE SYDNEY TOKYO

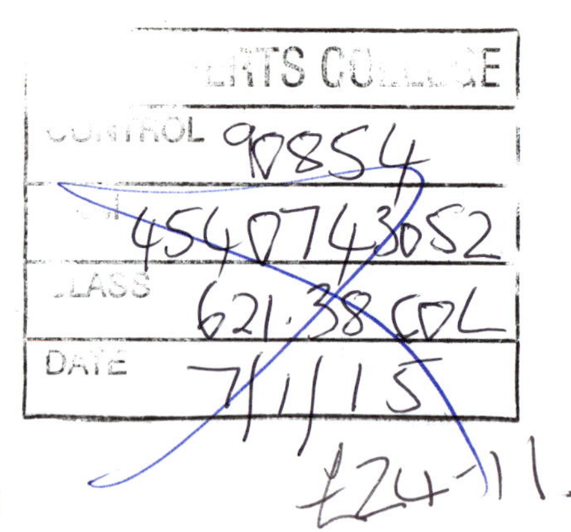

ERTS COLLEGE
CONTROL 90854
454074305Z
CLASS 621.38 COL
DATE 7/1/15
£24-11.

Focal Press
An imprint of Elsevier
Linacre House, Jordan Hill, Oxford OX2 8DP
200 Wheeler Road, Burlington MA 01803

First published 2003

Copyright © 2003, Mike Collins. All rights reserved

The right of Mike Collins to be identified as the author of this work has been asserted in accordance with the Copyright, Designs and Patents Act 1988

No part of this publication may be reproduced in any material form (including photocopying or storing in any medium by electronic means and whether or not transiently or incidentally to some other use of this publication) without the written permission of the copyright holder except in accordance with the provisions of the Copyright, Designs and Patents Act 1988 or under the terms of a licence issued by the Copyright Licensing Agency Ltd, 90 Tottenham Court Road, London, England W1T 4LP. Applications for the copyright holder's written permission to reproduce any part of this publication should be addressed to the publisher

Permissions may be sought directly from Elsevier's Science and Technology Rights Department in Oxford, UK: phone: (+44) (0) 1865 843830; fax: (+44) (0) 1865 853333; e-mail: permissions@elsevier.co.uk. You may also complete your request on-line via the Elsevier homepage (www.elsevier.com), by selecting 'Customer Support' and then 'Obtaining Permissions'

British Library Cataloguing in Publication Data
Collins, Mike
A professional guide to audio plug-ins and virtual instruments
1. Sound - Recording and reproducing - Digital techniques
2. Computer music
I. Title
621.3′893

Library of Congress Cataloguing in Publication Data
A catalogue record for this book is available from the Library of Congress

ISBN 0 240 51706 7

For information on all Focal Press publications visit our website at www.focalpress.com

Typeset by Keyword Typesetting Services
Printed and bound in Italy

Contents

About the author

Mike Collins is a music technology consultant and writer who has been making music in London's recording studios variously as a MIDI programmer, session musician, recording engineer, producer and arranger since 1981. From 1988 through to the present time, Mike has also been writing articles about music technology and the music industry in general, as well as post-production, film, broadcast, mastering and multimedia. These have appeared in a wide range of magazines around the world, including *Macworld (UK), Pro Sound News* Europe, *Sound on Sound* and *AudioMedia*, *Electronic Musician* and *MIX* in the USA, and so forth. A listing of the more than 500 articles and reviews published in more than 20 magazines is available on Mike's website at www.mikecollinsmusic.com.

Since the beginning of 2001, Mike has been writing for Focal Press and his first book, entitled *Pro Tools 5.1 for Music Production* has been available since December 2001. This book is Mike's second title for Focal and a third is planned for 2003 - about the range of professional software for music and audio applications. Other titles are at the planning stage, but will be in keeping with these.

While writing the books, Mike has continued to offer freelance Pro Tools engineering, consultancy, troubleshooting and personal tuition. This has led to presenting various seminars and lectures on Pro Tools and related music technology and audio recording topics at the Sibelius Academy in Finland, at Glamorgan University in Wales, at Staffordshire University, and at various universities and colleges in and around London.

Typical clients who have called Mike in to configure the latest software, or help them make the transition to using it, include producer Phil Harding, musician Errol Kennedy, film composer Simon Boswell, songwriter Michael Kirkham, radio DJ Allister Whitehead, and World Music studio- and label-owner John Michaelson.

Ongoing musical projects include guitar duets with top jazz guitarist Jim Mullen, more duets with ex-Soft Machine saxophonist/flautist Lyn Dobson, and a 'smooth jazz' album with Beautiful South keyboardist Damon Butcher. Mike plays acoustic and electric guitars, bass guitar and hand percussion and records using Pro Tools, Logic Audio, Cubase VST, Nuendo and Digital Performer as appropriate - and, of course, using lots of plug-ins and virtual instruments.

Background information

Mike was an active member of Re-Pro, the Guild of Recording Producers, Directors and Engineers, between 1995 and 1998, acting as Executive Consultant at policy meetings, and editing and programming the Re-Pro website.

Similarly, between 1999 and 2001, Mike worked alongside Phil Harding (Chairman of the Music Producers Guild during this period) as Head of the MPG Technical Special Interest Group. In this role, Mike organized and presented a series of technical seminars for the MPG at various London venues, dealing with issues of the day. A highly valued feature of these was the lively Q & A sessions moderated by Mike.

For example, the 'TC Bites Back' seminar at Media Tools in January 2000 started with a technical presentation by Thomas Lund of TC Electronic followed by a debate with a panel of industry professionals. The main topic was 'Uses and Abuses of the Finalizer' multi-band compressor. The panel included producer Phil Harding, Focusrite designer Rob Jenkins, recording engineer Steve Parr, Sony Studios mastering 'guru' Ray Staff, and mastering engineer Duncan Cowell. A surprising outcome for Thomas Lund was the realization that so many of the engineers present, particularly those working in post-production and mastering, were extremely wary of using the Finalizer and blamed it, often angrily, for reducing audio standards.

In July of that year, William Henshall and his team came over from the USA to demonstrate their Rocket Network system to the MPG members - 'live' at Media Tools' central-London demonstration rooms. Henshall used a PowerMac laptop running Logic Audio and overdubbed bass to a drum track created by Tim Bran in deepest Harlesden (a residential district in the heart of London) using an ADSL line. Over in Chicago, keyboardist Zoggo added a Rhodes keyboard part using a DSL line, while Jay added guitar parts from New Orleans using a 56k modem - and a couple of others contributed trumpet and keyboards using dialup connections. Interest was definitely stirring among the audience - although some said they would wait for Pro Tools to become Rocket Powered before using the system in earnest. Rocket Networking is now available in Pro Tools as well as Cubase and Logic.

In January 2001, at Ray Staff's suggestion, an invitation-only meeting was held to discuss 'Mastering and Pressing Quality Control' at Olympic Studios in Barnes. Around 60 mastering and recording engineers, producers, record company A & R personnel and staff from various pressing plants attended. At this meeting it became clear just how important liaison about technical and other issues can be if audio quality is to be maintained and expensive problems are to be avoided. In February, DTS presented a seminar to the MPG at Metropolis Studios, discussing applications of DTS encoding for DVD-Audio. The relevance of the DVD-Video format for audio releases was pointed out here – and, again, there was plenty of interest from both studio staff and independent engineers and producers.

Then, in March, a major DVD Forum event for 100+ delegates was held at Angel Studios. This was jointly organized with Dr Graham Sharpless of Disctronics and the DVD-A Association. The subject was the new DVD-Audio format. Panel speakers included Bill Foster from Understanding & Solutions, Peter Cobbin from Abbey Road, Bob Stuart from Meridian, Andy Murray from Warner Brothers, Rolf Hartley from Sonic Solutions and Andy Evans from The Pavement. The highlight of this event was the spontaneous applause for David Gray's 5.1 DVD-Video of White Ladder – which simply sounded great!

In April 2001, the 'Valves versus Plug-ins' seminar held at Planet Audio studios compared and contrasted a selection of guitar amps with Amp Farm – stimulating the debate about which is best. The Focusrite D2 and Bomb Factory LA2A and 1176 plug-ins were also compared with the 'real thing'. For this event, Mike helped to organize a group of musicians including Mike on rhythm guitar, Paul 'Wix' Wickens on keyboards, John McKenzie on bass and Shane Meehan on drums accompanying vocalist Emma Robbins with ex-Status Quo producer Pip Williams on lead guitar. Mick Glossop engineered the sessions, recording the band directly to Pro Tools, then playing back instruments through different amplifiers, speakers and effects units and comparing these played back through the software simulations. The consensus of opinion was that the plug-ins and virtual instruments acquitted themselves extremely well, with some people preferring the generally cleaner sounds and appreciating the ease of use of the software, while a minority felt that they would always prefer the hardware as they could hear the differences (while agreeing that these were small) and thus preferred the original equipment.

To complete the programme of events, in August, Rob Jenkins of Focusrite and Rob Kelly of Digidesign presented the Control|24 hardware control surface and the latest Pro Tools 5.1 software at Albert's Studios - followed by another stimulating Q & A session. The importance of using good analogue microphone pre-amplifiers was discussed, along with the practicalities of using a hardware control surface with Pro Tools rather than a conventional external mixer.

Personal statement

My involvement in organizing and presenting the various MPG Technical SIG events really brought to the foreground of my attention just how fast music and audio technologies are being developed these days. The pace of development is definitely 'hotting up', and this is particularly true when it comes to software plug-ins and virtual instruments. These include everything from 5.1 surround processing, to suites of plug-ins for recording and mastering engineers, to simulations of the Vox AC30 and Hammond Organ. If you go back even just two or three years, there were nowhere near as many of these available, and nowhere near as many people were using them. Now, 'everyone and his dog' is using plug-ins - or planning to with their next upgrade. The level of interest from MPG members at the 'Valves versus Plug-ins' and 'Control|24' events was extremely high and confirmed for me that plug-ins and virtual instruments were definitely 'coming of age' and being accepted for professional work - just as Pro Tools 5.1 had broken through the year before to become the major digital audio workstation used in professional recording. By the end of 2001, I was clear about what I wanted to do for the next nine months - namely, to write a book to tell everyone (who doesn't already know) all about this stuff. Focal Press agreed that this was a good idea - so here it is! I hope you like it.

Contact details

The author may be contacted at mike@mikecollins.plus.com, or by phone +44 (0)20 8888 5318.

Acknowledgements

I would like to thank all the manufacturers and publishers who have supported me by supplying software and hardware for review in connection with this book. It was a little like painting the Forth Bridge up in Scotland – I would just get finished and there would be a new software release or update and I would need to start over again. So there came a point where I had covered all the major plug-ins and lots of the more peripheral ones, and decided to stop. At this stage, I had had to go back and rework the entries for several plug-ins that had been substantially updated during the nine-month period in which the book was mostly written. (And, yes, before you ask, it did feel like 'giving birth' – at least metaphorically speaking!)

I would also like to thank Beth Howard and Jenny Welham at Focal Press for believing in this project and backing it so effectively. Again, thanks to Jenny for accepting my argument that not only would readers appreciate colour, but that it would be essential to allow a full appreciation of the user interfaces being described. Otherwise the marketing people would have prevailed with their view that if a more cheaply produced black-and-white book were offered at a lower price then more people would be tempted to buy it.

Edited versions of some of the plug-in overviews have previously appeared in the UK edition of *Macworld* magazine. A special thankyou goes to editor-in-chief Simon Jary, deputy editor David Fanning and all at *Macworld*.

Many thanks are due to my girlfriend Athanassia Duma, my parents, Luke and Patricia Collins, and my brother Anthony Collins – all of whose support was invaluable while writing this book. Special thanks also to my friends Barry Stoller, Damon Butcher and Jim Mullen who were so supportive when I spent some time in hospital during this period.

Introduction

'Plug-ins and virtual instruments' – so what's this all about? In case you haven't heard, there has been a quiet revolution going on over the last few years in the world of recording. Computer programmers are taking over from hardware engineers as the stars of this show and are busying themselves emulating in software what previous generations of audio engineers created using analogue and digital hardware technologies. Some new designs are emerging that are unique to the software environment, while others are aiming to replace classic hardware for music and recording. So today you can get simulations of everything from Fender guitar amplifiers and speakers, Hammond organs and Rhodes pianos, Prophet 5 and Moog synthesizers, to reverb units, delay processors, compressors, expanders, equalizers and just about any audio signal processors that exist.

Today's recording engineers, producers and musicians are increasingly working with computer-based digital audio workstations, such as Pro Tools, Digital Performer, Logic Audio, Cubase VST (Virtual Studio Technology) and Nuendo. These all provide highly integrated software environments that allow you to record, edit and mix audio and MIDI data and apply suitable signal processing and effects. Virtual instruments can also be integrated into these software environments so you can work with synthesizers, keyboards, drum-machines and samplers, using MIDI to create original parts, then record the audio outputs of these instruments to hard disk.

So how did this all start? In some ways we have the world of graphics software to thank. Companies such as Adobe came up with the brilliant idea of having a software 'plug-in' architecture that would allow not only themselves but, most importantly, third-party developers to produce additional software programs that would instantly 'plug in' to the host environment by placing the relevant files in a Plug-in Folder. Photoshop was one of the first programs to use this system to provide additional processing for graphic images. This was quickly followed by a plug-in format offering video (and audio) effects for Premiere – Adobe's video editing software. Of course, video usually has audio and it was not long before a company called Cybersound developed a range of audio-only plug-ins for Premiere. Almost immediately, the Adobe Premiere plug-in architecture was adopted by a number of audio software manufacturers and incorporated into products such as Deck (now published by BIAS, Inc.).

Digidesign were one of the first audio software manufacturers to develop their own plug-in architecture, and Waves software, based in Israel, became the first major developer of audio signal-processing software plug-ins for Digidesign systems. Waves originally developed their C1 compressor, L1 limiter, S1 stereo imager and Q10 equalizer as plug-ins for Sound Designer II software, and companies such as Antares and Arboretum quickly joined in to offer signal processors such as the Multiband Dynamics Tools and special effects generators such as Hyperprism. At this time, around 1995, emulations and simulations of actual hardware devices had not yet appeared, and the unique benefits of digital software processing, the integration with the host software, and the large screen-based user interface were being touted as the advantages of these.

Development quickly shifted to the Pro Tools TDM platform from around 1996 onwards, as Digidesign decided not to continue development of Sound Designer II in favour of concentrating all their efforts on the Pro Tools platform. Cubase VST also appeared around this time – complete with its own VST plug-ins format. Developments were initially slow, but inexorable, as more and more developers came on board. By 1997, Steinberg, for example, was developing plug-ins in both TDM and VST format, although these were relatively low-cost products offering relatively low quality at first. In 1997, Propellerhead Software developed ReBirth – a simulation of the Roland TB303 and TR808 analogue synthesizers. Although there had been previous software synthesizers, such as Digidesign's TurboSynth and others which appeared around the end of the 1980s, these had not been particularly successful commercially – possibly as they were not easy to operate and the factory sounds provided were not very exciting compared with the wide range of sounds on offer from hardware synthesizers. The great thing about ReBirth was that the sounds were excellent simulations of well-known hardware instruments and the operation of the software was similar to the way the hardware worked. A couple of years went by before other developers caught onto this new way of providing 'virtual instruments' in software, and, in the meantime, OMS compatibility was implemented in ReBirth so that it could be run on the computer at the same time as a sequencer – with its timing synchronized with that of the sequencer and with the possibility of controlling the parameters in ReBirth from the sequencer. (Opcode's Open MIDI System, OMS, is software for the Mac that provides robust MIDI input and output and inter-application communication features, with timing routines that were jointly developed with Steinberg.) Steinberg, distributors of the ReBirth software, subsequently developed a technology, jointly with Propellerheads, called ReWire. This allows the audio outputs of ReBirth to be played directly into Cubase VST audio tracks and recorded to hard disk – further broadening the possibilities for closer integration. Subsequently, ReWire compatibility was added to Logic Audio – and others are likely to follow suit.

By this time, Waves were also developing for the Premiere format, the VST format and for Mark of the Unicorn's MAS format. These formats all use the computer's CPU rather than the processors on the Pro Tools cards used by the TDM versions. Direct-X format was added to cater for PC-users, and, more recently, AudioSuite for Mac-based Pro Tools systems and Real Time AudioSuite for both Mac- and PC-based Pro Tools systems have been developed. Other developers, such as DUY in Spain and TC|Works in Hamburg followed suit, and Steinberg developed their VST 2.0 Interface to allow software synthesizers to be controlled directly from MIDI within VST-compatible host applications with the audio returning directly into the host application's mixing environment. Native Instruments in Germany, probably taking their cue from ReBirth, developed a range of virtual instruments that emulate classic keyboards such as the Hammond organ. Similarly, Waldorf developed a software version of the PPG Wave synthesizer, while BitHeadz in the USA developed their AS-1 analogue synthesizer and DS-1 digital sampler plug-ins as unique designs – not based on any actual hardware.

Over the last couple of years, many more developers have partnered with Digidesign, Mark of the Unicorn and Steinberg to offer an extremely wide range of both signal-processing and special effects plug-ins and virtual instruments – as a glance at their 'Development Partners' catalogues will show. There are also many shareware and freeware plug-ins being developed – especially for VST – and an increasing number for Direct-X on the PC. Plug-ins are also being developed for hardware platforms from Yamaha, Mackie and others. However, this book will focus on the leading software developments for use with Pro Tools, VST and MAS systems. Signal-processing plug-ins for Pro Tools TDM systems will be covered in depth. Various DSP cards on which plug-ins can be run alongside other audio hardware or 'native' audio applications will be featured in a separate chapter. Steinberg VST format plug-ins and virtual instruments, MOTU Audio System (MAS) and other similar formats are covered as well, so the book will also be of great interest to users of Steinberg Cubase VST and Nuendo, Emagic Logic Audio and MOTU Digital Performer.

What's in the book

Chapter 1, Overview, explains the differences between the various plug-in systems for Pro Tools, VST, MAS and so forth. The next four chapters focus on plug-ins and virtual instruments for Pro Tools systems - as Pro Tools is the leading system used in professional recording applications. Chapter 6 features Waves plug-ins - one of the leading suites of plug-ins available in just about every format. Chapter 7 introduces VST plug-ins, the next most popular format, followed by MAS plug-ins in Chapter 8. Chapter 9 covers virtual instruments for all formats (apart from the Pro Tools-only instruments already covered in Chapter 4). Similarly, Chapter 10 covers software samplers of various types while Chapter 11 features BitHeadz software and Chapter 12 features Propellerhead software. Chapter 13 covers miscellaneous software which doesn't fall neatly into any of these categories, but which is likely to be of interest. Chapter 14 introduces the two main DSP cards currently available which offer customised DSP processing for various plug-ins. Chapter 15 offers a brief guide to developing your own plug-ins in various formats while Chapter 16 offers a discussion of various key issues to be aware of, and includes a recommended portfolio of essential plug-ins for professional work. The Appendix lists websites for all the manufacturers mentioned in the book.

This book is aimed at a professional audience working on high-end systems - for example, recording engineers with many years of experience working with traditional studio equipment who wish to gain more understanding of the software versions of this traditional equipment, as well as to appreciate the possibilities now available with plug-ins and virtual instruments. This 'target' audience is largely Mac-based and, say, 80% are likely to be working with Digidesign hardware but may have three or four different 'host' software applications that they may be required to use at some point - or that they may wish to use as alternatives. This book is also intended as a companion volume to my first book for Focal Press, *Pro Tools for Music Production*.

This book is not intended as a replacement for the manuals that come with the products described, but many of the chapters do draw heavily on the equipment manuals - presenting the information supplied by the various manufacturers in a summarized or more efficiently organized manner. The intention is that this book will not only provide an overview of the range of products available with appropriate tips, hints and tutorial material, but that it will also serve as a very convenient reference to keep at hand when working with any of these products. In some instances the plug-ins or instruments are so easy to use that there is not much that can be said about these. In other instances, there is far more information than can be included in a book of this sort - so a compromise has had to be made. Generally speaking, in this book you will find most of what you want to know about these products to understand basically what they do. If not, you will simply have to delve into the manuals on your own. Wherever possible, I have tried to add comments or tips from my own personal experience. However, no-one in the world could be using all of these every day of the week - there are just too many. And many of these products offer similar or identical features, so you would choose one or other depending on price, quality, availability and so forth.

I would like to say a special thankyou to the legions of manual and brochure writers whose words I have had to wade through while summarizing, editing and reorganizing the information supplied by the manufacturers to present in this book. Thankfully, the level of readability is far better in these manuals than it used to be 15 years ago when I first had to deal with this type of information, and I must refer you to the original manuals for more detailed instructions and explanations about these products. Nevertheless, I believe that you will find it much easier to access much of this information here in this book, where I have been able to distil the information from the manuals, clarify many points, and, wherever possible, present the information in a more logical and friendly manner.

1 Overview

This section will introduce the concepts of plug-ins and virtual instruments, explaining how these can be categorized and how they integrate into the various host platforms.

The distinction between virtual instruments and plug-ins is subtle - in the sense that they are all software applications that 'plug in' to their host software. The term 'virtual instruments' applies to any software that acts as a musical instrument of some type. Synthesizers, samplers and drum-machines are all considered to be musical instruments, and there are increasing numbers of these available in software along with simulations of conventional instruments such as the Hammond organ, Wurlitzer piano and suchlike. The first plug-ins available were for signal processing rather than sampling or synthesis, so when most people talk about 'plug-ins' without mentioning the instruments in any way, they are probably talking about signal-processing plug-ins.

So what are the pros and cons of plug-ins and virtual instruments versus their hardware equivalents? One advantage is that it is often possible to achieve even better performance using software - most of which operates at 24, if not 32 bits these days. Compare that with the first, second and third generations of digital hardware samplers and synthesizers that used 8-bit, then 12-bit, then 16-bit architectures, and consequently suffered from noise problems - particularly with the 12- and 8-bit systems, of course. And a Yamaha DX7 cost a lot of money when it first appeared - much more than today's Native Instruments FM7, for example. Of course, many older hardware instruments are becoming increasingly hard to find, now that they are no longer manufactured - so software simulations provide a valid substitute when the real thing is not available.

Also, software can often do much more than hardware - especially if there is ongoing development of the product and if the capabilities of the computer are put to good use. For example, you can automate everything easily via the host application, and you are not stuck with just one device, either. You can usually have multiple instances of plug-ins, so you could have 6 Drawmer Compressors, four SampleCells, a bunch of Prophet 5s or FM7s, or whatever.

The way plug-ins work is very similar in all the host software environments. Typically, there is a popup selector associated with each mixer channel where you can choose the plug-in you want. Once this is 'instantiated', i.e. an instance of the plug-in is created within the host environment, you can click on the plug-in to open its window and access various controls and parameters. Non-realtime plug-ins let you select an audio region and process this using a plug-in to produce a new file that you can then bring back into your project to replace the region you just processed.

Choice of host platform is an important issue. If you want to use TDM plug-ins you will, of course, need TDM hardware - but there are options when it comes to software. If you want to use VST format plug-ins, Cubase VST is the obvious choice, but Logic Audio offers fair compatibility with these - although some VST instruments do not work correctly in some versions of Logic. MOTU have attracted almost as wide a group of third-party manufacturers to produce MOTU Audio System (MAS) plug-ins as Digidesign have persuaded to produce TDM plug-ins, with many companies now producing simultaneously for both platforms. Probably the greatest numbers of plug-ins are available in VST format, although many of these are of the 'cheap and cheerful' variety. Nevertheless, many VST plug-ins offer very high standards of performance, and plenty of users want to have access to these when using MAS or TDM systems. There are various 'VST-wrapper' plug-ins appearing that can host VST plug-ins from within MAS-compatible software, and even from within TDM software. As demand builds up, the companies producing the various plug-ins often make these available in multiple formats, so, one way or another, you can now use most plug-ins within most of the host environments.

Pro Tools is the leading digital audio system used in professional studios today and there are several different versions available currently - from the humble MBox hardware and LE software right up to the mighty new HD systems. As you might expect, there are several plug-in formats that you can use with these different 'flavours' of Pro Tools.

Steinberg are also undoubtedly a force to be reckoned with, because of the widespread use of Cubase VST and the growing numbers of professional users turning to Nuendo software. VST plug-ins also work with Logic Audio. On the Mac, Digital Performer software uses Mark of the Unicorn's MOTU Audio System plug-in format. On the PC, software plug-in formats include Steinberg's VST format, Windows DirectX format and the new DXi instrument format.

Pro Tools plug-in types

TDM

Pro Tools TDM systems include the Pro Tools HD, PT|24, and MIX systems. All of these have DSP chips on PCI cards that are used by the plug-ins to carry out the processing. This contrasts with 'native' plug-ins that run on the computer's CPU or 'powered' plug-ins that run on a dedicated DSP card such as the Universal Audio UAD-1 or TC|Works PowerCore.

Note TDM plug-ins cannot be used on the lower cost Digidesign systems that use the Pro Tools LE software, such as the Digi 001 or Mbox.

All TDM plug-ins work in real time and you insert these into Pro Tools mixer channels to provide immediate effects, rather than selecting regions and then processing these 'off-line' as with AudioSuite plug-ins. Of course, some plug-ins are extremely demanding of processing power. With these, it makes sense to bounce tracks to disk, complete with the processing, then use the processed tracks from that point on. This way you can remove the processor-intensive plug-in from your session, or use it for another track.

Pro Tools HD TDM systems

The latest Digidesign Pro Tools HD|Process systems use a very different design – so plug-ins designed to work with PT|24 and MIX systems will not work with HD systems!

This is less likely to be a problem for you if you are buying a new HD system today, particularly if it is your first Pro Tools system. However, it is a major issue for people with existing systems who want to upgrade, as it means they have to think about upgrading all their existing plug-ins if they want to transfer ongoing projects to the new hardware and have the same plug-ins in operation.

Because of the programming effort involved in porting plug-ins to the new hardware, it is almost certain that some of these will simply not 'make it' over to the

HD systems. Suppose your 'hit' vocal sound rests on that lovingly crafted patch for TC|VoiceTools? This was not available for HD systems at the time of writing (Summer 2002) and TC|Works had no current plans to port it to HD systems at that time either. So, if your production work is dependent on any particular plug-ins, make sure that that plug-in is available before taking your project to an HD-equipped studio, or you may get an unpleasant surprise!

It will be interesting to see how this situation develops - how quickly people will migrate to the new HD systems, and how many will continue to use the older MIX systems, and for how long!

AudioSuite

AudioSuite plug-ins are available for both TDM and non-TDM systems such as the Digi 001. These are non-real-time plug-ins which use the computer's CPU for processing. The way these work is that you select an audio region, apply the AudioSuite plug-in to this, preview and adjust the settings, then process the region to produce a new audio file to replace the original region. One advantage of this method is that once you have carried out this operation, the CPU is left free for other tasks. Another advantage is that it uses the computer's CPU rather than requiring TDM hardware resources. The disadvantage, of course, is that it is a non-realtime process.

RTAS

Realtime AudioSuite (RTAS) plug-ins use the host computer to carry out the processing rather than using the dedicated DSP chips on the Pro Tools TDM cards. Many TDM plug-ins are now supplied with RTAS versions that can be used when you run out of DSP power on your Pro Tools cards.

For example, say you are happily using a compressor on each new track you are overdubbing. Things are going well, and the creative juices are flowing. Then, just at the crucial moment when the guitarist is about to lay down the ultimate solo, you try inserting another TDM compressor and you get the warning message that there are insufficient DSP resources on your TDM card(s). Now if there is an RTAS equivalent of this compressor, no problem, just use this instead and the chances are that there is still plenty of DSP power available using your computer's CPU instead of the TDM card(s). And even if your favourite compressor is not yet available in RTAS, there is probably one that will do the job. So, RTAS is a very welcome format for running plug-ins within Pro Tools TDM systems.

Bear in mind, however, that there is also a limit to the processing power available in your computer. This means that you are trading off between how many tracks you are using, the density of your edits, the amount of mix automation you are using and the number of host-based plug-ins you are using.

HTDM

HTDM stands for 'Host TDM' - a hybrid of TDM and RTAS technologies. HTDM plug-ins, as with RTAS plug-ins provide all the functionality of standard TDM plug-ins but all the processing is done using the host CPU instead of using the DSP chips on your TDM hardware.

Both RTAS and HTDM plug-ins can be automated using standard Pro Tools automation, and both ProControl and Control|24 are supported for use with these.

HTDM and RTAS plug-ins can be used with Pro Tools Mix (DAE 5.1 on) and Pro Tools HD (DAE 5.3 on). The RTAS plug-ins can also be used with Pro Tools LE (DAE 5.01 on) and Pro Tools Free and can be used with all types of tracks.

Note With DAE version 5.1.1 you should not mix RTAS and HTDM plug-ins in the same session. If RTAS and HTDM plug-ins are used at the same time, occasional clicks may occur.

The first HTDM plug-ins and 'virtual instruments' for RTAS are available from Native Instruments and include the B4 Organ, the PRO-52 vintage synth, and the Battery drum sampler bundled as the Native Instruments Studio Collection Pro Tools Edition (PTE) - along with the NI-Spektral Delay PTE as an RTAS/HTDM plug-in. These plug-ins use the native power of the host computer and work as RTAS/HTDM plug-ins on Pro Tools TDM and as RTAS plug-ins on Pro Tools LE systems.

HTDM plug-ins appear in the list of TDM plug-ins and can be inserted into audio, aux or master tracks just like normal TDM plug-ins - but they actually run on the computer's CPU rather than on the DSP chips on your Pro Tools TDM cards. They can be inserted before, in-between or after any normal TDM plug-ins you are using - unlike RTAS plug-ins which can only be used with audio tracks and which must be inserted before any TDM plug-ins you are using.

Although the processing is carried out in the host computer's CPU, HTDM plug-ins do need to use an SRAM chip on a Pro Tools TDM card to transfer the audio data between the host and the TDM bus. Up to 16 different HTDM stereo (or 32 mono) plug-ins can be used on one SRAM chip. The audio connection is actually made using DirectConnect, although this is handled 'invisibly' to the user, so if any DirectConnect plug-ins are used in the session, these have to be taken into account as well. RTAS plug-ins, on the other hand, do not occupy any DSP chips on the Pro Tools TDM cards.

Note There is also the question of voice allocation to consider. RTAS plug-ins used on audio tracks will occupy one Pro Tools voice, or two voices in the case of a stereo plug-in such as those from Native Instruments, while HTDM plug-ins don't use any voices when inserted onto an audio track.

Now with DirectConnect plug-ins, you normally have to make sure that OMS is correctly configured for your plug-in, typically using the OMS IAC bus to send MIDI data from Pro Tools to the plug-in. This is no longer the case with the HTDM plug-ins. For example, when you insert any of the Native Instruments (NI) plug-ins into Pro Tools, the instrument will automatically appear as a MIDI destination in any MIDI tracks you are using in your session.

Another great feature of this new format (compared with using the standard DirectConnect format for linking virtual instruments into Pro Tools) is that it provides total recall of the plug-in settings. When you save your Pro Tools session, all the plug-in parameter settings, including the entire preset bank, are saved. So, for example, with the NI Pro-52 all the settings of its 512 presets are recalled when you start the next session. With NI Battery, all the samples you had loaded come back up. Even the positions of all the plug-in's knobs and faders come back exactly as you left them. This all helps immensely when you are loading a session onto another system, of course. For instance, you could create a session on a laptop using Pro Tools Free and load it onto a Mix system later - without any problems. You would find any changes you made to the plug-in's presets on the laptop waiting for you on the Mix system. So you wouldn't have to worry whether you'd have the same Pro-52 preset bank on the other system - as you would with a VST plug-in.

One thing to be aware of when using RTAS or HTDM plug-ins is the latency delay associated with the different types. For example, the audio buffer used with an RTAS plug-in is 32 samples. A buffer of 32 samples at a sample rate of 44.1 kHz produces a latency delay of 0.6 ms, which you are unlikely to be able to detect. On the other hand, the audio buffer used with HTDM is either 128 or 512 samples, depending on which version of the Digidesign StreamManager is installed in the Extensions folder within the System folder. Two alternative versions of this extension are supplied with the Native Instruments HTDM plug-ins - one that uses a 128-sample buffer, the other that uses a 512-sample buffer. A buffer of 128 samples at a sample rate of 44.1 kHz produces a latency delay of 2.8 ms, which is difficult to perceive, and a buffer of 512 samples produces a latency delay of 11.6 ms which is much easier to hear, and can be quite off-putting when playing a virtual instrument from your MIDI keyboard.

Tip If you want to work with minimum latency, it is quite clear from the above that you should use the RTAS plug-ins rather than the HTDM plug-ins.

MOTU Audio System (MAS) plug-ins for Digital Performer

Mark of the Unicorn have developed their own audio system (the MOTU Audio System, or MAS) which works on the built-in audio hardware available on Apple Mac computers, on many popular third-party audio cards, or on MOTU's own hardware systems - which provide a viable, lower-cost alternative to Pro Tools hardware systems.

MOTU's Digital Performer software, naturally, uses the MOTU Audio System software, but can also use Pro Tools TDM hardware, either via MAS or via Digidesign's DAE software. The MAS software will actually let you use Pro Tools TDM hardware for basic audio input and output - but MAS does not provide access to the TDM plug-ins. Alternatively, Digital Performer can use Digidesign's DAE software with Pro Tools TDM hardware to allow full access to TDM plug-ins. What would be much better is if Digital Performer would allow you to use both TDM and MAS audio tracks at the same time. After all, Logic Audio can do this trick!

Digital Performer comes with a basic set of MAS plug-ins, and these will certainly get you going if you are on a budget. Professional users will quickly find that they will need to buy additional high-quality plug-ins. Most of the third-party developers making TDM plug-ins are now developing MAS plug-ins as well - and quite a few VST plug-ins are also available for MAS.

VST plug-ins

Steinberg's Cubase and Nuendo software packages come with a basic set of Steinberg VST plug-ins, and Steinberg and other manufacturers offer a wide range of additional plug-ins and virtual instruments in VST format. Some VST plug-ins are available in TDM and/or MAS formats as well - although many are not.

Steinberg has developed the ASIO software interface for use with a wide range of popular audio hardware - including basic input and output via Digidesign hardware using the ASIO Digidesign DirectIO drivers, but there is no TDM compatibility. Neither Cubase VST nor Nuendo can use Digidesign's DAE for audio input and output, so they cannot access TDM plug-ins even when you have Pro Tools TDM hardware installed.

Logic Audio plug-ins

Emagic's Logic Audio software, like Digital Performer and Cubase VST, comes with a basic complement of proprietary plug-ins. Like Digital Performer, Logic Audio has a good selection of standard plug-ins, but for anything really special you will have to look elsewhere.

Logic Audio can also use VST-format plug-ins, and it works with both ASIO and DAE, and various other systems, for interfacing with the widest possible range of hardware.

Uniquely, Logic Audio can use DAE at the same time as it uses ASIO or other software interfaces. So Logic can use TDM plug-ins on selected tracks/channels while using VST plug-ins on others.

Windows plug-ins

On the Windows platform, there are two main types of plug-ins - DirectX and VST. As on the Mac, it depends on which host software you are using as to which of these formats you will use. Similarly, the various host software packages typically include a set of proprietary built-in plug-ins. CoolEdit Pro has an excellent selection of its own plug-ins, as does Sonic Foundry. Cubase VST and Logic Audio use the VST format while Cakewalk Sonar uses DirectX for signal-processing plug-ins and uses the DX instrument format for software synthesizers. Some software packages support all these formats. For example, Steinberg's Wavelab can use VST or DirectX or DX plug-ins.

Just as the plug-in manufacturers are all rushing to make their plug-ins available in all formats on the Mac, the same trend is observable on the PC. As far as features are concerned, whether you are concerned with the user interface or the resulting sound quality, there is little or nothing to choose between Mac and PC versions of these plug-ins. The crucial issue really is whether the plug-ins you want to use are available for your host software or not. If not, you may want to consider changing your host software - or even changing computer platform if necessary!

Managing plug-ins

If you have a lot of plug-ins, you will quickly realize that it can be a good idea to temporarily de-install any that you are not working with on your current project. The reasons for this are simple: it will take you much longer to launch your host software application if a long list of plug-ins has to be loaded. When you want to select a plug-in for use within your application, you will have to look through a long list - which will inevitably slow you down. Also, you need to allocate more RAM to your application (and/or to the DAE if you are using this) for each additional plug-in you load - which can be a problem if you have limited RAM.

A good way to handle this is to remove the plug-ins you are not using and place them in a nearby folder called 'Unused Plug-ins' or something similar. For example, Waves provide a folder of this name inside the Waves Plug-ins folder that comes with their bundled software.

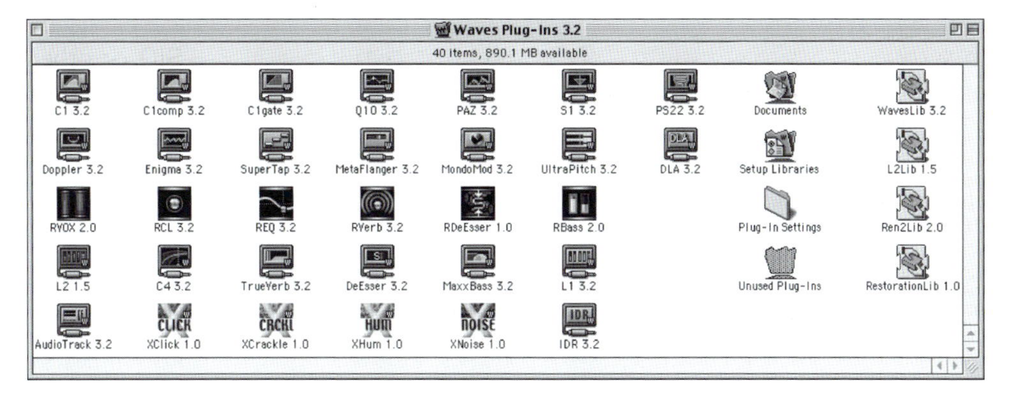

Figure 1.1 Waves Plug-ins folder.

Just drag the plug-ins you are not using into this Unused Plug-ins folder and you will speed up your loading time, reduce your RAM requirement, and make your plug-ins menus manageably shorter.

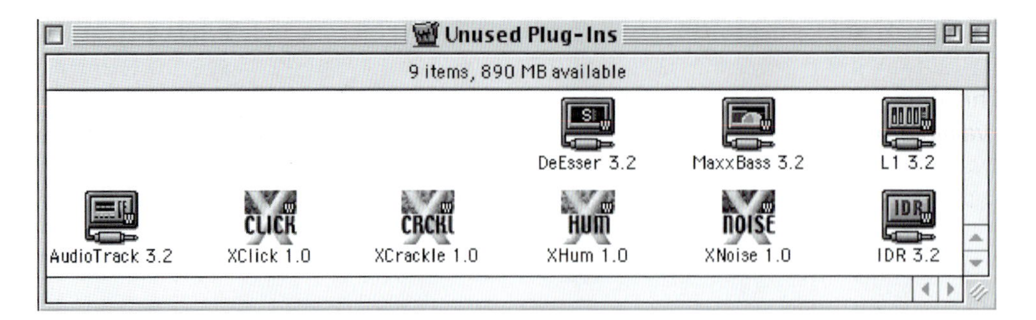

Figure 1.2 Waves Unused Plug-ins folder.

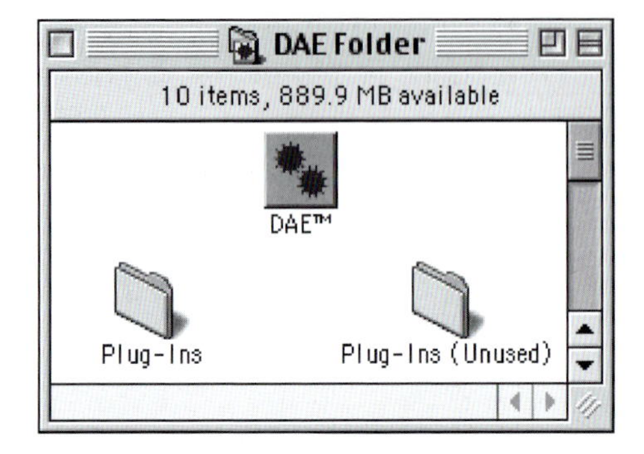

Figure 1.3 Digidesign DAE folder.

Similarly, the Digidesign Audio Engine folder that you will find inside your System Folder has a Plug-ins (Unused) folder set up ready for this purpose.

With Cubase VST, Nuendo, Logic Audio and Digital Performer, if there is not already a folder with this name, you can simply create a new folder within the application's folder, name it Plug-ins (Unused) or something similar, and drag any plug-ins that you want to de-install to this folder.

Freeing up your DSP resources

Once you have recorded a MIDI track using a virtual instrument, or if you are using an audio plug-in with an audio track, you can always use the Bounce to Disk feature in Pro Tools or the Export Audio feature in Cubase VST - or the equivalent feature in whichever host software you are using - to turn this into a new audio file. Mute all the other tracks in your song so that just the track (or tracks) you want to bounce are playing and so that you are using all the resources of your CPU to create the new bounced track (or tracks) at maximum quality. Once this is complete, you can import this back into the host software and use it instead of the original version. The important thing here is that you can now remove the virtual instrument or the audio plug-in that you were using to free up your DSP resources - and you can still keep your original tracks in case you want to change these later.

In summary

This book covers virtually all the major plug-ins and virtual instruments that were available up to August 2002. Two major developments that are not covered are any plug-ins specifically developed for Pro Tools HD systems that were not available on any previous systems and any plug-ins specifically developed for Mac OSX that are not available for other formats. At the time of writing, there were few (if any) plug-ins for which this is the case - so this book covers the bulk of what is 'out there'.

The chances are that some of the plug-ins covered in this book will never be made available for Pro Tools HD systems, or for Mac OSX. Nevertheless, it is a fair bet that the most important ones will - and many people will be running older hardware and older software for years to come. So this book will remain relevant, despite changes in hardware and software and in the host environments, for quite some time. After all, a Prophet 5 or a Mini-Moog, or an LA-2A or an 1176 simulation is not really ever going to change too much - especially as far as the user interface is concerned - as these are pretty much 'set in stone' by the original hardware. Other plug-ins, such as the Unity Session, are undergoing rapid ongoing development, while yet others are being dreamt up and coded for the first time right now.

Future revisions of this book will keep abreast of new developments, while continuing to offer a comprehensive guide to the majority of what is available.

2 Optional Digidesign Plug-ins

Apart from those supplied with the Pro Tools software, Digidesign also offer several additional plug-ins as options, including the Realverb and D-Verb reverbs, the DPP-1 pitch-processor, DIN-R noise reduction, the Maxim limiter, Bruno and Reso for synthesizer effects, Sound Replacer for sample-replacement, D-Fi for creating low-bandwidth 'grunge' effects and D-FX for Audio Suite. Some of these are great for everyday use, while others will only be used on occasion to create special effects. It all depends on the kind of projects you are working on.

My aim here is to provide useful overviews of these, simplifying and clarifying some of the information provided in the manuals for these plug-ins, and drawing this information together so that you can conveniently 'get a handle on' what each plug-in is like: what each can be used for, what unique design features are incorporated, what kind of controls are available, what presets are available, and how well they work.

Tip When TDM plug-ins are used as pre-fader inserts, their input levels are not affected by a track's volume fader - except when used on a master fader. Many of these plug-ins have no input gain controls, so clipping will sometimes occur on tracks recorded at high amplitudes. You should watch the on-screen metering closely to spot clipping if it occurs. To avoid clipping, don't insert the plug-in directly onto an audio track. Instead, insert the plug-in on an auxiliary input and attenuate the level of audio routed to it using the faders in the Send Output windows for each track you are routing to the plug-in.

Reverb One

Reverb One is the 'flagship' reverb available from the Digidesign range - with the friendliest user interface, the highest-quality reverb sounds, and a very usable set of presets.

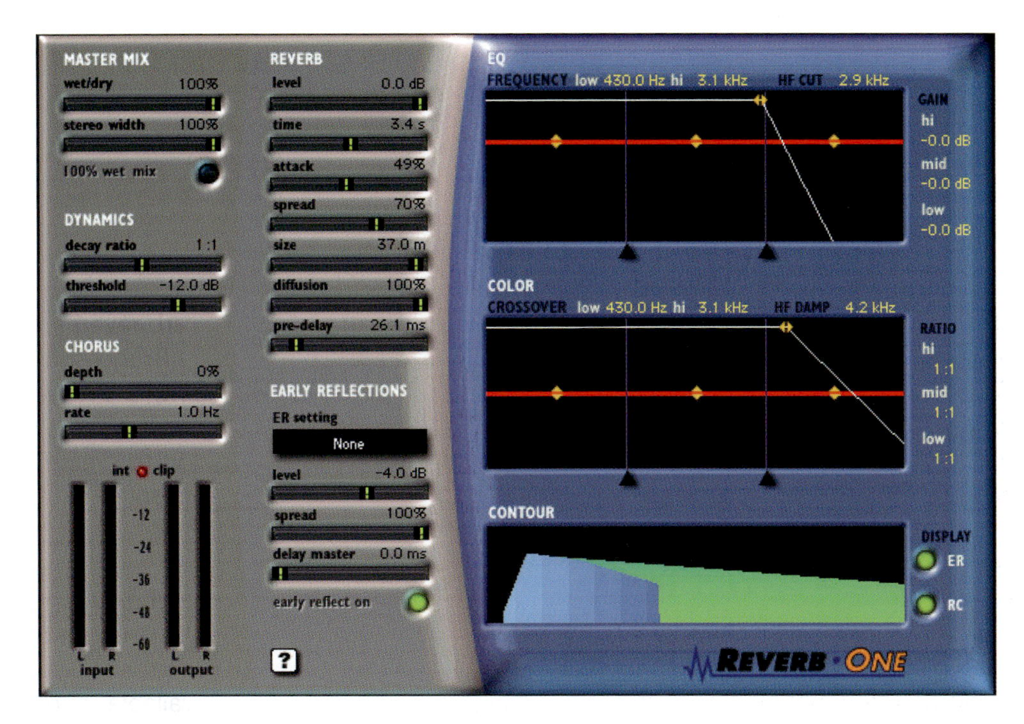

Figure 2.1 Digidesign Reverb One.

The Reverb section has controls for the various reverb tail parameters, including level, time, attack, spread, size, diffusion and pre-delay. These determine the overall character of the reverb. The Early Reflections section has further controls for the various early reflection parameters, including ER setting, level, spread and delay. Dynamics and Chorus controls let you refine the reverb decay and reverb effects. The Chorus section is straightforward enough, letting you set the depth and rate of the chorusing effect that you can use to thicken and animate the sound of the reverb tail. The Dynamics section demands a more detailed explanation, as this is not available as a standard feature on every reverberation device.

With Reverb One, the dynamics controls can be used to give a reverb a shorter decay time when the input signal is above a particular threshold, and a longer decay time when the input level drops below that threshold. This produces a longer, more lush reverb tail and greater ambience between pauses in the source audio, and a shorter, clearer reverb tail in sections without pauses. On a vocal track, for example, you can use Dynamics to make the reverb effect tight, clear and intelligible during busy sections of the vocal (where the signal is above the Threshold setting), and then 'bloom' or lengthen at the end of a phrase (where the signal falls below the threshold). Similarly, Dynamics can be used on drums tracks to mimic classic gated reverb effects by causing the decay time to cut off quickly when the input level is below the threshold.

The Decay Ratio parameter lets you set the ratio by which the reverb time is increased when a signal is above or below the threshold level. Dynamics behaviour differs when the Decay Ratio is set above or below 1. A ratio setting of greater than 1 increases reverb time when the signal is above the threshold. A ratio setting of less than 1 increases a reverb's time when the signal is below the threshold. For example, if Decay Ratio is set to 4, the reverb time is increased by a factor of 4 when the signal is above the threshold level. If the ratio is 0.25, reverb time is increased by a factor of 4 when the signal is below the threshold level. The Threshold parameter lets you set the input level above or below which the reverb decay time will be modified.

Plenty of visual feedback is provided with individual graph displays for Reverb EQ and Color and a third display showing the Reverb Contour. The graphs can be edited directly within the displays and you can use the EQ graph to control the 3-band equalizer to shape the tonal spectrum of your reverb. The Reverb Color graph lets you set decay times for the different frequency bands for additional tonal shaping. The Reverb Contour graph displays the envelope of the reverb, displaying early reflections and reverb tail in different colours to give you lots of useful feedback when setting these parameters. Finally, when you have adjusted all your settings, the Master Mix section lets you balance the relative levels of the source signal and the reverb effect and control the width of the reverb effect in the stereo field.

With Reverb One all the presets are of high quality and instantly usable - with everything from large arenas and cathedrals to halls, theatres and small wood rooms plus a fair selection of special effects such as gated reverbs and imaginary spaces.

In my opinion, Reverb One competes very favourably with the Lexiverb plug-in - sounding better at times (although Lexiverb has an even wider selection of presets) - and Reverb One blows the humble D-Verb right out of the water!

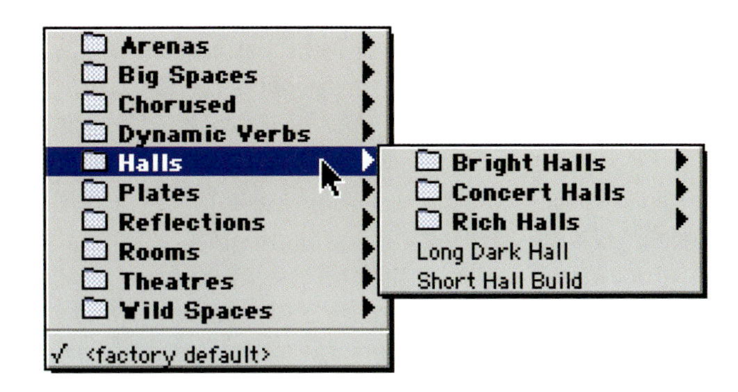

Figure 2.2 Digidesign Reverb One presets.

D-Verb

The D-Verb plug-in is a basic reverb with most of the controls that you would expect. There are no presets provided, but you can always create your own by adjusting the parameters and saving the settings.

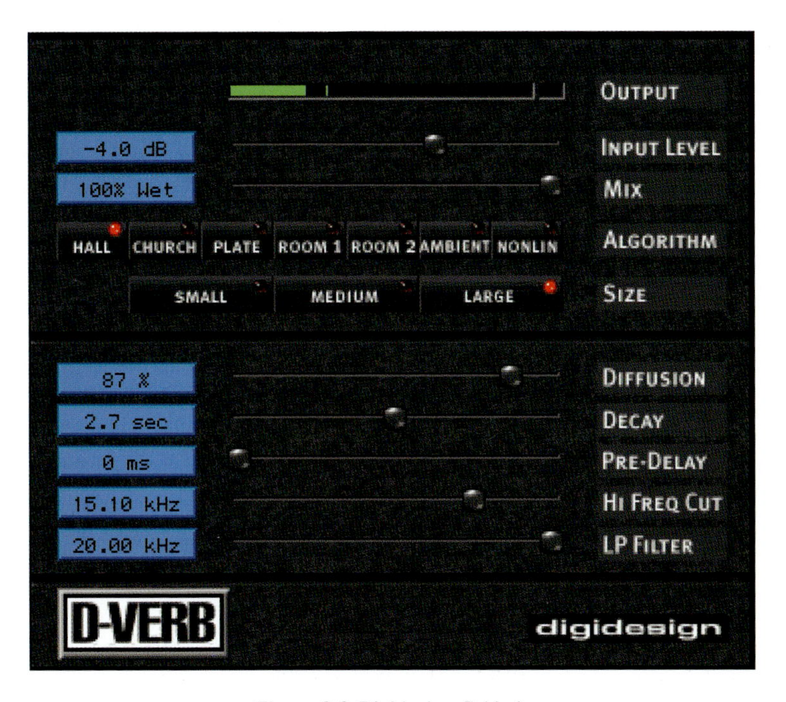

Figure 2.3 Digidesign D-Verb.

Looking at the user interface, you will see the Output Meter at the top of the window showing the mono output level of the processed signal, or the summed stereo output level in the stereo version. The Clip Indicator shows whether or not clipping has occurred. It is a clip-hold indicator, so if clipping occurs at any time during playback, the clip light stays on. To clear it, just click it. With longer reverb times there is a greater likelihood of clipping occurring as the feedback element of the reverb builds up and approaches a high output level. It is also possible that the signal gets clipped on input, so you need to watch for this. You can use the Input Level fader to reduce the input level to prevent clipping, and the Mix fader lets you adjust the balance between the dry signal and the effected signal, giving you control over the depth of the effect. The seven Algorithm buttons let you select which of the reverb algorithms to use - Hall, Church, Plate, Room 1, Room 2, Ambience or Nonlinear. Three buttons below these allow you to set the Size parameter for the reverb effect. The Diffusion parameter lets you control how the initial echo density increases over time. High settings result in high initial buildup of echo density while low settings cause low initial buildup. This control interacts with the Size and Decay parameters to affect the overall reverb density. High settings of Diffusion can be used to enhance percussion, for example, or you can use low or moderate settings for clearer and more natural-sounding vocals and mixes. The Decay parameter sets the rate at which the reverb decays after the original direct signal stops. This parameter is influenced by the Size and Algorithm parameters and can be set to infinity with most algorithms.

The Pre-Delay control lets you set the amount of time that elapses between the input of a signal and the onset of reverberation - up to a maximum setting of 370 ms. You can use Pre-Delay to create a sense of distance and volume within the simulated acoustic space, with longer settings placing the reverberant field further behind the source material. The Hi Frequency Cut control lets you set the decay characteristic of the high frequency components of the reverb. It acts in conjunction with the Low Pass Filter control to set the overall high frequency contour of the reverb. If you set this relatively low, for example, the high frequencies will decay more quickly than low frequencies - simulating the effect of air absorption in a hall. The Low Pass Filter control lets you choose the overall high frequency content of the selected reverb by allowing you to set the frequency above which D-Verb's 6 dB per octave filter attenuates the processed signal.

D-Verb is available in TDM, RTAS and AudioSuite formats. Note that D-Verb AudioSuite appears as 'D-FX D-Verb' in the AudioSuite menu of Pro Tools.

DPP-1

The DPP-1 Realtime Pitch Processor offers up to four octaves of pitch transposition and manipulation at 24-bit resolution. Applications include chorusing, doubling and delay effects, pitch correction, and extreme pitch shifting for special effects.

Note When the processed signal is transposed in pitch, it still keeps the same overall length as the unprocessed signal - i.e. the pitch is changed while the time is kept constant. This is in contrast to the way audio can be changed in pitch by speeding up or slowing down the speed of a tape recording - in which case the audio lasts a shorter or longer time, depending on the direction of the pitch change.

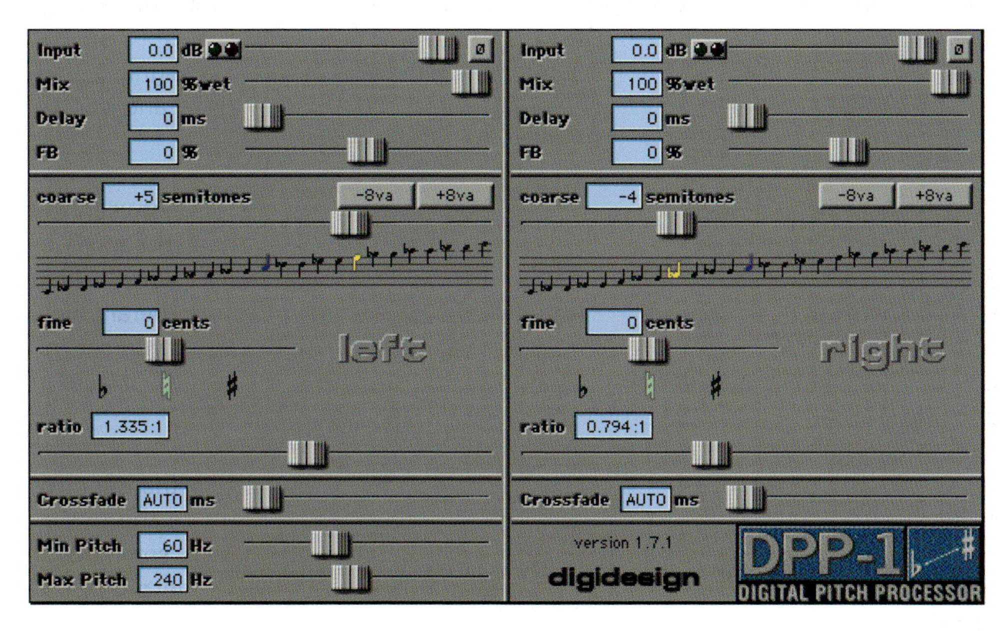

Figure 2.4 Digidesign DPP-1.

The DPP-1 is best used on an auxiliary return track if you want to apply pitch processing to more than one track at a time, as you need a full DSP chip each time you insert a DPP-1 plug-in. You can use DPP-1 as mono in/out, mono in/stereo out, or stereo in/out, and clicking on any note in the ascending musical staff lets you quickly select a relative pitch transposition value. Alternatively, you can use the Coarse pitch slider which lets you adjust the pitch in semitones over a two octave range.

With version 1.0 of DPP-1, several interesting templates were supplied to get you started. Unfortunately, they are no longer supplied with the plug-in, but I will describe some of these briefly here as ideas for how to use the DPP-1. These templates were configured for stereo operation and included Thickening, Bass Octave Doubling, Falling Feedback, and Low Warble. The Thickening template used pitch-shifting and delay to thicken and spread the audio in stereo. The positive-going pitch was panned to one side and the negative-going pitch was panned to the other side. This setup works well with guitar and vocals and you can experiment using a full wet mix or a mix between dry and wet. A special version of this was provided for Bass guitars with the Minimum Pitch slider set at 28 Hz. Another template for use with bass guitars, the Bass Octave Doubling Mix template, doubled the input with an octave above this. Octave-doubling is a useful technique for working with any low-frequency audio material. The Falling Feedback Pitch Wet template illustrated how you can use pitch change with feedback and delay to achieve an effect where the pitch drops and changes. When delay and feedback are used together, the delay continues to change pitch, resulting in a climbing or falling effect. The Guitar Low Warble template illustrated a simple semitone pitch transposition. It also provided optimal Crossfade, Minimum Pitch and Maximum Pitch settings chosen for pitch transposing guitars with minimal artefacts. The Crossfade parameter's Auto setting has a relatively long crossfade value which can adversely affect audio with sharp percussive attacks - as with guitar - so you do need to experiment with the Crossfade setting to find what works best for your particular program material.

To sum up, the DPP-1 is a useful pitch-processing tool for the more musically inclined engineer, featuring an effective and intuitive interface.

DINR

The Digidesign Noise Reduction (DINR) software actually comprises two plug-ins - Broadband Noise Reduction and Hum Removal. The Broadband Noise Reduction (BNR) plug-in provides broad and narrow-band noise reduction for suppressing such unwanted elements as tape hiss, air conditioner rumble, and microphone preamp noise.

Note BNR is available as a real-time TDM and as an AudioSuite plug-in. The Hum Removal plug-in removes pitch-based noise such as guitar pickup buzz, AC line noise, and fluorescent hum - but only works with DSP Farm (not MIX) cards.

DINR works with either mono or stereo audio files and at resolutions ranging from 16 to 24 bits, with sample rates of up to 48 kHz. DINR processing can be applied in real time or by permanently processing your audio file on disk using the AudioSuite versions.

Broadband Noise Reduction

The Broadband Noise Reduction module is best suited to reducing noise whose overall character doesn't change very much, like tape hiss, air conditioning rumble and microphone pre-amp noise. If your audio contains several different types of noise, sections of the recording can be processed differently and independently to remove the different types of noise.

To quote from the manual: 'The Broadband Noise Reduction uses a proprietary technique called Dynamic Audio Signal Modelling. This uses the audio information in the file itself along with settings made by the user to build a dynamic model or parametric description of what the noise "looks like". At the same time it also creates a model of what the "desired audio" looks like. DINR then attempts to pull apart these two models to separate the noise from the wanted audio. There are practical trade-offs between how much noise you remove from the signal and the amount of unwanted artifacts introduced into the signal.'

What this means is that sometimes you will barely achieve any worthwhile noise reduction before such unwanted artifacts are produced to spoil the sound. DINR can sometimes appear to work miracles - rescuing otherwise unusable material, but it is not a panacea that can fix all ills. For example, DINR cannot deal with noise amplitudes less than −96 dB in 24-bit files (not such a big problem), and you can't expect miracles if the noise components in the audio are masking the original signal (which can often be a problem).

You may get better results using two successive passes with smaller amounts of noise reduction. You should also remove any obvious pops and clicks first, using a waveform editor, as these can cause problems. It is also better to make any EQ changes you want prior to processing via DINR, and roll off any low-end rumble or sub-sonic frequencies – all of which will help make the noise reduction more effective.

In action

The way you use DINR is that you select a segment of noise in your audio file containing just the noise, not the signal, so that DINR can analyse this to create a model, or 'signature', of the noise. With this in place, you use the Fit button to make DINR fit an editable envelope called the Contour Line to the noise signature. This line represents an editable division between the noise signature and the desired audio, and has a series of 'breakpoints' that you can move up or down to make a better fit.

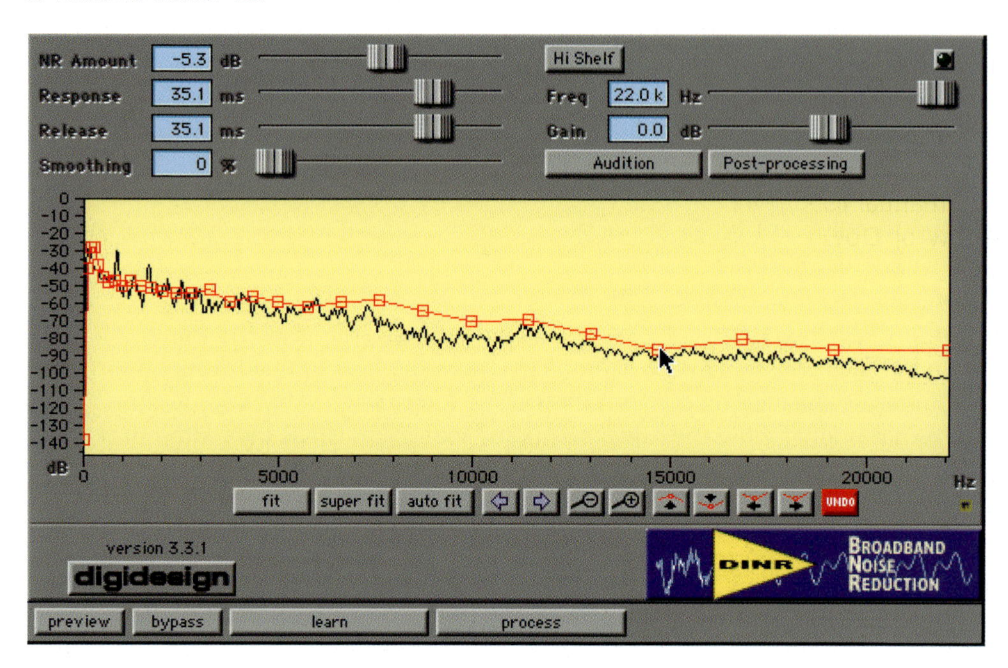

Figure 2.5 Digidesign DINR BNR.

The user interface features a frequency response graph with associated controls for Noise Reduction Amount and to help minimize audible artefacts in the processed file. You can make settings for smoothing, slow or fast response, slow or fast release, and a Hi Shelf filter can be used to restore some of the higher frequencies that are inevitably lost in the noise-reduction process.

There are two Learn modes: Learn First Audio Mode and Learn Last Audio Mode. Learn First Audio Mode is the default mode for use with audio that has an identifiable noise-only section that you can pre-select. In this mode, BNR builds a noise-signature based on the first 16 ms of audio playback. Learn Last Audio Mode is designed to allow you to locate and identify a segment of noise on-the-fly as you listen to audio playback. In this mode, you need to Option (Alt)-click the Learn button, before starting audio playback. When you hear the section of audio containing the noise you want to identify and remove, click the Learn button a second time. In this mode, BNR builds a noise signature based on the last 16 ms of audio playback. The frequency response graph displays data in real time in Learn Last Audio Mode.

The fit button computes a noise Contour Line with approximately 30 breakpoints to fit the shape of the current noise signature - enough to work well in most situations. Of course, if you are feeling very determined, you can use the Super Fit button. This creates a noise Contour Line consisting of over 500 breakpoints - so it will follow the shape of the noise signature much more accurately. Once you have your Contour Line, you should take a look at this to see how well it fits. You can often improve the fit manually - by dragging, adding or deleting breakpoints.

The manual lists several useful ways to edit the Contour Line: 'Pressing the Up Arrow or Down Arrow keys on your computer keyboard allows you to raise or lower the entire Contour Line, or a selected portion of the Contour Line. The Left/Right arrows allow you to move a selection left or right. To select a portion of the Contour Line with multiple breakpoints, Command-drag (Macintosh) or Control-drag (Windows) to highlight the desired area. After you use the fit function, BNR will automatically boost the entire Contour Line 6 dB above the noise signature so that all noise components of the audio file are below the Contour Line. You may want to adjust the Contour Line downwards as needed to modify the character of the noise reduction.'

So what if the audio that you want to perform noise reduction on does not have a noise-only portion at the beginning or end of the recording that DINR can analyse and learn? Obviously, analysing signal mixed with noise will produce a noise signature that is based partially on signal. Nevertheless, you can still obtain reasonable results using Auto Fit. With this, it is best to select a part of your audio that has a relatively low amount of signal and a high amount of noise, such as in a quiet passage, for best results here. The Auto Fit feature computes a generic noise curve based on the points contained within the currently selected audio, then fits the Contour Line to it. To use this, you first make a selection in the frequency response graph by Command-dragging (Macintosh) or Control-dragging (Windows). If the selected audio has both noise and wanted sound components, you can generate an approximate noise-only Contour Line by selecting a frequency range that appears to be mostly noise, then pressing the Auto Fit button. You will almost certainly need to edit the Contour Line for a better fit here.

Hum Removal

If you have problems with mains hum, light dimmer buzz, generator noise, guitar pickup noise or computer monitor emissions, then the Hum Removal module is the one to use.

Note Hum Removal is only available as a real-time TDM plug-in - and only for the Macintosh. Also, it is not compatible with Pro Tools|24 MIX cards. Instead, it requires a Pro Tools DSP Farm to operate.

As in the BNR module, you must have Hum Removal 'learn' the noise in your audio first. When you hit the Learn button, this sweeps through the material's frequency spectrum, scanning the noise to find the most prominent frequency characteristic of the audio signal. When this is completed, it automatically configures itself for maximum hum removal by setting its centre-frequency parameter to the fundamental frequency of the most prominent hum.

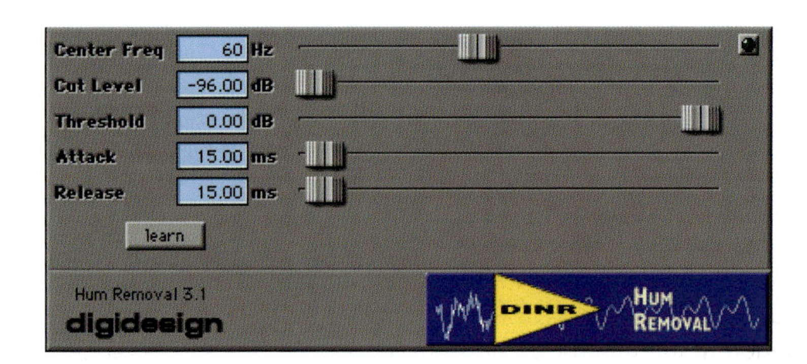

Figure 2.6 Digidesign DINR Hum Removal.

The Hum Removal module provides harmonic filters specifically designed to remove harmonic pitched noise. Harmonic noise not only contains a fundamental frequency component, such as the 50 Hz component of AC mains electricity line noise, but also contains significant components at the odd or even harmonics of this fundamental frequency. If you were to remove only the 50 Hz component of AC line noise, you would still hear significant amounts of noise at 100 Hz (the second harmonic), 150 Hz (the third harmonic), and so on. Harmonic filters not only remove the fundamental frequency they are set to, but they also remove all the odd and even harmonics of that frequency. The advantage of this type of filter is that it creates hundreds of digital notch filters at exactly the frequencies needed to remove certain types of noise. The disadvantage is that Harmonic filters can alter the sound, producing a hollow quality. Again there is a trade-off between noise-reduction amount and the amount of sound degradation.

The controls provided have familiar-sounding names, but need some further explanation to clarify exactly how these work – and when you should use them. For example, the Center Frequency slider sets the fundamental frequency of the Harmonic filter. Although the Learn function automatically sets the centre frequency to the fundamental frequency of the learned noise, this parameter can also be manually edited if necessary. The other notches in the harmonic filter are automatically set to the related harmonic frequencies. The Cut Level slider lets you set the depth of the notches within the Harmonic Filter – which determines the amount of hum removal that takes place when the signal falls below the threshold. As you increase the dB value, deeper notches are created and more noise reduction is achieved, but you will hear more artefacts. As you decrease the dB value you will hear less artefacts, so you have to choose the most acceptable setting on a case-by-case basis. The Threshold slider sets the signal level at which the harmonic filtering is activated. To minimize unwanted artefacts, you also need to set the Threshold slider carefully, so that it is just above the level at which the noise is apparent. Don't forget – loud signals will mask the noise components anyway! You can also control how long the Harmonic Filter takes to reach its full effect once the Threshold has been crossed, and how long it takes to switch off once the signal exceeds the threshold using the Attack and Release sliders.

Summary

To understand what DINR is capable of dealing with, you have to appreciate the trade-offs inherent in any type of noise-reduction system. Noise reduction always involves a 'juggling act' where you are trying to achieve a certain amount of noise reduction – the more noise reduction you apply, the more artefacts are added to the sound and the more the original signal is degraded in quality.

In practice I have used both DINR modules extremely successfully with some old material recorded to low-quality cassette – getting results fit to use for commercial release on CD. On other occasions I have been able to successfully reduce the noticeable background hiss from a close-miked guitar combo amplifier that was too audible on quiet jazz guitar passages. Generally speaking, the results you can achieve using DINR are extremely dependent on the nature of the audio you are processing, and in some cases I have found that it is not possible to remove anywhere near as much noise as I would have liked without introducing unwanted artefacts and unacceptable degradation of the sound quality. Overall, I can recommend DINR highly.

Maxim

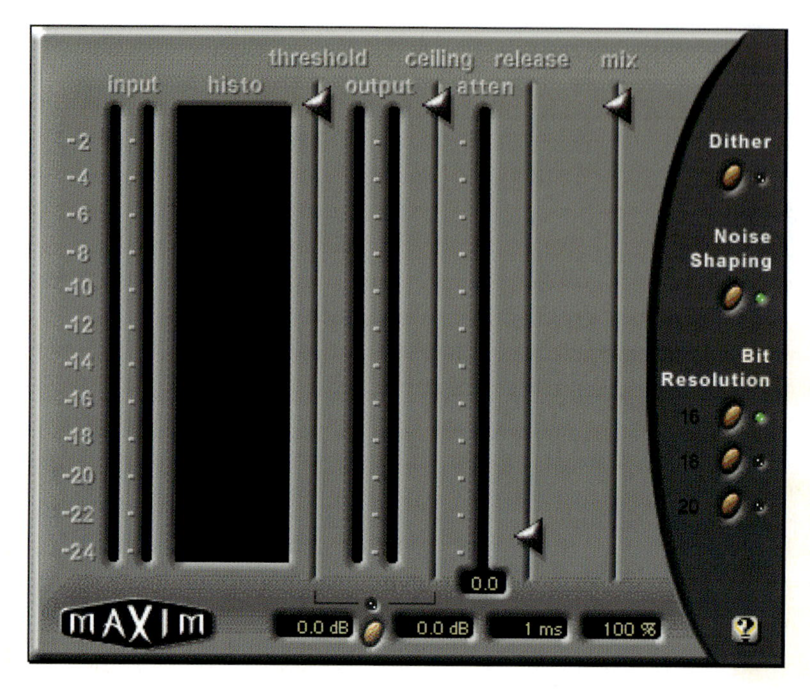

Figure 2.7 Digidesign Maxim.

Maxim is a peak limiter and sound maximizer that lets you raise the overall level of your tracks without causing any distortion. And how does it do this? It's quite simple, really. Maxim uses a look-ahead design to avoid clipping any peaks, which would introduce distortion. Remember - the effect of limiting the output is to prevent high signal levels from reaching high enough levels to become clipped. So if you apply limiting to your final mixes to remove any occasional large peaks in the music then you are free to raise the overall level of your mix to much higher levels without risk of clipping. Limiting can also be used creatively with individual instruments to smooth out peaks or to change the sound of the instruments radically. Using radical settings with drums, for instance, you may come up with that 'hit' snare sound you are after!

The user interface is easy enough to understand and operate. To provide visual feedback while you are setting your threshold levels, a histogram plots a history of the input levels. You can set an absolute ceiling that the dynamic range cannot exceed using the peak limiter, or set a threshold for the limiter beyond which gain reduction is applied according to the settings chosen. The look-ahead design allows Maxim to anticipate signal peaks and respond instantaneously - preserving the attack transients as transparently as possible while preventing clipping of the peaks.

Tip Like all look-ahead designs, Maxim introduces a relatively long delay to allow the processing to take place after the 'looking ahead'. To preserve phase relationships between two or more tracks when processing is applied to only one of these, you will need to use the DigiRack Time Adjuster plug-in to compensate for the 1024-sample delay introduced by the Maxim plug-in.

Note Maxim should be used on final mixes when you want these to sound their best on 16-bit DAT or CD-R versions that you intend to use as 'listening' copies - which are not going to be sent for mastering. For this application you can use Maxim to maximize the levels so that your mixes sound louder, and add dither so that you hear the widest dynamic range without distortion. On the other hand, if you intend to deliver 24-bit audio files on disk to a professional mastering studio, it is better not to use level optimization or dither on your final mixes. The mastering engineers will handle this at the mastering stage - so you should not remove their options during your mixing stage. Of course, you can always apply some creative limiting to individual tracks without causing problems later on during mastering.

Maxim is available in TDM, RTAS and AudioSuite formats and can be a very useful tool to have - particularly for your 'listening' CDs.

Bruno/Reso

Bruno and Reso are a pair of real-time TDM plug-ins that process audio using a sound generation technique known as cross-synthesis. Cross-synthesis generates complex sound textures by using an audio track as a tone source then applying a variety of synthesizer-type effects to that tone source.

Figure 2.8 Digidesign Bruno.

Bruno uses time-slicing for tone generation, extracting timbres from the audio track during playback and cross-fading them together at a user-selectable rate. This cross-fading creates a rhythmic pulse as the timbre of the sound changes. Consequently, Bruno is ideal for creating tonal effects with a continuously shifting timbre - similar to the wave sequencing found on synthesizers such as the PPG, Prophet VS, Korg Wavestation, and Waldorf XT. Time-slice switching can be controlled using envelope triggering or using a MIDI beat clock, and velocity-sensitive gain and detuning are also provided. The 11 intriguingly named presets include 1/16th MIDI clock with slow x-fade, Basic Unison Drone, Gnat Drone, Quick x-f Mono Vocody Drone, Quick x-f Vocody Drone, Spooky Drone, Sustained 6-voice drone, Sustained 6-voice drone Thru Trigger, Sustained 6-voice drone Trigger with MIDI clock, Trigger Quick x-fade-voice and Trigger Vocody drone.

Figure 2.9 Digidesign Reso.

Reso uses a resonator, which adds harmonic overtones to the source audio through a short signal delay line with a feedback loop. It also features velocity-sensitive resonance, damping, gain, and detuning and allows harmonic switching using envelope triggering or a MIDI beat clock. Six factory settings are provided with Reso, including Center Cluster Drone with Filter, Cluster Drone with Threshold, Deep Drone 24 stack, Hi Cluster + Envelope FeedBack, Lo Drone Envelope Wah with negative FeedBack, and my favourite, Thwonk cluster + Wah!

Features common to both Bruno and Reso include up to 24 voices of polyphony, multi-voice detuning, an editable ADSR envelope generator, and voice-stacking and portamento capabilities. Both plug-ins also have a side-chain input to allow control from an external audio source. For example, you can use side-chain processing to control the rate at which Bruno performs sample switching or the rate at which Reso toggles its harmonics back and forth using the dynamics of another signal - the so-called 'key' input. Typically, a rhythm track such as a drum kit is used to trigger these parameters. This lets you create rhythmic timbral changes that match the 'groove' of the key input.

To use a key input for side chain processing you must first decide which audio source to use as your trigger. Once this is chosen, click the plug-in's Side Chain

Input selector (just above the plug-in window) and choose the input or bus with the audio you want to use to trigger the plug-in. Then click the Key Input button (the button on Bruno or Reso with the key icon above it) to activate side-chain processing. To hear the audio source you have selected to control the side-chain input, click the Key Listen button (the button below the ear icon in Bruno or Reso). Now when you playback audio through the plug-in, it will use the input or bus that you chose as your side-chain input to trigger whatever effect you have set up.

These effects are unlikely to be needed for everyday use with your Pro Tools system. However, Bruno and Reso really 'score' when you want some extra-special creative effects to spice up your music - and the rhythmic timbral effects that you can create are ideal for music genres ranging from 'dance' to 'ambient'.

Sound Replacer

A typical task in both music and post-production is to replace a particular drum sound or sound effect. To go to each instance of an original sound you have used in the Edit window in Pro Tools and manually replace this can be very laborious. In the past, engineers and producers had to use audio delay lines with sampling capabilities, such as the popular AMS or TC2290 units, or MIDI triggered audio samplers from Akai, Emu and others - methods that had distinct disadvantages. Delay units, for example, will only hold a single replacement sample, and while they can track the amplitude of the source events, the replacement sample itself remains the same at different amplitude levels. The result is static and unnatural. In addition to these drawbacks, sample triggers are notoriously difficult to set up for accurate timing. Similarly, with MIDI triggered samplers, MIDI timing and event triggering are inconsistent, resulting in problems with phase and frequency response when the original audio is mixed with the triggered replacement sounds.

Now there is a solution. Sound Replacer matches the original timing and dynamics of the source audio with phase-accurate peak alignment while providing three separate amplitude zones per audio event. This allows you to trigger different replacement samples according to performance dynamics. Each replacement sample is assigned its own adjustable amplitude zone and you can adjust the threshold for each amplitude zone using the zoomable waveform display. You can cross-fade or hard-switch replacement audio using the different amplitude zones for soft, medium and loud drum hits, for example - and Sound Replacer will intelligently track the source audio dynamics to match the feel of the original performance. Variations in amplitude within the source audio can be used to determine which sample is triggered at a specific time. For instance, you could assign a soft snare hit to a low trigger threshold, a standard snare to a medium trigger threshold, and a rim shot snare to trigger only at the highest trigger threshold.

Operation is very straightforward. First select an audio region in Pro Tools' edit window that you wish to replace. When you open Sound Replacer, the audio waveform that you have selected will be automatically displayed.

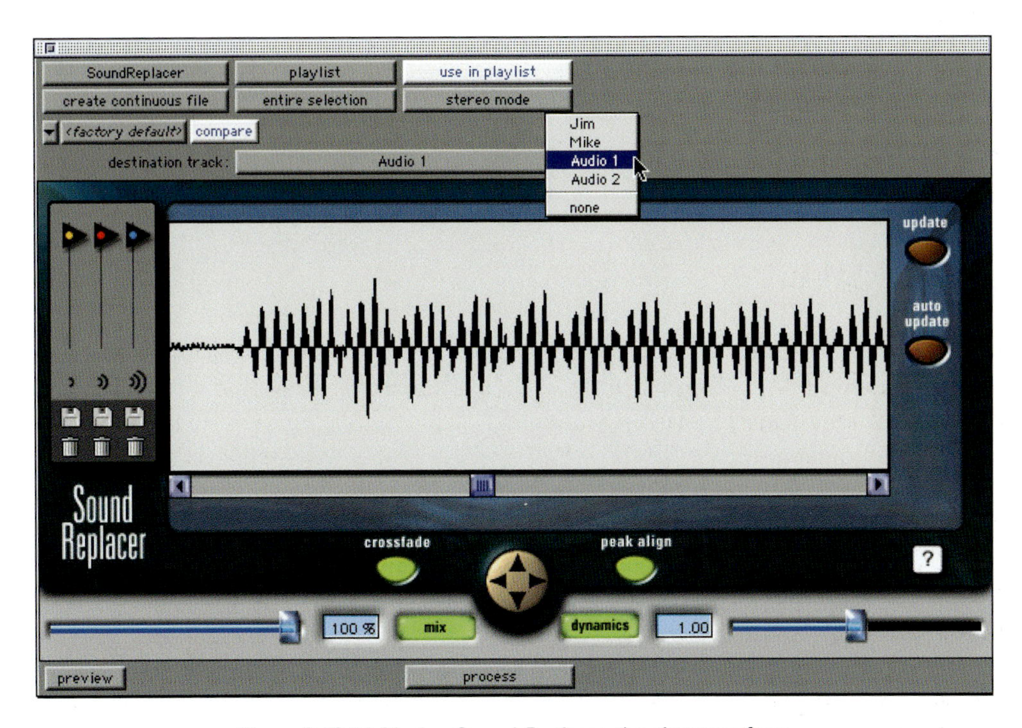

Figure 2.10 Digidesign Sound Replacer showing waveform.

After you select the region you want to replace, hit the Update button so you can see the waveform in the display, and choose the destination track where you want the replacement files to be placed using the popup menu above the display. This could be the same track if you want to replace what you have, or could be to a different track if you want to keep what you have but create an alternative.

Once your audio selection is displayed, you can load the files containing the replacement audio by clicking on the three diskette icons at the left of the plug-in window. A click on any of these brings up an 'Open File' dialog box that lets you navigate through your disk drives and folders to find the files you want to use. When you have loaded one or more of these, you can adjust their trigger thresholds while viewing the waveform peaks. Trigger markers then appear in the

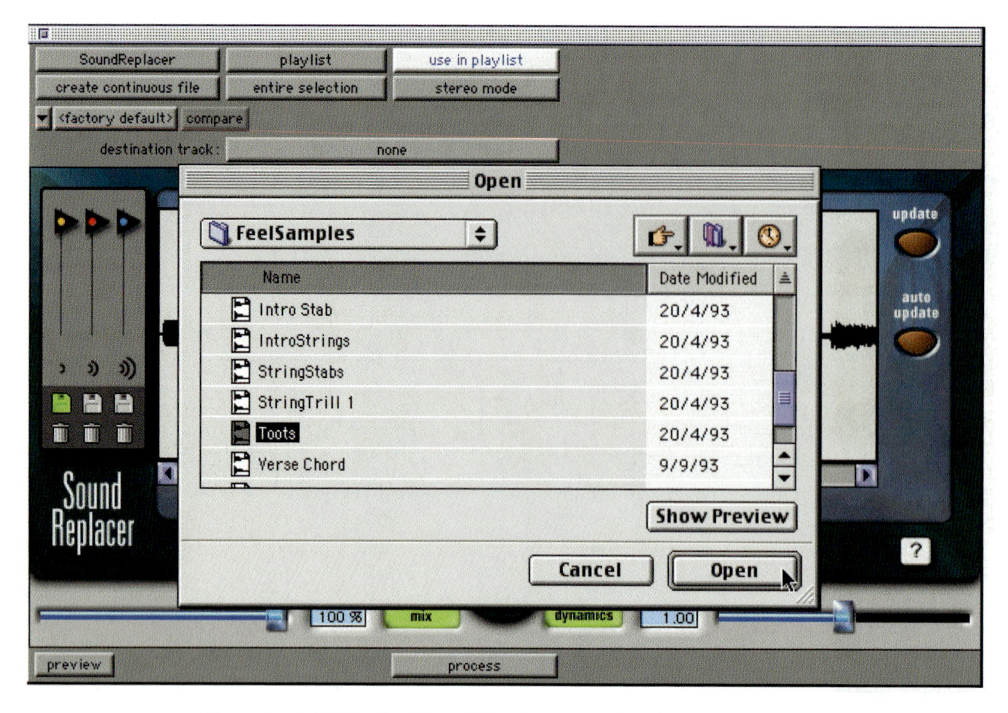

Figure 2.11 Digidesign Sound Replacer showing Open File dialog.

waveform, indicating the points at which the samples will be triggered. The colour of each marker indicates which threshold/replacement sample will be triggered at each of the three dynamic levels.

Move the sliders at the left of the display to set the actual level thresholds at which your loaded files will replace the audio in your source region. The display becomes shaded in three different shades of grey - light, medium and dark - that indicate which amplitudes in the original file will trigger which replacement files. The vertical lines that appear in the display are each coloured to correspond to the colour marked on one of the three sliders - to indicate which peaks in the file will trigger which sample.

The Zoomer tool at the bottom centre of the plug-in window allows you to increase or decrease waveform magnification here to help set trigger thresholds accurately. This can be just a little fiddly to get used to at first, but once you have got the hang of it, it will be plain sailing from then on.

When Peak Align is on, Sound Replacer aligns the peak of the replacement file with the peak of the source file while taking account of any 'pre-hit' sounds that might trigger the replacement sample too early. For example, if you look at the

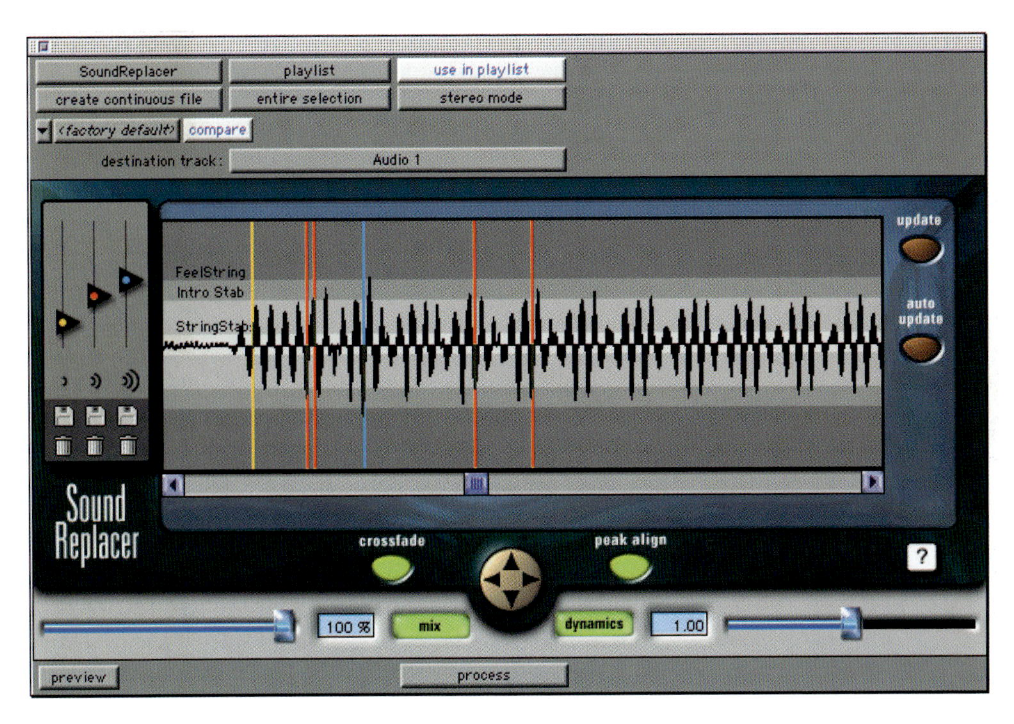

Figure 2.12 Digidesign Sound Replacer showing coloured trigger markers.

Note If you are only using a single replacement sample, you should still set all three amplitude zones for best results. Load your replacement sample into the first amplitude zone (the one with the yellow colour identifier) and set its threshold, then use the red and blue trigger threshold sliders to set amplitude zones 2 and 3 – without loading samples into either of these.

waveform of a kick drum, you will often see a 'pre-hit' portion of the sound that occurs as soon as the ball of the kick pedal hits the drum. This is rapidly followed by the denser 'attack' portion of the sound, where most of sound's 'weight' really is. If you replace this with a kick sample that does not have a pre-hit, or whose pre-hit is shorter or longer, then the timing of the replacement audio will be off. So you should turn Peak Align on if you are replacing drum or percussion sounds whose peak level occurs exactly at the initial attack and turn it off if you are replacing sounds whose peak level occurs somewhere after the initial attack. Normally, Peak Align should be turned off if the sounds you are replacing are not drum or percussion sounds.

Sound Replacer also lets you blend the replacement audio with the original audio if you want to do this. This is what the Mix control is for - to let you balance the mix between the replacement audio file and the original source file. Higher percentage values weight the mix toward the replacement audio. A value of 100% will replace all the original audio, while lower values will blend in progressively more of the original.

You can use the Dynamics control to decide how closely the audio events in the replacement file track the dynamics of the source file: setting the ratio to 1.00 simply follows the original dynamics of the source file; decreasing the ratio below 1.00 compresses the dynamic range so that there is less variation between loud and soft hits. This is useful if the loud hits are too loud and the soft hits are too soft. Increasing the ratio above 1.00 expands the dynamic range so that softer hits are softer, and louder hits are louder. This is useful if the loud hits are too soft and the soft hits are too loud. You can use the Dynamics button to switch this on and off to quickly check how your settings sound with and without Dynamics control while you are setting this up.

Finally, you can use the Preview button to let you hear your source audio and replacement audio together so that you can check how accurately the replacement triggering timing is set. Once everything sounds OK, just hit the Process button and sit back and wait for SoundReplacer to do its thing!

Producers, engineers and remixers often decide to change the sound of the bass drum, snare drum or whatever in the search for the ultimate mix. Sound Replacer has become an indispensable tool for me when I am working on drum tracks. Many of you will find this indispensable too. Sound Replacer is also extremely useful for sound design and post-production, where morphing gun shots, changing door slams or adding a Doppler effect can be accomplished in seconds, with sample-level precision.

D-Fi

D-Fi is a set of four individual plug-ins, each of which purposely degrades the audio quality of the processed sound to suit contemporary music styles such as hip-hop and rap that make extensive use of vintage drum machines, samplers, and analogue synthesizers. These older machines often used 8-bit or 12-bit digital processing or relatively low-quality analogue circuitry to produce 'grungy' sounds.

Lo-Fi

Figure 2.13 Digidesign Lo-Fi.

Lo-Fi down-processes audio by reducing its sample rate and bit resolution. It is ideal for emulating the grungy quality of 8-bit samplers. Other effects include soft-clipping distortion and saturation; an anti-aliasing filter; and a variable-amplitude noise generator.

Note Lo-Fi can be used as either a real-time TDM or RTAS plug-in or as a non-real-time AudioSuite plug-in.

You can use the Sample Rate slider to adjust the audio file's playback sample rate in fixed intervals from a high of the session's current sample rate (44.1 kHz or 48 kHz) down to a low of 700 Hz. Reducing the sample rate of an audio file has the effect of degrading its audio quality – the lower the sample rate, the grungier the audio quality, because there is less high-frequency content in the processed audio.

The Anti-Alias filter works in conjunction with the Sample Rate control. As you reduce the sample rate, aliasing artefacts are produced in the audio. These produce a characteristically dirty sound. Lo-Fi's anti-alias filter has a default setting of 100%, automatically removing all aliasing artefacts as the sample rate is lowered. This parameter is adjustable from 100% to 0%, allowing you to add precisely the amount of aliasing you want back into the mix. This lets you mimic the sound of older equipment that had poor-quality (or no) anti-alias filters, so there was always a certain amount of aliasing audible.

Note This slider only has an effect if you have reduced the sample rate with the Sample Rate control.

The Sample Size slider controls the bit-resolution of the audio, allowing you to set this anywhere from as many as 24 bits down to 2 bits. Like sample rate, bit resolution affects audio quality and clarity. The lower the bit-resolution, the grungier the quality because the detail in the sound is lost due to the lower resolution. Two buttons are provided to let you choose either linear or adaptive quantization to be applied when Lo-Fi processes the target audio signal at the selected bit resolution. These types affect the character of the audio signal in different ways. Linear quantization abruptly cuts off sample data bits in an effort to fit the audio into the selected bit resolution. This imparts a characteristically raunchy sound to the audio that becomes more pronounced as the sample size is reduced. At extreme low bit-resolution settings, linear quantization will actually cause abrupt cut-offs in the signal itself, similar to gating. Thus, linear resolution can be used creatively to add random percussive, rhythmic effects to the audio signal when it falls to lower levels, and a grungy quality as the audio reaches mid-levels. Adaptive quantization, on the other hand, reduces bit depth by adapting to changes in level by tracking and shifting the amplitude range of the signal. This shifting causes the signal to fit into the lower bit range. The result is a higher apparent bit resolution with a raunchiness that differs from the harsher quantization scheme used in linear resolution.

The Noise slider mixes a percentage of pseudo-white noise into the audio signal. Noise is useful for adding 'grit' into a signal, especially when you are processing percussive sounds. This noise is 'shaped' by the envelope of the input signal, again allowing you to mimic yet another deficiency of older equipment. The Distortion and Saturation sliders provide clipping control. The Distortion slider lets you set the gain - allowing clipping to occur in a smooth, rounded manner. The Saturation slider lets you set the amount of clipping - simulating the typical high-frequency roll-off effect of tube saturation.

To give you some inspiration as to how to use these effects, four presets are provided as settings files, including Bass Dirty Amp, Lo-rate Distorto Kit 1, Ring-Moddy Trash Bass, and Trash Bass.

To summarize, Lo-Fi has all the tools you need to effectively simulate the sound of older samplers and synthesizers that suffered from a range of deficiencies in the sound. These deficiencies, nevertheless, gave the sounds of these instruments a very unique, instantly recognizable character. Today's synthesizers and samplers are 'clean as a whistle' by comparison - yet the sounds these make are often thought of as less interesting as a consequence! Lo-Fi is simple to operate, once you understand what the controls do, and produces effective results.

Sci-Fi

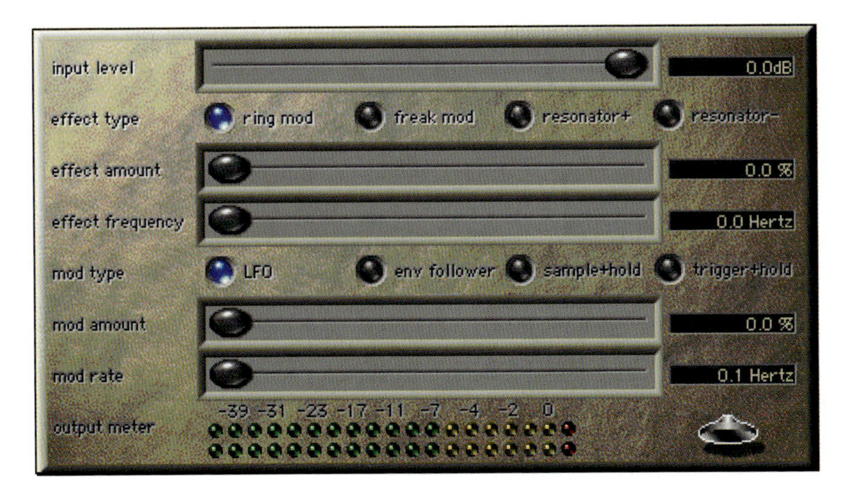

Figure 2.14 Digidesign Sci-Fi.

Sci-Fi makes your audio sound as though it has been synthesized - by adding effects such as ring modulation, resonator, and sample & hold that are typically found on older, modular analogue synthesizers. It is ideal for adding a synth 'edge' to a track.

Note Sci-Fi can be used as either a real-time TDM or RTAS plug-in or as a non-real-time AudioSuite plug-in.

The Input Trim control is typically used to achieve unity gain between input and output. Some of the effects that you may use in Sci-Fi, such as the Resonator, can cause very high signal levels to be reached within the processor. Reducing the input gain can compensate for this to make sure that the processor's output level is the same as the input level.

Sci-Fi provides four different types of effects. The Ring Modulator modulates the signal amplitude with a carrier frequency, producing harmonic sidebands that are the sum and difference of the frequencies of the two signals. The carrier frequency is supplied by Sci-Fi itself, while the modulation frequency is determined by the Effect Frequency control. Ring modulation adds a characteristic hard-edged, metallic sound to audio. Freak Mod is a frequency modulation processor that modulates the signal frequency with a carrier frequency - producing harmonic sidebands that are the sum and difference of the input signal frequency and whole number multiples of the carrier frequency. Frequency modulation produces many more sideband frequencies than ring modulation and imparts an even wilder metallic characteristic to the processed sound.

The Effect Amount parameter controls the mix of the processed sound with the original signal, while the Effect Frequency parameter controls the modulation frequency of the Ring modulators and Resonators. With the Ring Modulator and Freak Mod effects, the frequency ranges from 0 Hz to 22.05 kHz. The Resonator – (minus) and Resonator + (plus) effects both add a resonant frequency tone to the audio signal, the frequency of which is determined by the Effect Frequency parameter. For Resonator +, the frequency ranges from 344 to 11 025 kHz, while for Resonator –, the frequency ranges from 172 Hz to 5512 Hz. The difference between these two modules is that Resonator – (minus) reverses the phase (polarity) of the effect, producing a hollower sound than Resonator + (plus). The Resonator can be used to produce metallic and flanging effects that emulate the sound of classic analogue flangers.

Four Modulation Type buttons let you select the type of modulation to apply to the selected effect. Depending on the type of modulation you select, the parameter sliders below will change to provide the appropriate type of modulation controls.

LFO modulation uses a low-frequency triangle wave as a modulation source, the rate and amplitude of which are determined by the Mod Rate and Mod Amount controls, respectively.

Envelope Follower modulation causes the selected effect to dynamically track the input signal by varying with the amplitude envelope of the audio signal. As the signal gets louder, more modulation occurs. This can be used to produce a very good automatic wah-wah type effect. When you select the Envelope Follower, the Mod Amount slider changes to a Mod Slewing control. Slewing provides you with the ability to smooth out extreme dynamic changes in your modulation source. This provides a smoother, more continuous modulation effect. The more slewing you add, the more gradual the changes in modulation will be.

Sample-and-Hold modulation periodically samples a random pseudo-noise signal and applies it to the effect frequency. Sample-and-hold modulation produces a characteristic random stair-step modulation. The sampling rate and the amplitude are determined by the Mod Rate and Mod Amount controls, respectively.

Trigger-and-Hold modulation is similar to Sample-and-Hold modulation, with one significant difference: if the input signal falls below the threshold set with the Mod Threshold control, modulation will not occur. This provides interesting rhythmic effects, where modulation occurs primarily on signal peaks. Modulation will occur in a periodic, yet random way that varies directly with peaks in the audio material. Think of this type of modulation as having the best elements of both Sample-and-Hold and the Envelope Follower.

The Mod Amount and Mod Rate sliders control the amplitude and frequency of the modulating signal. The modulation amount ranges from 0% to 100%. The modulation rate when LFO or Sample-and-Hold are selected ranges from 0.1 Hz to 20 Hz. If you select Trigger-and-Hold as a modulation type, the Mod Rate slider changes to a Mod Threshold slider, which is adjustable from −95 dB to 0 dB. It determines the level above which modulation occurs with the Trigger-and-Hold function. If you select Envelope Follower as a modulation type, the Mod Rate slider changes to a Mod Slewing slider, which is adjustable from 0% to 100%.

Note If the Mod Amount is set to 0%, no dynamic modulation is applied to the audio signal. The Effect Frequency slider then becomes the primary control for modifying the sound.

Seven cryptically, yet descriptively, named presets are provided, including Freq Mod Env. F. Kit, Freq Mod Env. F. Wah, Res.-1/16 note Trig. & Hold, Res.-1/4 note Trig. & Hold, Res.-Env. Follower, Ring Mod Trig. & Hold Kit, and Wah Res-LFO Faux Flange.

Some, or all, of these effects can be found on typical modular analogue synthesizers. You can use Sci-Fi to add these characteristic effects to any audio track.

Recti-Fi

You can use Recti-Fi to 'beef up' the harmonic content of your audio tracks by adding subharmonic or superharmonic tones to the original signal.

Note Recti-Fi can be used as either a real-time TDM or RTAS plug-in or as a non-real-time AudioSuite plug-in.

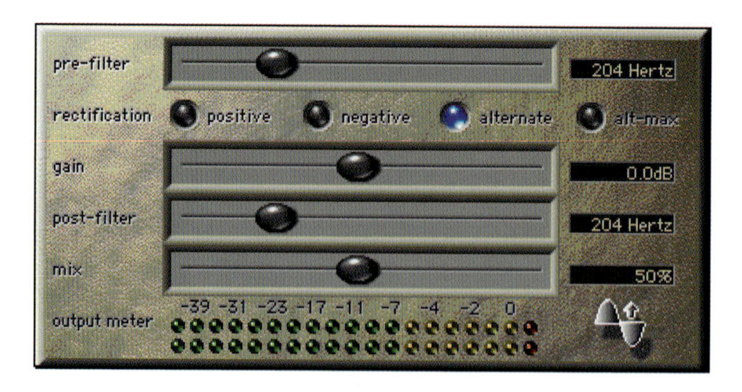

Figure 2.15 Digidesign Recti-Fi.

The incoming signal is treated as follows: the signal is split into two paths - one of which passes the input signal through unchanged to the mixer prior to being output. The other passes the input signal through the processor where it is rectified before being passed to the mixer, then mixed with the desired amount of the original signal prior to being output.

In the processor path, the Pre-Filter is first used to filter out high frequencies from the incoming audio signal prior to rectification.

Note Technically, this is desirable because the rectification process can cause instability in waveform output - particularly in the case of high-frequency audio signals. Filtering out these higher frequencies prior to rectification can improve waveform stability and the quality of the rectification effect.

Once the signal has been filtered, the rectification process is used to create a processed version of the signal whose frequency is either doubled, to create a new superharmonic frequency, or halved to create a new subharmonic frequency that will subsequently be mixed together with the original signal before output. To create classic subharmonic synthesis effects, you should set the Pre-Filter and Post-Filter to a relatively low frequency, such as 250 Hz. Two rectification options are provided to let you halve the audio signal's frequency to create the subharmonic tone. If you choose Alternating Rectification, this alternates between rectifying the phase of the first negative waveform excursion to positive, then the next positive excursion to negative, and so on, throughout the waveform. The result is that the frequency of the processed audio is halved. If you want to create a subharmonic tone with a hollow, square wave-like timbre, you should use the Alt-Max Rectification option. This alternates between holding the maximum value of the first positive excursion through the negative excursion period, switching to rectify the next positive excursion, and holding its peak negative value until the next zero crossing. Again, the audio signal is halved in frequency, but in this case a hollower sound is produced. To create a superharmonic tone, you need to double the audio signal's frequency. In this case you can use either Positive or Negative Rectification. These options rectify the waveform so that its phase is 100% positive or negative – with the result that the frequency of the processed signal is doubled.

The Gain control allows you to adjust signal level before the audio reaches the Post-Filter. You can use this to restore unity gain if you have used the Pre-Filter to cut off high frequencies prior to rectification. Waveform rectification, particularly alternating rectification, typically produces a great number of harmonics – and there may be far more than you want. You can use the Post-Filter to filter out some of these harmonics to create a smoother sound. For example, to create classic subharmonic synthesis effects, you should set the Pre-Filter and Post-Filter to a relatively low frequency. Finally, the original and the processed signals are combined using the Mix fader to produce an output signal that now contains the new harmonics you have created.

Six presets are provided, including Noise Hat, Sub-Kit, Sub-Oct. Bass, Sub-Oct. Heavy Bass, Trasho Kit, and Up-Oct. Wah.

Recti-Fi is the tool to use if you want to simulate the sounds of analogue equipment such as tape recorders and tube-based, transformer-coupled outboard devices that typically add harmonic distortion components that 'fatten up' the sounds of drums, guitars and so forth.

Vari-Fi

Vari-Fi is an AudioSuite-only plug-in that provides a pitch-change effect similar to a tape deck or record turntable speeding up from or slowing down to a complete

Figure 2.16 Digidesign Vari-Fi.

stop. Vari-Fi preserves the original duration of the audio selection. This plug-in just couldn't be easier to use! There are just two settings to choose from: you can Speed up from a complete stop to normal speed or Slow down to a complete stop from normal speed. Simple - but effective! If you want it - you got it with Vari-Fi. Say no more!

D-FX

D-FX is a set of non-real-time AudioSuite plug-ins that you can use for thickening, spatializing and adding a sense of space and depth to elements in a mix - using your choice of the five types: Chorus, Reverb, Flanging, Multi-tap delay or Ping-Pong delay. The D-FX plug-ins can be used in either mono or stereo.

Tip If you want to process a mono track and obtain a stereo result, first select the desired track or region plus an empty track or region, then select the stereo version of the D-FX from the pop-up menu. When you process the audio, the result will be two tracks or regions that represent the right and left channels of the processed audio. You should then pan these tracks hard right and hard left in your mix.

Note If you choose to use a D-FX plug-in in stereo mode, and then select an odd number of Pro Tools tracks for processing, D-FX will process the selected tracks in pairs, in stereo. However, the last odd (unpaired) track will be processed as mono, using the left channel settings of the stereo D-FX plug-in. If you want the last track to be processed in stereo, you must select an additional track to pair it with an empty one if necessary.

D-FX Chorus

The D-FX Chorus plug-in combines a time-delayed, pitch-shifted copy with the original signal. This produces an effect that is ideal for thickening and adding a shimmering quality to guitars, keyboards, and other instruments. The Input Level slider is set to a default of +3 dB. If your source audio has been recorded very close to peak level, this +3 dB default setting could cause clipping so you should use this control to reduce the input level.

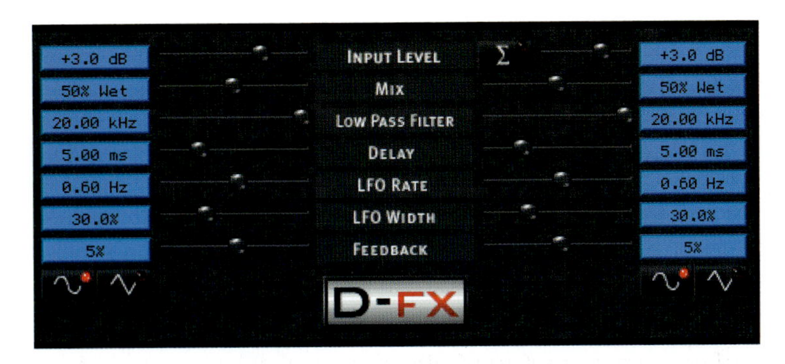

Figure 2.17 Digidesign D-FX Chorus.

Note When you use the Chorus plug-in in stereo mode, a Sum Inputs button appears next to the right channel's Input Level slider. Selecting the Sum Input button will sum the dry input signals before processing them. The dry signal will then appear in the centre of the stereo field, and the wet, effected signal will be output in stereo. When the Sum Inputs button is selected, the LFO waveform on the right channel is automatically phase-inverted in order to enhance the mono-stereo effect.

The Mix control lets you set the balance between the effected signal and the original signal to control the depth of the effect. The Low-Pass Filter control lets you set the cutoff frequency of the low-pass filter to attenuate the high frequency content of the feedback signal. The Delay control lets you set a delay time of up to 20 milliseconds between the original signal and the chorused signal. The higher, longer delay settings produce a wider chorusing effect. You can also adjust the rate of the low frequency oscillator (LFO) that can be applied to the delayed signal to produce modulation effects, using the LFO Rate slider. The LFO Width control lets you set the depth of the LFO modulation and at the bottom-left of the window you can select either a sine wave or a triangle wave as the modulation source. Finally, a Feedback control is provided to let you set the amount of feedback applied from the output of the delayed signal back into its input. Negative settings here provide a more intense effect. These controls let you quickly alter the effect to suit your requirements, and four presets are provided to help get you started, including Deep Chorus, EvenSlo Flange, Stereo Deep Chorus ad Stereo Flange 1.

The D-FX Chorus works well enough, although it doesn't have as interesting a character as some of the vintage designs which have been more closely modelled by some of the third-party developers.

D-FX Flanger

Figure 2.18 Digidesign D-FX Flanger.

The D-FX Flanger is ideal for thickening and adding a swirling, moving quality to guitars and other instruments. Like the Chorus, the D-FX Flanger combines a time-delayed, pitch-shifted copy of an audio signal with itself. The D-FX Flanger differs from other digital flangers in that it uses a thru-zero flanging algorithm that results in a truer tape-like flange. This technique delays the original dry signal very slightly (approximately 256 samples), then modulates the delayed signal back and forth in time in relation to the dry signal, passing through its zero point on the way. Looking at the controls, at the top of the plug-in window you will see the Input Level slider. This is set to a default value of +3 dB. If your source audio has been recorded very close to peak level, this +3 dB default setting could cause clipping so you should use this control to reduce the input level.

Note When you use the D-FX Flanger plug-in in stereo mode, a Sum Inputs button appears. Selecting the Sum Input button will sum the dry input signals before processing them. The dry signal will then appear in the centre of the stereo field, and the wet, effected signal will be output in stereo. When the Sum Inputs button is selected, the LFO waveform on the right channel is automatically phase-inverted in order to enhance the mono-stereo effect.

The Mix slider lets you set the balance between the effected signal and the original signal to control the depth of the effect. The High-Pass Filter control lets you set the cutoff frequency of the high-pass filter to attenuate the frequency content of the feedback signal and the frequency response of the flanging. The higher the setting, the more low frequencies are removed from the feedback signal. The LFO

Rate control lets you set the rate of the low frequency oscillator (LFO) applied to the delayed signal as modulation, while the LFO Width control lets you set the depth of the LFO modulation. You can select either a sine wave or a triangle wave as a modulation source, using the LFO Waveform selector at the bottom-left of the plug-in window. Finally, the Feedback control lets you set the amount of feedback applied from the output of the delayed signal back into its input. Negative settings here provide a more intense effect. Five presets are provided with this effect, including a Factory Setting, a Fast Leslie, Slow 'n' Deep, Vocal Flange and Whacked AM Flange – which sounds 'whacky' enough to me.

Again, the D-FX Flanger works well enough, although it doesn't have as interesting a character as some of the vintage designs which have been more closely modelled by some of the third-party developers.

D-FX Multi-Tap Delay

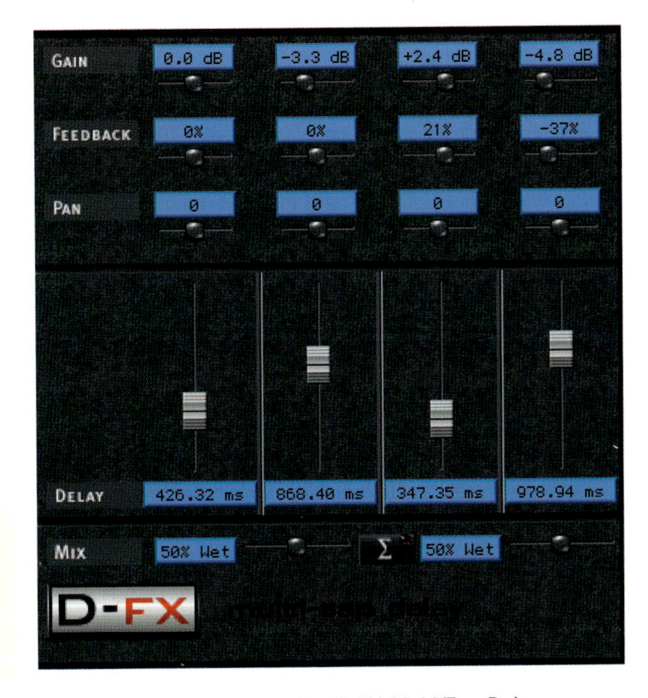

Figure 2.19 Digidesign D-FX Multi-Tap Delay.

The D-FX Multi-Tap Delay is ideal for adding complex rhythmic echo effects to your audio. Multi-tap delays were originally created using tape-recording machines with multiple playback heads spaced at appropriate distances from the record head to produce multiple delayed versions of the original input signal.

The different delay paths were often referred to as delay 'taps'. The D-FX Multi-Tap Delay is much more flexible than these original devices, allowing you to individually control the delay time and number of repetitions of each of its four delay 'taps'. At the top of the plug-in window, a Gain slider for each of the Multi-Tap Delay's delay lines controls the input level of that individual delay tap.

Note When you use the D-FX Multi-Tap Delay in stereo mode, a Sum Inputs button appears. Selecting this button will add the dry input signals together before processing them. The dry signal will then appear in the centre of the stereo field, and the wet, effected signal will be output in stereo.

Immediately below the Gain controls, four Feedback controls let you set the amount of feedback applied from the output of the delay into its input to control the number of repetitions of the delayed signal for each delay line. When you use the Multi-Tap Delay in stereo mode, a Pan slider appears below each Feedback slider, allowing you to control the apparent locations of the four delayed signals in the stereo field. A Delay control for each 'tap' lets you set the delay time between the original signal and the delayed signal of up to 1.5 seconds in millisecond increments. Finally, at the bottom of the window, the Mix control lets you set the balance between the effected signal and the original signal to control the depth of the effect.

Multi-Tap Delay is a very useful effect to have available in your selection of effects-creation tools and this D-FX plug-in operates faultlessly to blend up to four delay signals in with your original. If you are looking for a 'clean' sound you will not be disappointed. On the other hand, unlike third-party TDM plug-ins such as the Line 6 Echo Farm, the D-FX Multi-Tap Delay will not let you simulate the quirkier sounds of older multi-tap delay designs, such as the Roland Re-101. What you do get are six great presets to get you started, including 4 against 3, Med Slap, Short Slap, Stagger 'n' Bounce, Stereo 'n' Slap and Thicken.

D-FX Ping-Pong Delay

The D-FX Ping-Pong Delay plug-in is mainly intended for use in stereo to add a panned echo to any sound. In stereo mode this plug-in feeds back delayed signals to their opposite channels, creating a characteristic 'ping-pong' effect. The Input Level control lets you adjust the input level to prevent clipping and the Mix control lets you set the balance between the processed and the original signals to control the depth of the effect. The Delay control lets you set a delay time between the

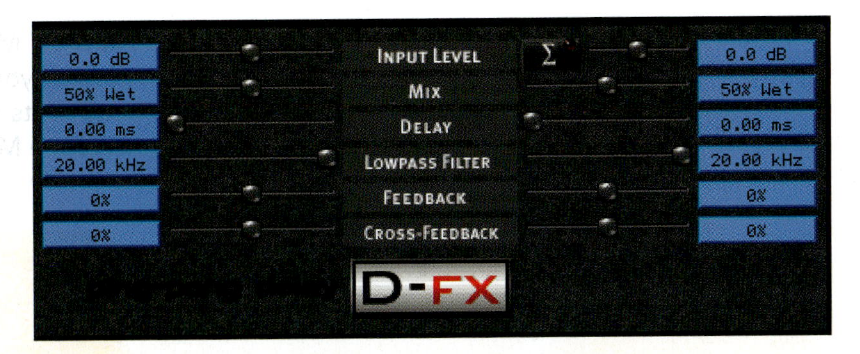

Figure 2.20 Digidesign D-FX Ping-Pong Delay.

original signal and the delayed signal of up to 1.5 seconds in millisecond increments. The Low-Pass Filter parameter controls the cut-off frequency of the low-pass filter, allowing you to attenuate the high frequency content of the feedback signal. Feedback controls the amount of feedback applied from the output of the delay into its input, thus controlling the number of repetitions of the delayed signal. When used in stereo mode, Cross Feedback feeds the delayed left channel signal to the right channel input and vice-versa. The result is a stereo echo that ping-pongs back and forth between the right and left channels. Presets provided include 4 against 3 Bounce Delay, I Looongg For You, Long Bouncing Step and Soft Bouncer – all of which serve to demonstrate the kind of effects you can achieve very successfully. Ping-Pong Delay is not an effect that you will use every day of the week. But when you need it, D-FX will deliver it – with no fuss, no muss.

D-FX D-Verb

This AudioSuite version of D-Verb is virtually identical to the TDM version previously described, although there is no Output Meter or Clip Indicator. Another difference is that when you use the D-Verb AudioSuite plug-in in stereo mode, a Sum Input button appears. Selecting the Sum Input button will sum the dry input signals (regardless of whether the input is mono or stereo) before processing them. The dry signal will then appear in the centre of the stereo field, and the wet, effected signal will be output in stereo. Yet another issue to be particularly aware of when using the AudioSuite version is that you need to take account of any reverb tails which ought to fade away after the last part of the selected audio has finished. If this is the case, you need to make a selection (containing blank space) that extends some way beyond the end of the actual audio source material. This extra time selection has to be sufficiently long to let Pro Tools record the complete decay of the reverb. Watch out for this! A suddenly truncated reverb

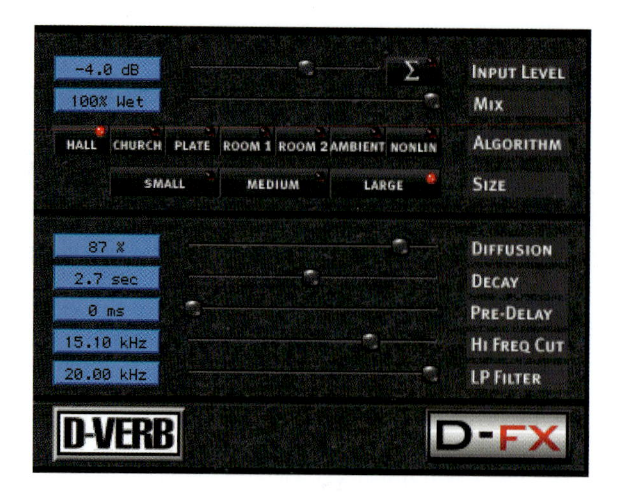

Figure 2.21 Digidesign D-FX D-Verb

decay can stick out like a sore thumb in some mixes. As with the TDM version, sadly, no presets are supplied. Although D-Verb can be a useful additional reverb processor to have around, I would definitely not recommend this as your main reverb processor.

Chapter summary

Realverb is a serious contender as the main reverb for your Pro Tools system, although it does use a lot of processing power. D-Verb, on the other hand, you can definitely live without. The DPP-1 is a useful tool for the sound designer, and DIN-R is a useful tool for the technical engineer - although this will be completely overshadowed by Sonic Solutions No Noise system which is due to be launched for Pro Tools systems soon. The Maxim Limiter does what it says it will - but there are better tools available from third parties such as Waves that do the same thing, and are better. Bruno and Reso are great tools for sound designers and remixers looking for 'fun' effects, as is D-Fi. On the other hand, you can probably give the D-FX a complete miss - especially compared with, say, the Waves Pro-FX which just blow these out of the water. The most essential of these tools for me is Sound Replacer, which is simple to use, effective in what it does and fills a niche that no other plug-in addresses.

3 Digidesign-distributed Plug-ins

Digidesign distributes a range of third-party plug-ins that has been developed to simulate conventional analogue, or digital, hardware devices. Often, a selection of these will be bundled with the system when you buy a Pro Tools system. I have regularly worked with all of these on commercial recording sessions and they have proven their worth time and time again - delivering consistent and excellent results in a variety of situations.

In this chapter I will provide short overviews to help guide you when making your own choices, followed by tips and hints culled from the manuals and from personal experience, along with other interesting information. For Drawmer Dynamics, which is the most complex plug-in here, I have presented more in-depth material drawn from the manual but reorganized and abbreviated to get the information across more succinctly.

Two of my favourites are the Focusrite D2 equalizer and the Focusrite D3 compressor/limiter which simulate the Focusrite Red 2 and Red 3 hardware units, respectively. As a guitarist, I find the Line 6 Amp Farm particularly interesting. This simulates valve guitar amplifiers and various speaker cabinets with presets for many popular combinations. And the Line 6 Echo Farm does the same trick for classic guitar effects units and pedals. If you are looking for vocal or 'talking' synthesizer effects, try the Orange Vocoder. Or you could go for the LexiVerb plug-in with its excellent selection of Lexicon digital reverb algorithms. If you are looking for effects to enhance the harmonic content of your audio, Aphex offer their popular Aural Exciter and Big Bottom processors. Finally, Drawmer Dynamics simulates the popular Drawmer expander, compressor, limiter and gate processors.

The burning question for some is how close these are to the original devices. Others will take a more pragmatic view and simply ask how usable these plug-ins are in their own right. I have had the opportunity to compare the Focusrite D2 with a Red 2. The D2 automatically filtered out some of the low frequencies even with the settings at zero boost/cut. Also, the frequencies were hard to match with the hardware unit – as this uses standard pots with no readout of the exact values you are setting. Despite this, the D2 performs extremely well, providing musically appropriate filter designs controlled by a first-rate user-interface which is greatly enhanced compared with the hardware unit by the repeatability of the settings, the graph of the frequency response, and the automation features.

When I compared Amp farm with several of the actual amplifier and speaker combinations simulated, I was amazed to discover how close the Vox AC30 and Fender Deluxe simulations were. Other combinations, such as the Fender Bassman and Marshall JCM800 were not as close – but the Amp Farm versions were extremely-useable in their own right. Echo Farm does not remain 100% faithful to the original effects designs, but does incorporate most of their quirks, while often adding useful features made possible by the software. The LexiVerb does not sound like any actual Lexicon reverb unit I have used, but does provide an extremely interesting and useable selection of presets. Demo versions of most of these plug-ins are widely-available, so my recommendation is to try before you buy and let your own ears be the judge.

Note Don't forget to allocate 1–2 megabytes of additional RAM to the DAE for each additional plug-in installed on your system.

Focusrite D2

The D2 was jointly developed by Focusrite with Digidesign – using the Red 2 equalizer as the model. For this digital version, new EQ algorithms were developed with all the 'warts' you would get with analogue gear.

The D2 has up to six simultaneous bands of EQ, including high-pass, low-shelf, low-mid peak, high-mid peak, high-shelf, and low-pass filters. A very useful graph is provided which displays the EQ curves in real time as you adjust the EQ parameters. There are actually three configurations of the D2 supplied – each of which will use a lesser or greater amount of your system's DSP. The 1–2 Band EQ provides two filters that you can separately enable or disable. You can enable both filters if you opt for low-mid peak, high-mid peak, or high shelf filter types. However, if you choose high-pass, low-shelf, or low-pass filter types, these use the entire module so you can only enable one filter. The 4-Band EQ lets you use up to four filters simultaneously, but uses more DSP, while the 6-Band EQ provides six filters, and uses the most DSP.

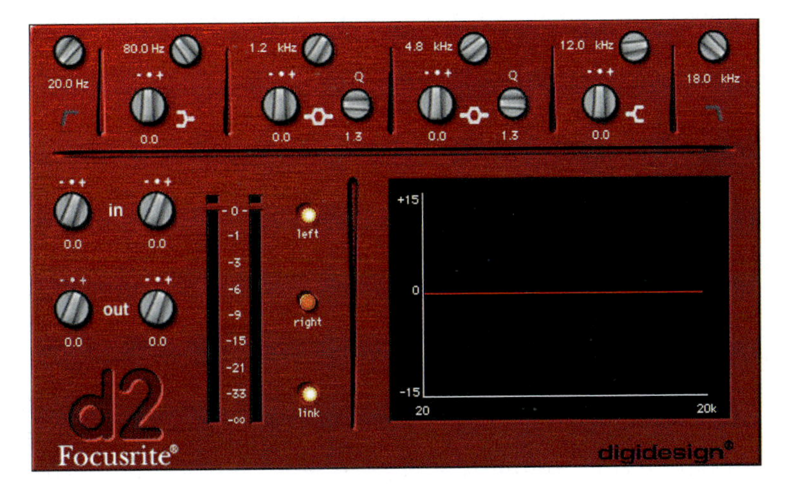

Figure 3.1 Focusrite D2.

If you know how to set the controls on the analogue EQ, then you know how to use the D2 – it's as simple as that! I EQ'ed a mix where the guitar sounded too cutting and the bass guitar sounded a little 'light'. Adding a dB of boost at 80 Hz and a couple of dBs of cut at around 4 kHz sorted this out in seconds flat – giving me exactly the results I wanted and expected to achieve. For my money, this is a much more musical-sounding EQ compared with any of the other EQ plug-ins available for Pro Tools.

The D2 plug-in vs the Red 2 hardware

Being of a curious nature, I wondered how close the plug-in would be to the real hardware. So I helped to organize a 'shoot-out' between these for the UK Music Producers Guild. The event took place at Planet Audio Studios in London a little before Easter 2001. A recording session was produced by Pip Williams, Pro Tools engineering was taken care of by Mick Glossop, and I documented the event.

Rob Jenkins, Director of Product Strategy for Focusrite, who designed the D2 in collaboration with a programmer from Digidesign, was invited to match the settings of the plug-in to those of the hardware so that playback of a vocal recording could be compared through both.

The consensus of the invited audience of professional engineers and producers was that the plug-in removed some of the low-mid frequencies from the recording even when the settings were 'flat'. It was also agreed that it was very difficult to match the settings on the hardware exactly because the hardware had switchable frequency range controls but no readouts of exact frequencies, while the D2 controls were continuously variable throughout their ranges and had numeric readouts of their settings. Everyone agreed that the D2 was an excellent EQ and worked well on the vocals – while performing slightly differently to the hardware version.

Rob Jenkins offered these comments afterwards: *'It is fair to say that it is pretty easy to design an EQ in software and to put a Red-style application window on the screen – but my intention was to go beyond that. Our main aim in developing the D2 was to give the purchaser of the software version the same pride of ownership as the Red 2 hardware user. What are the elements that make the Red 2 special and instill that pride of ownership? The fact that Rupert Neve put 30 to 40 years of audio experience into designing the Focusrite EQ hardware. It sounds special and like nothing else. It looks fantastic.*

'The software development process began by first measuring the Red 2 and graphing the audio performance and secondly we derived transfer functions for each circuit element from the circuit diagrams. These transfer functions were then used to construct the first basic code and its performance measured against the Red 2 hardware graphs. The second stage involved many iterations of the code as we moved the code performance closer and closer to the actual performance of the hardware. In phase 3 the code complexity increased as we concentrated more and more on the non-linear and Red 2 unique elements of the Red 2 hardware performance. The software you hear today is the output of this process and I think we succeeded in our original aim of transferring the original Rupert Neve analogue design techniques into software and capturing the unique details that make a Red 2 sound like a Red 2. It is these subtle details, the non-linear elements, that you won't find in any analogue or digital design book that we worked so hard to capture.

'Now in terms of the big question "Is the D2 really the same as a Red 2?" I have to say that we stand by the design process and we are very proud of the outcome. Do they sound the same? That is a truly contentious question for me. There are issues that impact upon the performance of the D2 to exactly replicate the Red 2 sound. Firstly, analogue electronics is a very imperfect science. Its components are both non-linear and bring inherent characteristics with them – and capacitors do what capacitors do in their own way. In the world of software, components and circuits that are developed from analogue circuits can only be modelled to a certain level of sensitivity, before the amount of code and processing becomes impossible to sustain. Therefore the model will always be different to the original.

But we are talking about degrees of difference - and to my ears the differences are very small. The D2 is as close to the original Red 2 design as we could make it. What differences exist are small, and at the end of the day the D2 is a great sounding software EQ which stands on its own merit. The D2 is a Focusrite software EQ and the Red 2 is a Focusrite hardware EQ with both conforming to the same design philosophy and values. I hope that clears up my thoughts on the subject.'

Focusrite D3

The Focusrite Red 3 dual mono/stereo compressor and limiter is a highly regarded hardware unit designed by Rupert Neve. Digidesign developed the D3 software in collaboration with Focusrite to provide a high-quality dynamics processing plug-in for Pro Tools TDM and AudioSuite systems.

Two configurations of the plug-in are supplied: Compressor + Limiter and Compressor/Limiter. The Compressor + Limiter lets you use both the compressor and the limiter at the same time, but uses twice as much DSP as the Compressor/Limiter. The Compressor/Limiter, on the other hand, restricts you to using either the Compressor or the Limiter, but uses half as much DSP as the Compressor + Limiter. The Compressor/Limiter is the one to use if you only need one or other of these processing types, as it leaves more of your precious DSP resources available for use with other plug-ins. By default, the Compressor is enabled. To enable the Limiter you need to Control-click its icon on the Mac or Start key-click the icon when using Windows.

Figure 3.2 Focusrite D3.

The D3 Compressor has the usual controls to let you set the threshold above which the dynamic range of signals will be reduced and to set the compression ratio by which the output level will be reduced compared to the input. The D3 Limiter uses a very fast attack time combined with a very high compression ratio to limit the dynamic range of signals in a musically appropriate fashion - rather than providing the 'brick-wall' limiting used in broadcast limiters, for example. The detected amplitude of the input signal is normally the control source used to trigger the compression or limiting once the threshold is passed.

Side-chain processing is also available. This allows the amount of compression or limiting to be triggered by an independent audio signal. In Pro Tools you can use the dynamics of one track as a control source, feeding this into the key-input to trigger compression or limiting for a different track being processed by the D3. Using this technique lets you compress your chosen Pro Tools track using the dynamics of the control track's audio. Another typical application is de-essing, where the compression can be triggered by selected frequencies.

Note When used in stereo configurations, levels for the two input channels can be set separately, while all the other D3 controls affect both channels equally. The single control signal that drives the dynamics processing is derived from the pair of input channels. This prevents stereo image shifting which would otherwise occur due to level differences between the two channels causing different amounts of compression or limiting to be applied to each channel.

The Focusrite D3 packs a lot into its deceptively simple interface and works well in practice. As with the D2, this plug-in is particularly effective for musical applications. Other compressors and limiters may be more appropriate at times, so I would not regard the D3 as the only compressor/limiter to have in my 'tool-kit'.

Line 6 Amp Farm

Line 6 originally developed their digital software modelling technology for a new type of combo amp for guitarists. This digital technology enabled Line 6 to successfully simulate the sounds of classic guitar amplifier and speaker combinations - which guitarists to use at the flick of a switch. Subsequently, this simulator was released as The Pod and The Pod Pro - stand-alone processor-only versions that could be plugged into any amplifier or recording system. Around the same time, Line 6 became Digidesign Third-Party Developers and developed a TDM plug-in using this technology. The Line 6 Amp Farm for Mac-based Pro Tools TDM systems uses the same 'TubeTone' software modelling technology used in the company's AxySys 212 guitar amp to model the sounds of a range of classic amplifiers and speakers. Amp Farm was the first plug-in to bring these 'virtual' guitar amps to the TDM system - something of a revolution for guitarists recording with Pro Tools.

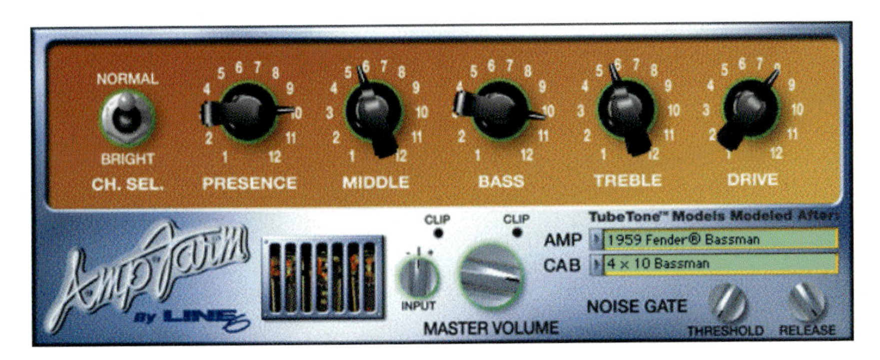

Figure 3.3 Line 6 Amp Farm.

The software couldn't be easier to use, featuring a user interface designed to look like the panels of the modelled amps. You choose your amp and cabinet from pull-down menus and the chosen amp's controls are replicated onscreen - so you see exactly what you would see on the control panel of the actual amplifier itself. For example, the Fender Twin has a Bright switch, Drive, Treble, Middle, Bass and Vibrato Speed and Intensity controls - no Reverb control as it doesn't have a spring reverb simulation, unfortunately. Amp Farm also has an input gain control and a master volume control along with threshold and release controls for a noise gate. Presets are provided to simulate the sounds of several of the most sought-after guitar amplifiers. The tone controls' centre frequencies and response curves, and the all-important interactivity between the controls were analysed and carefully emulated for each amplifier - with surprisingly usable results. Several popular speaker cabinets were also analysed, along with the results of recording these using various microphone positions.

Using the Amp popup menu at the lower-right of the plug-in window, you can choose which amp 'head' you want to use. The list includes a 1959 Fender Bassman, a 1964 Fender Blackface Deluxe, a 1967 Fender Twin, a 1960 Vox AC30, a 1966 Vox AC30 with Top Boost, a 1965 Marshall JTM45, a 1968 Marshall 'Plexi', a 1986 Marshall JCM800, a 1995 Mesa Boogie 'Recto' Head, a 1994 Mesa Boogie Trem-O-Verb, a 1989 Soldano SLO Head, a 1987 Soldano X-88R Preamp and a 1996 Matchless Chieftain.

Using the Cab popup menu underneath this, you can select the type of cabinet you want to use - and the type of sound that would be achieved using different microphone positions. Here, the list includes 4 × 10 Bassman, 4 × 10 Bassman (near), 4 × 10 Bassman (off-axis), 4 × 10 Bassman (near/off-axis), 4 × 12 Marshall, 4 × 12 Marshall (near), 4 × 12 Marshall (off-axis), 4 × 12 Marshall (near/off-axis), 2 × 12 Twin, 2 × 12 Twin (near), 2 × 12 Twin (off-axis), 2 × 12 Twin (near/off-axis), 2 × 12 AC30, 2 × 12 AC30 (near), 2 × 12 AC30 (off-axis), 2 × 12 AC30 (near/off-axis), 1 × 12 '52 Deluxe, 1 × 12 '52 Deluxe (off-axis), 1 × 12 '64 Deluxe, 1 × 12 '64 Deluxe (off-axis), Big Cab, Big Cab (off-axis) - or 'none' if you simply want to simulate the sound of the amplifier head on its own.

A total of 36 preset combinations are provided as plug-in settings accessible from the standard popup menu just above the plug-in window. These include AC 30 Non-Top Alternative, AC 30 Non-Top Driven, AC 30 Top-boost, AC 30 Top-Boost Crunch, AC-30 Top-Boost Beatles, Bassman Blues, Bassman Semi-Clean, Boogie Recto Rocks, Boogie Recto Scooped, Boogie Recto Thermometer, Boogie TOV Crunch, Boogie TOV Grunge, Boogie TOV Insane Gain, Deluxe Blackface Clean, Deluxe Blackface Driven, JCM-800 Extra Heavy, JCM-800 Heavy, JCM-800 Hollow, JMP 100 '76 50-Watt, JTM-45 '65 Crunch, JTM-45 Clean '65, Matchless Fat Clean, Matchless Growl, Matchless Long Drive, Plexi Light Crunch, Plexi Maxed Out, Soldano SLO Blues, Soldano SLO Crunch, Soldano SLO Scoop, Soldano X88 Big Hair, Soldano X88 Grit, Soldano X88 Lead, Super Diddley, Twin Bright, Twin Dark, and Twin Spy Soundtrack.

Ever wished you could record guitar without having to mic up an amp and blast at full volume? Or recorded some guitar and then wished you could go back and adjust the tone? Or try out a part with an amp you don't even own? With Amp Farm, you can record direct, add the amp and/or cabinet sound later, choosing the best mike position without having to go out into the studio to move microphones around, and then change all this any time you like. You can save your favourite setups for instant recall, and all the controls are all automatable - so you can even make the settings change continuously while the guitar is playing back!

I had the opportunity to compare the simulations of the 1960 AC30, the 1964 Fender Blackface Deluxe, the 1959 Fender Bassman and the 1986 Marshall JCM800 at a 'shoot-out' session at London's Planet Audio Studios. The Bassman and the JCM800 simulations were not close enough for my liking - although they were somewhere 'in the right ballpark' and very usable-sounding in their own right. But I was amazed at how close the AC30 and the Fender Deluxe simulations were - especially on lead guitar parts where I found it very difficult to tell the simulations and the real amps apart. OK, so I wouldn't really swap my blackface Vibrolux Reverb for this software. But if you have guitar parts recorded into Pro Tools, Amp Farm is to die for - giving the very next best results to having the real thing!

Line 6 Echo Farm

After Amp Farm, Line 6's next move was to use software to model the various 'stomp boxes', pedals and other effects units popular with guitar players everywhere – many of which are no longer in production. Now this software is available as a Pro Tools TDM plug-in called Echo Farm.

Note Echo Farm requires MIX hardware, and will not run on older NuBus or pre-MIX PCI TDM systems. Echo Farm also will not run as an RTAS or AudioSuite plug-in and cannot be used, for instance, with a Digi 001 system, which relies on 'native' processing for its plug-in power.

The interface is very 'clean' and dead simple to use (it would have to be for guitarists, wouldn't it!), with several controls common to each effect, while others change when you select a different effect. At bottom-left of the window a popup menu lists all the available effects, including classics such as the Maestro EP3 Echoplex, the Roland RE101 Space Echo, the TC2290-based Dynamic Delay, the Electro-Harmonix Deluxe Memory Man and the Boss DM-2 Analog Echo.

At the top left of the window there is a silver Delay Time knob, with a display in milliseconds next to this. To the right there is a BPM display that you can use to set the delay by entering the tempo in BPM if you prefer. The associated Note Value buttons lets you select a note value for this, and a Tap Button is provided for you to tap the tempo if you favour this method. Next to this is a button marked Time Ramp. This lets you choose how Echo Farm will respond when you change the delay time. With a tape-based delay, for example, changes in delay time are effected by moving the tape head or by varying the speed of the tape. The delay time does not change instantaneously to the new value with these machines, so you can distinctly hear it ramp up or down to the new value – while software can switch delay times virtually instantaneously. With this button on, the software mimics the way a tape-based system would change the delay time.

The rest of the controls that appear on-screen represent the similarly labelled knobs and switches on the original effects that are being modelled. For example, the Maestro EP3 effect simulates the classic Echoplex delay unit which uses a special cartridge of looped 1/4-inch tape that wraps past separate record and playback heads. With this design, the position of the playback head can be moved to adjust the delay time from 60 to 650 ms. The software gives you up to 2.5 seconds of delay - and no worn-out tapes! Just as on the original, controls are provided for echo Repeats and Mix between original and effect, and the software adds Bass and Treble controls.

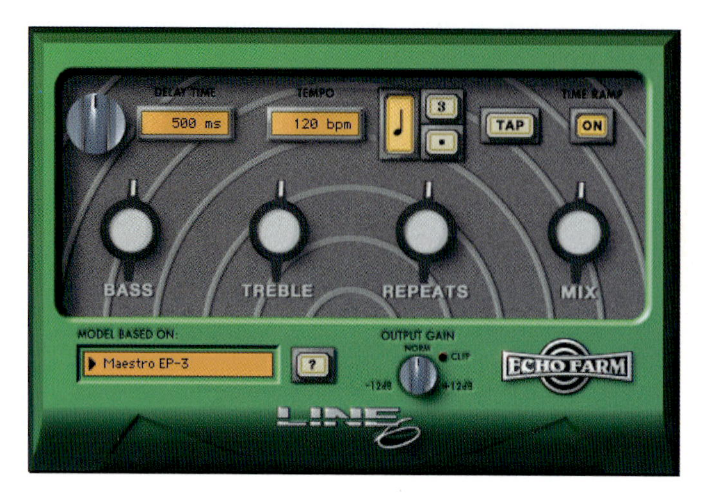

Figure 3.4 Echo Farm Maestro EP-3 effect.

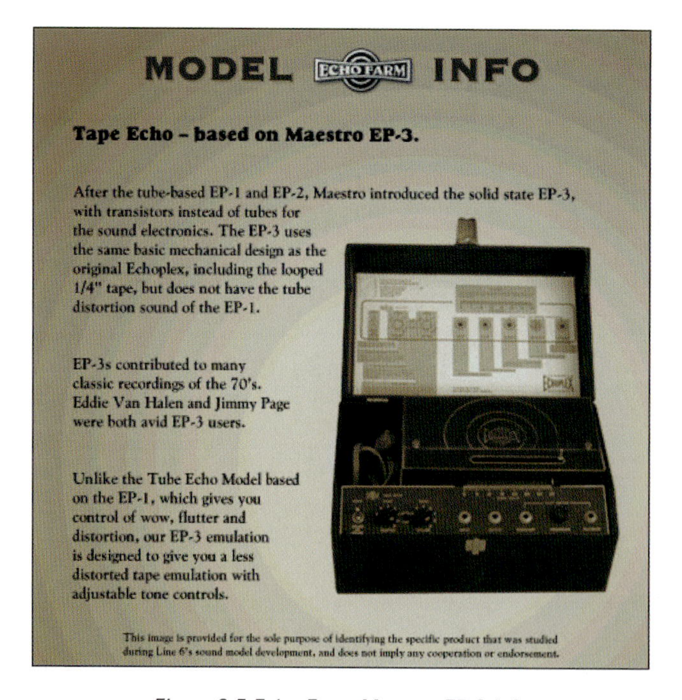

Figure 3.5 Echo Farm Maestro EP-3 info.

The Roland Space Echo was always one of my favourite effects, and I bitterly regret selling the one I owned when I was a starving songwriter. Instead of having one movable playback head (like the Echoplex) this machine has three stationary

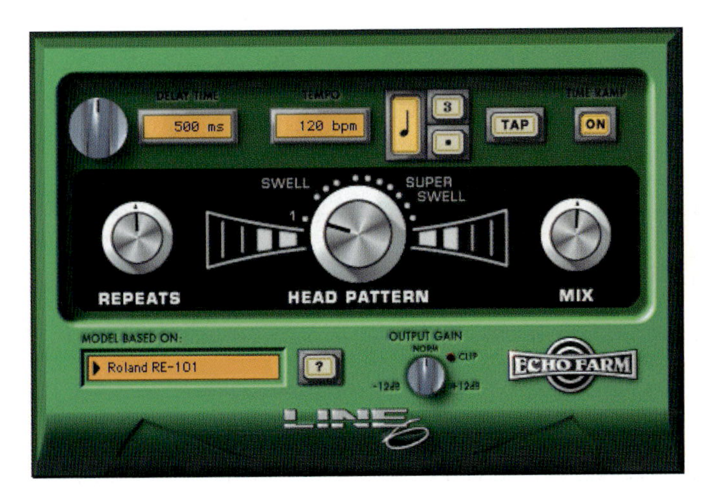

Figure 3.6 Echo Farm Roland RE-101 effect.

MODEL ECHO FARM INFO

Multi-Head – based on Roland RE-101 Space Echo.

Long before Boss pedals, the Space Echo was Roland's first venture into the world of effects processing. Instead of having one movable playback head (like the Echoplex) this machine has multiple stationary heads. You change delay times by switching amongst these heads, and then fine-tune delay time with a motor speed control. The groovy part is that you can play back on multiple heads at the same time to get multi-tap delay effects, all with the great tone that only tape delay (or Line 6 emulation) can deliver.

The Space Echo allowed six choices of from the "mode selector" knob: you could run any of the three heads (allowing one of three delay time ranges), plus three "Swell" modes which used heads in combination.

Echo Farm's "Head Pattern" knob lets you choose from a single head, one of the three Space Echo "Swell" patterns, or seven Line 6-crafted combinations of up to *four* emulated heads.

SPACE ECHO RE-101

This image is provided for the sole purpose of identifying the specific product that was studied during Line 6's sound model development, and does not imply any cooperation or endorsement.

Figure 3.7 Echo Farm Roland RE-101 info.

heads. You change delay times by switching amongst these heads, and then fine-tune delay time with a motor speed control.

The software version allows you to choose various combinations of heads via the multi-position Head Pattern dial which corresponds to the original Space Echo's head selection dial. On the original Space Echo, this head selection control had six positions: each of the three heads solo'ed, plus three combinations of the heads labelled 'Swell'. The Space Echo's designers had to give access to the three individual heads in order to provide the various delay times; the software is free of this limitation, so there is a single one-head selection as the first position on this control. The next three selections correspond to the Space Echo's three 'Swell' settings. A further seven selections that are not available on the original use various combinations of up to four modelled tape heads - arranging them from less to more as you turn the dial toward its four-heads-at-once setting.

So - if you play guitar and you have a Pro Tools TDM MIX system, you will want this plug-in. And you can get great results using Echo Farm on vocals, keyboards and many other instruments. These effects work brilliantly, and Line 6 has chosen the selection wisely.

Aphex Aural Exciter

The Aphex Aural Exciter has become a standard effects unit in studios around the world since its introduction in 1975, allowing engineers and producers to bring out more detail or add more 'air' to their mixes. Basically, the Aural Exciter processes the audio to recreate and restore missing harmonics to revive the natural brightness, clarity, presence and intelligibility of an original performance or to enhance the sound of any recorded audio. This is now available as a Pro Tools TDM plug-in - based on the Aural Exciter Type III technology.

The user interface offers a wide range of controls including faders and switches, along with accurate metering of 'drive' and output levels. You set the Drive switch to control the input sensitivity of the processor and use the Level fader to avoid clipping. The fader control section is where the real action takes place. Here the user can control the bandwidth of the side-chain at the point where the harmonic excitement effect is taking place - using a high-pass filter. The Tune fader lets you choose the range of frequencies in the side-chain that will be enhanced and then mixed in with the original signal at the output. You can adjust the high-pass filter to accentuate the response and you can also adjust the amount of harmonics being generated and the balance between odd and even harmonics.

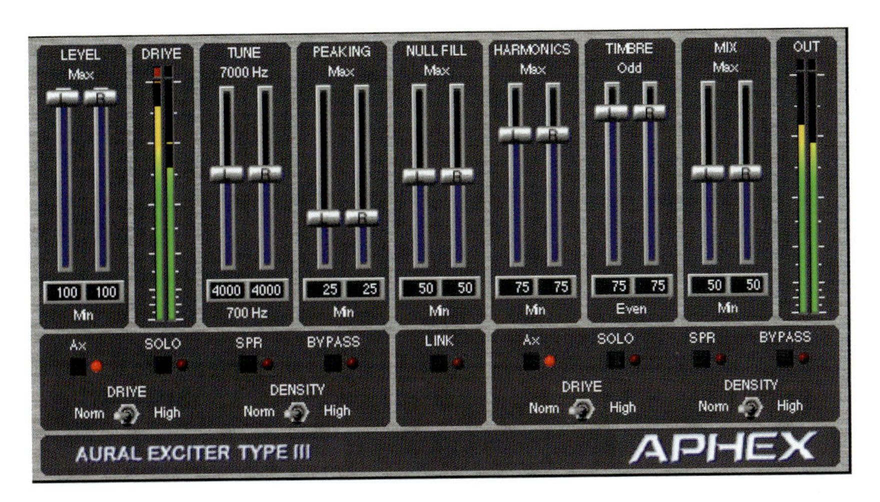

Figure 3.8 Aphex Aural Exciter.

In action, the Exciter makes things sound brighter and harmonically more interesting – and without increasing the peak level of the audio material. I tried the Exciter on a solo electric guitar which was not 'jangling' enough. Lo and behold – the Exciter brought the 'jangle' back, like magic! And after fiddling with the controls for a few minutes the sound got even better, making the guitar arpeggios 'ripple' much more delicately while they had sounded a little clumsy beforehand. So should you buy the plug-in or the hardware version? Well, according to Aphex, 80% of the beta testers think it sounds as good or better than the hardware version – and the software version is claimed to handle sibilant frequencies much better. As a Pro Tools user, I would definitely prefer to use the Aural Exciter plug-in on my mixing sessions – both for the convenience and for the automation possibilities.

Aphex Big Bottom

'Beefing up' the bottom end (i.e. enhancing the low frequencies) in your sound effects and music can be particularly important for multimedia applications to counteract the effects of small speakers and low listening levels, which both tend to make the bass sound weak. The primary function of Big Bottom is to help create stronger, more powerful bass along with bass density and 'endurance', without significantly increasing the peak level of your audio material. This is a different process from the one used in Waves MaxxBass that generates additional harmonics at lower frequencies – which Aphex claim can result in boosted peak levels that may lead to distortion.

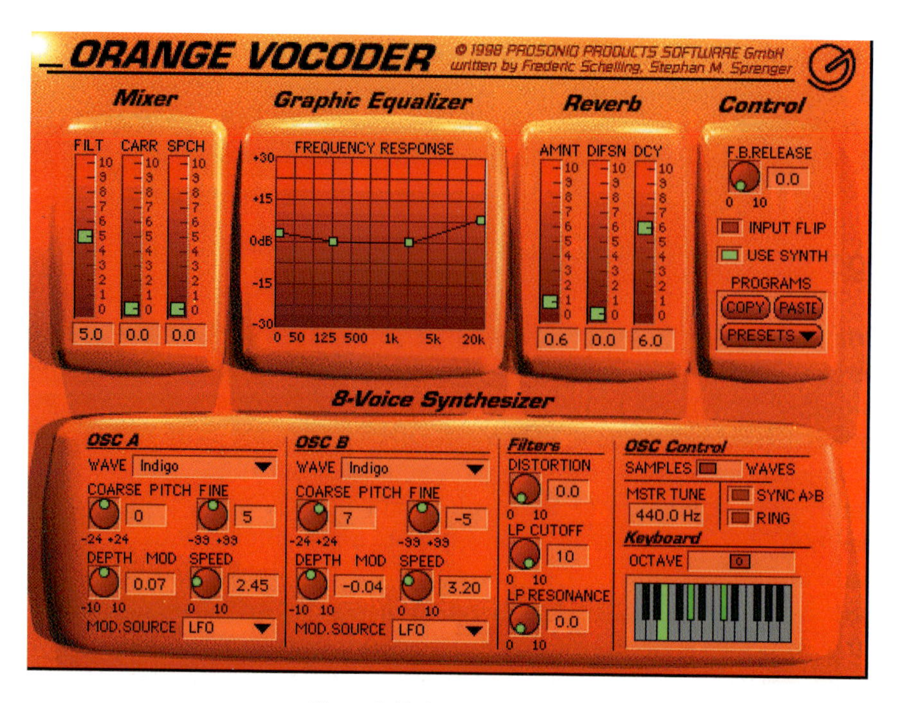

Figure 3.10 Orange Vocoder.

There are 32 preset effects accessible from the Presets popup menu at the middle-right of the plug-in window. These include Default, Full Chord, Rotating Robot, Upset, Robot Voice, One Note Voc, Ethereal Voices, fmaj Vocoder, Weird Talk, Jazz Vocoder, Ring Mod Vocoder, Weird Ring & Sync, Talking Voices, Synthetic Speech, Narrator II, Rubber Tongue, AM Radio, Synth 1, 1000 Flutes, Talking of Flutes, String Arpeggio, Crystal, Phat, Synth 2, Squared Voc, Dephective, Noise Only, Mosquito, Radar Pulse Train, On The Bridge, Sync'n'ring and Prutz.

Vocoder effects are coming back into fashion these days, especially for 'pop' and 'dance' tracks where producers and remixers are always looking for those special sounds to 'hook' the audience. The Orange Vocoder plug-in gives Pro Tools users access to these sounds, with plenty of presets provided to get you started.

Lexicon LexiVerb

Lexicon's hardware reverb units are the most sought-after units in professional studios worldwide, so this plug-in has a lot to live up to. The LexiVerb's main window has a large 3D frequency response graph in the central position which you can drag laterally to view from any perspective – a useful practical application of Apple's QuickDraw 3D. This graph has three bands which show the reverb's amplitude and decay time in the low, mid and high frequency bands – a great help when setting your parameters. One word of warning here, though. Quickdraw 3D allocates its memory from the System's allocation, so you will need extra RAM to accommodate this. Immediately below the frequency response graph there is a 'soft' row of popup controls for the extended parameters of some of the algorithms, while at the bottom of the window there is a diagrammatic representation of the signal flow through the selected processor algorithm, again with popup controls to let you set the parameters visually. You can also switch the display to show the Macro Editor. A Macro is a fader that can simultaneously control as many as four parameters – changing these independently under control of the one fader – and you can create up to six macros per preset, which will then appear in the 'soft' row of controls.

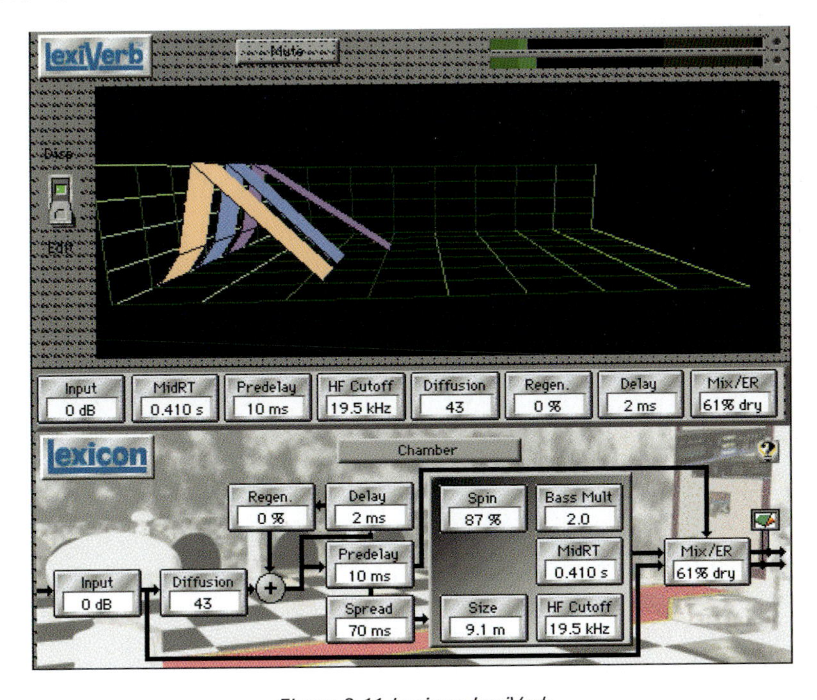

Figure 3.11 Lexicon LexiVerb.

One of the beauties of TDM plug-ins is the high level of automation which plug-in designers are able to incorporate into their software. The LexiVerb is no exception here – letting you create dynamic automation for virtually all LexiVerb parameters. You choose which of these parameters will be recorded by Pro Tools, enable one of Pro Tools' automation modes, hit Play in Pro Tools to begin recording – then start making your moves. You can only move the plug-in parameters you have enabled for automation one at a time using a mouse, of course, so you may find yourself wanting to buy a control surface of some sort - ProControl, Control|24, or the Mackie HUI, for example.

Of course, what really matters are the sounds you can get out of the LexiVerb. You will not be disappointed here! With careful programming you can get great effects with LexiVerb, although it could take newcomers quite some time to learn how to use all the parameters to their best effect, so you want to know about the presets. OK, the presets menu has folders for Chambers, Gates, Inverses, Plates and Special Effects - and each of these contains 20 to 30 descriptively named presets such as Drum Cave, Vocal Gate, Stutter Echo, Clean Guitar Plate and Chaos Engine. This is a much better selection than you will get with any other reverb plug-in for Pro Tools TDM right now - I know, I've tried them all! And the great thing about the LexiVerb presets is that so many are usable, just as they are, for a wide variety of applications.

I tried various of the presets on a fairly dry recording of a drumkit to see how the LexiVerb would handle percussive sounds - always a good test! I immediately found several presets with very usable reverb settings for my dry kit. Then I started on the rest of the presets ... and then it was the morning and I realized how much fun I had been having! The only thing I didn't like is that it is too easy to accidentally set a different value to any of the popup controls when you click on these and change your mind. This is undoubtedly the 'Rolls Royce' of third-party reverb plug-ins for Pro Tools TDM systems - with great reverbs and effects to suit most applications.

Drawmer Dynamics for TDM

The Drawmer Dynamics plug-in, based on the Drawmer DS201 and DL241 hardware units, provides first-rate limiting, compression, expansion and gating for Pro Tools TDM systems. The user interface emulates the controls on the original Drawmer models - which should appeal to existing users of these.

When you install the software, there are two configurations of the plug-in provided. The DrawmerECL features an Expander, a Compressor and a Limiter. The DrawmerGCL features a Key section, a Noise Gate, a Compressor and a Limiter. The Compressor sections common to both plug-ins combine both traditional compressor styles with a 'soft-knee' approach to provide flexible creative control. A Peak Limiter is also included to catch any peaks a slow compressor attack might miss - particularly useful for digital recording. The Expander module allows a wide range of adjustment and will automatically vary ratio and release times depending on the dynamics of the input signal - making it easier to avoid cutting off quiet word-endings, for example. The Key section lets you designate any track as an external trigger source for the Gate module. A key is often used with a gate to achieve tighter synchronization between two parts played 'live'. The gate only opens to let you hear the audio when audio is playing on the key track. So, if the bass drum is played very accurately, but the bass guitar less so, you can gate the bass guitar track and key it using the bass drum track so you only hear bass guitar notes play while the bass drum plays - not starting just a little before or ending just a little after as you would hear this if the bass player was playing slightly off-time.

Three presets are provided for the Expander/Compressor - BassGentle, CleanVox and ExpandUp. Drawmer has been more generous with presets for the Gate/Compressor, with eight presets provided here: 1000Drums, Drive, KeyFilterNoGate, Looney, SnareAbuse, SqrSnare, Stoccato and VoxOver. Some of these are intended to clean up the sound without altering it too much, while others are intended for use as creative effects to deliberately change the sound. Using the 'Drive' preset on a bass guitar, for example, makes the notes punch through much more aggressively, adding an exaggerated 'click' to the initial attack. Bass Gentle, on the other hand, cleans up any spill between the wanted notes while taming any unevenness in the level without affecting the sound noticeably. Several instructive demo sessions are included with Drawmer Dynamics. These illustrate how to use the plug-in in typical recording scenarios. For example, the 'Wrong & Right' Demo illustrates the right and wrong way to use a Gate and the 'Noise Removal' Demo illustrates how to use the Expander to effectively remove excessive background hiss from a short drum loop.

Note Drawmer Dynamics functions as a pre-fader insert, so its level is not affected by a track's volume fader (except when used on a master fader). For this reason, clipping can occur if you boost its gain to extremes. This is particularly true on tracks recorded at high amplitude. Watch the on-screen metering carefully to identify and rectify clipping if it occurs.

Drawmer Dynamics provides several types of LED-type signal meters and indicators. With the mono versions of these plug-ins, a Peak Input LED positioned above the main controls near the top left of the plug-in window indicates peaks in the input signals. At 1 dB below digital full scale the LED is amber. At -0.01 dB, just below full scale input, it is red.

Note If the Peak Input LED frequently flashes red, the input signal is clipping and should be reduced.

When you insert a stereo Drawmer Dynamics plug-in, you will see a Stereo Input Balance control at the top of the plug-in window. This lets you correct any imbalance between the two sides of the stereo input - without introducing any gain. You should make sure that your mix is properly balanced before applying dynamics processing - as stereo dynamic control always works best if the mix is equally centred.

Figure 3.12 Drawmer Dynamics Input Balance control.

A row of LEDs above the Input Balance control shows the relative input balance and how far off-centre the incoming stereo mix is. A perfectly matching stereo input should produce virtually no LED activity, regardless of audio level. When centred, the Stereo Input Balance control has no effect. Moving the control to the left attenuates the right input signal by up to 6 dB and pans the image to the left. Moving the fader to the right attenuates the left input signal by up to 6 dB and pans the image to the right. Note that Peak input LEDs for the left and right channels appear at the left and right ends of this meter.

The 16-element LED VU Meter at the far right of the plug-in's window shows the output levels after processing. The metering ranges from 60 dB below full scale up to 0 dBfs. Green LEDs indicate levels below −4 dB. Yellow LEDs indicate levels from −3 dB to −1 dB. An amber LED indicates −0.5 dB and turns red for peaks held at 0 dBfs.

The Expander module

The Expander section of the Drawmer ECL Plug-In is modelled on the Drawmer DL241 and DS201 hardware processors. The controls are grouped into two sections containing the Compressor controls and the Expander controls. The Compressor section is identical to the Compressor section in the Drawmer GCL plug-in and will be described separately. The Expander has controls for Threshold, Ratio, Attack, Release and Range. The Threshold control sets the level below which expansion starts to take place. Threshold range is from −70 dB to +12 dB. Setting this control above 0 dB permits upwards expansion, whereby over-compressed material can be re-expanded. The Ratio control sets the amount of attenuation applied to the signal as it decreases below the Threshold level. The Attack control sets the rate at which the Expander opens from a closed state and the Release control determines the speed at which the Expander closes to the Range setting once the input signal has fallen below the Threshold level.

Figure 3.13 Drawmer ECL Dynamics.

A unique feature is the Auto-Adaptive Expansion. The Expander's Ratio control ranges from 1.1:1 up to 50:1 and then has two further settings - Soft and Softer. Engaging either of these causes the ratio and release times to be automatically varied according to the dynamics of the signal being processed. Low-level signals will be processed with a lower ratio of expansion while the residual noise during pauses will be processed with a higher ratio of expansion - resulting in greater attenuation. This makes the Expander easier to use effectively and more likely to preserve desired audio signals that are only slightly above the residual noise floor.

The manual contains plenty of tips on using the Expander section. For example, it recommends that you 'Set the Attack and Release controls based on the character of the audio material you want to expand. For material with long, legato release, set a long Release. For material with lots of low frequency content, set a slow Attack. For sharp, percussive sounds, set a short Attack. Set the Ratio control based on the character of the audio material. Try a maximum setting of 2.5:1 for vocals, and higher ratios for dynamic, full mix material. Adjust the Threshold control until the vertical VU meter and the Expander Gain Reduction (GR) meter begin to light. As a rule, the Threshold should be between 6 dB to 10 dB below the average input level. Use a section of audio material with pauses and adjust the threshold to be as low a dB level as possible while still attenuating the noise during pauses. Listen carefully to how the sounds come in after the pauses and how cleanly they fade away again. If the Expander changes the sound in an unacceptable way, the Threshold is probably set too high.'

The Gate module

Figure 3.14 Drawmer GCL Dynamics.

The Gate is part of the Drawmer GCL plug-in. This is based on the Drawmer DS201 hardware dynamics processor – widely used in recording studios to create effects such as Gated reverb and enhanced percussive kick-drums. It features variable high-pass and low-pass filters for frequency-conscious gating and has comprehensive envelope control with attack, hold, release and range controls. A ducking mode is provided for voice-overs, along with comprehensive side-chain filtering and a Key listen facility to allow monitoring of the Key Filter effect.

The controls are arranged into three sections – one each for the Key controls, the Compressor controls and the Noise Gate controls. The Compressor section is identical to the Compressor section in the Drawmer ECL plug-in and will be described separately.

The Key controls

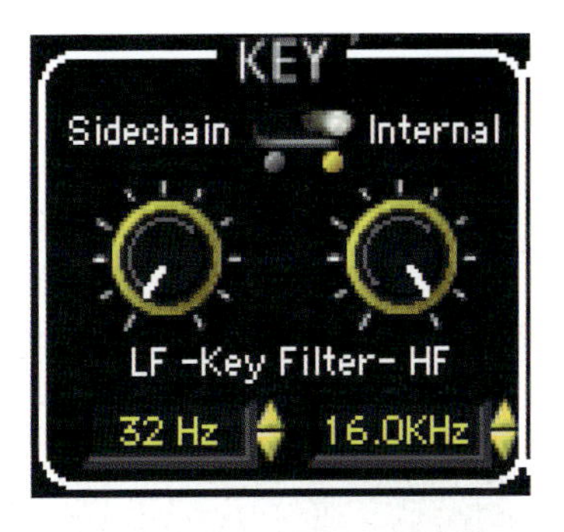

Figure 3.15 Drawmer Dynamics Key section.

The Key section has controls for the HF and LF Key Filters and a switch to enable the side-chain.

With side-chain processing enabled, you can control gating of a mono or stereo audio signal using the dynamics of a different mono signal - the key input. The idea here is that you set the key input filters to define a specific frequency range in the key input signal to trigger the gating action.

A typical application of this technique is to use a bass drum track to trigger a bass guitar track that is intended to play in step with the bass drum. Gating the bass guitar track so that it is only heard when the bass drum plays ensures that they both sound exactly in step. Similarly, a rhythm guitar track can be used to gate a keyboard pad so that the pad sound plays with the rhythm of the guitar - almost as though the guitarist is strumming the keyboard.

To set this up you simply choose the input or bus carrying the signal you want to use to trigger your gate from the Side Chain Input pop-up menu just above the main plug-in controls. In Pro Tools, for example, this is identified by a key icon. Now, when you play back, gating will be controlled by the signal from the input or bus you have chosen as your key input.

Note If you want to listen to the audio source you have selected to control the side-chain input to make sure that you have the correct source, enable the Key Listen switch in the Noise Gate section.

You can filter the Key Input so that only certain frequencies trigger the plug-in using the LF Key Filter and HF Key Filter controls to adjust the roll-off frequencies of the Key filters. The LF or low frequency filter ranges from 32 Hz to 4 kHz and attenuates signals below the cut-off frequency at 12 dB per octave. The HF or high frequency filter ranges from 250 Hz to 16 kHz and attenuates signals above the cutoff frequency. When both filters are used, only the frequency band between the two settings remains. A common technique is to use these controls to filter a drum track so that only specific high frequencies (a hi-hat, for example) or low frequencies (a tom or a kick, for example) trigger the gate.

Figure 3.16 Drawmer GCL showing the side-chain input popup menu.

> **Tip** Conventional equalizers seldom have a sharp enough response to remove unwanted noise without also removing some of the wanted audio signal. The Drawmer GCL Key input filters can be used instead of conventional equalizers in situations where the frequency range of the audio signal you are working with does not occupy the full audio spectrum. For example, electric guitar produces little below 100 Hz or above 3 kHz, so setting the Gate to Key Listen mode will enable you to use the filters to exclude much of the amplifier hum at the low end and hiss at the top end while having little effect on the sound of the guitar. The same is true for acoustic guitar, so these filters can also be used to reduce fret noise or a player's breathing.

The Noise Gate

The Noise Gate section has a Key Listen switch which you can use to monitor the Key input signal during setup, along with the usual controls for Threshold, Attack, Hold, Release and Range. Switches are provided to enable the ducking mode and to bypass the Gate section. In Gate mode, a signal above the Threshold will cause the Gate to open. In Duck mode, the audio passes unattenuated until the signal exceeds the Threshold. Ducking is mainly used for ducking the music behind voice-overs, although it can also be used to remove pops and clicks.

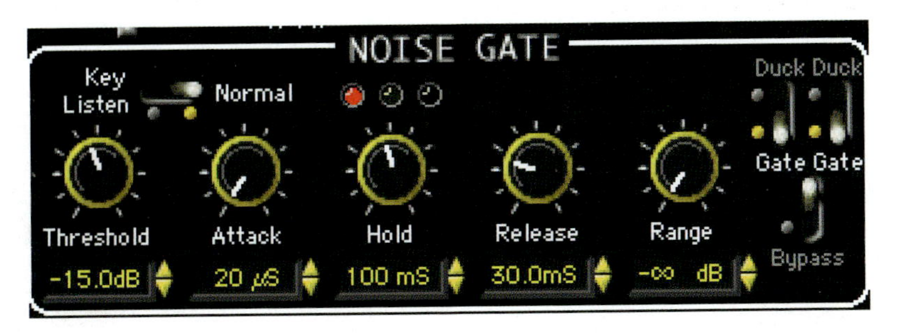

Figure 3.17 Drawmer GCL Noise Gate.

The three LEDs positioned above the Hold control in the Noise Gate section show the active state of the Gate Envelope: when the Gate is closed, the red LED is lit. When the Gate is open, only the green and yellow LEDs are lit. When the input signal falls below the threshold, the yellow LED will light, then fade over the duration of the release time. In Duck mode, the green LED indicates an un-triggered state.

The Threshold control lets you set the level below which gating takes place (the level at which the output of the Key Filter triggers the Gate). The Attack control lets you set the rate at which the Gate opens from a closed state. The Hold control lets you set how long the Gate is held open after the signal falls below the Threshold setting. The Release control lets you set the speed at which the Gate closes to the Range setting after the input signal has fallen below the Threshold level. The Range control lets you set the level to which the Gate closes after the input signal has fallen below the Threshold setting. When Ducking is enabled, the Range control sets the level to which the signal will drop when the Ducker is triggered from the key input.

The Compressor

The Compressor section appears in both the Dynamics ECL and Dynamics GCL plug-ins. This features a 'soft-knee' design offering subtle level control where the sound is changed as gently as possible as the signal first exceeds the threshold. If the input signal increases further, above the knee the compressor's transfer characteristic becomes a straight line once more. The Compressor also features Auto-adaptive Attack and Release and has Automatic Gain Make-up which calculates the most suitable output Gain position for maximum dynamic range. In short, there are a heck of a lot of features packed into this section. Let's take each control in turn and describe its action.

Figure 3.18 Drawmer Compressor.

The Threshold control lets you set the input level above which gain reduction will be applied. You can judge this partly by ear and partly by observing the compressor gain reduction meter - the row of LEDs above this.

The Ratio control lets you set the compression ratio (from 1.1:1 to infinity) applied after the 10 dB soft-knee region of the Threshold is exceeded. Higher ratios are more audible in operation, especially when using high levels of gain reduction. The soft-knee feature helps to minimize this, allowing higher ratio settings to be used than on conventional compressors.

The Attack control sets the rate at which the Compressor will respond to input signals that exceed the Threshold level setting. A slow attack time is often used to accentuate the beginning of percussive or plucked sounds such as drums, basses and guitars. This is safe to use because the Limiter will prevent problems occurring due to any peaks that get past the Compressor.

The Auto switch disables the Attack and Release rotary controls and instead automatically optimizes the Attack and Release times to suit the dynamics of the material being processed. This is the best option to use to control signals with widely varying dynamics or when you are compressing the whole mix.

The Release control sets the rate at which gain returns to normal after the input signal level has fallen below the Threshold. Release time should be set short enough so that the gain has returned to normal before the next signal peak occurs - to avoid audible gain 'pumping'.

The Gain control has two functions, depending on the current state of the Auto Gain Make-up switch. When Auto Gain Make-up is off, up to 36 dB of gain can be added to compensate for level changes caused by Compression. With Auto Gain

Make-Up on, Gain Make Up is adjusted automatically when the Compressor's Threshold, Attack and Ratio or Limiter Threshold are adjusted – keeping the output level as close as possible to the Limit Threshold. In this case, the gain control becomes a +/− 9 dB trim with a centre-zero, intended for use as a fine trim to optimize the output level.

If the Compressor module is used with higher ratios, normally greater than 20:1, it will function as a Limiter. There is also a separate auto-adaptive Peak Limiter. This not only catches any peaks that a slow Compressor attack might pass through unprocessed, but also allows the user to set an absolute output signal level that will not be exceeded.

The Limit Gain Reduction Meter, a horizontal row of five LEDs above the Limit threshold control, indicates the gain reduction in dB. The Limit control lets you set the level threshold that the output signal is not allowed to exceed. Normally, you should leave this at its maximum setting – so that it just catches the highest peaks. Limiting is extremely valuable in digital recording where an absolute maximum recording level exists. On the other hand, some engineers deliberately set the Gain sufficiently high that level-pumping effects are clearly audible – for creative effect.

Finally, there is a Bypass switch for the Compressor and Limiter section that can be used to compare the unprocessed signal with the processed signal.

Note Setting up the Compressor is simpler if the Expander or Gate is initially set to Bypass and the Peak Limiter threshold set to maximum. This allows you to adjust the Compressor in isolation.

Summary

Drawmer Dynamics packs all the complexity and flexibility of the original hardware units that it emulates into a pair of compact and powerful software plug-ins for Pro Tools TDM systems. Professional engineers who already know the Drawmer user interface will get along with these straight away. If you are new to Drawmer, it will take you a little while to get used to all the controls and to understand what they do – but it is definitely worth the effort.

Chapter summary

The Focusrite D2 and D3 plug-ins are the kind of standard outboard units that every studio should have, and the same goes for the Drawmer Dynamics plug-ins.

Depending on the kind of work you do, the others are more optional. If you record guitars you are going to want Amp Farm and Echo Farm for sure - and these are great for mixing and remixing as well. For acoustic guitars and bass instruments, and even for finished mixes, the Aural Exciter and Big Bottom effects can be very useful. The LexiVerb is great when you are looking for that special reverb preset to jazz up your mix, and, if you need a Vocoder, the Orange will do nicely . . .

4 Digidesign Virtual Instruments

Digidesign has followed the trend toward 'virtual instruments' by making its SampleCell sample playback device available as a software plug-in for Pro Tools TDM systems. This chapter contains a detailed review of the software.

The hardware version, SampleCell II Plus, is still available (priced at $1295), in the form of a PCI card containing DSP chips and RAM. A 'daughtercard' that mounts on this card is available (for another $395) that feeds SampleCell's eight 20-bit outputs directly into the Pro Tools TDM mixer. It can still make sense to buy this rather than the software, especially if you are running out of processing power on your CPU.

Digidesign also distributes a software simulation of the popular Access Virus analogue synthesizer along with a synthesizer/sampler/drum-machine package, the Studio 9000 from Koblo in Denmark.

Access Virus

Access Virus is a real-time TDM plug-in version of Access Music's Virus analogue modelling synthesizer. Access Virus lets you reproduce the sounds and performance features of the classic ARP, Moog, and Oberheim analogue synthesizers from the 70s – with the added benefits of digital control and stability.

The Virus synthesizer features Analogue, FM and Wave-shape synthesis and has a Matrix Modulation section, a Ring Modulator, Chorus and Delay effects, an Arpeggiator and a Vocoder. You can also use Virus to process external inputs from Pro Tools tracks. The arpeggiator syncs to MIDI beat clock so that it can be made to follow the tempo of your sequence.

Note Virus uses one DSP chip for each 16 voices (with a maximum of 128 voices) and because it operates in stereo internally, it requires the same amount of DSP whether you use it as a mono insert or a stereo insert.

Each Virus master plug-in provides 16 voices of polyphony and these 16 voices can be divided among up to eight 'virtual' Virus plug-ins for multi-timbral operation. Virtual plug-ins can be created by assigning the same Virus plug-in to multiple inserts. For example, assigning Virus 1 to three different inserts will create three virtual plug-ins, all of which share the same DSP and use the same pool of 16 voices. Virtual inserts are numbered a–h, and once created, appear in the MIDI Device/Channel Selector as Virus 1a, Virus 1b, Virus 2a, Virus 2b, and so on. This arrangement makes efficient use of DSP resources, since you could potentially create a synth bass, lead, pad, drums, and more, all from a single master Virus plug-in – using a single DSP.

You can insert a Virus plug-in on an audio track or master fader if you want to use its filters or other parameters to process that track's audio, but, normally, you will insert it on an auxiliary input and create a new MIDI track to control this. You can then play it using its on-screen keyboard, a MIDI keyboard controller, or a MIDI playback track.

Tip To play Virus, your MIDI keyboard or other MIDI controller must be enabled in the Input Devices dialog – and to be able to hear your MIDI tracks while recording, you should enable MIDI Thru in Pro Tools and disable Local Control on your MIDI keyboard controller.

The Virus plug-in has six 'pages' – the Oscillator page, the Filters and Envelope page, the LFO page, the EFX page, the Miscellaneous page and the Modulation Matrix page – containing the various controls. These can be accessed (hmm . . .?) by clicking on the name of the page you want, at the left of the keyboard that runs along the top of the plug-in's window.

The Oscillator page has controls for the various tone generators – the oscillators, the noise generator, the frequency modulator, and the ring modulator.

Figure 4.1 Virus Oscillator page.

The Filters and Envelope page has controls for the two filters and the two envelope generators. Here you will find all the usual controls for filter cutoff and resonance, and envelope attack, decay, sustain and release.

Figure 4.2 Virus Filters/Env page.

The LFO page has controls for the three Low Frequency Oscillators. LFOs generate low-frequency waveforms that can be used to change or modulate pitch, filter cutoff, volume and many other parameters. The most basic pitch or volume modulations can be used to create vibrato or tremolo effects, and, with three extremely flexible LFOs like these to play with, many more effects are possible.

Figure 4.3 Virus LFO page.

The EFX page has controls for Chorus, Arpeggiator, Vocoder and Delay. Here you can quickly get your sounds 'vibey' and 'happening' - a touch of flanging, automatic arpeggios - effects that will make your synth parts blend with or stand out in the mix.

Figure 4.4 Virus EFX page.

The Miscellaneous page has controls for global tuning, key mode, pitch bend, unison and other parameters. Here you can add a touch of portamento to make the notes in your solo glide between each other or detune an oscillator for a fatter sound in unison mode.

Figure 4.5 Virus Miscellaneous page.

The ModMatrix page has controls for routing the sources, destinations, and amount of modulation in a patch. Three separate routing matrixes allow virtually any control source to be routed to virtually any Virus parameter. For example, you can control the filter cutoff frequency with the mod wheel of a MIDI keyboard or control an oscillator's waveshape with a breath controller. Other possibilities include controlling filter or amplifier envelope parameters with key velocity or controlling LFO frequency with a foot pedal.

Figure 4.6 Virus Modulation Matrix.

The simplest way to play Virus is to use its on-screen keyboard. You can click one note at a time or use keyboard latch to hold multiple notes. You can also play Virus 'live' using a MIDI keyboard controller or use a Digidesign ProControl, Mackie HUI, or CM Automation MotorMix as a MIDI control surface for Virus. Virus comes with a library of over 500 preset sounds, so you are spoilt for choice here, and if you understand how to program an analogue synthesizer you will be 'in business' right away tweaking the sounds or creating your own custom patches.

Note The + Presets in the Input and Vocoder folders are designed for external signal processing and vocoding, so they require an audio input.

Now you may be thinking of Access Virus as a total replacement for or alternative to the hardware version, but it actually makes lots of sense to use the plug-in version alongside the actual hardware. For a start, you can swap single patches between the software and hardware versions. But, more importantly, you can use the hardware version as a control surface for the plug-in parameters. Simply route the MIDI data from the hardware virus through to the software virus. Then, moving a control on the Virus hardware will move the corresponding control on the Virus plug-in. So why bother with the software if you've got the hardware? Well, what's particularly good about the software version is the total recall and automation that you get - and the convenient integration into your Pro Tools environment.

Soft SampleCell

Soft SampleCell is Digidesign's first software sampler offering, based on the original SampleCell hardware that was provided on a PCI card. The original SampleCell card had a maximum of 32 Mb of RAM, while Soft SampleCell can use however much free RAM is available on your Mac - up to a limit of 1 Gb. In most other respects, Soft SampleCell is identical to its hardware forerunner. Samples are organized into Instruments, which are then loaded into Banks for playback using the SampleCell Editor application. Instruments may be mono or stereo and can contain a single sample or multiple samples playable across adjustable key zones - with up to six velocity zones that can be velocity-switched or cross-faded. Once loaded into a Bank, MIDI channels and audio outputs can be set up. Samples cannot be recorded directly into the SampleCell Editor, but audio samples recorded using any other application can be loaded into SampleCell

Instruments. Only one disc of samples is provided with the software, but sample libraries in SampleCell format are widely available. There is no import facility for other formats such as Akai or Emu, but third-party conversion tools, such as Osmosis from Bitheadz, are available that can convert to SampleCell format.

Setting up Soft SampleCell

Before you can use Soft SampleCell, you must configure several parameters such as the Sample Memory, the Number of Voices, and the Sound Output using the Soft SampleCell Setup dialog.

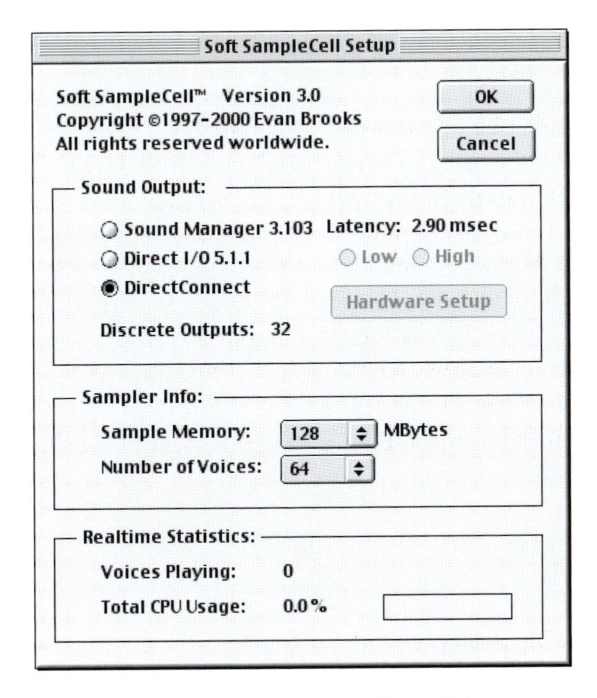

Figure 4.7 Soft SampleCell Setup dialog.

Soft SampleCell plays its samples back from your computer's RAM, so the more RAM you have installed, the better. You can set the amount of RAM to make available to Soft SampleCell using the Sample Memory popup. Of course, you have to make sure that you leave memory free for your other applications to use, so always leave at least 10% free to avoid problems. To check how much RAM you need for the Soft SampleCell Editor and all other audio applications you are likely be using at the same time, open all these applications, switch to The Finder (click on the desktop) then choose About This Computer from the Apple Menu.

The amount of available RAM will be shown as the 'Largest Unused Block', and you should set a maximum of, say, 80–90% of this amount in Soft SampleCell's Sample Memory popup.

Note Although Soft SampleCell supports up to 64 voices for sample playback, the actual number you can work with depends on the speed of your computer's CPU. With slower computers you won't get this maximum, so you should just allocate as many voices as you will need.

Soft SampleCell will play back 16-bit or 24-bit Sound Designer II files at either 44.1 or 48 kHz and can output audio using one of three options: using Apple's Sound Manager, or via Digidesign hardware using Direct I/O or DirectConnect. By default, Soft SampleCell uses the Apple Sound Manager with the sound output jack from your Macintosh for its audio connections. Unfortunately, you are restricted to just two outputs from Soft SampleCell using Sound Manager, so for professional work you will almost certainly need to use Digidesign audio hardware. If you run Soft SampleCell as a stand-alone application with your Digidesign hardware, the Direct I/O option lets you access up to 16 analogue or digital outputs from Soft SampleCell. If you are using Pro Tools or another DAE-compatible audio application such as Logic Audio or Digital Performer, then you should use DirectConnect. This allows up to 32 independent Soft SampleCell outputs to be routed within DAE-compatible applications.

DirectConnect works well with Soft SampleCell. When you save, close and then re-open your Pro Tools session, DirectConnect launches the Soft SampleCell Editor and prompts you to load the bank of sounds you last used with your session. DirectConnect enables third-party software synthesizers, samplers and drum-machines on the Macintosh platform to connect their audio outputs directly into Pro Tools mixer channel inputs. DirectConnect works either with Pro Tools MIX cards or with Pro Tools LE systems with software versions 5.01 or greater. DirectConnect also works with DAE-compatible applications such as Digital Performer, Logic Audio or BIAS Peak with version 5.01 or higher of DAE and the DigiSystem INIT. The specification allows up to 32 separate audio channel outputs connected from compatible host-based applications to be routed into Pro Tools so you can record, edit and mix the audio from these using Pro Tools. Unfortunately, there is no ASIO support, so you cannot use Soft SampleCell with Cubase VST, Nuendo or other applications that do not support the DAE.

Samples, Instruments and Banks

SampleCell organizes samples into Instruments that can then be loaded into Banks for playback. You cannot record audio into SampleCell - it is a playback-only sampler. Instead you import SDII files that you have recorded or sourced elsewhere, such as from the wide range of SampleCell CD-ROM libraries available from third-party sound developers. When you work with your own samples, it is a good idea to create folders for Banks, Instruments and Samples to keep everything organized - as these libraries are.

Samples are mapped to appropriate keys within each Instrument where the sound can be further enhanced using synthesizer-style controls. Once an Instrument has been set up, or selected from a library, you can open this in a Bank and choose the audio outputs and MIDI channels. Banks are typically used for organizing all the Instruments needed by a song or MIDI sequence into a single document. You can have just one Bank open containing all the Instruments for your session, or you can have multiple Banks open and active at the same time with different Banks containing Instruments for different songs or sets. Or you could use one Bank for a rock drum set, while another contains a jazz set. Whichever you prefer - it's a flexible system.

> **Tip** When you make up a working Bank, it can be useful to include the Stick Click and Tuning instruments supplied with Soft SampleCell. The Tuning instrument includes sine wave samples at 220, 440 and 880 Hz, while the Stick Click Instrument contains a side-stick sample for use as a click.

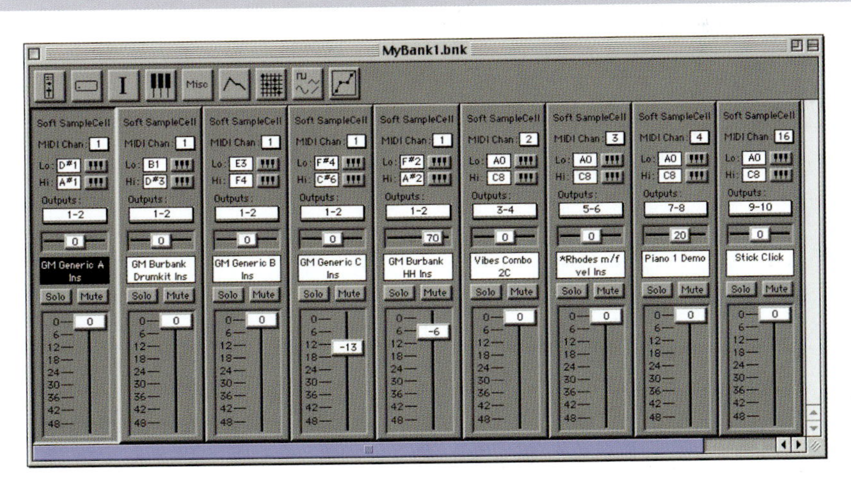

Figure 4.8 A SampleCell Bank showing nine Instruments.

Assigning audio outputs

Once you have set up a Bank with the Instruments you want to use for your session, you need to assign the MIDI channels for each Instrument so that they play back from the correct sequencer tracks, and set up the audio outputs, levels and pans. Each Instrument control panel looks something like an audio mixer channel – so you should get the idea easily enough. The outputs are selected in pairs so that each Instrument can be panned as necessary. The actual number of outputs available will depend on the sound output option chosen in the Soft SampleCell Setup dialog – according to which hardware you have.

Tip I strongly recommend that you run checks during setup using the SoundCheck Bank. This includes instruments set up to play through the left channel, the right channel, low frequencies and sample and hold. It is all too easy to get some of your cables round the wrong way and find that what should be in the left speaker comes out of the right speaker!

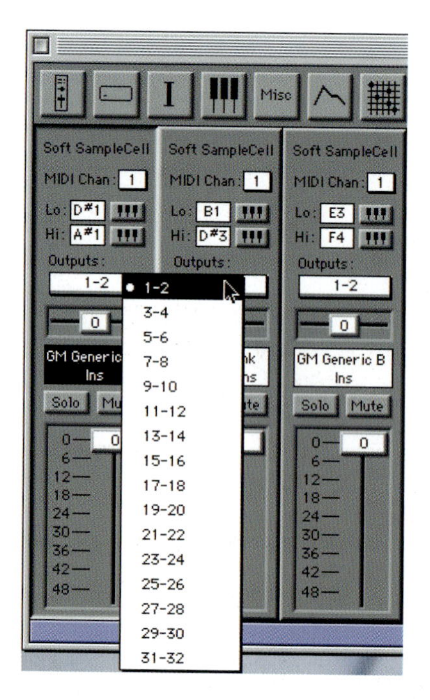

Figure 4.9 Setting SampleCell audio outputs.

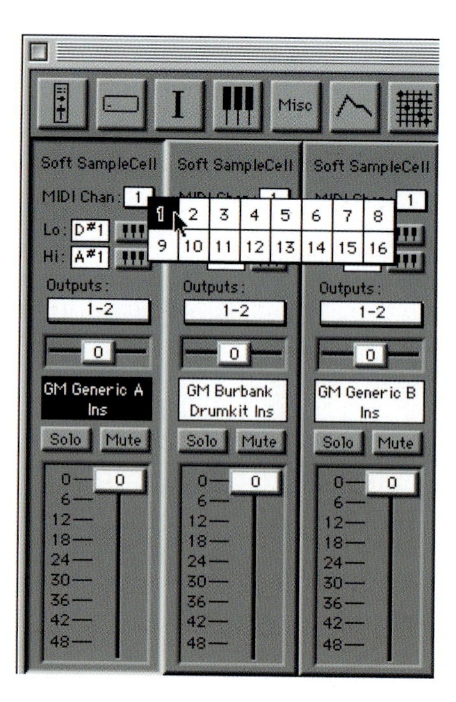

Figure 4.10 Setting SampleCell MIDI channels.

Setting up the MIDI

You can choose a MIDI channel for each Instrument using the MIDI channel pop-up menu on the Instrument strip.

When an Instrument is first loaded into a Bank, it plays back over the entire range of MIDI notes. This range can be changed for each Instrument. For example, to create a split keyboard you might want a bass sound to play only on notes below C2 and an organ to play only on notes C2 and above. The Hi/Lo Note popup keyboards provide quick ways to define split points by setting the low and high note range for the Instrument.

> **Tip** A quick way to set the ranges is to Control-click the Hi or Lo value and enter a new value by playing a note on your MIDI keyboard.

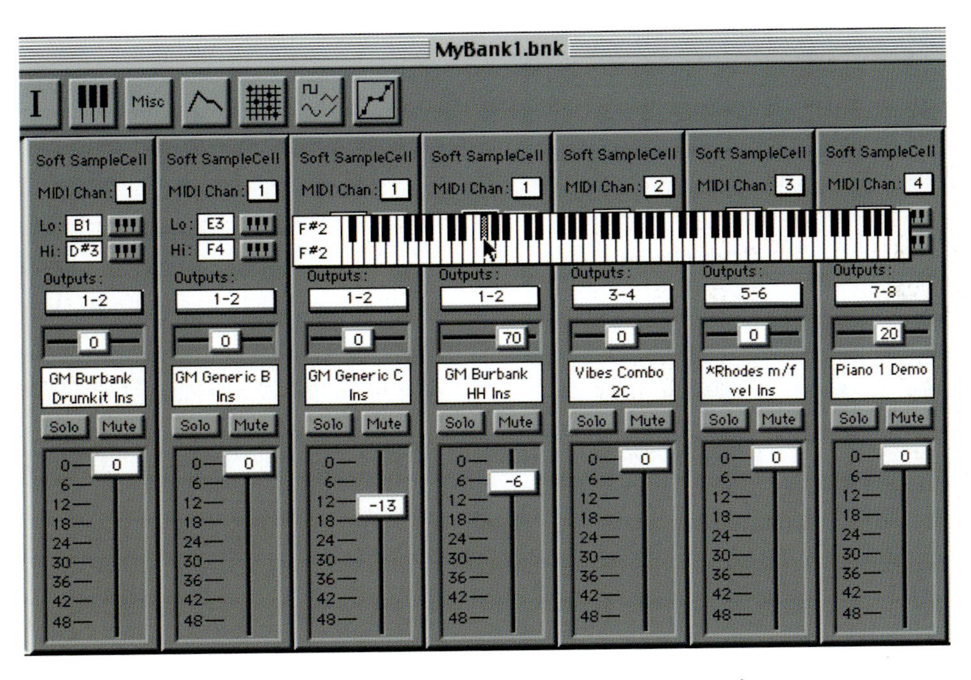

Figure 4.11 Setting SampleCell Hi/Lo MIDI note range values.

Instrument menu

You can use the Instrument menu to load samples into a new or existing Soft SampleCell Instrument.

The New Velocity Zone command creates a new Velocity Zone within the currently selected Key Group and loads a selected sample into that Velocity Zone. You can create a maximum of 6 Velocity Zones per Key Group.

The New Key Group command creates a new Key Group for a selected sample and loads the sample into one of up to 60 Key Groups. This new Key Group is automatically placed at the rightmost position of the Sample Map, with a default range of one note.

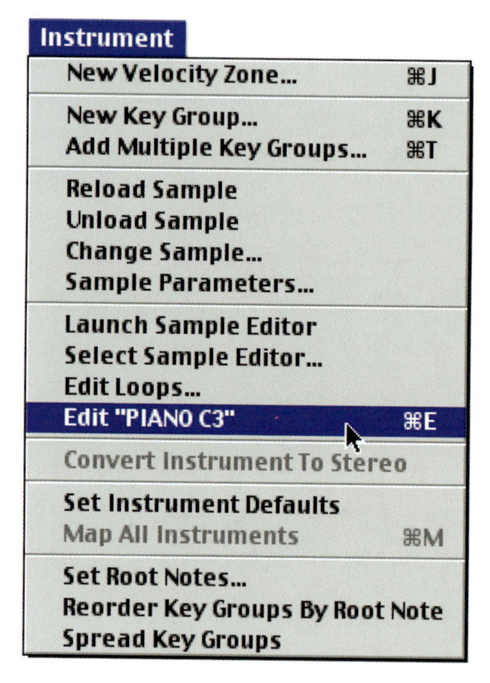

Figure 4.12 SampleCell Instrument menu.

The Add Multiple Key Groups command lets you select multiple samples and load them into an Instrument – useful for quickly building multi-sample Instruments.

The Load Sample command loads a sample into the currently selected Key Group or Velocity Zone. The Unload Sample command unloads the currently selected sample from an Instrument while retaining its Key Group/Velocity Zone mapping and file referencing.

Reload Sample reloads a sample or region that has been unloaded or modified and re-saved. Change Sample loads a new sample into the currently selected Key Group or Velocity Zone. Any parameters you have set for this sample (such as Start Point, Tuning, and so on) will be applied.

Sample Map window

Once you have set up a new Instrument or chosen one from an existing library, you can use the Sample Map window to map the samples to various ranges of

MIDI notes and velocities called Key Groups. The number of Key Groups can be anywhere from a single Key Group spanning the entire keyboard, up to 60 different Key Groups - memory permitting. Velocity mapping lets you assign different samples that have been recorded at varying intensities to corresponding MIDI velocity ranges. You can set up as many as six different Velocity Zones for each Key Group, each of which will play a different sample according to the velocity of the incoming MIDI notes. This is useful when emulating the sounds of acoustic instruments and other natural sounds that have different timbres depending on their volume. Velocity Zones can also be used to switch between completely different sounds. A classic example of this would be a bass guitar with a fingered bass in the lower Velocity Zone, and a slap sample in the upper Zone. Notes played louder would trigger the pop samples, while softer MIDI velocities trigger regular finger-plucked bass.

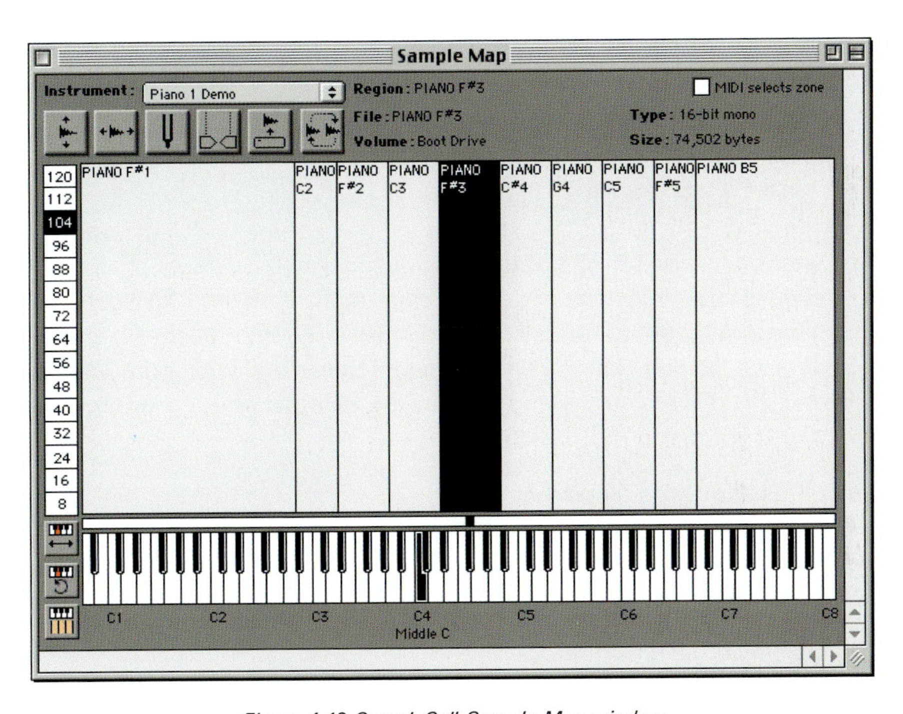

Figure 4.13 SampleCell Sample Map window.

Loop window

Many samples contain playback loops that allow them to be sustained as long as a note is held. In the Sample Loop window you can create and edit loops for a specific sample - adjusting loop start points, loop type, and other parameters.

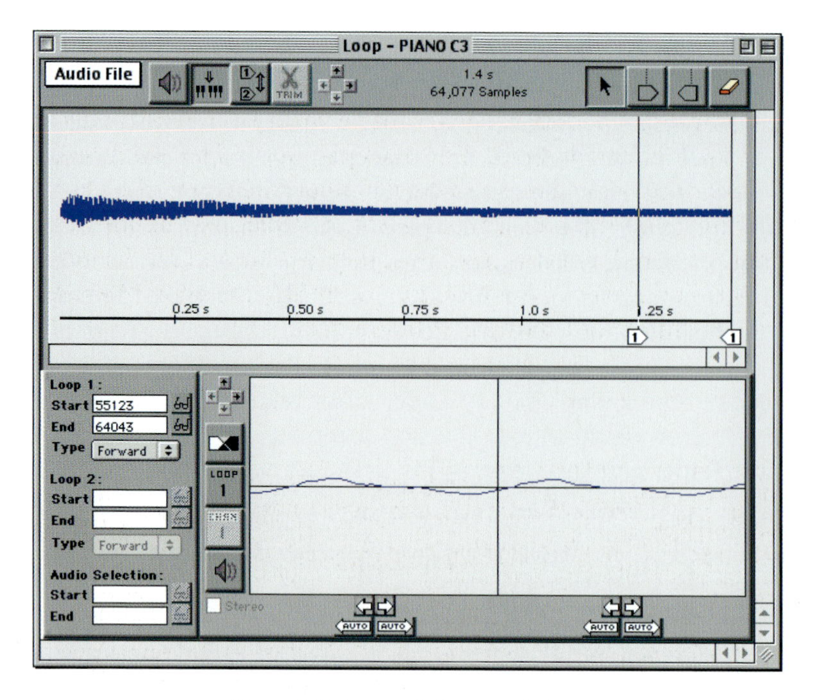

Figure 4.14 SampleCell Loop window.

The Loop window is fine for straightforward looping tasks. However, for detailed editing of samples you will have to use another application – such as BIAS Peak. For convenience, a menu option is provided in Soft SampleCell that lets you sub-launch whichever editing software you have installed on your computer, automatically bringing the selected sample up in the external editor's window.

SampleCell Modulators window

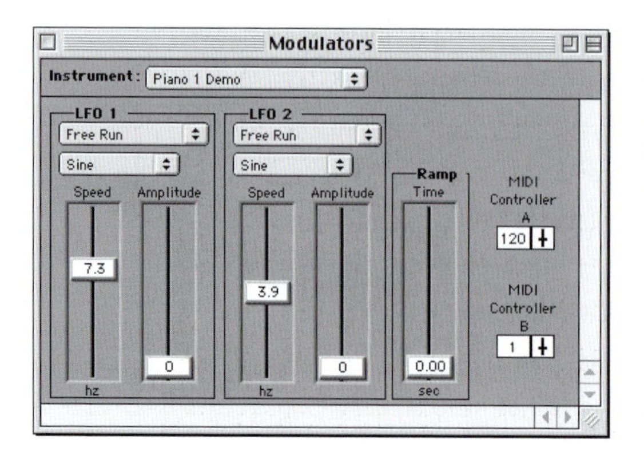

Figure 4.15 SampleCell Modulators window.

The Modulators window contains parameters for five different modulators: LFO 1 and LFO 2 (each with variable speed, amplitude, shape and trigger mode), Ramp Generator, MIDI Controller A and MIDI Controller B. These can be used as both modulation sources and destinations in the Matrix Modulation window.

Matrix Modulation window

The Matrix Modulation window lets you route a variety of modulation sources to a variety of modulation destinations. For example, you might want to control the filter cutoff frequency using the mod wheel of your MIDI controller. Similarly, you could use an LFO to vary the volume of a Soft SampleCell Instrument to create a tremolo effect, or vary the filter cutoff frequency with an LFO for an automatic wah-wah effect. For a 'wackier' effect, try controlling the pitch of a sample using the Random Generator for a sample & hold effect.

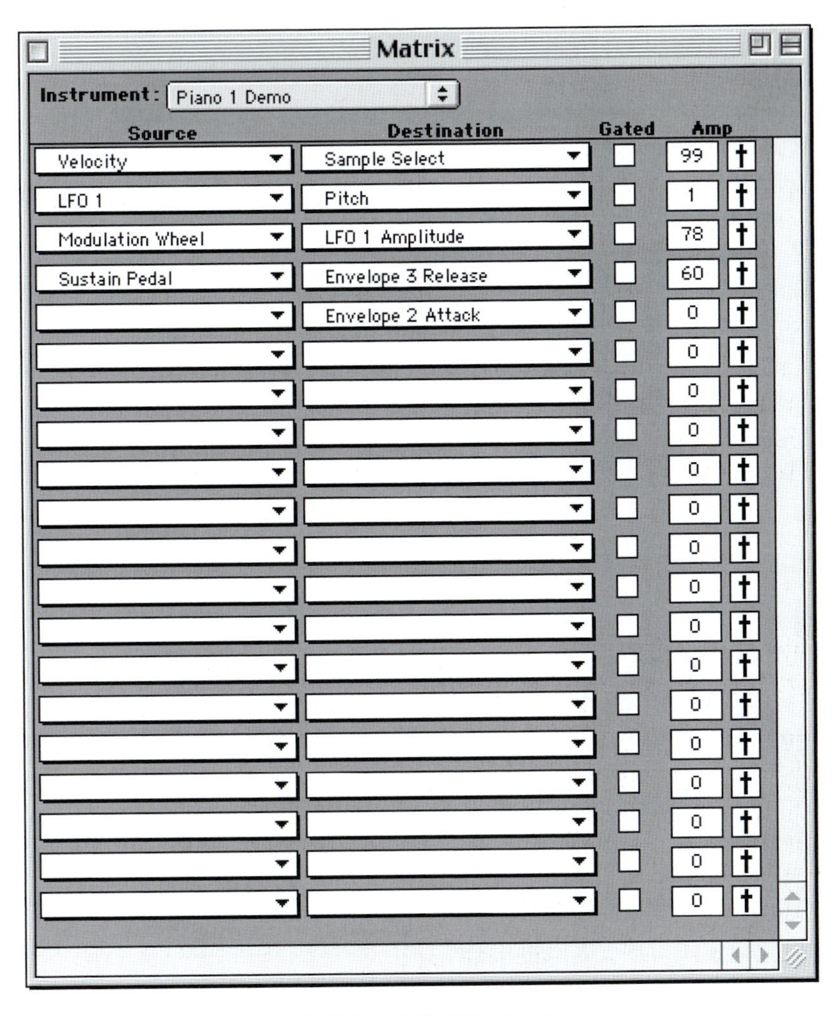

Figure 4.16 SampleCell Matrix window.

Trackers window

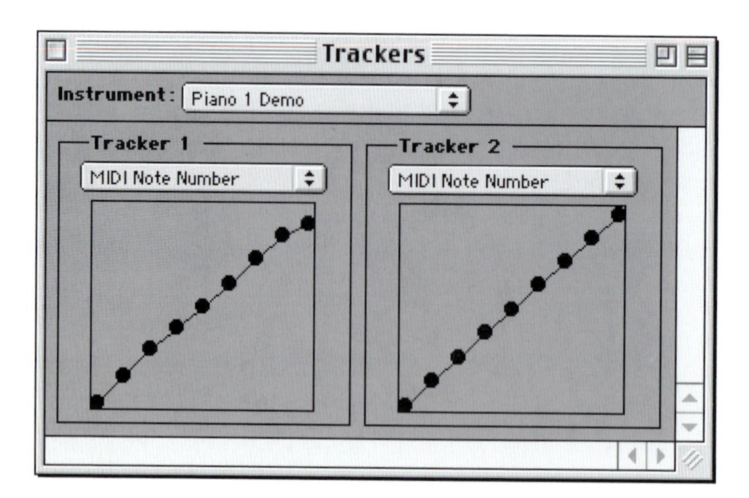

Figure 4.17 SampleCell Trackers window.

You can use the Tracking Generator to vary the modulation signal using a graph or 'Tracker' curve with nine movable 'breakpoints' that let you define the shape of the curve. The popup menus above each Tracker curve let you select the modulation source that will be affected by the Tracker. When that Tracking Generator is then used in the Modulation Matrix, its normal behaviour will be modified according to the curve you have created.

MIDI Note Number
Velocity
Release Velocity
Pressure
Modulation Wheel
Pitch Wheel
Sustain Pedal
Envelope 1
Envelope 2
Envelope 3
LFO 1
LFO 2
Ramp Generator
• Random Generator
MIDI Controller A
MIDI Controller B
Tracking Generator 1
Tracking Generator 2

Figure 4.18 SampleCell Trackers popup.

For example, you could modify your pitch wheel or mod wheel responses with an exponential tracking curve to produce a much wider bend at the end of the bend range or a much 'wider' modulation at the end of the mod wheel range. The manual also explains in detail how to use the Trackers to add random pitch variations to an instrument to produce a more realistic simulation of a non-Western instrument such as an African Kora. The Tracking Generator can also be used to vary the width of a vibrato depending on the relative pitch of the note to create a dynamic vibrato effect, or to delay the onset of the vibrato effect, as singers often do.

Envelope window

Envelope generators let you 'shape' the volume, filter cutoff and other parameters of an Instrument each time a note is played. Soft SampleCell provides three separate Envelopes with controls for attack, decay, sustain and release and additional controls for Key Track, Envelope Amount and Gate Time. Envelopes are primarily used to control the volume and harmonic content of a sound, so Envelope 3 is permanently assigned to amplitude and Envelope 1 is assigned to Filter Frequency by default. Using the Modulation Matrix you can use any of the three Envelopes to control parameters such as pitch, panning, LFO speed, and so forth.

> **Note** Envelope controls can be displayed as line segments or sliders. Use the SampleCell Preferences dialog to choose the display format you prefer.

Envelopes can track the keyboard so that envelope times are progressively shortened as MIDI notes get higher, and lengthened as they get lower. This is useful for simulating acoustic instruments such as brass, where higher-register instruments such as trumpets have much faster attacks than low-register instruments such as tubas.

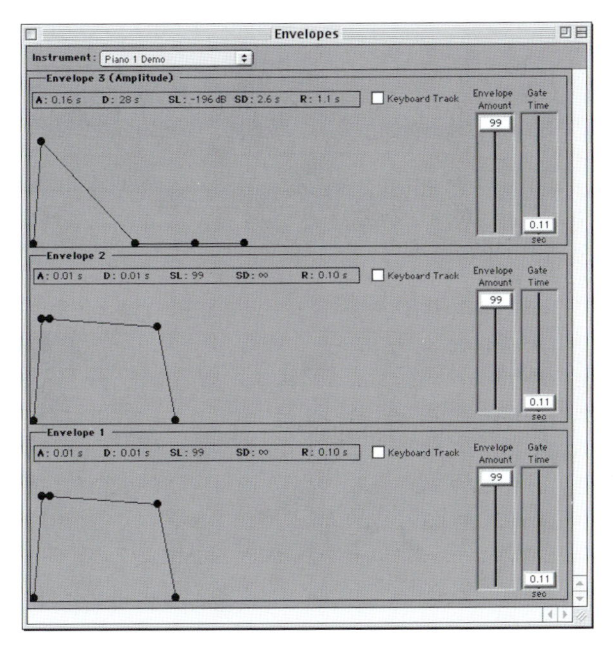

Figure 4.19 SampleCell Envelopes window.

Miscellaneous Parameters window

SampleCell doesn't have any global Bank tuning controls. Instead, each Instrument has its own individual tuning parameters – located in the Instrument's Miscellaneous Parameters window. Here you will find controls for pitch, filter frequency, cross-fade, key track, pitch tracking, intonation, MIDI filtering, voice overlap and priority, and velocity sensitivity. These parameters apply to Instruments as a whole, so adjusting the Filter Frequency, for example, adjusts it for all samples within the Instrument.

If you play many Instruments simultaneously, it is possible to run out of voices. When you attempt to play more voices than are available, Soft SampleCell will 'steal' voices from some Instruments and allocate them to others as needed. You can prioritize Instruments using the Sound Priority parameter so that their voices are more likely or less likely to be stolen when necessary. Typically, lead Instruments should be given high priority while background or pad-type Instruments are given low priority.

To prevent sounds such as ride or crash cymbal from being artificially choked off when played repeatedly you can set a voice Overlap Amount. To set this correctly, adjust the slider while repeatedly playing the same note on your MIDI controller.

You can also adjust the low-pass filter settings to change the harmonic content and character of instruments as you would with the filters of an analogue synthesizer to create wah-wah effects and synthesizer-like filter sweeps.

The Filter Frequency control specifies the cutoff frequency – the frequency below which harmonics in a sampled waveform are removed. The higher the cutoff frequency, the brighter the sound; the lower the cutoff frequency, the duller the sound.

The Filter Resonance control determines how much frequencies near the cutoff are emphasized, and how much those that are farther away are suppressed. High resonance settings add a nasal, ringing quality to sounds.

The Filter can operate in either one-pole, 6 dB/octave mode or four-pole, 24 dB/octave mode. The four-pole filter is better suited to creating synthesizer-type resonant filtering effects.

Note Be aware that excessive use of the Overlap feature may steal voices from other Instruments – and Filter Resonance controls are not available when the Filter is set to one-pole mode.

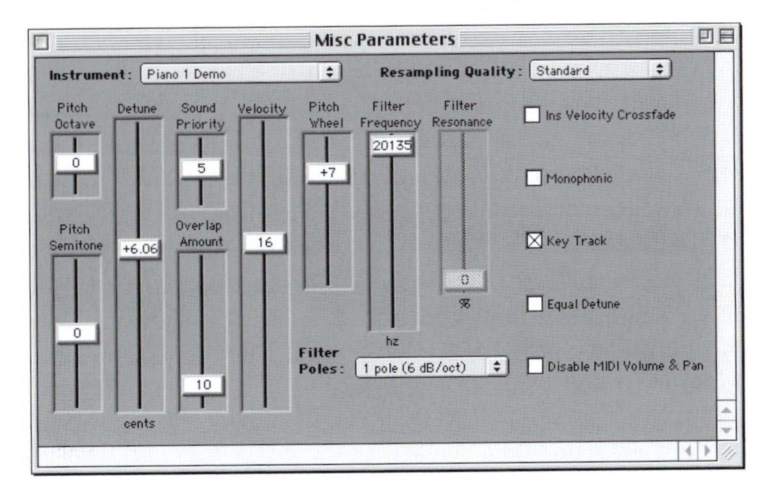

Figure 4.20 SampleCell Miscellaneous Parameters window.

Using Soft SampleCell with Pro Tools

If you are running Pro Tools on a Mac, you can use the DirectConnect plug-in to route the audio from Soft SampleCell directly into the Pro Tools mixer. Up to 32 separate audio channel outputs from Soft SampleCell can be independently routed, recorded, processed and mixed within Pro Tools. So you can apply plug-in processing, automate levels, pans and plug-in parameters, and record the results to disk.

First you need to make sure that you have a connection set up in OMS for the IAC (Inter Application Communication) Driver. This is a feature of OMS that lets two or more applications running on the Mac communicate with each other – sending timing messages, note data or other MIDI information. The IAC Driver is an optional install from the OMS Installer that comes with the Pro Tools software, so, if in doubt, install this using the Custom Install option in the OMS Installer and restart your computer. When you have the IAC Driver installed correctly, double-click on its icon in the OMS Studio Setup document to bring up the dialog box. Here you can name one of the four available IAC buses with the name of the device you want to connect – in this case SampleCell. This way, Pro Tools or any other DAE-compatible application can send MIDI data via OMS IAC to SampleCell, and you can choose this OMS IAC connection as the input device in SampleCell to receive the MIDI output from Pro Tools or your other application.

To get Soft SampleCell working you will need to use one or more Auxiliary tracks to get the audio into Pro Tools and you will need to set up corresponding MIDI tracks to play SampleCell. For example, you might use a stereo Auxiliary track to accept the stereo output from one SampleCell instrument and a mono Aux track to accept the output from another. To set this up, first create a new Auxiliary Input for each Soft SampleCell Instrument you want to add. In the Mix window you need to insert the DirectConnect plug-in into each of these Auxiliary Input tracks, then open the DirectConnect plug-in window and set its input to the appropriate SampleCell outputs. Create a MIDI track for each Soft SampleCell Instrument you want to use, click the MIDI track's Output Device/Channel Selector, then select Soft SampleCell and the desired MIDI channel. Normally, you will set the MIDI track to play back using the OMS IAC connection that will appear in the list of available MIDI output devices under the usual popup menu on the channel strip - using, say, MIDI Channel 1.

Tips Don't forget to enable the SampleCell as an Input Device under the MIDI menu in Pro Tools - or else it won't work! Also, the DirectConnect input channel assignment must match the output channel assignment of the desired Instrument in Soft SampleCell Editor or you will not hear any output.

Keep in mind that each channel of audio transmitted through DirectConnect uses the same amount of PCI bus bandwidth as a single audio track on Pro Tools TDM systems. If, for example, you are playing 48 disk tracks on a Pro Tools MIX System, that will leave just 16 channels of audio (16 mono or 8 stereo) that can be used with DirectConnect - because the maximum number of audio channels is never more than 64.

Note On Pro Tools LE-based systems, performance is determined by several factors, including host CPU speed, available memory, and buffer settings. As the manual points out, Digidesign cannot guarantee 32 simultaneous audio channel outputs with DirectConnect on all computer configurations.

Once you save and re-open a session that uses Soft SampleCell with DirectConnect, Pro Tools will automatically recall all settings, launch Soft SampleCell Editor, and open the appropriate Banks and Instruments for you. This provides convenient total recall of your Pro Tools and Soft SampleCell set-ups.

If you are using the MIDI capabilities of a Digi001, you must launch Pro Tools before launching Soft SampleCell Editor (or any other OMS application), or Pro Tools will not open. In addition, you must quit the other applications before quitting Pro Tools, or MIDI will be inactive on the Digi001 until all OMS applications have quit.

Tip If you aren't using the MIDI capabilities of the Digi001, remove the Digi001 OMS driver from your OMS folder to avoid the above restrictions.

Koblo Studio9000

The Koblo Studio9000 packs four powerful music production tools into one versatile software bundle. The system includes the Vibra1000, Vibra6000 and Vibra9000 monophonic analogue-modelling synthesizers, the Stella9000 polyphonic sampler, and the Gamma9000 multi-timbral drum-machine. You can install all of these, or just the ones you want to use. The Studio9000 uses Koblo's Tokyo 'engine' to handle the MIDI and audio and to provide the user interface for the system.

Studio9000 works seamlessly with Pro Tools via DirectConnect, which allows up to 16 audio outputs from the Studio9000 synthesizers to appear as separate inputs in the Pro Tools mixer. It supports AIFF and Sound Designer II file formats and works with VST-2, ReWire, MAS, OMS and FreeMIDI for integration with most major third-party sequencer manufacturers. It can also work in stand-alone mode using Digidesign's Direct I/O or Apple's Sound Manager.

All four Studio9000 synthesizers can be played in real time via an external MIDI controller and triggered from Pro Tools or a third-party MIDI sequencer. Most parameters can be manipulated via MIDI controller messages. You can also use the Pro Tools automation – giving you total recall of all the synthesizer settings within your projects.

Figure 4.21 Koblo Vibra9000.

The Vibra is a 'virtual analogue' synthesizer. It is totally digital but simulates the classic analogue synthesizers, which are valued for their flexibility and unique sounds but disliked for their disadvantages – noise, high price, lack of MIDI and inability to store presets or keep their tuning.

The Vibra9000 has two stereo oscillators with five waveform choices: square, sine, triangle, saw and noise. It has three syncable LFOs, four assignable ADSRs, and eight filter types, and allows pitch adjustment in octaves, semitones and cents. Other features include portamento, oscillator mixing controls, FM effects and chorus effects.

The Vibra6000 and Vibra1000 are cut-down versions of the Vibra9000 that let you conserve processing power when a quick and simple sound such as a synthesizer bass is all you need.

Stella9000 is an eight-voice polyphonic sample playback synthesizer that lets you load samples and adjust parameters in real time. It supports SampleCell key maps – making it a great partner for Digidesign's SampleCell hardware or software. Presets include acoustic instruments, voices, pads, strings and so forth, and you can tweak these to your taste using the powerful synthesizer

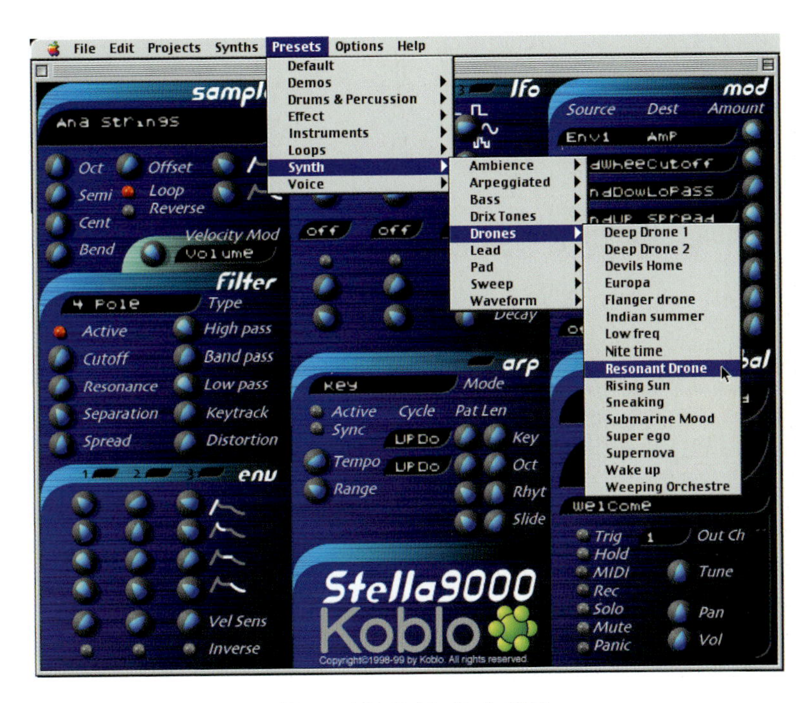

Figure 4.22 Koblo Stella9000.

controls. Like the Vibra900, it also has three syncable LFOs, a modulation section, four assignable ADSR envelope generators, and eight filter types including two, four, and eight-pole designs with notch, saw, comb and square comb characteristics.

The Gamma9000 is a multi-timbral drum-machine with a built-in step sequencer. It is a type of sample playback synthesizer that supports SampleCell key maps and has separate output channels for each track. There are six 'sample' tracks, represented by horizontal strips at the upper part of the window. The sample tracks allow precise control of individual sounds like the bass drum and the snare. Each track uses a single sample for sound synthesis and has settings such as pitch, envelope and tone to let you further modify the sound. There is also a multi-timbral 'keymap' track that allows quick access to a complete set of sounds. This has the same settings as the single sample tracks, but uses a keymap instead of a single sample. By simply pointing to a folder, you can quickly load in a set of sounds. The track settings will affect all samples in the keymap. The Gamma9000 also has a monophonic filter section that you can use to further shape your percussion sounds - and each track can be routed to this filter. The factory presets give you a great selection to get you started, or you can load in any sounds you like in AIFF or SDII format. Once you have chosen the sounds you

Figure 4.23 Koblo Gamma9000.

want for your 'kit', the Step Sequencer can be used to program your own rhythms and combine them into songs.

Chapter summary

Access Virus is a useful synthesizer to have available within your Pro Tools setup. The Virus has a very distinctive sound. It is a 'big' sound that cuts through on mixes with lots of the attributes of the classic analogue synths combined with an indefinable 'digital' edge.

Soft SampleCell makes a good choice as a sample-replay device for use within Pro Tools systems. It takes some time to become familiar with all the windows, but each of these is easy enough to understand and use in itself. There are samplers that have more sophisticated features, but SampleCell, despite showing its age, can still play a very useful role in any professional system - and there are plenty of CD-ROM libraries available in this format.

Koblo's Studio9000 system gives you a full 'kit' of synthesizers, sampler and drum-machine with a wide enough range of presets to let you work on most types of musical arrangements.

5 Third-party Plug-ins

Various third-party companies also produce TDM plug-ins, most of which are listed in the Digidesign Third-party Developers Catalogue. Many of these are also available as VST and/or MAS plug-ins.

My current favourite is the Sony Oxford EQ plug-in with the George Massenburg option – giving you two of the best equalizer designs around in one convenient package. If you have to choose just one high-quality third-party EQ for your Pro Tools system, then this should probably be the one.

Native Instruments Spektral Delay is a tool for the creative sound designer, as are the plug-ins that you will find in the GRM Tools bundles. These GRM Tools plug-ins are also available for VST, for example, although the implementations are significantly different on the different platforms. The DUY suite of plug-ins has various creative sound modifiers, including as Shape, Wide, Valve and Tape, ReDSPider to replay preset effect types created using DSPider, and SynthSpider, which does for 'virtual instruments' what DSPider does for plug-ins – it lets you create them!

Metric Halo Labs offer their Channel Strip – an 'all-in-one' plug-in that provides all the features you would expect in a mixing console channel in one convenient plug-in – and SpectraFoo which provides professional metering facilities for Pro Tools systems. Wave Mechanics offer UltraTools and Speed. UltraTools is actually three plug-ins – Pure Pitch, Pitch Doctor and Sound Blender – while Speed, as its name suggests, lets you speed your audio up (or slow it down). Serato's Pitch 'n Time is similar to Speed, but has more features.

For the more technical engineer, Apogee MasterTools provides their proprietary UV22 dither algorithm and has various features that technical engineers will appreciate. Similarly, Cool Stuff Labs Generator X lets you create a range of

WATFORD LRC

test signals for your Pro Tools system. Also aimed at the more technical user, McDSP offer an excellent range of plug-ins including their Filter Bank, Compressor Bank and MC2000 multi-band compressor, along with their AC1 and AC2 Analog Channel emulators.

One of the original TDM plug-in developers, Andy Hildebrand at Antares, offers a range of software including MDT Multiband Dynamics, Jupiter Voice Processor, Auto-Tune and Microphone Modeler. Another early plug-in developer, Georges Jaroslaw at Arboretum, offers a suite of plug-ins called Hyperprism aimed specifically at sound effects designers along with Realizer Pro, Ray Gun, Harmony and Ionizer plug-ins.

Sony Oxford OXF-R3 EQ

The Sony Oxford mixing console's EQ and filters have been modelled in software and made available as a plug-in for Pro Tools TDM systems – handsomely priced at around £550.

The plug-in is also available for Cubase, Nuendo, Logic 5 and Digital Performer 3 as a VST plug-in on both Windows and Mac platforms for the TC|Works PowerCore PCI card – priced around £350.

Figure 5.1 Sony Oxford Plug-in.

The design of these EQ filters is state-of-the-art. Total dynamic range is greater than 138 dB (RMS unweighted) for any single section, including the filters.

Harmonic Distortion is undetectable using FFT analysis down to −160 dB. Control-induced noise is less than −95 dB (peak) during adjustment of any single control. The user interface is simple and effective, with controls laid out similarly to the way they would be on a hardware unit, and with the benefit of a detailed frequency response graph display to show the effect of the processing. The user interface has separate sections for the EQ and Filter functions and the plug-in can be instantiated with just the EQ active, or just the filters active - or with both EQ and filters active. Note that the filter or EQ section controls are greyed-out when either section is not in use.

The EQ function offers five bandpass sections, each with rotary controls for Frequency, Q and Gain. The LF Peak/Shelf section covers the range between 20-400 Hz; LMF covers the range between 30-600 Hz; MF covers the range between 100-6 kHz; HMF covers the range between 900-18 kHz; and HF Peak/Shelf covers the range between 2-20 kHz. The highest and lowest sections can be independently switched to shelving functions. Separate variable-slope low pass and high pass filters are provided and the LF Filter cutoff ranges from 20-500 Hz, while the HF Filter cutoff ranges from 1-20 kHz.

Tip The A and B buttons let you set up two different sets of settings and quickly switch between these.

The EQ also features four different selectable EQ types that cover most of the EQ styles currently popular amongst professional users, including some older styles that are highly valued for their creative applications.

Most analogue EQ circuit designs show some dependency between the Gain and Q settings. (Remember that the Q control lets you adjust the bandwidth of the filter - i.e. the range of frequencies that will be boosted or cut using the Gain control.) To emulate these designs, the Oxford EQ provides three different styles of EQ - each of which provides different amounts of Gain/Q dependency.

EQ Type 1 has minimal Gain/Q dependency and is the most flexible type, although you do need time and patience to make use of this flexibility. The nearest comparison is to the original 4000 series SSLs and similar types of EQ that were popular in the 1980s. With this type, the user has to adjust the Q control to maintain an effect whenever the gain is changed.

EQ Type 2 is similar to type 1, but has asymmetrical response curves because the Q is held constant when the EQ is cut (as opposed to boosted). As the manual explains, 'this EQ type lends itself well to resonance control for percussion instruments such as drums, since relatively high Q is available at low gain settings, whilst fairly subtle "fill EQ" can be achieved in boost settings at the same time.'

EQ Type 3 has a moderate amount of Gain/Q dependency such that the Q reduces as you lower the gain. Referring to the manual: 'This provides the EQ with a softer characteristic as EQ is progressively applied and since the effective bandwidth is increased for low gain settings, it sounds louder and more impressive when used at moderate settings. The gentler Q curve also lends itself better to overall EQ fills and more subtle corrections in instrument and vocal sources. Turning the Gain control seems to produce the effect that the ear is expecting without needing to adjust the Q control too often. Therefore EQ's of this type are often dubbed as "more musical sounding".' This EQ most resembles the older and well-loved Neve types, their modern derivatives and later SSL G series. Also many of the more popular outboard EQs have this dependency to some extent.

EQ type 4 builds on type 3, using a far greater Gain/Q dependency that maintains an almost equal Q in the boosted region. This EQ type is extremely soft and gentle in use making it particularly useful where subtle changes to completed mixes are required. It is also useful in mastering situations or whenever you need to match the sounds of tracks from different sources.

When the shelving function is enabled, the Q control provides control of the 'overshoot' at the crossover frequencies. Note that the curves are symmetrical in cut and boost gain settings. With the Q set to minimum there will be no overshoot and the EQ will act as a standard shelving type similar to the original 'clean' SSL 4000 EQs. Although these are still very popular, some engineers complain that they can sound harsh and overbearing in comparison with 'vintage' EQs. As the Q is increased the overshoot amount is increased. As the manual explains it, 'This has the effect of suppressing the perceived mid-range boost that occurs with the previous "clean" variety, reducing the apparent "hardness" of the sound. This, along with the increased slope rate, provides more apparent definition to the EQ in the band of interest. With the Q control in mid position, this produces a gain loss of around 10% (of the boost gain) in the overshoot region and further defines the effect described above. This setting provides a response most like the older Neve designs and their derivatives and the later SSL G series EQs. This response produces an optimum effect by providing what the ear expects to hear as the gain control is operated and can explain the enduring popularity and renowned musical qualities attributed to EQs of this type.' Finally, when the Q control is set to maximum, it produces an overshoot of half the total boosted gain, i.e. 10dB.

Sony claim that 'The use of novel coefficient generation and intelligent processing design provides unparalleled performance that surpasses analogue EQ in both sound quality and artistic freedom. This plug-in may well provide all the EQ you will ever need.' What I can definitely confirm is that this plug-in produces extremely smooth and effective results. I have found myself automatically using one of these across my main Master Fader whenever I am using Pro Tools so that I can tweak the sound of the whole mix whenever I like. The graph display provides excellent visual feedback to help guide EQ choices and the controls are logically arranged.

Sony Oxford GML 8200 option

For an additional £220, you can add the GML 8200 EQ emulation option to the Sony Oxford EQ plug-in. The GM designation refers to George Massenburg, the American designer of the highly acclaimed EQ units that bear his name. This Sony software emulation of Massenburg's hardware is reputed to be a highly accurate simulation. It is fully endorsed as such by George Massenburg himself, who collaborated on the design of the software – originally intended for use with the Oxford OXF-R3 console. The ranges of the controls and the responses of the original EQ have been faithfully reproduced – even to the point of producing centre frequencies up to 26 kHz while running at 44.1 or 48 kHz! The intention here has been to match the operational characteristics of the original analogue unit as closely as possible by providing this highly accurate control range and control law matching between the software and the original hardware.

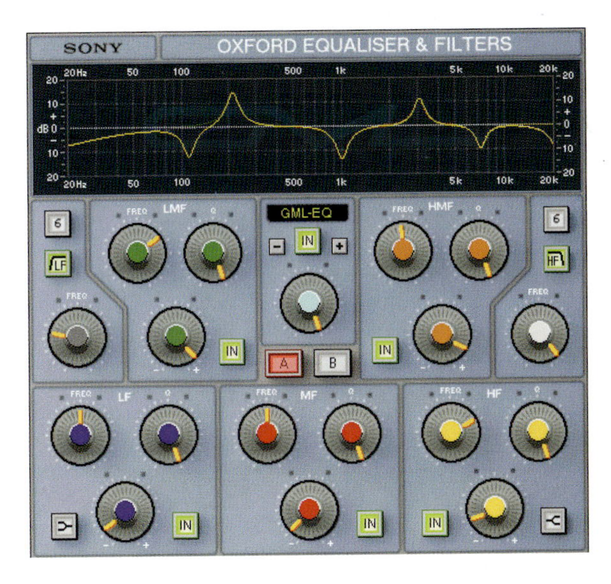

Figure 5.2 Sony Oxford GML option.

When you install this option, the plug-in looks identical to the standard Sony Oxford plug-in. In fact, the filter sections are identical - they are the Oxford EQ filters. The GML8200 band pass section responses, on the other hand, cover a significantly wider range of frequencies than those of the Sony Oxford - from 15 Hz to 26 kHz. These responses have been very accurately matched to those of the original analogue unit, giving all the character of the original GML8200 EQ but with the added bonus of full +/− dB gain ranges. It is also worth noting that the extremely low noise and distortion performance of the plug-in delivers performance levels many times greater than those attainable from the analogue unit.

The user interface follows the style of the original Oxford EQ plug-in, which is clear and consistent. To access the GML 8200 bandpass response curves, you can use the +/− buttons underneath the EQ-type field in the plug-in's window. Alongside the four standard EQ types, a fifth type will be available - the GML-EQ. When you select the GML-EQ type, the Q controls of the LF and HF sections are made inoperative when the shelving function is enabled and the EQ control ranges change to accurately match those available on the outboard unit. Now, the LF Peak/Shelf ranges from 15–800 Hz; the LMF ranges from 15–800 Hz; the MF ranges from 120 Hz–8 kHz; the HMF ranges from 400 Hz–26 kHz and the HF Peak/Shelf ranges from 400–26 kHz.

Note With a Pro Tools system operating at 44.1 kHz or 48 kHz sampling rates it has an upper frequency limit of around 20 kHz. When centre frequencies above 20 kHz are selected, the EQ provides a response that is accurate for any portion of the curve that lies within the legal bandwidth limit of the system.

Native Instruments Spektral Delay

Native Instruments' Spektral Delay plug-in lets you create unique delay effects ranging from subtle colourations to rhythmic effects and atmospheric sound textures. Spektral Delay takes a radical new approach by splitting the audio into 1024 different frequency bands using a real-time Fast Fourier Transform (FFT) analysis. It then allows you to set levels, delays and feedback individually for each band.

Spektral Delay's FFT effects, including frequency band rotation, exotic filters and spectral gating, provide plenty of scope for sound sculpting. All the parameters can be controlled from the graphical user interface using simple pencil and line editing tools - you just draw in the amplitudes, delay times and feedback levels for the frequency bands.

Each of the bands for each side of the stereo input can be delayed separately and fed back to the delay input with an individual amount. The maximum delay time is 12 seconds and individual delay times can be set freely - or according to the rhythm via a selectable note grid. Any frequencies within the signal can be attenuated or filtered completely and the Spektral Delay's filter curves can have any shape, allowing for extreme filter settings and sweeps. To help you to visualize the way the processing will affect the sound, a pair of stereo sonograms - one for input and one for output - constantly display the frequency content of the input and output signals. All the parameters such as filter, delay and feedback settings can be modulated by an integrated tempo-syncable LFO - or via MIDI. When the Sync button is activated, the Spectral Delay will slave to the tempo of the host program. When activated, the LFO will also sync to the external tempo. The Spektral Delay is also completely MIDI-controllable, so it can be used for live performance with a compatible MIDI control device.

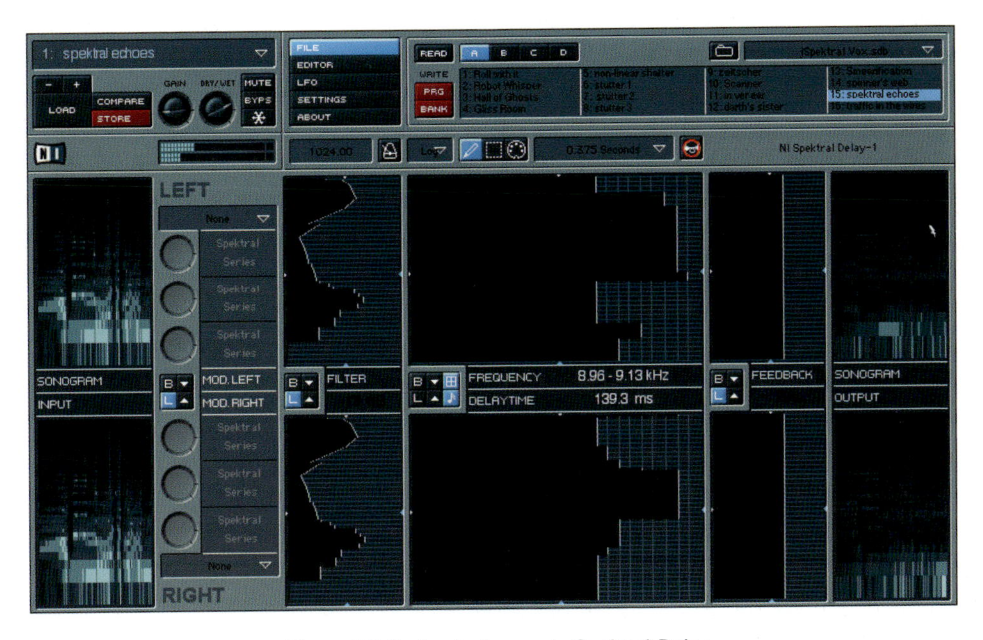

Figure 5.3 Native Instruments Spektral Delay.

The upper part of the plug-in's window contains controls for the presets and lets you access settings for the LFO and so forth. Below this, you can think of the user interface as revealing the signal flow through the plug-in - with the input signals displayed on a sonogram at the left, split into the 1024 bands, followed by modulation effects, followed by filtering, delay settings, then feedback settings, before the recombined signal is passed to the output. The displays showing the settings for the Filter, Frequency/Delay Time and Feedback sections can be altered using a pencil tool – making the Spektral Delay quick to use once you have familiarized yourself with these displays.

Native Instruments offer Spektral Delay in both Host TDM (HTDM) and RTAS formats for Macintosh-based Pro Tools TDM 5.1 or Pro Tools LE 5.1 or higher systems.

Note 'Host TDM' was developed by Digidesign as a hybrid of TDM and RTAS technologies. HTDM plug-ins provide all the logistical functionality of standard TDM plug-ins, but, like RTAS plug-ins, they allow for all the processing to be done on the host instead of the DSP chips found on your TDM hardware.

The Spektral Delay is also available for Macintosh and Windows, either standalone with MME, Direct Sound, Sound Manager and ASIO or as a plug-in with a VST or DirectX-compatible host program.

An extensive library of presets including guitar, vocal, drums, ambience, reverberation and spatial effects is provided. These will not disappoint. You will immediately find creative uses for many of these – whether for music production or for sound effects work. Creating effects from scratch will take some effort on your part to learn how to get the best out of the user interface – but the effort will certainly be worth it. Highly recommended for creative sound designers.

DUY

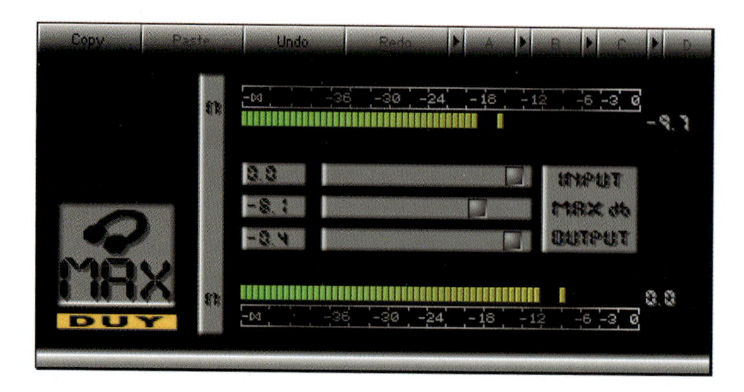

Figure 5.4 DUY MAX.

DUY offer a suite of interesting plug-ins originally designed by Rafael Duyos in Madrid, Spain. Most of these will let you 'pump up the volume' or otherwise enhance the sound of your tracks.

DUY Max TDM

DUY's Max plug-in lets you 'maximize' your audio tracks using DUY's Intelligent Level Optimization algorithm. This increases the perceived level of your audio without the pumping and breathing effects sometimes heard in compressors or limiters, and it doesn't distort or harm low-frequency signals. So you can process any kind of sound - from individual tracks to final mixes - and it even works on tracks that have already been compressed or limited.

It is dead easy to operate - you just move the middle 'Max' fader back towards the left of the display until the effect sounds good to you. The more you move the fader to the left, the more the 'maximizing' effect, so you will hear a greater loudness level for a given peak reading. Because the music gets louder you may need to reduce the input level or back off the output level a little to prevent distortion. The best bit is that this perceived level increase leaves the dynamics and spectral balance of the original signal sounding unaltered. So that's Max - and it does work!

DUY Shape TDM

Shape is a multiband dynamics processor based on DUY's proprietary Frequency Dependent WaveShaping algorithm. The software features three independent user-defined 'Shapers' operating on the Low, Mid and High frequency bands.

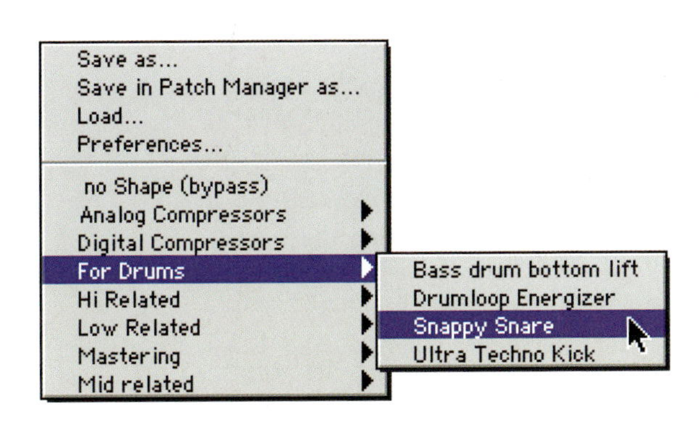

Figure 5.5 DUY Shape presets.

Each Shaper has eight different dynamics curve types, including Linear, Log and Cosine functions selected by a button above each display; a button marked E for the linear Expand function which you can use to optimize dynamics; a button marked I to initialize the curve; five factory presets for typical applications; and a pair of simultaneous Input and Output plasma-like meters.

The Mix switch to the left of the meters lets you turn off any or all of the Shapers so that you can adjust each of these independently and two buttons to the left of this let you save and load your settings. Three sliders are provided to control Input level and Low Cut and High Cut filter frequencies.

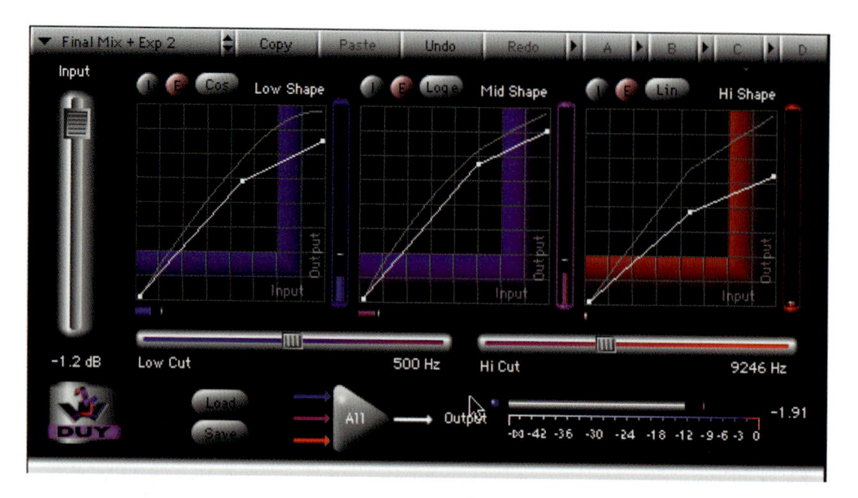

Figure 5.6 DUY Shape.

Setting up is a two-stage operation. First you adjust the cut-off frequencies of the filters to split the audible range into three bands, choosing these according to the frequency content of the audio to be processed. This stage produces three (or six in stereo) split bands of audio material ready to be processed by their corresponding Shaper. Then you create the shapes for each band that best suit your application. To create a curve you just click and drag its points to the desired parts of the grid - and any changes you make will be heard in real time. For example, you can compress the Low band while leaving the Mid and Hi bands untouched. In this case, just draw the low Shaper and leave the Mid and Hi ones untouched, pressing the Initialize button to make sure they are completely inactive. You can also eliminate whole bands by drawing a horizontal line over the X-axis thus making all points equal to zero - with the curve type set to linear and the Expand button deselected. Using this approach you can adjust the filters to find the cut-off points of any band you want to eliminate from the audio path.

Shape is suitable for a wide range of applications, from simple compression or expansion to frequency enhancement and revitalization of any kind of sound, whether music, musical instruments, voices or effects. And, as with Max, Shape will allow you to process any kind of sound - music, musical instruments, voices or effects either on individual tracks or final mixes.

DUY Wide TDM

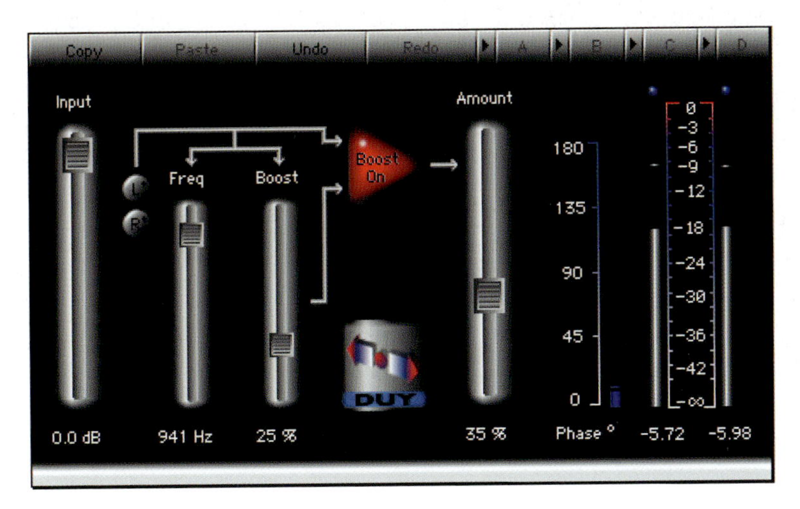

Figure 5.7 DUY Wide.

Wide is a spatial enhancer which lets you place sound elements from within your mix outside the normal stereo field, without adding unwanted colorations. Again, you can process any kind of stereo or multi-channel sound source - music, voices or effects - on each track of your recording, or you can process final mixes. It has a wide range of applications, from music and film soundtracks to multimedia and virtual reality. Re-mastering, re-balancing and spatial enhancing of multi-channel mixes is also possible. Two toggle switches permit left and right channels to be independently phase-inverted and a 'Boost' control is provided to let you compensate for any low frequency losses due to the spatial processing. The amount of effect can be adjusted subtly or dramatically - allowing a wide range of applications. As you turn up the 'Amount' slider the stereo image is progressively widened. Settings up to about 40% are quite subtle while settings more than 80% produce a very dramatic effect. As with Max, it is extremely simple to operate and produces instantly usable results. Ideal for film soundtracks and multimedia applications, Wide can also be used for rebalancing and spatial enhancing of stereo and multi-channel mixes.

DUY DaD Valve TDM

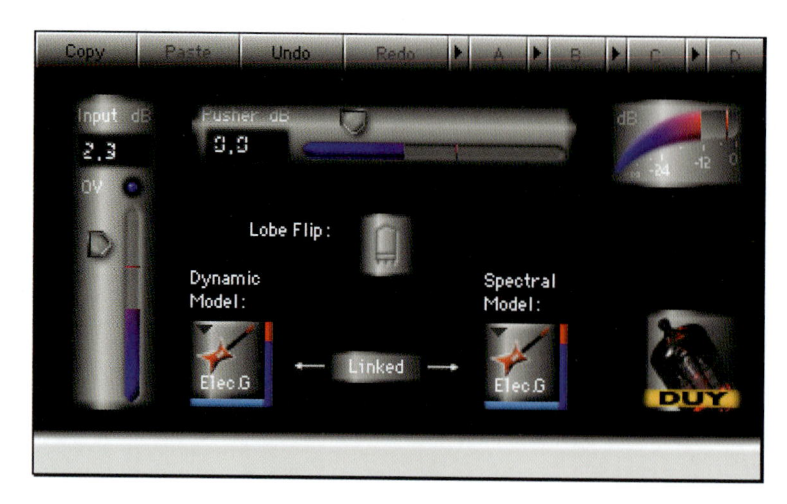

Figure 5.8 DUY DaD Valve.

DaD Valve, as its name suggests, lets you apply the characteristics of a valve amplifier circuit to your audio. To design DaD Valve, DUY modelled a complete valve chain, its associated circuitry and all the significant parameters such as noise, temperature, frequency response, impedance and so forth – using a Silicon Graphics workstation. The software was then ported to the Pro Tools platform. Separate modelling sections are provided for 'Spectrum' and 'Dynamics'. The Spectrum model emulates the frequency and transient response of a real valve circuit while the Dynamics model produces harmonic distortion identical to that of a real valve. This both complements and interacts with the Spectrum model and both sections are required to produce a complete simulation indistinguishable from a real valve. The characteristics for each section can be chosen individually from the presets provided.

Figure 5.9 DUY DaD Valve presets.

Rather than providing individual user controls for the myriad of parameters which have been modelled, such as valve type (triode, tetrode or pentode), bias, de-coupling, component non-linearities, drift, noise, and so forth, DUY have simply opted to provide 40 of the best presets their experts can come up with, with just one control to allow tweaking - the Pusher slider which lets you set the amount of feedback to emulate the microphonic effect of a real valve.

I tried the Electric Guitar preset on a mix of two electric guitars playing rhythm and lead. These had already been played through valve amps but with a very clean sound. The DUY software fattened up the sound in a very pleasant way - and sounded very similar to the way I would have expected a valve amplifier to affect the audio if I had processed the mixed tracks through one. Overall, I found that most of the presets were very convincing.

DUY DaD Tape TDM

Figure 5.10 DUY DaD Tape.

To design DaD Tape, DUY modelled five of the most representative tape recorders available along with examples of the best 1/4-, 1/2- and 2-inch tape types. When driven into overload, analogue tape recorders produce 'rounded', 'warm' sounding distortion rather than the harsh clipping produced by digital equipment. Another unavoidable aspect is noise. DaD Tape lets you simulate both these characteristics - so you can have that tape hiss back again if you really want it! Three popup menus are provided - one each for Machine Type, Tape Speed and Noise Reduction Type - plus input and output level controls. Make any Noise Reduction Type choice, or Noise Reduction 'off', and you will hear typical tape noise and hiss, even without playing any audio through the software! Of course, there is a 'noiseless' setting if you just want to simulate the other tape characteristics without adding hiss.

Simulations of the two most common noise reduction systems are provided, along with a proprietary noiseless-tape mode. Tape machines modelled include a latest generation state-of-the-art 1/4'' analogue recorder, a professional 2-inch 24-track machine from the 1970s with op-amp circuitry and low-noise transistor input stages, and a late 60s transistor-based professional stereo machine. The other two machines are an old valve model and a cheap 1/2'' multi-track recorder. The vintage valve machine produces a subtle asymmetrical distortion that can be appreciated even with medium-level signals and delivers a very 'rounded' sound – ideal for voices and guitars. The cheap 1/2'' multi-track has a relatively limited dynamic range that can be extremely effective on drums, drum loops and other percussive sounds.

A tape speed switch is available for each of the tape brands and you can choose between 7.5 ips, 15 ips and 30 ips. In the real world not all speeds are available to all tape brands, but DUY decided to remove this limitation in order to provide maximum control and allow unusual combinations of tape recorders and speeds. Thankfully, an 'Ideal' tape mode is also provided with all the desirable features of an analogue tape machine, but with none of the aspects usually considered as faults. DUY took out the 'wow and flutter' and unwanted high frequency non-linearities – just leaving the desirable tape characteristics. As with the other DUY plug-ins, I found this software to be extremely easy to operate and effective in use. You can use it to round off peak transients and add warmth to electronic and acoustic instruments and it works well with percussive sounds. And if you really want to simulate the sound of analogue tape, complete with all its faults, you can use Tape to process your final mix.

DUY ReDSPider

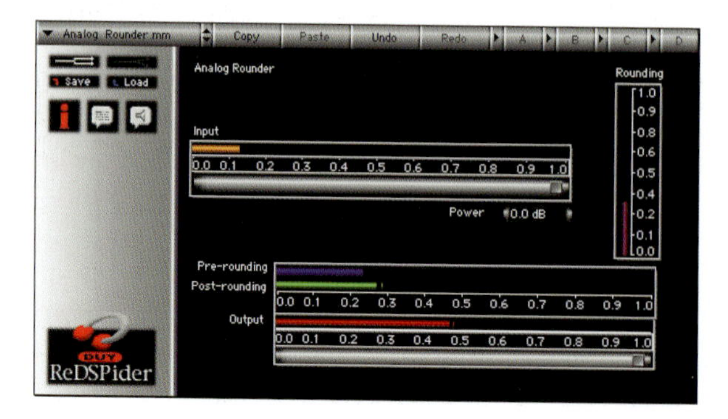

Figure 5.11 DUY ReDSPider.

ReDSPider features over 200 plug-ins for TDM which include compressors, limiters, reverbs, noise gates, expanders, equalizers, de-essers, panners, levellers, flangers, delays, multi-band and multi-effect processors, 3D effects, synth sounds, sound enhancers, de-noisers, and so forth. This is the same selection that you get with DUY's DSPider package. In fact, ReDSPider is a 'replay-only' version of DUY's DSPider; in other words, you can use the presets, but the editing capabilities are deliberately restricted. Users can edit visible module parameters and subsequently save them as new plug-ins - but DSPider's advanced plug-in creation facilities are not available. ReDSPider will appeal to anyone who prefers using presets, perhaps because they don't have the time to develop their own.

DUY DSPider; DUY SynthSpider

See Chapter 15, 'Roll Your Own . . .'.

Metric Halo

Channel Strip

Figure 5.12 Metric Halo Channel Strip.

Metric Halo's Channel Strip provides all the typical signal processing functionality that you might expect a high-end mixing console to have and packs this into one DSP-efficient plug-in. You get a Gate and a Compressor, both with side-chain features, and six bands of fully parametric EQ. Extensive metering features include individual gain reduction meters for the Gate and Compressor sections, input and output metering for the EQ section, I/O transfer function displays for the Gate and Compressor, and frequency response displays for the Gate and Compressor side chains and for the EQ section. Channel Strip can be used as a fuller-featured alternative to the standard DigiRack plug-ins for gating, compression and EQ. It uses comparable amounts of DSP to these, or maybe a little more, but provides a much more attractive user interface with better performance. And it uses a lot less DSP than you would need for three individual higher-quality plug-ins for gating, compression and 6-band parametric EQ.

SpectraFoo TDM

Metric Halo's SpectraFoo Visual Audio Monitoring System includes a suite of analysis and measurement tools for Pro Tools TDM systems. The analysis engine uses a Fast Fourier Transform (FFT) to convert a continuous record of amplitude vs time into a record of amplitude vs frequency. SpectraFoo can perform multi-channel FFTs at up to 84 times a second, allowing you to see spectral features as small as 5 Hz wide while staying absolutely synchronized with the source material. SpectraFoo also provides a full suite of precise quantitative analysis tools, including a capture and storage system that is seamlessly integrated into the monitoring environment.

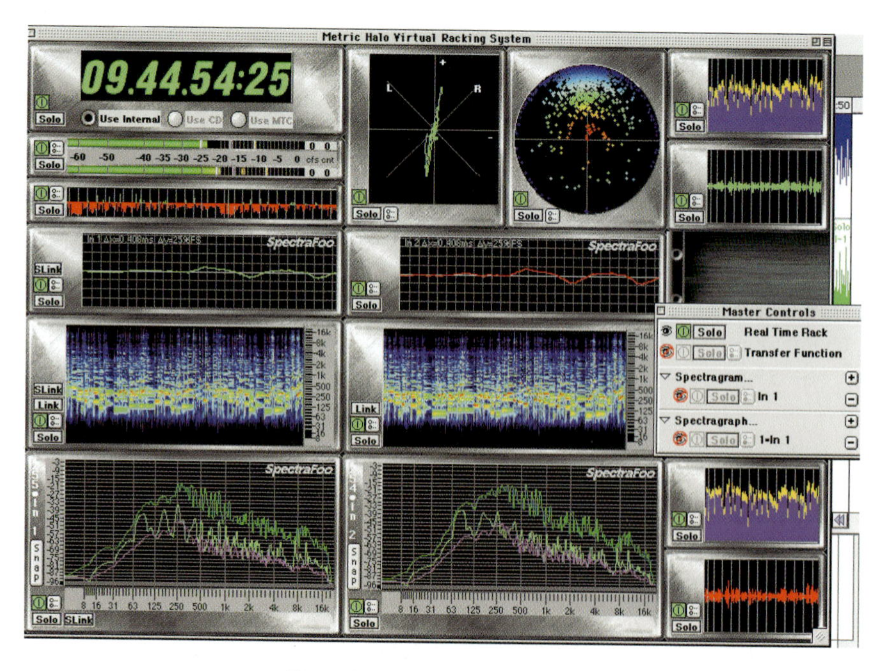

Figure 5.13 SpectraFoo windows.

SpectraFoo's instruments include Transfer Function, Level Meter, Oscilloscope, Spectragram, Spectragraph, Power History, Phase Torch, Lissajous Phase Scope and several others. The Master Controls are provided in a floating window that 'sits' on top of Pro Tools' windows. You can simply click on the icon next to any of SpectraFoo's 'instruments' to open a window for just that instrument.

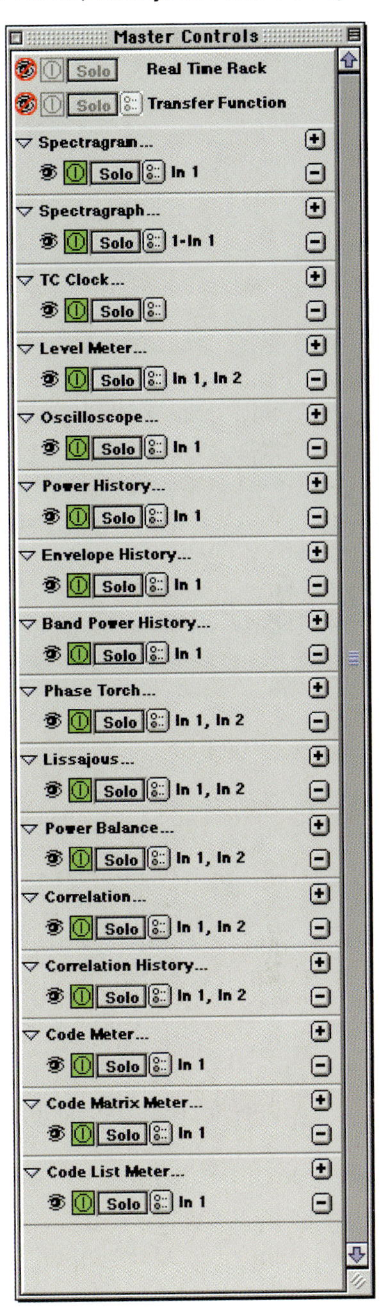

Figure 5.14 SpectraFoo Master Controls.

However, even with just a few windows open, the screen gets very cluttered. I found myself preferring to switch to the consolidated 'Real Time Rack' view which contains two channels of most of SpectraFoo's instruments neatly arranged in one screen - although this almost completely occupies the area of a 17" monitor. Still, two screens are virtually mandatory with Pro Tools anyway.

The Transfer Function window displays graphs of Power vs Frequency and Phase vs Frequency - both of which will be appreciated by the more technical engineers who will be familiar with these.

Every engineer will appreciate SpectraFoo's two-channel Digital Level Meters, which offers much greater accuracy and detail than you will find elsewhere. The yellow part of the meter indicates the instantaneous peak-to-peak level of the program material while the purple part indicates the RMS level. The Peak and RMS meters have independent, peak hold functions represented by triangular carets of the same colour as the meter to which they correspond. The VU meter is an average indicator with adjustable ballistics and is represented by the grey triangular carets. Detailed meter readings - to 1/100 dB accuracy - and scaling options can be accessed by clicking on the Show Details button. The meter also contains a numerical counter that indicates the largest number of consecutive samples surpassing 0 dBfs.

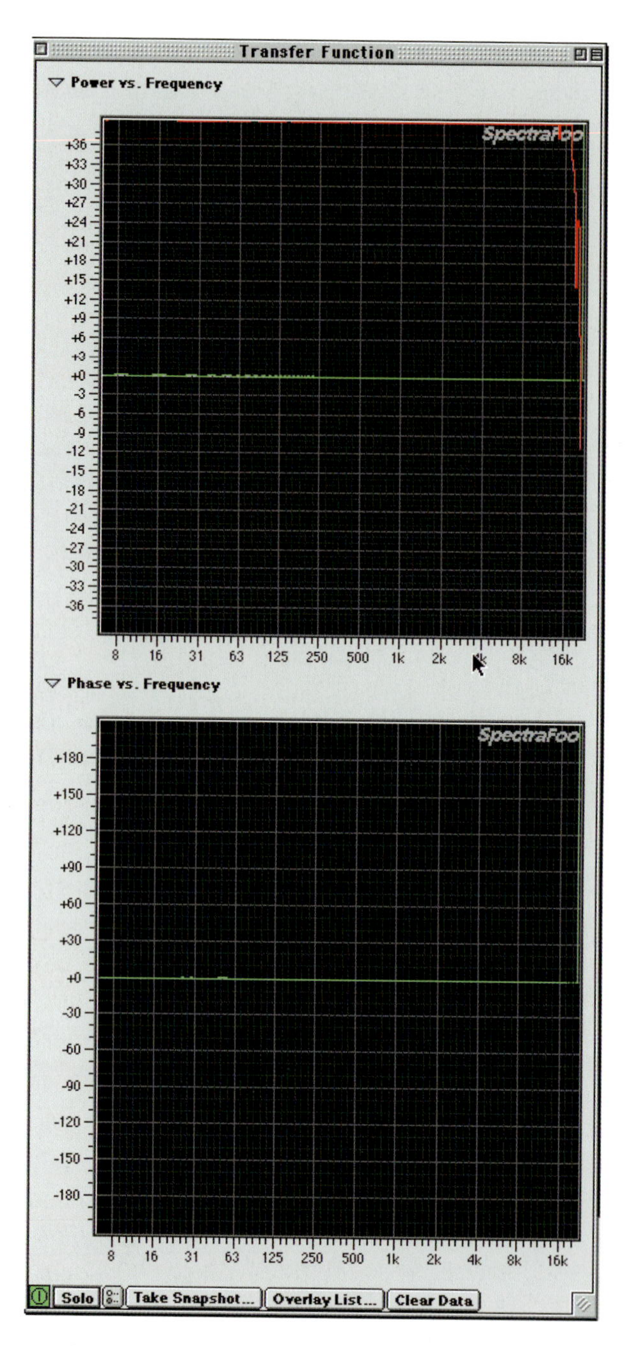

Figure 5.15 SpectraFoo Transfer Function window.

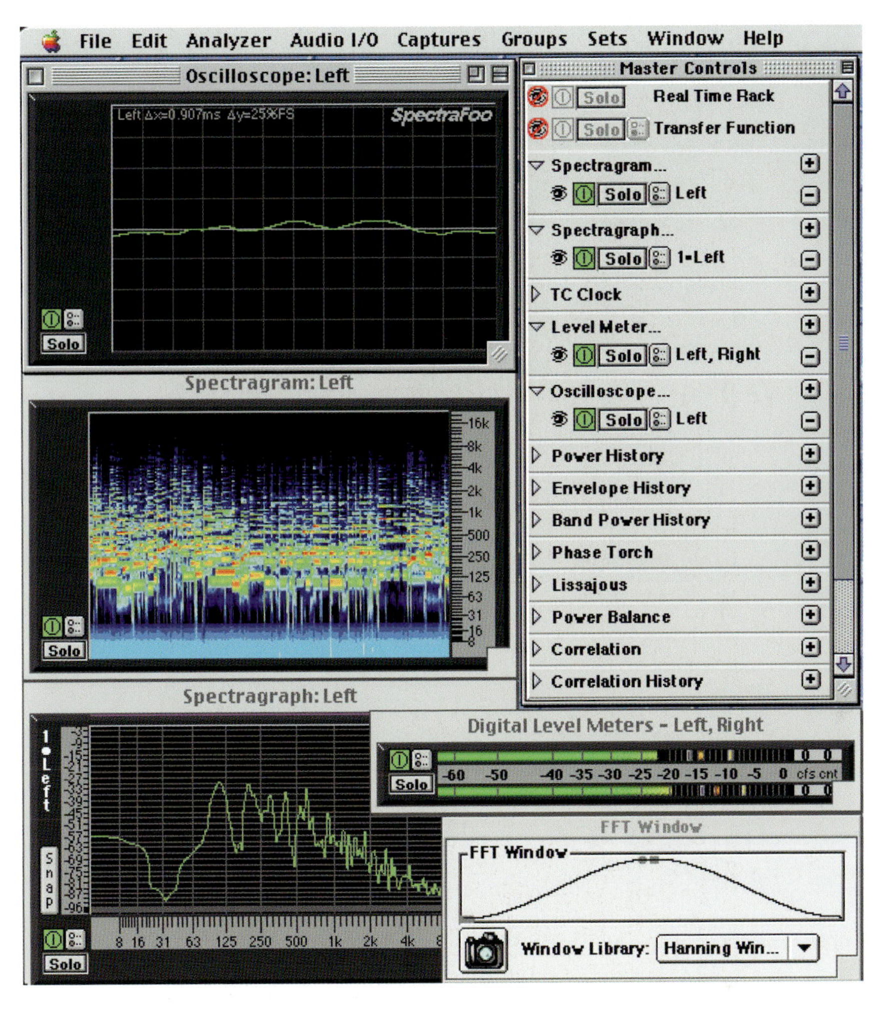

Figure 5.16 SpectraFoo windows in detail.

The Triggering Oscilloscope provides an amplitude vs time waveform display of the assigned input channels. It can often be more informative to look at amplitude versus frequency using the Spectragraph. This is a real-time spectrum analyser which displays three different-coloured traces. The bright green trace provides instantaneous spectral analysis while the purple trace shows the average spectral content referenced over a user-specified length of time, and the dark green trace represents the peak spectral power monitored since the peak was last reset. You can alter the Spectragraph's frequency and power scaling as well as its averaging characteristics in real-time, allowing you to focus on very fine details of the recordings such as phase structure, overall spectral balance, dynamic range, frequency range, and low frequency roll-off. Using this, for example, I was able to quickly identify the 'boomy' frequencies in my acoustic guitar recordings and then reduce the level of these frequencies using one of my EQ plug-ins.

To see how the spectral response changes over time, you can use the Spectragram. This represents the power of frequency events through a colour scale and the display medium correlates not only frequency and power, but also time. The result is a visual 'sonic fingerprint' which is very effective for precisely identifying frequency overlap and masking effects – such as a bass drum and a bass guitar occupying the same frequency range and obscuring each other. It gives you a good picture of the rhythmic aspects of program material and is very helpful in pinpointing timing problems in dense arrangements.

The Peak and RMS Power History instrument allows you to see the history of the information displayed by the Peak and RMS meters. The ratio between these two levels is responsible for the 'perceived volume'. With a large difference between Peak and RMS levels, perceived volume will be low compared to the peak levels on the tape. Conversely, when the difference is small, perceived volume will be louder – so this instrument can help you determine how much compression and limiting is needed for specific program material.

Although level metering tells you plenty about the behaviour of your audio signals, phase metering can also be extremely important when using multiple microphone setups during recording sessions as well as for checking final mixes. SpectraFoo has a couple of very powerful tools for checking phase – the Phase Torch and the Lissajous Phase Scope.

The Phase Torch compares the phase difference between two channels as a function of frequency, independent of power. A mono (in-phase) signal is indicated by what looks like a torch precisely aligned with the Y axis of the scope while an out-of-polarity signal appears as a negative version of this. Delays appear as spirals within the scope. This meter is very useful when recording a musical instrument with multiple microphones. You can use the Phase Torch to identify the frequency ranges in which phase cancellations are occurring as a result of the comb filter created by the use of multiple mics on a single source. Then you can set up multiple microphone placements that create nulls which are outside of the critical frequency range of the instrument you are recording. The Phase Torch can also be used for identifying phase anomalies in studio wiring; setting azimuth on analogue tape recorders; checking for time alignment in complex PA systems; identifying frequency dependent phase and polarity problems (such as an out of polarity high frequency driver in a sound reinforcement system); and frequency-sensitive mono compatibility analysis.

The Lissajous Phase Scope shows you the amplitude of the first input signal versus the amplitude of the second. This instantly lets you know if a mix has polarity problems. It also allows you to see the width of the stereo field of the material being monitored. With the scope in vector mode, in-phase material appears as a vertical line while out-of-phase material appears horizontal. When the scope is in X–Y mode and the activity of the trace occurs mostly in the lower left and upper right quadrants of the display, the left and right channels are predominantly in polarity (in-phase). When the activity of the trace occurs mostly in the lower right and upper left quadrants of the display, the left and right channels are predominantly out of polarity (commonly referred to as out-of-phase, or mono-incompatible).

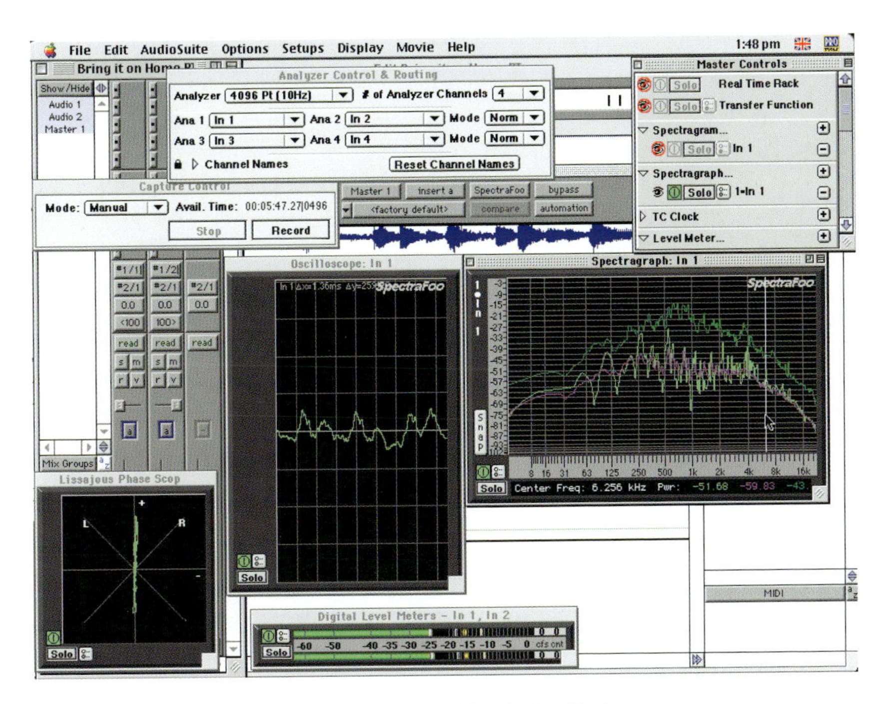

Figure 5.17 SpectraFoo in Pro Tools.

SpectraFoo is capable of monitoring up to 24 separate channels (depending on the speed of your Mac and your input hardware). However, there are probably very few situations in which you would want or need to be analysing 24 channels at the same time. More typically you will want to look at 4–8 channels. For example, you might use two channels to monitor the main stereo mix and another two channels to switch on the fly between the individual channels of the mix. This way, you could check out, say, the kick drum and the bass to see if there is a masking effect inducing frequency overlaps that could be cured with EQ.

SpectraFoo's greatest strength lies in its ability to provide quantitative measurements that can be synchronously referenced against the subjective experience – what Metric Halo refer to as the 'Foo factor'. Good specs and performance 'on the bench' do not necessarily guarantee exceptional sound quality as these don't take into account the dynamic behaviour of electronic systems. Previously available test equipment simply was not fast enough to analyse and display measurements that provide any meaningful information about the dynamical behaviour of audio devices in the presence of real musical signals. SpectraFoo will allow you to make dynamic measurements that are difficult or impossible to make with other analysers due to their inherent speed limitations.

Wave Mechanics

UltraTools

Wave Mechanics offer a bundle of three TDM plug-ins called UltraTools including their original Pure Pitch processor for vocal doubling or creating harmonies or detune effects; Pitch Doctor to correct out-of-tune vocals and instruments; and Sound Blender, which is actually two plug-ins – Pitch Blender and Time Blender.

You can use Pure Pitch to create realistic harmonies from a solo performance, deepen a vocal part without changing its pitch, or create extremely realistic double-tracking effects. It can even alter formants and pitch inflections of voice-over tracks, producing subtle character alterations or more extreme gender and species morphing effects. Over 50 preset patches are included for music production, sound design and post-production.

Pitch Doctor lets you enter a key and scale of your choice and then automatically adjusts the tuning of any out-of-tune notes to conform to your settings. Sound Blender combines synth-like filtering and modulation with pitch-shifters, arpeggiators and digital delays with which you can create zillions of effects.

Pure Pitch

Let's take a closer look at Pure Pitch first. Independent controls are provided for both pitch and formants, enabling you to create truly realistic vocal doubling effects – or to make your vocalist sound like a creature from another world. Using the built-in delay and LFO you can create modulation effects such as vibrato or tremolo, and if you are working with dialogue the 'expression controller' can be used to alter or remove the pitch inflections from your dialogue recordings, or to create instant robot voices using the more extreme settings.

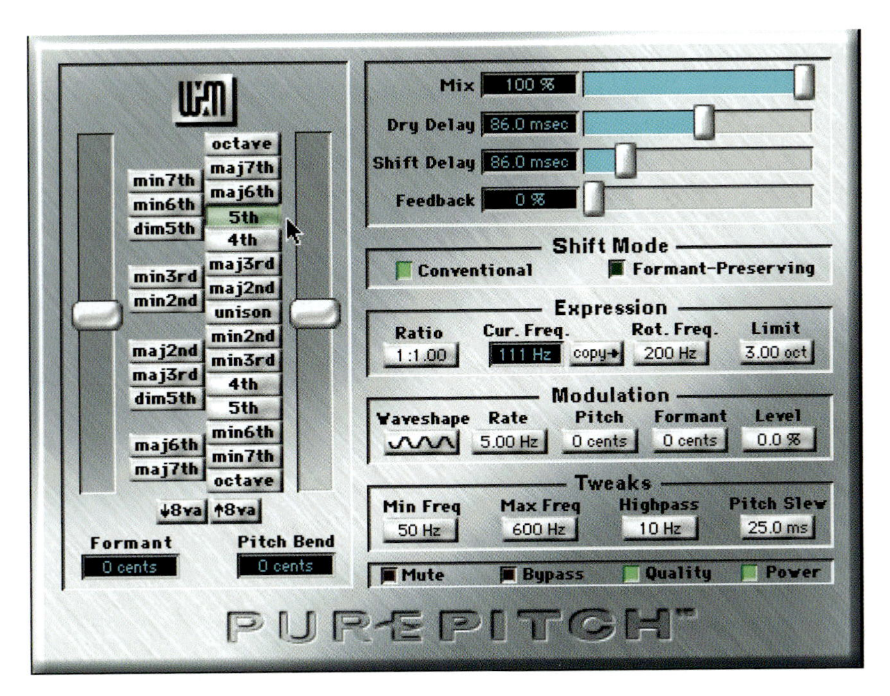

Figure 5.18 Pure Pitch.

The ability to shift the pitch of vocals without affecting the formant structure is where much of the power of Pure Pitch lies. Formants are what gives each vocal sound, and each person, a unique character. A formant is simply a resonance that appears as a narrow peak in the frequency spectrum, similar to the peak created by a parametric equalizer with a narrow bandwidth and high gain. Each vocal sound has several of these 'formant' resonances, all at different frequencies and with different bandwidths and it is these characteristics which differ from person-to-person which help us to identify any individual by their voice. Most pitch processors, like those in many popular samplers, will distort the location of these resonances – resulting in the familiar 'chipmunk' effect when you shift pitch by any more than about a semitone. Pure Pitch lets you avoid this problem but still allows you to use this effect if you want to. So, although you would normally want to preserve the formant structure when changing the pitch, the Formant Shift slider gives you the ability to change the formant structure independently of the pitch. For example, small amounts of formant shift can be used to create more realistic harmonies or double-tracking, by subtly altering the character of the original voice. And more extreme changes in the formant can be used to deepen voices – or to perform digital gender-changes!

Pure Pitch presets are organized into categories including Chorus for thickening effects, Dialogue for inflection manipulation, Harmony for musical harmony effects, Spacey for long delay effects, Tremolo/Vibrato for level and pitch modulation effects, Vocal Transforms for manipulating the character of speech and singing, and Wacky effects for 'when you've been staring at the computer screen too long'! Using these you can double vocals, give dialogue more expression, create natural-sounding octave doubling, turn your lead singer into an operatic soprano, and make the backing singers sound like a bunch of 'space monkeys' or evil aliens!

Pitch Doctor

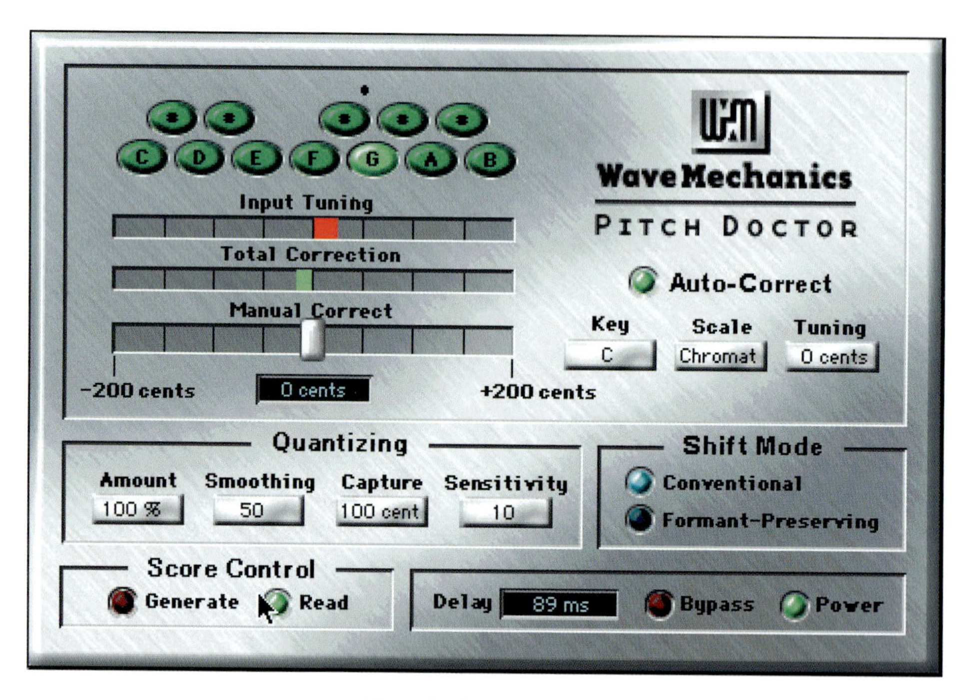

Figure 5.19 Pitch Doctor.

Pitch Doctor, as its name suggests, is designed specifically for fixing out-of-tune vocal and instrumental tracks. You can also use it to retune or transpose vocal or instrumental tracks over a wide pitch range while still maintaining the original, natural sound. It provides precise manual control over the amount of pitch shift, making it easy to fine-tune intonation, for instance. Pitch Doctor also offers both a semi-automatic note-based pitch-correction mode and an automatic mode that you can use as a quick fix for anything slightly out-of-tune. Manual pitch correction is extremely easy to operate – you simply dial in the amount of pitch shift you want using the Manual Correct slider. The downside of this is that if you have a vocal that needs lots of attention you could be at this for a very long time! If you want to correct a note which is wildly off-pitch, or even to change the melody of the original, the note-based mode is the one to use. To set this up, you just click on one of the keys in the pitch keyboard while your track is playing. Then when the pitch of any note in your recording falls within the Capture range of that key, it will be pulled in towards that note. To set up automatic mode you just click on the Auto-Correct enable button, choose a key and scale, and Pitch Doctor will automatically modify the pitch of any notes falling outside of the selected scale tones.

As enticing as this sounds, however, there are some major caveats. In general, automatic pitch correction will only work well on tracks that are slightly out of tune. For more serious intonation problems, Pitch Doctor (and any other automatic intonation correction tool) will not be able to work out what the intended pitch is and will often try to 'correct' the pitch to the wrong note. However, for the right kind of track, this mode can save an incredible amount of time. For example, I used automatic mode to correct the pitch of a guitar recording where the instrument had gone slightly out of tune towards the end of the recording. It worked pretty well on steady notes, but gave a sort of vibrato effect to some of the bent notes which was not ideal.

One of the best features in Pitch Doctor is the Correction Score parameter. This is a special 'automation' parameter available in the Plug-In Automation window which can be set to No Correction or Auto-Correct or to any of the notes of the chromatic scale. By editing the automation graph associated with this parameter, different regions of a track can have note-based or automatic pitch-correction as appropriate - a brilliant idea!

You can also use the 'Generate' button to automatically generate correction score automation data by analysing the pitch of your audio track. This data can then be edited and used for automation-driven note-based pitch correction. This feature might be used in cases where automatic pitch correction is having difficulty accurately tracking the pitch.

Sound Blender

Sound Blender is a pair of plug-ins - Pitch Blender and Time Blender - designed for creating radical sound modifications.

Pitch Blender

Figure 5.20 Pitch Blender.

Pitch Blender is a two-channel effects processor that combines pitchshifting, delay, filtering, panning and modulation. Virtually every parameter of the signal processing can be modulated by a wide range of modulation sources to create evolving effects that respond to changes in the input signal. The Pitch Mapper section adds intelligent harmony and arpeggiation features, allowing you to dynamically adjust the pitch shift interval depending on the detected pitch, the selected key, and the pitch shift interval. Plenty of exotic scales are provided as pitchmaps, including Pelog, Siamese, Indian and Iranian. This plug-in gives you lots of control over the effects and you will need to spend some time with it to get to know it thoroughly. Fortunately, Wave Mechanics have provided an extremely comprehensive set of presets, including the wonderfully named 'Deep Throat' (sub-octaves with resonant filtering for solo instruments), 'Hip-Hopper-izer' (makes a sleazy sax even sleazier), and 'Switchy-Twitchy Fifths' (up and down fifths with switched resonant filters and panning).

Time Blender

Figure 5.21 Time Blender.

Time Blender is also a two-channel effects processor that combines reverse pitch-shifting with delay, filtering, panning and modulation. Reverse pitch shift lets you create backwards tape effects – similar to those used by Jimi Hendrix, for instance. As with Pitch Blender, you get lots of instantly usable presets such as 'Power Tools' (which sounds sort of like the name suggests) and 'Randomania' (which gives you completely wild panning and modulation).

The control panels of the two plug-ins are quite similar, featuring input and output level controls, a BPM control for rhythmical delay and modulation effects, a Trigger Control for the trigger source and threshold, and a modulation control panel where you can choose a modulation waveform, set a rate in Hertz, set threshold and output levels and set pitch parameters. The main control section lets you set feedback, modulation depth and rate, mix, pitch and delay amounts. A SideChain feature is also provided to let you dynamically alter the sound of the effects processing. This side-chain is simply an alternate audio source – other than the track being processed – that can be patched into the modulation section to drive an envelope detector or a gate, which in turn can be patched to any of the modulatable parameters.

Speed

Speed is an AudioSuite plug-in that lets you change the pitch or the length/tempo of your audio files with ease when you are working in Pro Tools.

To use the plug-in you just select some audio in the Pro Tools Edit window and choose Speed from the AudioSuite menu. The Simple control window, which appears by default, appears with two rotary controls - Speed, which controls tempo without affecting pitch, and Pitch, which controls pitch without affecting tempo. Two buttons at the bottom of the window - Preview and Process - let you preview then process your selected audio and save the results to disk. You can also switch the display to Calculator mode or to Graphical mode - but more on these later.

Simple control panel

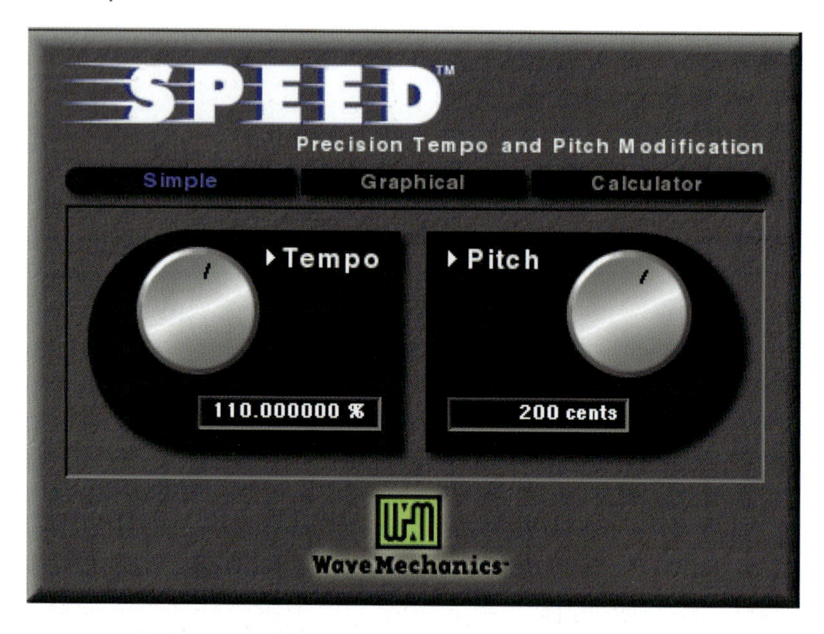

Figure 5.22 Speed window showing Simple controls.

Let's take a closer look at the Simple control panel. The knob on the left controls speed without affecting pitch, while the knob on the right controls pitch without affecting speed. Both controls have different modes that can be changed by clicking on the small triangle associated with each control. Values can also be entered directly by clicking on the displayed value and typing in a new value - useful when you know exactly what amounts of tempo or pitch change you want.

The speed knob has two modes - Tempo and Length. Increasing the Tempo above 100% produces a corresponding increase in tempo while decreasing the setting below 100% reduces the tempo, so a Tempo setting of 200% will be twice as fast as the original and a setting of 50% will be half as fast.

When the mode is set to Length, the effect will be the opposite. Higher settings will increase the length of the selection, resulting in a slower tempo, while lower settings will shorten the selection, resulting in a faster tempo. Like the Tempo mode, a setting of 100% will produce no effect on the audio. So, a setting of 200% will produce a segment twice as long as the original, and a setting of 50% will produce a segment half as long.

The pitch knob has three modes - Cents, Semitones and Percent. In Percent mode, the adjustment is given as a frequency ratio in percent, where 100% produces no pitch change. This is similar to the pitch ratio control of older hardware pitch change devices and can be useful for compensating for the pitch change produced by sample rate conversion or tape varispeed. Talking about varispeed, this is an effect that I have often wished I could achieve in Pro Tools - and using Speed this is now possible. Just set the Speed control to Tempo mode and the Pitch control to Percent mode and set both to identical values; the effect will be identical to that of varispeed.

Calculator control panel

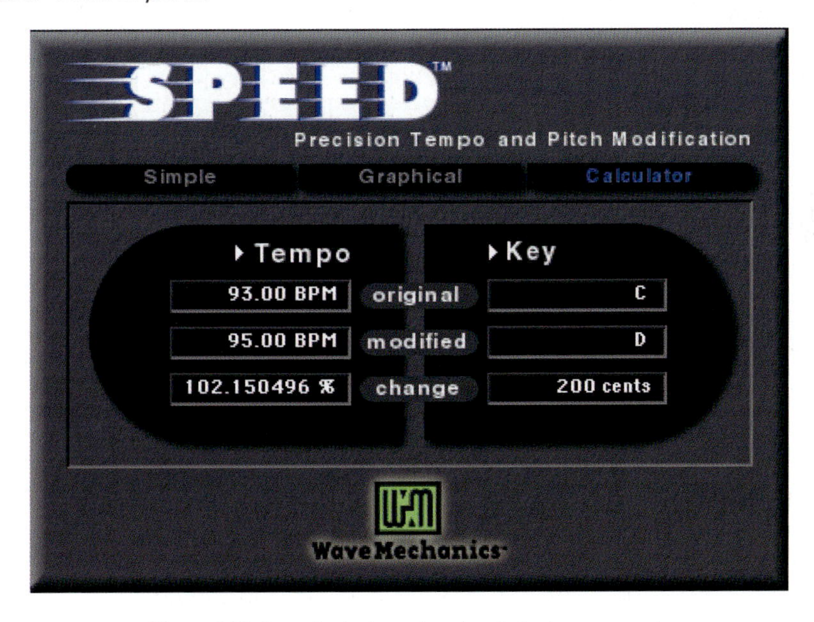

Figure 5.23 Speed window showing Calculator controls.

You can use various modes available in the Calculator panel instead - select a new key in Key mode, type a new BPM in Tempo mode, or enter a new length in Length mode.

If you prefer working with numerical values or are converting from a known speed or pitch to a known new speed or pitch you should use the Calculator panel rather than the Simple controls. This has two calculators - one for speed and one for pitch - both of which have different modes that can be changed by clicking on their triangle control. You can use the pitch calculator's Key mode to transpose music between specific keys: just enter the original key in the original key field and the desired key in the modified key field. The Tuning mode is useful for converting a piece of music tuned to one standard reference pitch (such as A435) to another (such as A440). As with Key mode, you just enter the original tuning reference into the original field and type the desired tuning reference into the modified field.

With the Speed calculator in Tempo mode, you type in the modified BPM then hit the preview or process buttons for this change to be applied to the audio. Working in Length mode, you can choose between Length in Samples and Length in Seconds. To stretch or squeeze the selected audio to a specific length, simply enter a new value into the modified length field, preview it to check, then process it.

The two Length modes have an additional feature called length Copy - shown as a curved arrow. Clicking on Copy transfers the original length value to the modified length field and puts the modified length into a 'locked' state - indicated by a red lock icon. When in the locked state, selecting a new piece of audio within Pro Tools will no longer change the value of the modified length field. Instead, the percentage change field will change to reflect the relative length difference between the new original length field and the locked modified field value.

You can use this Copy and Lock feature to make the lengths of several audio regions identical - which you often need to do when you are working with drum loops. With the Speed plug-in open, just select the region with the desired length, then select one of the Length modes in the speed calculator. Hit the Copy arrow to copy the original length into the modified length then select the target region whose length you want to change. The original length now shows the length of the currently selected region - but the locked modified length remains the same as before. Just hit the Process button and the target region will have the same length as the first region.

Graphical control panel

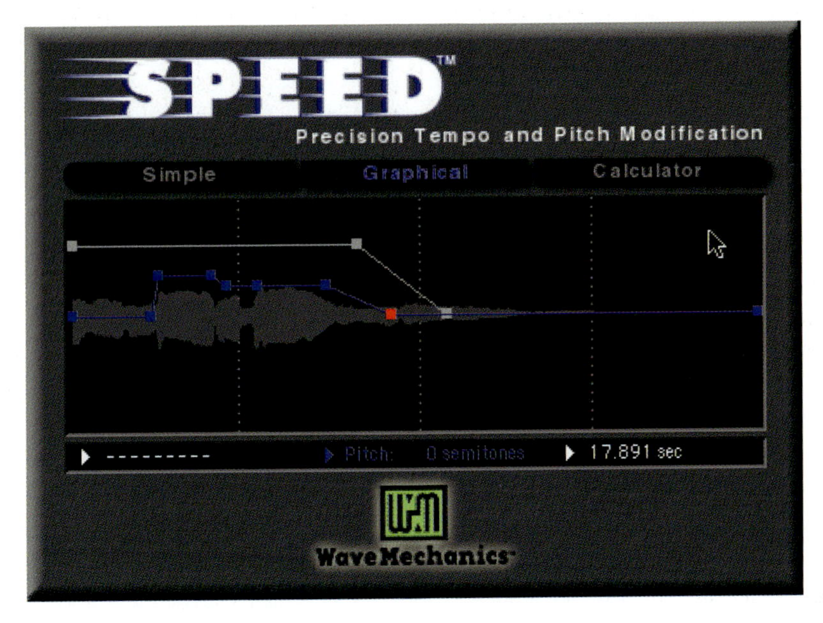

Figure 5.24 Speed window showing Graphical controls.

Speed's 'secret weapon' is the Graphical control panel. This lets you see a graphical representation of the audio waveform in the background and provides two horizontal graph lines that you can use to shift the pitch or the tempo. You can use these to correct pitching or timing or to create accelerandos or decelerandos within your music tracks – inserting changes wherever they may be needed. The white graph line controls speed while the blue graph line controls pitch.

Again, there are multiple display modes, accessible by clicking the small triangle controls. Both graphs are initially set to have no effect on speed or pitch. Editing the graphs works very much like editing automation data in Pro Tools. Clicking on an existing point (shown as a tiny square) will allow you to drag that point to a new location. Clicking on a graph where there is no point will create a new point that can be dragged to a different location. Option-clicking on an existing point will delete that point unless it is either the left or right endpoints. If one graph lies on top of the other at any point (obscuring it from view and mouse clicks), Control-clicking at that location is equivalent to clicking on the obscured graph (possibly creating a point) and moving that graph to the top view. Shift-clicking or Command-clicking points will limit the range of movement for the point to either vertical or horizontal, respectively. As a point is selected or moved around, the speed or pitch change value of that currently selected point (in terms of the selected mode) is displayed below. Also, the time location (in either seconds or samples) of the currently selected point is displayed below at right. The currently selected point is displayed as a red square dot. To enter a precise pitch or speed value for the currently selected point, click in the pitch or speed field and enter a new value using the keyboard. Don't worry if this all sounds a bit complicated when explained in words - it is actually very straightforward and intuitive in practice.

In action

Speed is not the tool to use for vocals. Speeding up or slowing down moves the vocal formants as well. And increasing or decreasing the tempo causes the vibrato in the voice to speed up or down in a telltale way. With solo acoustic guitar (my old 60s Epiphone Texan) when I moved the pitch down by one semitone from F to E, with the guitar playing arpeggios from the bottom F upwards, the sound was perfect when it played back in E. When I moved to E flat I could tell that the instrument did not sound completely natural, although it could fool some listeners. Same thing but more so when I dropped to D - and when I got to D flat I felt that it sounded too unnatural to use. Although there were no glitches in the sound, the characteristic 'formant' resonances were giving the game away here. Going up in pitch I managed three or four semitones with it sounding like I was using a capo on the guitar, while with five or more semitones, the sound became too unnatural again.

For my next test I used Carol King's original recording of 'It's Too Late'. I had been asked to make a new arrangement of this classic song for a singer who wanted to sing it in a key one semitone lower. The idea was to base the new arrangement very closely on the old one, using the original as a 'template' to work to in Pro Tools. Using Speed, it was very easy to make a new audio file to work to in the new key. However, something must have been running at the wrong speed when this song was transferred from vinyl to tape before I put it into Pro Tools, so I had to shift the new file up again by about 20 cents to make the music in tune with A = 440 Hz. This time, I could definitely hear the result of the pitch-shifting (a sort of warbling-as-though-under-water kind of sound) on the voice and on the saxophone, although the piano still sounded good. I decided to try shifting the original down two semitones - and this time the warbling sound was quite audible on the lead vocal. Definitely not good enough to inflict on the record-buying public, although it was good enough to let me rehearse the singer in another key as a trial.

On the other hand, Speed is great with drum loops! I changed the tempo of a drum pattern all the way from 50% right up to 200% and it sounded perfect! When I tried pitch-shifting, I discovered that I could select, say, two bars of drums, hit preview and then move the pitch control in between repeat plays - very convenient - and, again, the sound was completely glitch free! Still, the drums did begin to sound unnatural if I shifted them more than a couple of semitones down or three or four semitones up. Nevertheless, the results were usable all the way from −12 to +12 semitones if you wanted a deliberately unnatural-sounding kit.

In conclusion

The Simple mode just could not get any simpler. It could not be any faster or more intuitive to use - exactly what I want on a busy Pro Tools session. Using the Calculator mode is also very useful when you know exactly what you want, and the Graphical mode is something else! For trying ideas out in Pro Tools with any kind of audio material, Speed is a great tool to use - much better than the built-in time-stretching and pitch-shifting algorithms. And when it comes to working on drum tracks, Speed is the best tool I have found yet.

Serato Pitch 'n Time 2

Everyone wants to stretch and warp digital media these days - and audio people are no exception. Serato's Pitch 'n Time is a feature-filled AudioSuite plug-in for Pro Tools that lets you alter pitch and tempo graphically or numerically - or 'varispeed' your audio the 'old-fashioned' way.

Professional tape recorders use a 'varispeed' control to vary the speed of the motors. Run it fast and voices sound like 'Pinky & Perky' cartoon voices. Run it slow for the 'voice of doom' - slow and deep like in a horror movie. The digital audio equivalent involves varying the samplerate. Pitch 'n Time lets you do this using its Varispeed Mode, where altering the tempo also alters the pitch. Simple - but effective. The clever trick is to change pitch without affecting tempo - or vice versa. This is what Pitch 'n Time is really all about. You can change the tempo from half through to double speed independently of pitch, and pitch shift up or down by up to 12 semitones independently of tempo.

Pitch 'n Time has one of the most responsive interfaces I have come across on this type of software. OK - it is a non-real-time AudioSuite plug-in, but the previewing is excellent and you can move the pitch or tempo sliders up or down while previewing and hear the audio change almost immediately in response

- with hardly any glitches. The Pitch 'n Time window has three sections: Tempo (time-stretching), Length, and Pitch (pitch-shifting). The Length section shows the start, end and length of your selection both before and after processing. To change the length of your selection, simply type what you want here. This section is located centrally, above the Pitch and below the Tempo sections. The Tempo section's time-stretching capability lets you change the BPM of your loops to match a given BPM, for example, or to convert audio recorded at one SMPTE frame rate to match another. The Pitch section lets you alter the pitch of, say, an out-of-tune instrument that is consistently flat or sharp.

The Tempo section

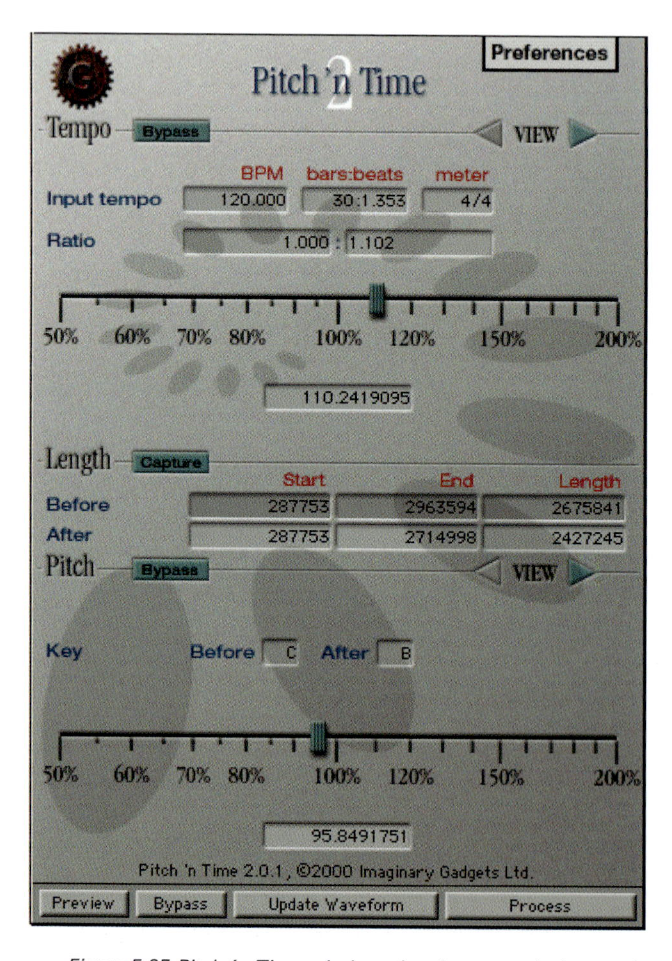

Figure 5.25 Pitch 'n Time window showing numerical controls.

The Input Tempo field shows the original tempo of the current selection in Beats Per Minute. If the tempo of the selection is defined within Pro Tools then this field is filled in automatically (if no value has been entered manually). When the input BPM is entered, the Bars & Beats field is calculated automatically. Entering the current time signature into the Meter field assists the plug-in with interpreting the Bars & Beats field, by nominating how many beats each bar contains. The Bars & Beats field shows the length of the current selection in Bars and Beats - if known. Entering this length will cause the plug-in to calculate the BPM of the selection.

Click the view arrow at the right of the Tempo section and the Variable Time-stretch Panel reveals itself - offering an editable tempo automation graph. Click on the red line to insert 'handles' then drag these up or down to 'draw' your tempo graph.

The time and tempo of the selected handle is shown in the two fields Time and Tempo below the graph. To select or move a handle, click within its box. To create a new handle, click anywhere on the line where there isn't already a handle. To delete a handle option, click it, drag it outside the graph, or move it on top of one of the handles to either side. The line always shows how the tempo will vary smoothly between the handles you have set.

The same feature is available for the Pitch section - just click the view arrow once to the right. This can be used to correct vocal or instrumental notes or phrases by moving the offending items to the correct pitch using the graph of pitch against time.

To help with detailed editing of automation data, the graphical editing modes include both horizontal and vertical zoom sliders to the left of the graph, and these always zoom in on the centre of the current view. The size of the scroll bars varies to reflect how much of the entire graph is visible.

By pressing the Update Waveform button at the bottom of the plug-in's window, you can have Pitch 'n Time display a waveform overview as a visual aid behind the automation graph.

The Tempo section offers a third panel called 'Morphing' Time-stretch. Here, you click on the display to insert vertical markers and then drag these horizontally to make the audio run slower or faster between pairs of markers. This lets you alter the timing of notes played within a musical phrase, for example, while keeping the overall length of the phrase the same - perfect for correcting slight performance mistakes.

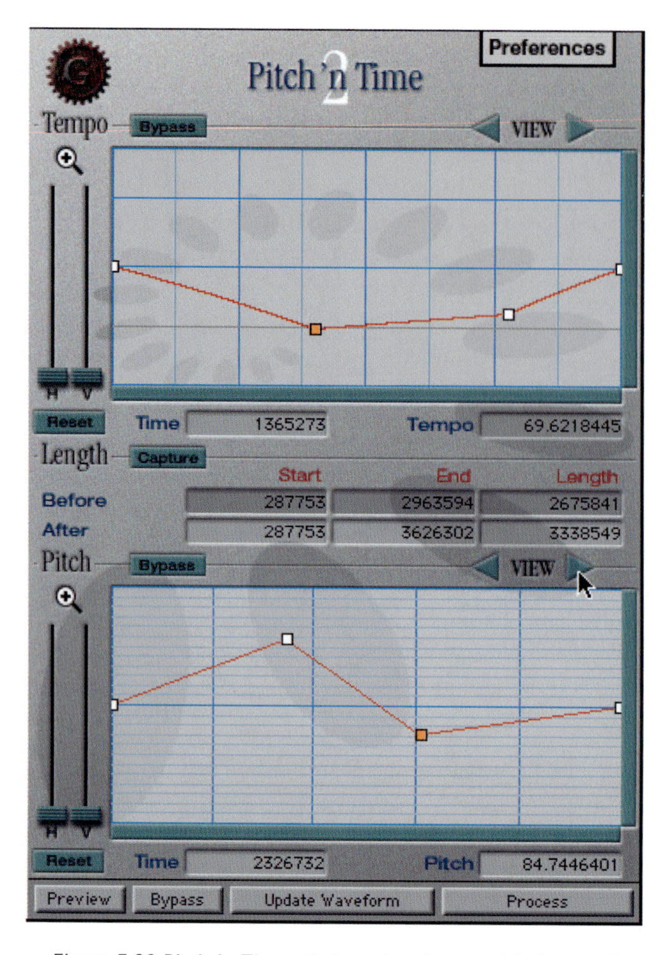

Figure 5.26 Pitch 'n Time window showing graphical controls.

This panel is useful if you need to control when things happen, rather than how quickly. Just drop markers on the audio and move them where you want them to go. Pitch 'n Time then stretches and squashes the audio to fit. There are two graphs on the morphing panel. The lower of these shows the selection and markers before processing. The upper graph shows where the markers will go after processing. The output graph also has a guide waveform overview. The guide waveform is available as a visual reference and appears beneath the output waveform, which is shown as a dark outline.

There are buttons labelled Guide and Source next to the graphs. Whichever one is pressed in controls which waveform overview will be updated when the Update Waveform button is pressed. So to load a guide waveform in, press Guide and then Update Waveform. To load the waveform for your source, just make sure Source is pressed in and press Update Waveform. It is recommended that you load your guide waveform first, to avoid any confusion.

Markers are shown as vertical lines with handles at the top and bottom. The selected marker's handles are shown on both graphs in orange, and the corresponding input and output times are displayed at the bottom of the panel. Markers can be selected and moved in either graph by clicking anywhere on the line. Option-clicking a marker deletes it, as does dragging it outside the graph. New markers can be created by clicking anywhere except on an existing one.

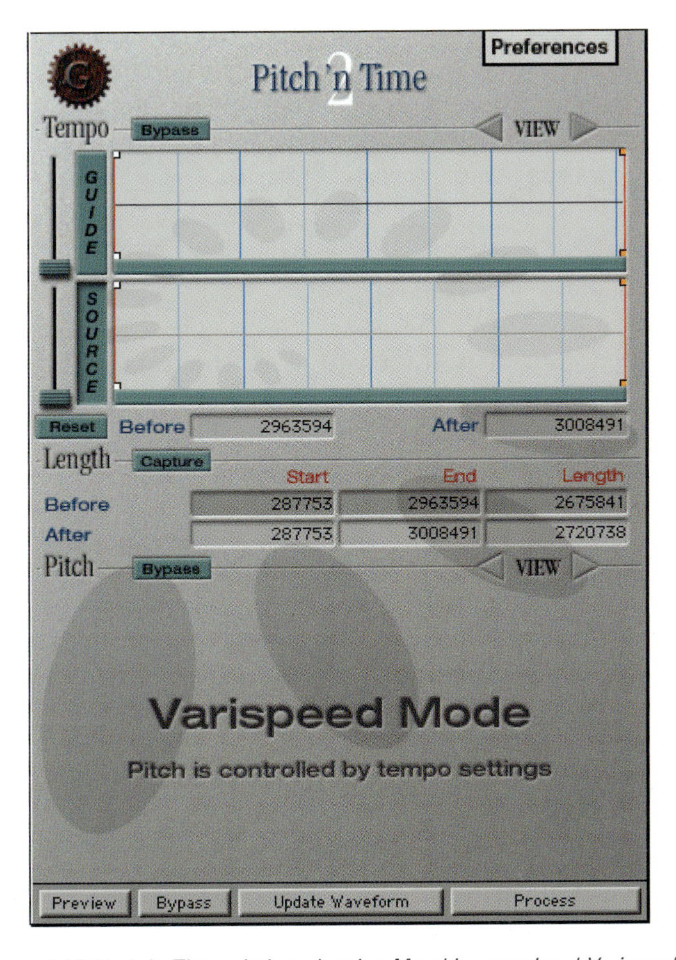

Figure 5.27 Pitch 'n Time window showing Morphing panel and Varispeed Mode.

The Length section

No matter which time-stretching mode you're in, the Length panel always shows the start, end and length of your selection both before and after processing. This panel is located centrally above the Pitch section and below the Time section, You can change the output values at any time by typing into those fields. The 'Capture' button here lets you easily match the lengths of two different audio regions. Pressing the capture button memorizes the length of the current selection. When a second region is selected, the current time-stretch panel is automatically adjusted to create that same output length.

You can use the Capture button to match several loops to the same tempo. This is very straightforward. Just select the loop with the correct tempo and press Capture. Then select a loop with a different tempo. The controls will automatically update to show the amount of time-stretch needed to match the tempo of the second file to that of the first, so all you need to do is to hit the Process button. You can then select another file to be matched to the first, the controls will update again and you can process this file - and so on.

The Pitch section

The Fixed Pitch-shift panel works similarly to the fixed time-stretch panel, applying a simple, single pitch-shift factor. In the preferences panel you can choose between working with percentage frequency change or semitones and cents. The two fields above the pitch slider provide a simple way to transpose your selection. Just enter the key of the source and the new key you would like and the plug-in will calculate the required pitch shift.

The Varispeed Pitch-shift Panel has no controls, and forces the pitch shift to precisely match the time-stretch at any point in time. This recreates a classic varispeed effect, allowing you to make great sounding record/tape speed-up and slow-down effects. To go really slow, use the variable time-stretch panel on a linear vertical scale.

Preferences panel

The Preferences panel can be accessed at any time by pressing the button labelled Preferences. Here you can change settings that apply globally to the plug-in.

For example, the Percent option lets you edit time-stretch values as a percentage of the original tempo while the BPM option lets you work in terms of the

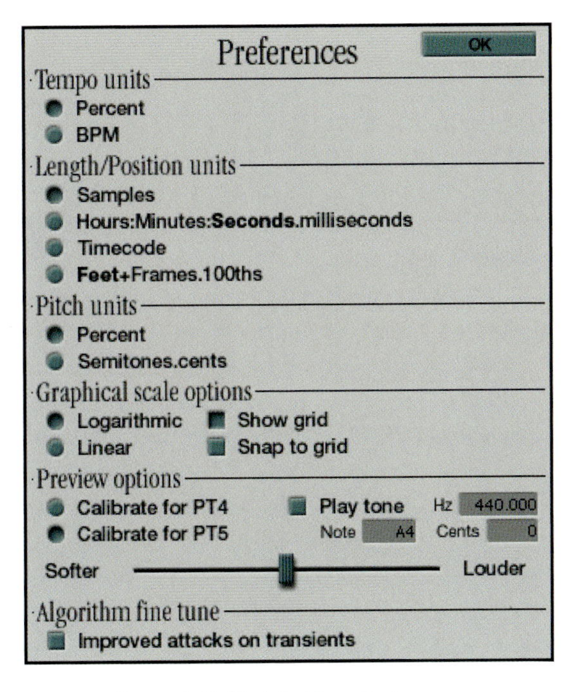

Figure 5.28 Pitch 'n Time Preferences.

output BPM. An input BPM must be specified for these units to make sense. You can make settings for how the length is displayed in samples, Hours, Minutes and Seconds, SMPTE time or Feet + Frames. Pitch-shift values can also be displayed as a percentage frequency change or in semitones and cents. The Graphical scale options include Show Grid Display to include grid lines in all graphical views and Snap To Grid which 'snaps' handles to nearby grid lines when editing these. You can also set the vertical scale to Log or Linear for the automation graphs. Interpolation between handles is not affected by this setting on the variable time-stretch panel, so in log mode lines will appear bent. The variable pitch-shift panel alters its interpolation to match this setting, so lines are always straight.

The Preview options let you choose whether to optimize for Pro Tools 4 or 5 software. Using the Play Tone option, you can play a reference tone along with the preview, to facilitate tuning by ear. The note to play and detuning can also be entered here, or the frequency can be entered in Hertz. Finally, the Algorithm fine-tune slider lets you bias the Serato Intelligent Sound algorithm towards more tonal sounds such as pipe organ or more transient sounds such as drums.

Note Pitch 'n Time always works non-destructively. That is, it always creates a new region file and never overwrites the source. The Use in Playlist button controls whether the processed audio should automatically be used in the Edit Window in place of the source. With this option selected with Playlist as the source, only the audio selected in the Edit Window is replaced. However if Use in Playlist is selected with region list as the source, every occurrence of the source region in the Edit Window is replaced - potentially a dangerous operation.

The Create Continuous File option is only available when Playlist is selected as the source. In this mode the entire selection within the Edit Window is treated as a single region. The option to the right indicates entire selection to reflect this fact. All of the audio - including the silent gaps between regions - is fed into the plug-in, and a single audio region created to hold the output from the plug-in.

Conversely the Create Individual Files option causes each selected region to be processed in turn, and corresponding regions are created from the output. When create individual files is used with Playlist selected as the source, Pitch 'n Time re-spots the new regions to new locations to match the selected time stretch.

Multi-channel mode enforces phase coherency between all the selected tracks, up to the AudioSuite limit of 48 channels. Use this feature only when absolutely necessary, as enforcing phase coherency unnecessarily can be detrimental to the sound quality.

The 'Bottom Line'

You can use Pitch 'n Time to change the key or tempo of your audio up or down by greater amounts than most rival products without rendering the processed audio unusable. Pitch 'n Time's closest rival is Wave Mechanics' Speed. Pitch 'n Time has the edge for me on account of its 'morphing' time-stretch and varispeed features. Speed is great for its simplicity, but Pitch 'n Time ultimately allows finer control of both pitch and time aspects of your audio - and often delivers superior results. Now if it would pitch-shift without shifting formants it could be a world-beating product!

GRM Tools

A standard interface is used for all the GRM Tools TDM plug-ins, with a set of controllers for adjustment and visualization of the processing parameters, plus a system for management of the presets. On the right of each plug-in's window there is a keypad with 16 buttons and a vertical slider for store and recall of presets, while at the bottom there is a horizontal slider that allows interpolation between the presets. Several of the plug-ins feature two-dimensional controls in a display area that allow the user to control two separate parameters at the same time - as you would do with a joystick. As well as using the standard Import and Export settings in the plug-in menu, there is also a simplified local management system making it possible to easily store and load preset configurations of parameters and interpolate between these to produce 'morphing'-style effects. You can store up to 16 preset configurations using the keypad: a Command + click stores the current preset configuration in the selected memory, while a simple click just loads the selected preset configuration.

Volume 1: Doppler, Comb Filter, EQ, Shuffling

Volume 1 of the GRM Tools plug-ins includes Comb Filters, Equalizer, Doppler and Shuffling effects.

Comb Filter

The Comb Filter amplifies the audio signal at a given frequency and at all the harmonics (integer multiples) of that fundamental frequency. The GRM Comb Filter plug-in is offered as banks of either one or five filters, in mono and stereo. These filters feature resonant filtering at specified frequencies, with controls for duration of the resonance and for the cutoff frequencies of the low-pass filters used to control the number of harmonics amplified. Each filter has three controllers - master frequency, master resonance duration and master filter cutoff. These master controls act as multiplier coefficients for the individual settings.

Comb filters can produce a wide range of effects, from the subtle to the outrageous. Low resonance values will produce the characteristic colorations that you would expect from valve (tube) amplification.

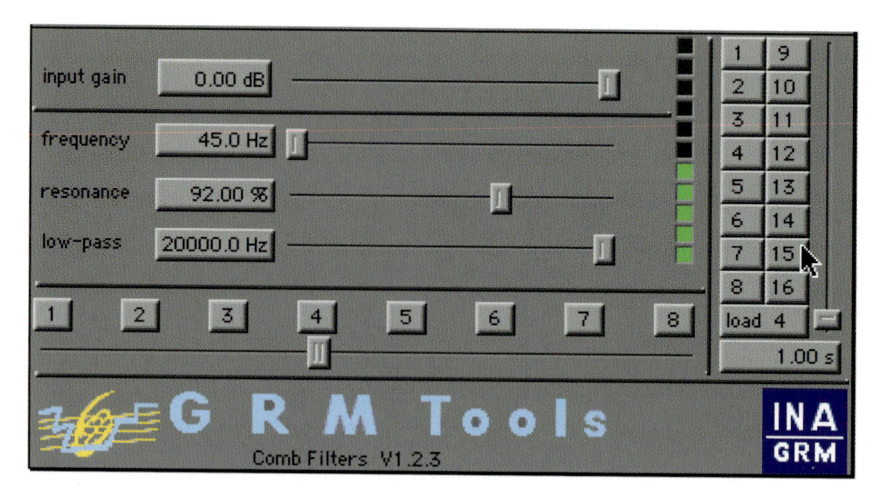

Figure 5.29 GRM Tools 1-Band Comb Filters.

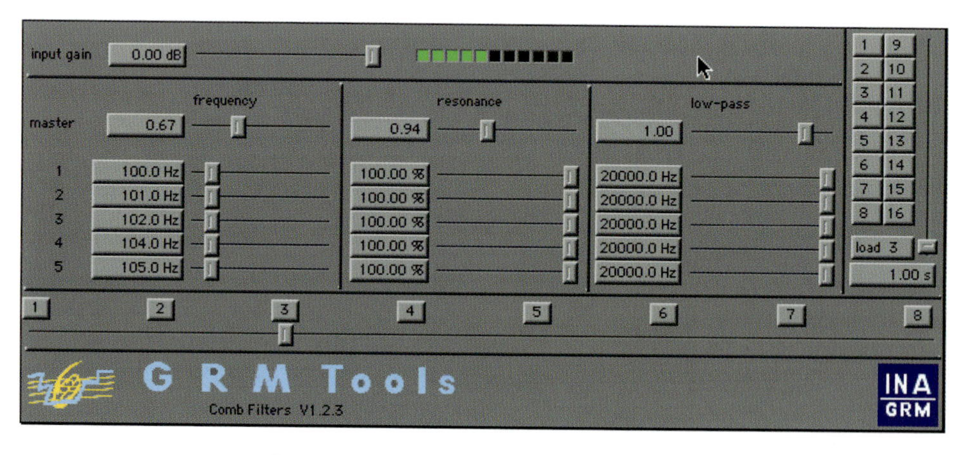

Figure 5.30 GRM Tools 5-Band Comb Filters.

Equalizer

The Equalizer features a bank of 23 fixed frequency filters equally spaced one third of an octave apart. The gain of each filter can be set from −100 dB to +12 dB to allow precise alterations of the spectral balance of the audio passing through. The high attenuation values available allow you, for example, to extract radical portions of the audio spectrum to create spectacular effects. Versions are available in mono or stereo. With the stereo version you can copy values from the left channel to the right channel and vice versa - and the controls for the two channels may be linked together.

By fading between presets, you can create some incredible timbral effects that change over time. This plug-in does take some effort to set up for a particular project, but your efforts will be well rewarded.

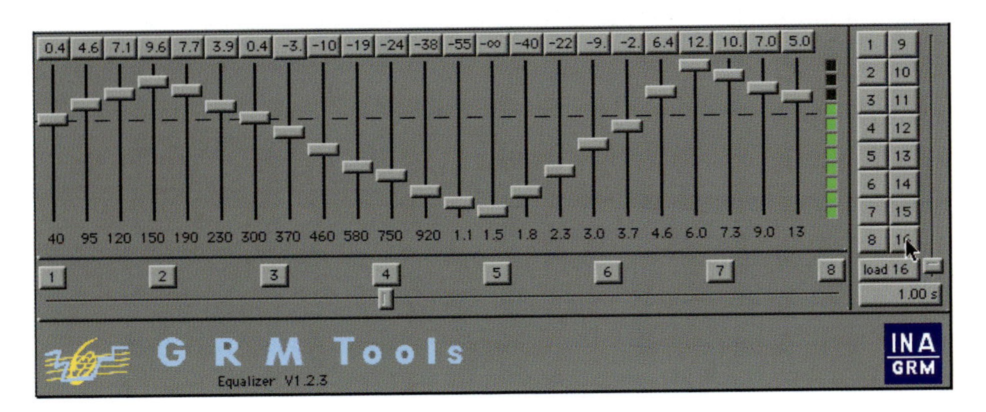

Figure 5.31 GRM Tools Equalizer.

Doppler

The Doppler effect can be used to create typical Doppler effects where the pitch of a sound source rises as it approaches the listener and falls as it travels away. The GRM plug-in has a wide range of controls that also allow you to create some very unusual rotating effects – in both stereo and mono versions. The smaller of the two squares in the display represents the sound source while the larger square represents the 'target' or 'listener'. The sound source reaches the target in a time that you can adjust using the Following Time parameter. The Amplitude Variation slider lets you set the maximum amplitude of the intensity variation with respect to distance and the Doppler Variation slider can be used to set the rate of the Doppler effect with respect to distance. The Circle Amplitude slider lets you set the amplitude of automatic rotation of the source around the target, and the Circle Frequency slider lets you set the rotational speed in the clockwise direction, or in the counterclockwise direction if you use negative values. In the stereo version, each channel becomes an independent source, so two small squares are provided in the display and two additional control sliders are provided to the right of the display – allowing you to control channel separation and phase during circular modulation. In this case, the Channel Separation slider lets you set the spacing between the source and the Phase Control slider lets you set the phase ratio between the two circular motions.

These controls let you make sounds 'fly' past you with accompanying natural-sounding pitch-shifting effects. Or you can create entirely new effects, with the source sounds rotating around the 'listener', for example. The effects that can be created using this plug-in will not only suit sound designers working to picture, but also provide an excellent selection of unusual spatial effects for use with any type of recording.

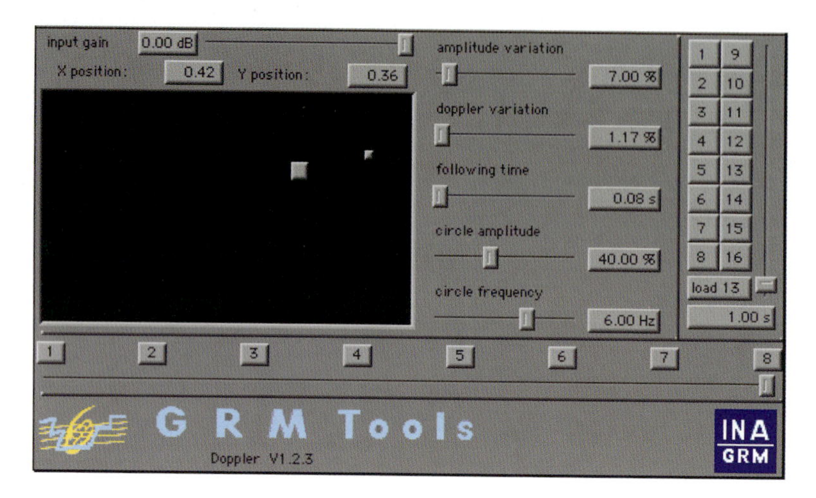

Figure 5.32 GRM Tools Doppler.

Shuffling

Shuffling allows random micro-splicing of audio fragments, with mono input and stereo output. This plug-in samples up to 683 ms of the audio playing through it, then takes fragments of specified duration from this and applies random delays to these. You can then adjust the envelope, delay time, initial and final pitch of these fragments. You can also adjust the density, which controls the way that some of the fragments are replaced by silence. The envelope and delay parameters of the fragments are set using the envelope and delay sliders and the pitch of each fragment may be modified either using the random pitch slider or in a linear fashion – in which case you use the initial and final pitch sliders. Feedback can also be applied.

The effects that this Shuffling plug-in creates simply have to be heard to be appreciated. Suffice it to say that sound designers will find this to be a very useful tool to have at their disposal for creating cartoon voice effects, space effects and unusual pitch transposition effects.

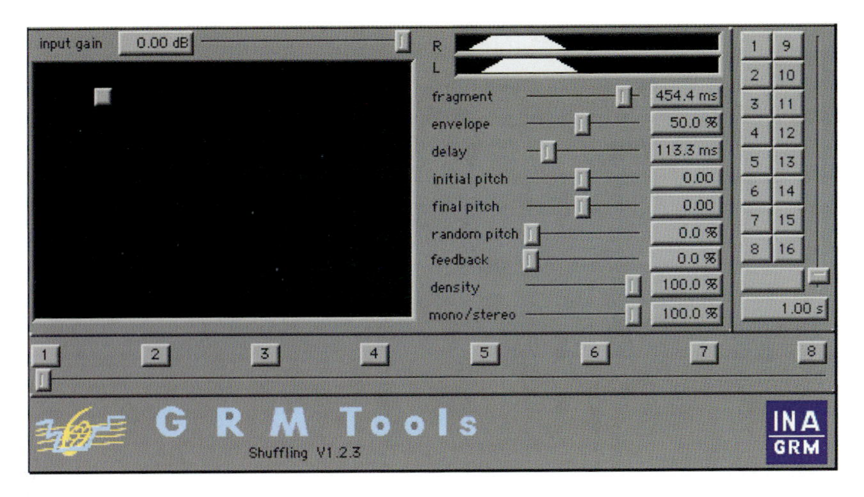

Figure 5.33 GRM Tools Shuffling.

Volume 2: Band Pass, Delay 24, Freezing, Pitch Accum

Volume 2 of the GRM Tools plug-ins includes four algorithms, each with mono and stereo versions. These include Band Pass filters, the Delay 24 multiple delay algorithm, a multiple loop replay device called Freezing, and a dual 're-injection' pitch shifter called Pitch Accum.

Band Pass

The Band Pass plug-in combines two filters, a high-pass and a low-pass, which together form a variable width band-pass or band-reject filter. The cut-off frequencies can be set individually, while the cut-off slope is fixed and set at a very high value of 560 dB per octave, so out-of-band attenuation is an impressive 60 dB. The High and Low sliders let you set the high and low cut-off frequencies. Dragging the square in the display area lets you set the centre-frequency and the bandwidth around this according to the horizontal and vertical co-ordinates of the square's position. In the stereo version, two display areas with two sets of high and low controls let you control the two channels. A Link button is provided to slave the controls of one channel to the other if desired. A pair of Left-Right and Right-Left buttons let you copy slider values from one channel to the other.

This plug-in lets you quickly cut away any frequencies in the audio spectrum that you are not interested in so that you can bring out the frequencies you want to feature. You could easily create a telephone type sound by restricting the pass band to between 300 and 3000 Hz - the bandwidth of a typical telephone system, for example. The presets include many useful settings that you can apply right away, some of which restrict the bandwidth to only allow a tiny portion of the audio spectrum 'through', while others take out the highs, or the lows. Switching from one setting to another while interpolating the in-between values produces some great effects.

Figure 5.34 GRM Tools Band Pass.

Delay 24

Delay 24 is a group of 24 variable delays featuring a multiple 're-injectable' delay algorithm. You select an initial delay and a number of delays are calculated from this with various spacings in time, amplitudes and feedback amounts.

The First Delay slider lets you set the initial delay up to a maximum of 683 ms. The Delay Range slider then lets you set the time interval from the first to the last delay and the Number of Delays slider lets you set the total number of delays. The two-dimensional control in the display window lets you simultaneously vary the delay distribution parameters. Horizontal movements set the delay distribution centre, i.e. the mid-point of the range given by the first delay value and the first delay + delay range sum. Vertical movements set the delay range value. The Feedback slider lets you feed signal from the output back to the input to produce delay repeats. The Amplitude Distribution slider lets you set the amplitude progression law for each delay - i.e. how much softer or louder each subsequent delay will be - linear to exponential or inversely exponential. Similarly, the Delay Distribution slider lets you set the delay time spacing law, again from linear to exponential or inversely exponential. The Random Delay slider offers random distributions of the delays in time, updated depending on the value of the Variation Rate. The Variation Rate slider lets you set the time interval between two updates of the delay amplitude distribution or variation. Lastly, clicking in the Delay Representation field brings up a popup menu to let you select the activation/deactivation status for each delay.

On the mono in/stereo out version, the mono/stereo slider sets the delay spacing at the output from fully monophonic to two-channel with odd delays on the left and even delays on the right. On the stereo in/out version, the Xfeedback slider cross-feeds the odd delays in the left-hand output to the right-hand input and the even delays in the right-hand output to the left-hand input.

With this impressive array of controls, the Delay 24 is capable of creating an extremely wide range of effects, including randomized delays, reverberation and reverse reverberation. This is definitely one for the creative sound designer working on film soundtracks and suchlike.

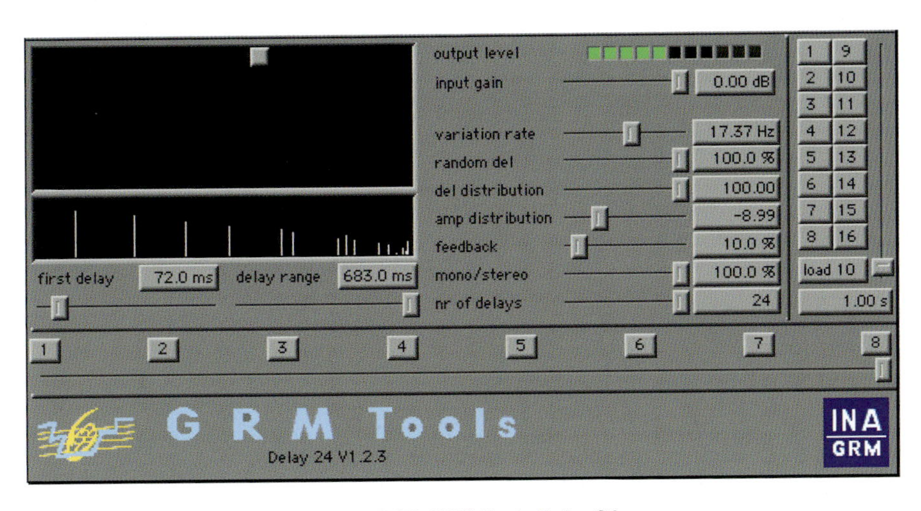

Figure 5.35 GRM Tools Delay 24.

Freezing

This plug-in basically samples a portion of the sound playing through it and then creates up to eight loops within the sample. The Freezing algorithm loops around the previous 683 ms of audio playing through the plug-in when you hit the Freeze button. The playback point, the time and number of simultaneous loops and synchronization can all be varied, along with the replay speed. This means that all the loops can start playing back at the same time, or at different times. You can also offset the pitch of the loops relative to one another. And the durations and pitches of the loops can be varied randomly.

A display screen at the top left of the window shows a representation of the signal elapsed since the current time, or, when you hit Freeze, of the signal held in memory. When the signal is 'frozen', i.e. looping round a captured section, you can adjust the loop points using the left and right pointer parameters or by dragging the two-dimensional square control, or by dragging the loop start and end points directly in the display. The Pitch slider can be used to transpose the loop from −4 octaves to +2 octaves by changing the replay speed. The Pitch Offset slide lets you offset the pitch transposition of loop pairs 3-4, 5-6 and 7-8. The Random Pitch slider lets you set an amount of random pitch deviation and the Random Duration slider lets you set a random loop range limited by the memory available around the loops.

On the mono in/stereo out and stereo in/out versions, the mono/stereo slider controls loop distribution at the output, from fully monophonic (centred) to two-channel (odd loops on the left, even loops on the right). If multiple loops are selected, two parameters determine the time setting of the loops relative to one another. The Reset Phases button resets all loops to the point of origin with zero phase difference between them. The Random Phases button resets all the loops to any point of origin with random phase, evenly distributed over the range of possibilities. These two parameters are most effective when used applied to loops that are nearly identical in length.

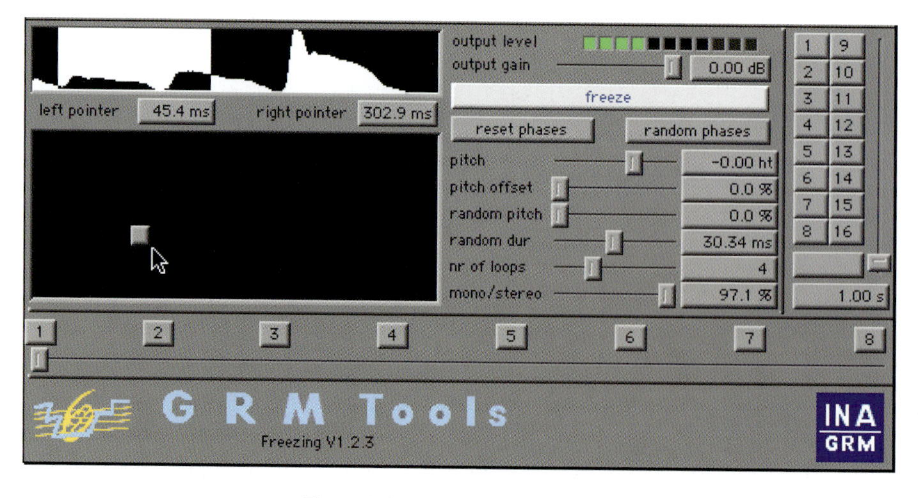

Figure 5.36 GRM Tools Freezing.

In action, this plug-in lets you grab source material from any audio playing back through the plug-in and 'mess' with this creatively to produce interesting effects. Often you can create rhythmic effects or textures that can act as interesting 'hooks' to feature or as beds to blend into the background of your music or sound effects. Useful for dance music producers as well as sound designers.

Pitch Accum

The Pitch Accum plug-in combines two transposition processors with a feedback delay section. The transposed audio can be modulated on a periodic or random basis and a pitch follower is provided to allow improved transposition of periodic signals.

Two transposition 'tracks' are provided, each with its own set of controls. The Transpose sliders let you individually set the transposition interval for each of these tracks from −24 to +24 semitones. The two-dimensional slider can be used to control both transpositions simultaneously. Horizontal movement produces identical transpositions of both tracks, while vertical movement produces a divergent offset of both tracks relative to one another, up to a maximum deviation of two octaves. The delay sliders set the delay of each track up to a maximum of 683 ms per track, and the Gain sliders control the input signal amplitude. A pair of left and right arrow buttons are provided to let you reverse the fragment replay direction for each track.

Control sections are also provided for two types of modulation that can be applied to the transposed audio - Periodic and Random. In the Periodic Modulation section, you can choose the shape of the modulation waveform from the popup menu at the top of this section. Shapes include sine, square, triangle, sawtooth, reversed sawtooth, and pulse. Sliders are provided for frequency and amplitude, and you can adjust the synchronization of the modulation applied to the two 'tracks' using the Phase slider. For example, at 180 degrees, the modulation shape applied to both transposers is in antiphase. In the Random Modulation section, you can set the sampling Frequency, the maximum Amplitude, and a Smooth value. This Smooth value is described in the manual as 'an inertia constant between two consecutive values' and serves to smooth out the effect. The feedback section lets you apply feedback to the delay of the two tracks using the Feedback slider. The Mono/Stereo slider controls the distribution of the transposition tracks at the output, from fully monophonic (centred) to two-channel. In the stereo version, the signal is summed into a single track before the transpositions are made. The Direct slider is used to mix the original signal, mono or stereo, at the output.

Underneath the Feedback section, controls are provided for the Windowing function and Pitch Detector. The size of the fragments used in the transposition algorithm is controlled by the Window slider. If the Pitch Detector is switched on, the size of each fragment depends on the detection of a signal zero-crossing. The Cross-fade parameter is used to set fragment overlap, from permanent fade-in/fade-out (100%) to no overlap (0%). Transposition quality depends heavily on the type of signals processed. For example, when using a simple voice signal, very good results can be obtained with the pitch detector switched on, coupled to a small fragment size of about 10 ms, i.e. a fundamental period window between 10 and 20 ms (100 to 50 Hz). When using complex signals, a larger fragment size, such as 50 ms, should be tried - with or without the pitch detector.

The best way to get started with this plug-in is to switch through the 16 presets and listen to the effects. Preset 1 is fairly subtle, providing a short delay with just a little modulation applied to make the sound more interesting. The presets become more radical as you step through these, progressively applying greater pitch shifts, longer delays and more complex modulations, with some of the higher-numbered presets providing reverse-sounding and tape-rewind effects. Preset 16 is totally far out and would be very interesting in an appropriate context!

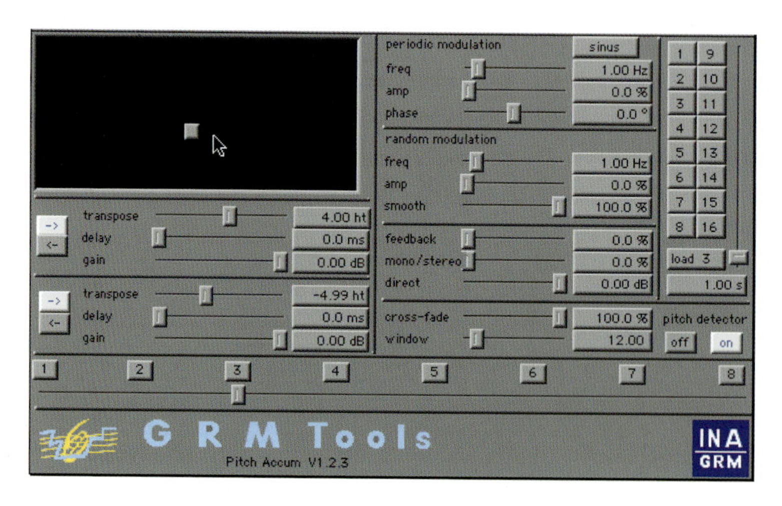

Figure 5.37 GRM Tools Pitch Accum.

Apogee MasterTools

Apogee MasterTools is a stereo TDM plug-in to use for Mastering or detailed editing of audio files. Apogee's proprietary UV22 process is provided as an alternative to the standard dithering available in Pro Tools and MasterTools also includes useful metering and other functions.

Once you have finalized your processing and mixing, you will often need to prepare a final output file that uses just the 16 bits which are available on DAT or CD rather than the 24 bits available within your TDM system. Apogee offers the highly regarded UV22 encoding process for this purpose. With standard noise-shaping and bit-mapping systems, users sometimes hear the noise floor changing with the music. This can happen because traditional dither compromises the 16-bit noise floor by adding noise – trading a reduced noise floor for a large boost at high frequencies. UV22 keeps the audible noise floor solid at the theoretical minimum for 16-bit systems, presenting a constant, smooth and stable noise floor through which can be heard undistorted detail up to 30 dB lower in level – thus extending the perceived dynamic range beyond that of 20-bit resolution.

With MasterTools you get most of the stuff you would have in a conventional analogue or digital mastering studio. So, for instance, you can mono the two channels to make sure nothing 'disappears' from the stereo mix due to phase cancellations when summed to mono. Or you can switch the polarity of left, right or both channels to make sure all your phase relationships are correct. And in what circumstances might they not be correct? Well, for example, if you make several routings in and out of Pro Tools to the DSP farms, there will be a small delay that upsets the phase relationships – or incorrect miking techniques might do this. You can use MasterTools to check whether these cumulative delays have audibly affected the phase relationships, so you can then slip a track or remix to correct these problems.

MasterTools also has extremely accurate meters to show level, phase and balance – presented both in real time and as a 'history'. The metering displays Peak and Average levels simultaneously along with Peak and Average Hold and you can use the Phase Correlation metering to check for mono compatibility. A five-second 'history' of Level, Overs and Phase Correlation is also displayed in colour. The OverLog displays how many 'overs' occurred during the session and logs them against timecode. A facility is provided to automatically normalize these peaks, but if there are too many serious 'overs', the only fix is to go back and remix more carefully.

Note In case you were wondering, an 'over' is the digital equivalent of clipping in the analogue world – where you have exceeded the available dynamic range of your system and the waveform becomes squared off at the maximum level. Rather than calling this 'Digital Clipping' or whatever, in digital systems this is usually referred to as an 'Over'. However, there is no universal agreement among manufacturers of digital audio equipment as to exactly what constitutes an 'over'. Apogee have chosen to regard four clipped samples in a row as being an 'over'.

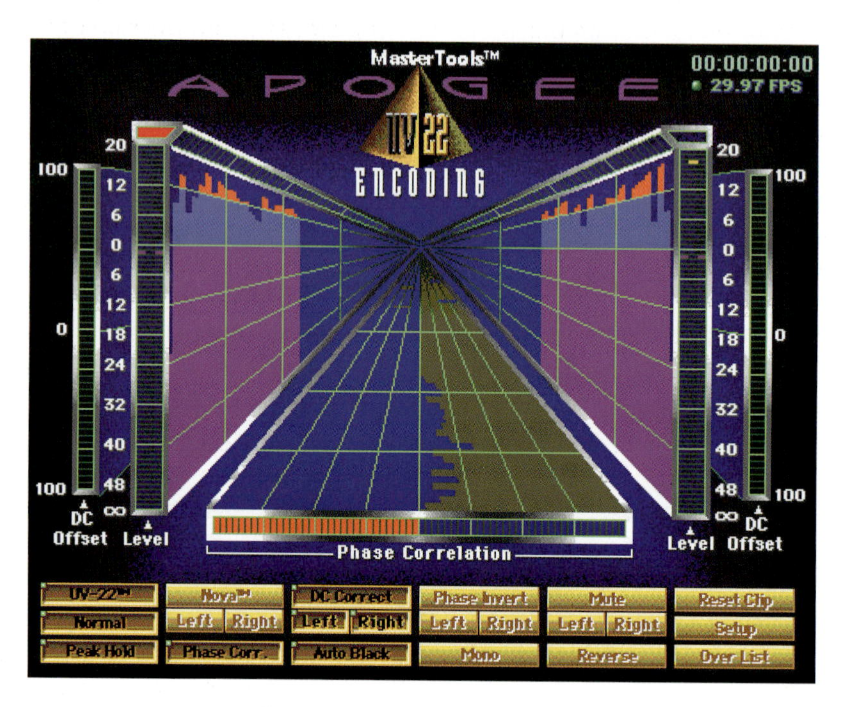

Figure 5.38 Apogee MasterTools.

When you playback audio through the plug-in, you will see the audio levels on the main metering display located to the left and right of the displays at the sides of the 'flying V'-shaped display. These levels march backwards down the 'V' as time passes, until they disappear over the 'horizon'. If you keep checking this display, it is easy to identify any problematic sections within your mix.

To the left and right sides of the main display, the Level displays show the signals at the left and right inputs of the MasterTools plug-in. Each channel has an 'over' indicator that turns red when an 'over' occurs and stays red until the Reset Clip button is clicked. The wide meter reflects the average signal over a period of time, while the narrow overlay meter displays the Peak Value. If peak hold is on, then the highest value that both the Peak and Average reach will be held for three seconds or until a higher level is reached.

Something to watch out for with digital audio is that if you edit signals with different levels of direct current (DC) on them, you can get clicks and pops at your edit points. MasterTools lets you automatically remove any such DC offset in real time using the DC Correct feature. The DC Offset displays to the left and right of the Level meters indicate the number of LSBs of DC Offset at the 16-bit level that are being removed from the left and right channels, with the sign of the offset indicated by the associated LED indicator.

Below these displays, there are various On/Off buttons for UV22, Normal/Low (for a lower level of UV22 encoding), Peak Hold, Phase Correlation, AutoBlack, Mono, Reverse, Overlog, Setup and Reset Clip. Selection buttons are also provided for Nova, DC Correct, Polarity Reverse and Mute. The AutoBlack functions like a gate. When the audio drops below a certain threshold, for a given time period, the UV22 signal is muted until activity at the input occurs again. The AutoBlack function allows automatic insertion of digital silence, by turning off the UV22 process. It triggers when no signal at the 20-bit level occurs for 128 consecutive sample pairs, left and right, and the signal is at least −42 dBFS. AutoBlack is removed and the UV22 process is reinstated when a signal on either left or right channel occurs. When AutoBlack is active, the UV22 control will flash indicating that the UV22 process has been overridden by Auto Black.

Note UV22 is expected to be the final step in the audio signal chain. No more processing of any kind should be performed on audio that has been processed using UV22.

In my experience, the UV22 processing is much better than the simple dithering available using Digidesign software, and the metering and other tools are extremely useful for mastering work. MasterTools is ideal for small project studios using Digidesign equipment, where people might otherwise be using a small Soundcraft or Mackie mixer – which would not have the meters and switching or processing facilities available in MasterTools.

Cool Stuff Labs Generator X

Generator X is a high-quality signal generator plug-in that produces all the test tones you may need for alignment, reference and so forth with your Pro Tools system. Generator X produces a wide variety of fully automatable waveforms and impulses, along with white and pink noise - which can also be used for creative sound design applications. An adjustable trigger allows the tone generator to be keyed from the input or from a side-chain, so you can add a sine wave tone to strengthen the sound of a kick drum, for example. And the single impulse mode provides an ideal way to audition reverbs without colouration. Priced at under $100, Generator X is simple to use, has no awkward copy protection, and works with all Pro Tools systems from PT|MIX back to 68k Nubus, with an AudioSuite version also provided.

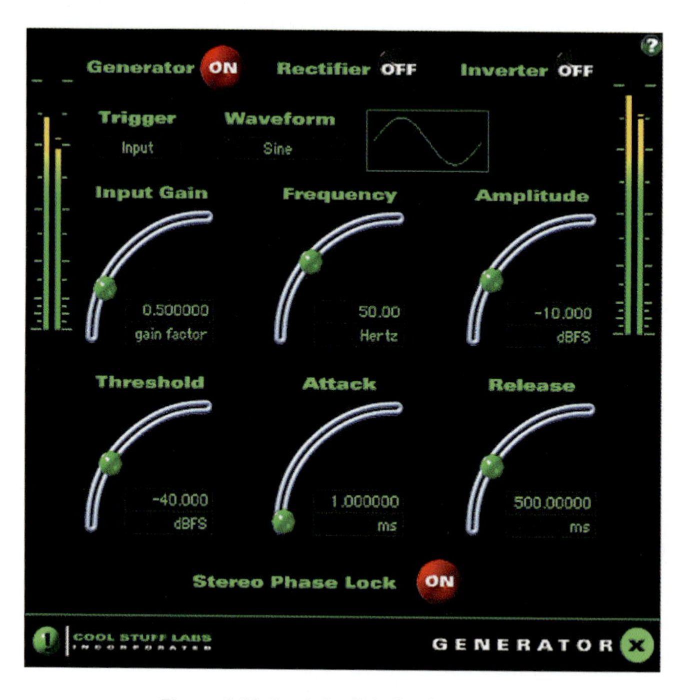

Figure 5.39 Cool Stuff Labs Generator X.

The user-interface is interesting - and efficient. Any text contained in a box reveals a popup menu when you click on it and numeric values are directly editable on-screen. The sliders are of a unique curved design. Because of the sliders' design, the further the cursor is from the lower right corner of the control being manipulated, the more movement is necessary to generate the same change in value - providing better resolution than with a normal control. The behaviour of the sliders also depends on how you use the mouse. If you move the mouse in an arc approximating that of the on-screen slider, the motion is linear and tracks your movement directly. On the other hand, if you move the mouse in a straight horizontal line directly to your right, the motion of the knob is approximately logarithmic - large initial change followed by smaller and smaller incremental change. And if you move the mouse in a straight vertical line directly up, the motion is logarithmic, but in the opposite sense - small initial change followed by incrementally larger and larger change. For example, with the Amplitude control set approximately in the centre of its range, when you click on the slider and drag the cursor inside the arc, towards the value readout for the control, small movements of the mouse will produce large movements along the arc. If you drag the cursor away from the control, towards the top left of the screen, the precision is now much higher, as much more movement is necessary to cause a change in the control value. For fine control, simply click on the value display for the control and use the up/down arrow keys to nudge the value up or down by very small increments.

The controls in Generator X are simple and can be divided into two main groups - the Basic Controls for Waveform, Frequency and Amplitude, and the Trigger Controls including Trigger input selector, Input Gain, Threshold, Attack and Release. The Basic Controls include on/off buttons for Rectifier and Inverter. The Rectifier button extends the existing waveforms by flipping the phase of any negative part of the generator output. (It doesn't affect any input signals.) The Rectifier can be useful if, for example, you want a train of positive-only impulses, or need to rectify a square wave to get a DC offset. The Inverter button, as with the Rectifier, extends the existing waveforms by flipping the phase of the generator output. (Again, it doesn't affect any input signals.) When used in combination with the Rectifier button, the Inverter will result in a waveform that is entirely negative. Whenever you use these buttons to modify the output, you will see the effect on the waveform in the small waveform display underneath the buttons. To the left of this waveform display, the Waveform pop-up menu lets you pick the waveform - sine, square, triangle, sawtooth or impulse - that will be shown in the display and form the basis of the generator output. The Impulse waveform consists of a train of single impulses, alternating positive and negative. Selecting Single Impulse in the waveform pop-up places Generator X in a special mode, displaying an additional button labelled SNAP. In this mode, the output of the signal generator is zero. Instead, pressing the SNAP button results in one single positive impulse - very useful for auditioning reverbs or checking absolute or relative phase. Other options are provided for White Noise and Pink Noise. White Noise can be useful for performing room measurements, EQ measurements, or adding some spice to snare or cymbal hits. Pink Noise is also useful for room measurements, EQ measurements, and acoustically resembles naturally occurring random broadband sounds such as ocean waves.

Note Various standard test tones are provided with the Generator X package as presets. Sine tones and noise at a number of standard levels are available, along with settings for Inter-Modulation Distortion (IMD) testing, alternating Least Significant Bit (LSB), and tape biasing.

The Input Gain slider controls the amount of signal present at the plug-in's input which is passed to the output. This allows for control over simple additive synthesis and other blending applications. The units are selectable for whatever you are most comfortable dealing with (dB or gain factor) and positive gain values are permitted. The Frequency control allows adjustment of the waveform frequency and the value may be specified either as a frequency in Hertz or as a wavelength in samples. The Amplitude control lets you adjust the amplitude of the generated waveform. As with Input Gain, the units may be adjusted according to your needs - in dBs below Full Scale, or as a gain factor, or as quanta (a hexadecimal value representing the number of quantization steps).

You can use the Trigger popup to select the source used for Generator X's triggering mode. Here you can choose whether to trigger the generator output from the input signal or from a side-chain input. If you choose side-chain, you also need to select the actual side-chain input that you will use from the standard Pro Tools plug-in side-chain input popup above the Generator X window. Any Pro Tools input or bus can be selected. The Input Gain slider lets you determine whether or not the Generator will create any output; it does not affect the level of the output signal. You use the Threshold control to set the level at which the trigger switches from OFF to ON, producing output from the generator and this value may either be specified either in dB or as a gain factor. The Attack slider lets you set the length of time it takes the trigger to go from completely OFF to completely ON, and the Release control lets you set the decay time of the trigger, i.e. how long it takes the gate to go from completely ON to completely OFF. Typically, you will use a fast attack with a longer release to achieve a natural sound.

Tip You can easily create a low frequency drone for a movie soundtrack or whatever using Generator X. Insert the plug-in on a new track, set the frequency to 50 Hz and the waveform to Sine with, say –6 dB amplitude. Now insert a second Generator X immediately after the first and set the frequency to 52 Hz, the waveform to Sine and the amplitude to –6 dB, with the Input Gain set to 0 dB. This will create a deep, pulsating 'drone' sound.

The bottom line? Generator X is an affordable, useful tool that ought to be included in every Pro Tools rig.

McDSP

FilterBank

Filters 'for all seasons!' That's the soundbite that comes to mind when I think about the McDSP FilterBank plug-ins. Filter types include low and high shelving EQ with independent peak, slope and dip controls; low and high pass filters with resonance controls; band pass and band reject filters; and parametric EQ with more than five octaves of bandwidth. These are supplied in stereo and mono versions in 2-,4- and 6-band configurations for a total of 17 different plug-in types. Despite the fact that there are so many, this doesn't mean the quality is low; all the filter designs use double precision 48-bit processing. We are talking high-quality technical tools here – with down-to-earth user-interface design that makes these filters 'a snap' to use!

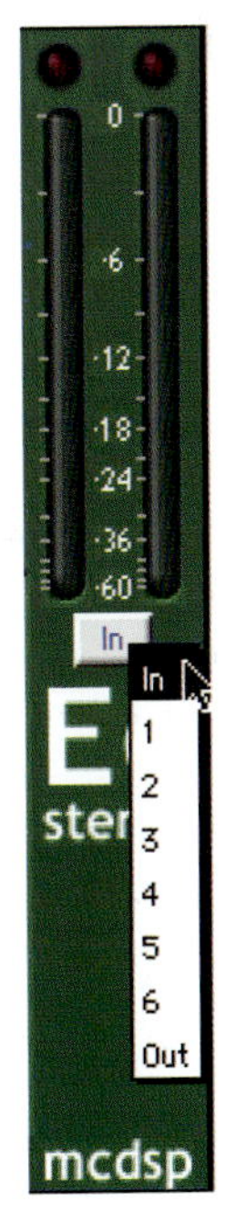

Figure 5.40 McDSP FilterBank Meter Input Selector.

The user interface has been carefully designed to provide flexible access to the wide range of features. Every FilterBank configuration has an input trim control and phase switch and a set of meters that can be switched to monitor the signal level at any stage as it passes through the plug-in. A popup menu lets you select any of the numbered EQ/filter sections as input to the meters – or you can check the input signal or the output signal.

All the FilterBank plug-ins, apart from the B1, feature a Display graph that shows the frequency response you get by applying the EQ filters. A grid indicating logarithmic frequency positions can be toggled on or off by clicking on the display while holding the Control key.

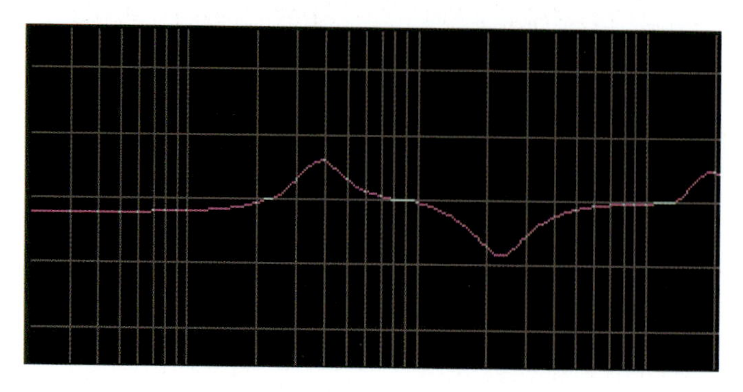

Figure 5.41 McDSP FilterBank Display with Grid Enabled.

You get a choice of user interface styles with FilterBank. The default is to use slider controls - or you can use rotary knobs instead. There is also an optional, space-saving, smaller user interface for stereo configurations in which a single slider or button is used to adjust both the left and right controls of each EQ/filter band. Input and phase controls are the exception, with individual left and right controls still provided. When FilterBank is being loaded as Pro Tools launches, a dialog appears asking you to choose which preference you want.

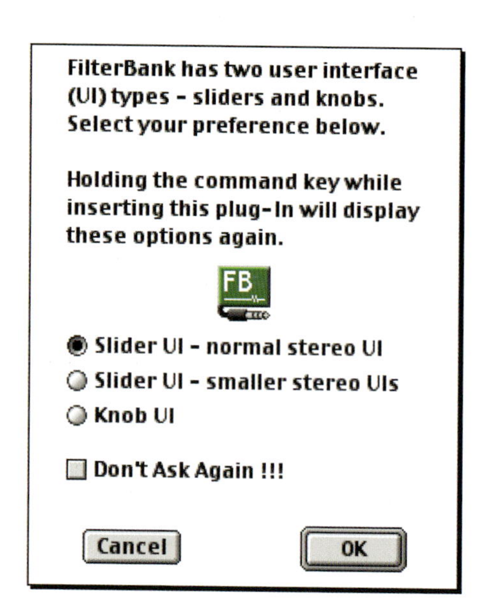

Figure 5.42 McDSP FilterBank user interface preferences.

You can change this preference while a session is running by holding the Command key on your Apple Mac keyboard as you insert the plug-in.

Band-stop and band-pass filters

The FilterBank B1 is a one-band filter, switchable between notch (band-stop) and band-pass operation. While this can be used as a notch filter with wider Q settings to create special effects, it is most typically used as a band-stop narrow-Q notch filter to remove 50 or 60 Hz hum or other noise frequencies. In the band-pass configuration, the B1 can be used to band-limit signal content - or to identify what the band-stop is removing.

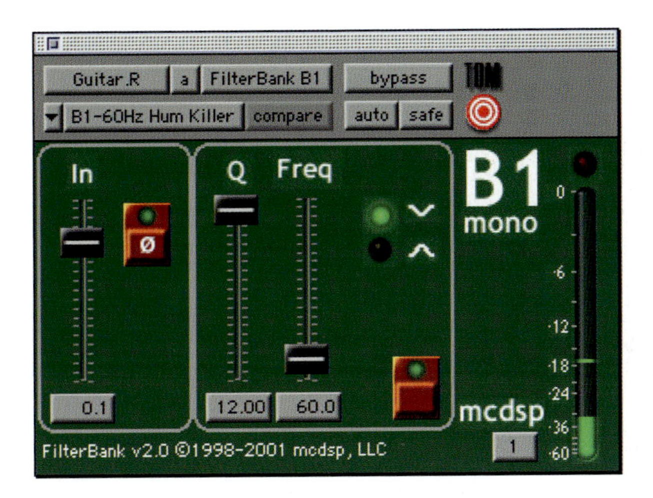

Figure 5.43 McDSP FilterBank B1.

Shelving equalization

Shelving EQ is provided in the 'E' FilterBank configurations, which can be used on a wide variety of recording projects.

FilterBank's shelving controls include a unique Peak-Slope-Dip control section along with the more usual Frequency and Gain controls. These controls allow the user to model a wide variety of vintage and contemporary shelving equalizers, or to create their own type of shelving response. Peak is the amount of emphasis in the shelved band. Increasing this adds 'punch' or 'brightness' to the sound. The Slope control lets you adjust the gradient of the shelved response in the transition band. The more gentle the transition between the shelved and non-shelved bands, the smoother the equalizer sounds. Dip is the amount of de-emphasis in the non-shelved band. Increasing this adds 'warmth' to the sound. Note that the peak and dip controls are interactive - as the peak control is increased, the overall dip in the shelved response is decreased, and vice versa.

The E2 has just two shelving equalizers while the E4 adds a high pass filter and a parametric equalizer. The E6 FilterBank configuration has six sections, including Band 1: High Pass Filter; Band 2: Low Shelf EQ; Bands 3&4: Parametric EQs; Band 5: High Shelf EQ; Band 6: Low Pass Filter.

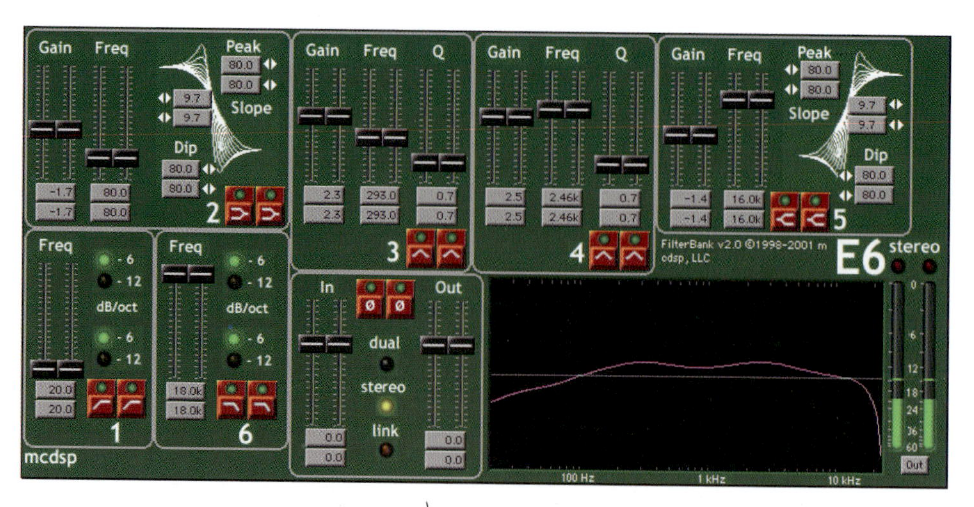

Figure 5.44 McDSP FilterBank E6.

Parametric equalization

Parametric equalization is provided in both the 'E' and 'P' configurations (E2, E4, E6, P2, P4, P6). The difference is that the 'P' configuration just contains parametric equalizers – no other types – providing the widest range of control. Each EQ/filter band has the usual parametric controls for Gain, Frequency and Q.

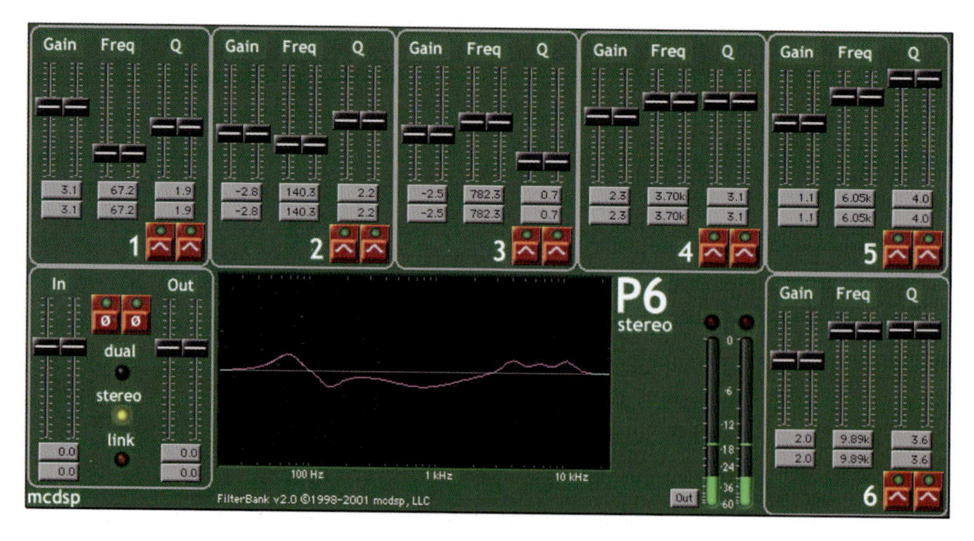

Figure 5.45 McDSP FilterBank P6.

Low and high pass filters

Low and high pass filters are provided in both the 'E' and the 'F' configurations (E2, E4, E6, F1, F2, F3). The filters in the 'E' and the 'F' configurations are different in a couple of respects. For example, the 'F' configuration high and low pass filters also have a peak (resonance) control. And the filter characteristics differ as well: the 'E' configurations have 2nd-order high and low pass filters capable of −6 and −12 dB/octave attenuation, while the 'F' configurations have 4th-order high and low pass filters capable of −6, −12, −18 and −24dB/ octave attenuation. The obvious application is to band-limit the audio material and these filters can be very useful when putting final touches to a track.

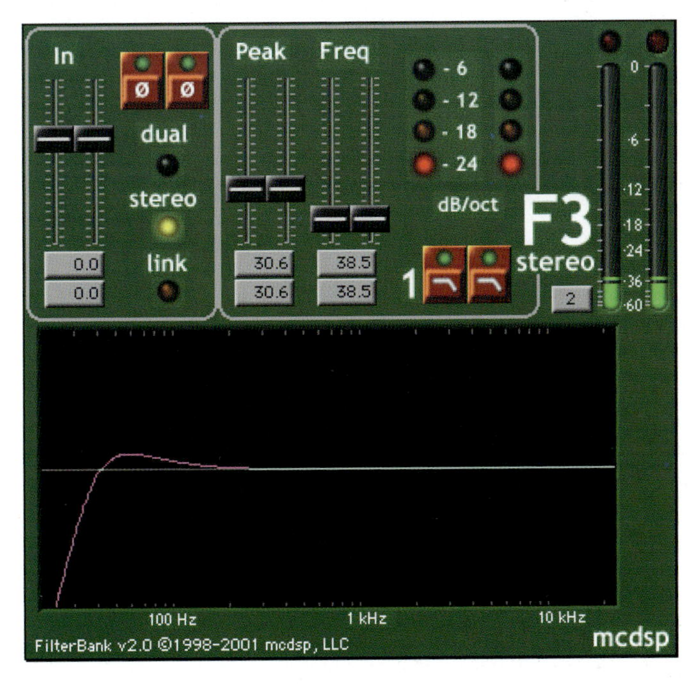

Figure 5.46 McDSP FilterBank F3.

FilterBank comes with a great selection of presets, especially if you are looking for classic sounds. The FilterBank presets are inspired by EQs such as the Neve 1084, Avalon 2055, GML 8200, Manley and Pultec and FilterBank's design allows for extremely faithful emulation of these classics. For example, the peak, slope and dip controls allow you to tailor the critical response characteristics of the low and high shelving EQ to replicate the sound of just about any other shelving EQ ever made. And all FilterBank's equalizer and filtering sections have an analogue

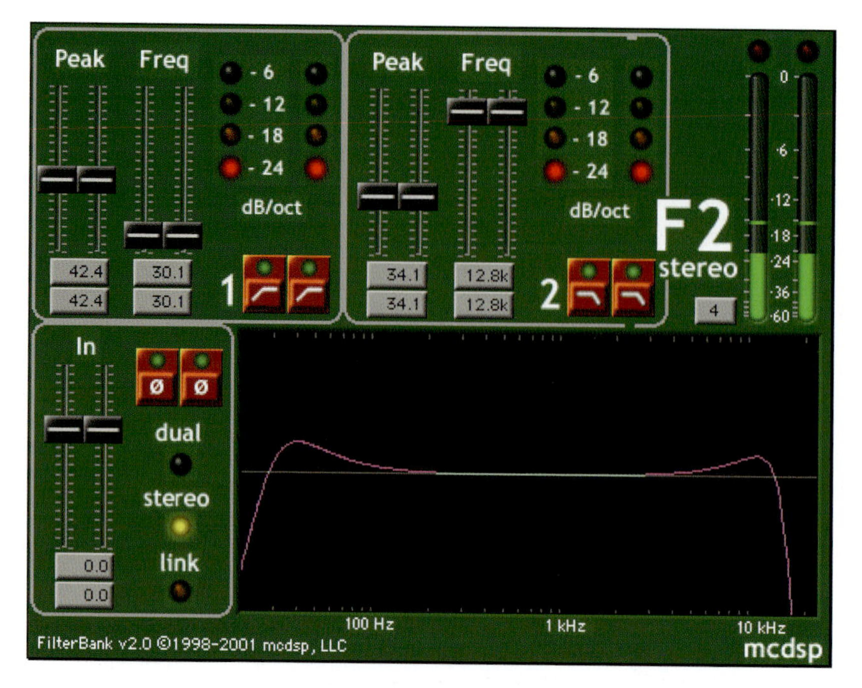

Figure 5.47 McDSP FilterBank F2.

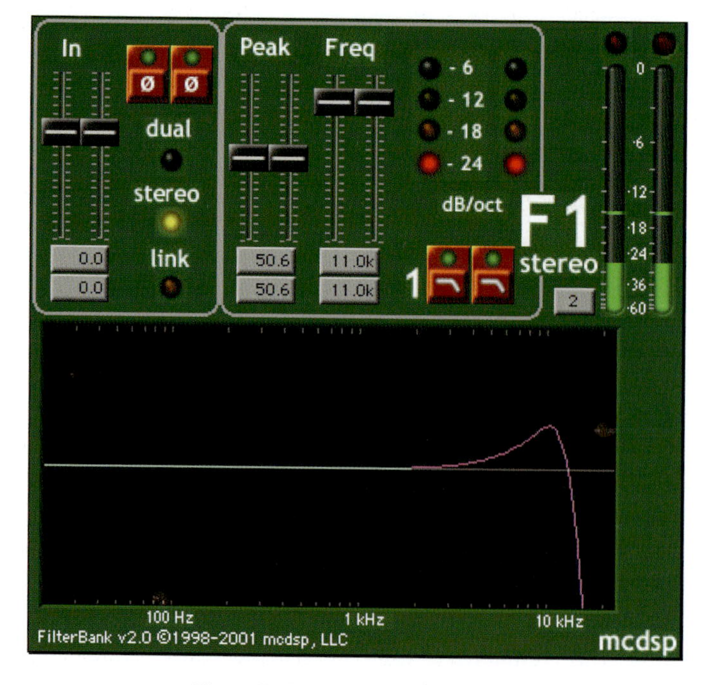

Figure 5.48 McDSP FilterBank F1.

saturation modelling circuit to prevent digital clipping from occurring – which also helps to simulate the over-driven sound of analogue gear.

CompressorBank

The McDSP CompressorBank plug-in is designed to emulate the sounds of both vintage and modern compressors. Six different compressor configurations are available – three mono and three stereo. The CB1 provides compression and side-chain processing, CB2 adds pre-filtering to the input or side-chain signal, and CB3 adds both pre-filtering and static/dynamic EQ.

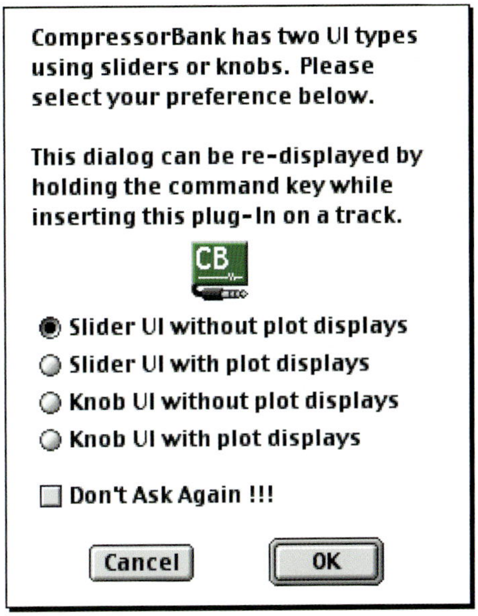

Figure 5.49 McDSP CompressorBank user interface preferences.

You get a choice of four user interface styles with CompressorBank. The default is to use slider controls. The other three preferences are an expanded slider-based user interface with graphic displays, a knob-based user interface, and a knob-based user interface with graphic displays. The graphic displays show the signal flow through the compressor, the compression curve, and the frequency response of the Pre-Filter and Static/Dynamic EQ section. When CompressorBank is being loaded as Pro Tools launches, a dialog appears asking you to choose which preference you want.

CB1, CB2, CB3

The CB1 has the usual controls for make-up gain (Output), Threshold, Compression ratio, Attack and Release, along with additional controls for 'Bite' and 'Knee' characteristics.

The shape and response of the compression curve can be adjusted using the Knee and Bite (Bi-directional Intelligent Transient Enhancement) controls. The Knee characteristic controls whether there is a sharp (angular) or gentle (rounded) transition between the two sections of the transfer characteristic. A 'soft' knee produces a smoother response. Bite gives the compressor the ability to allow signal transients (rapidly changing signals – i.e. high frequency data) to pass uncompressed, while the overall compression response is unchanged. These controls allow the user to emulate the responses of any type of vintage gear.

Every compressor uses a particular model to detect and track signal peaks and then apply dynamic compression. Three basic models are available in Compressor Bank: Type 1, pure peak detection; Type 2, pure peak detection combined with adaptive release times; and Auto, in which signal levels are automatically tracked.

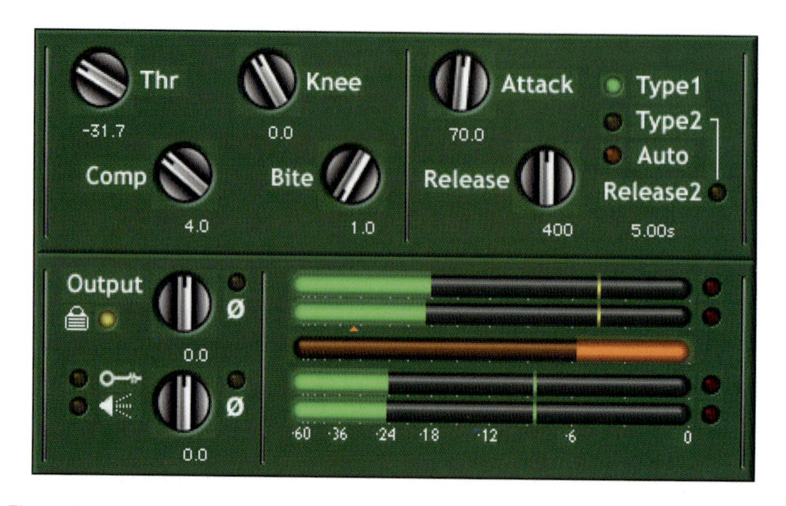

Figure 5.50 McDSP CompressorBank CB1 with knobs but without the plot display.

The Compression Curve Graph shows the transfer characteristic of the compressor – graphing the input on the x-axis against the output on the y-axis.

All the CompressorBank plug-ins have a Side Chain input that can be enabled by clicking on the small key icon. The CB2 configurations of CompressorBank also have a Pre-Filter section and a Frequency Response Graph that shows the frequency response of the pre-filter as a white line. The Pre-Filter section can be used to provide high pass, low pass, band pass and parametric filtering. This pre-filter can also be placed in-line and applied to the compressed signal and it uses the same technology as in the FilterBank plug-in with the Analog Saturation Modeling to prevent digital clipping.

This Analog Saturation Modeling is also used in the Static/Dynamic EQ section to prevent the digital clipping that can ruin your audio. When the input signal is too 'hot' to handle without distortion, the CompressorBank will not produce digital clipping, but will instead saturate and sound like an overloaded analogue EQ.

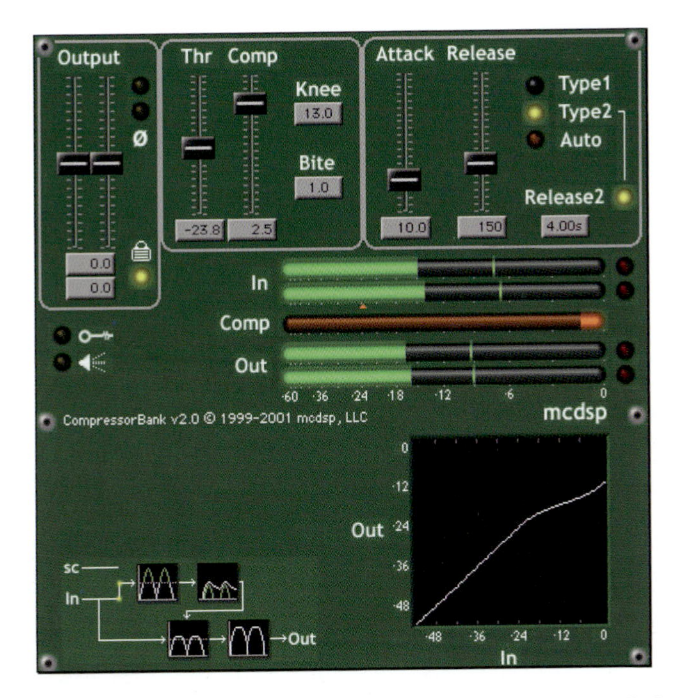

Figure 5.51 McDSP CompressorBank CB1 with sliders and plot display.

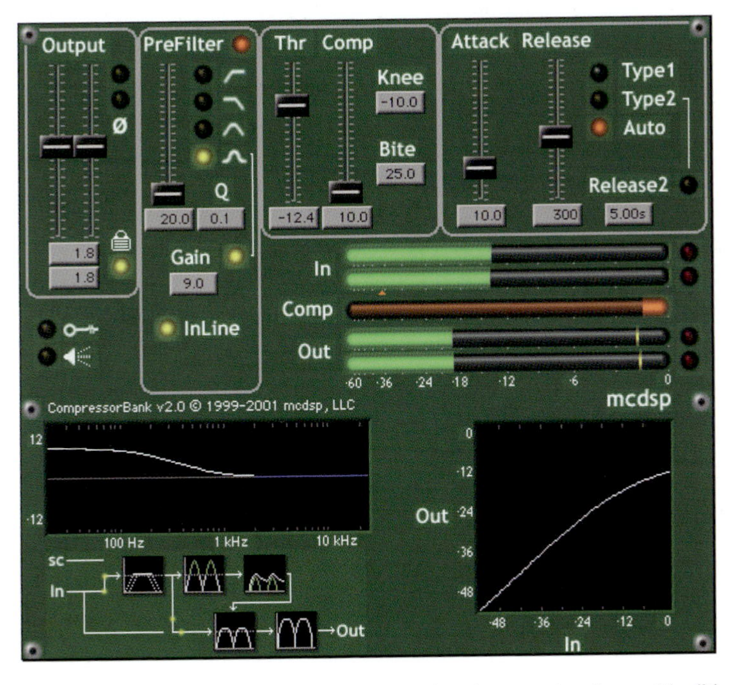

Figure 5.52 McDSP CompressorBank CB2 with PreFilter showing user interface with sliders and plot displays.

The CB3 configurations of CompressorBank have both a Pre-Filter section and a Static/Dynamic EQ section. With these configurations, the Frequency Response Graph shows the frequency response of the pre-filter as a white line and the response of the static/dynamic EQ section as a purple line.

The Static/Dynamic parametric EQ section can be applied to the compressor output. Here, the user can select a fixed amount of gain or a dynamic gain that tracks to the attack and release settings of the compressor. Suggested applications include signal enhancement, noise reduction and harmonic distortion.

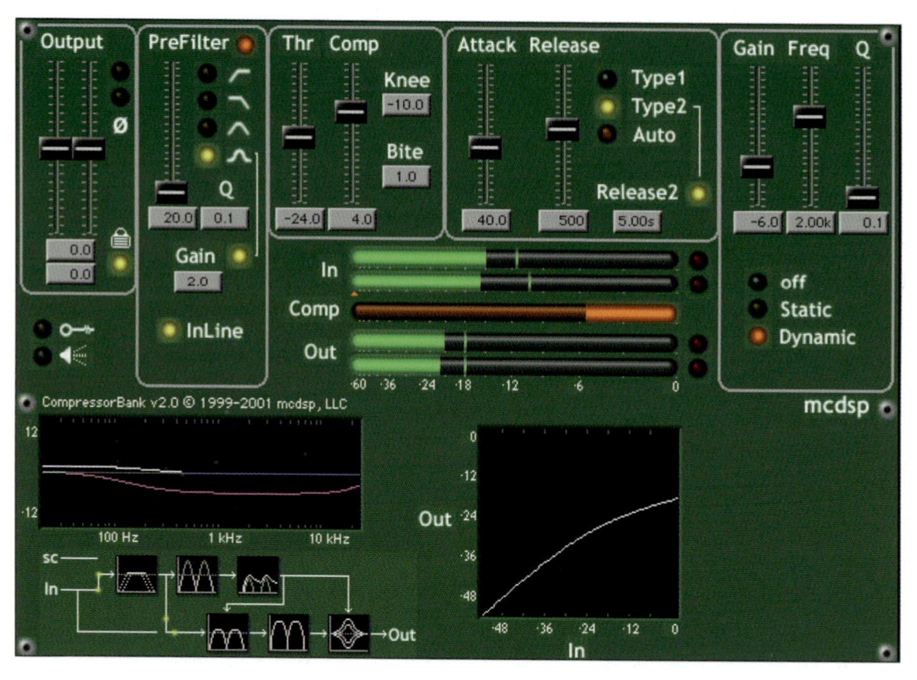

Figure 5.53 McDSP CompressorBank CB3 with post-compression EQ section showing user interface with sliders and plot displays and the standard Digidesign plug-in controls at the top.

Note Stereo versions of the CompressorBank configurations have two input and two output meters, but still only one compression gain meter. Stereo versions of CompressorBank base their compression gain on a composite signal derived from the current left and right input signal levels.

The CompressorBank presets are inspired by classic compressors such as the UREI 1176 LN, Teletronix LA2A, Neve 2254E/33609, Avalon Designs 2044, Empirical Labs EL8 Distressor, dbx 165, and Altec Lansing 9473A. A variety of other presets are named for their application ('vocal', 'drums', 'guitar'). In action the presets provide extremely useful starting points which you can tweak till you achieve your purpose. I had particularly good results from the MacValve and LA3, eh?* CB3 presets when I used these to process some finished mixes of some smooth jazz tracks I was recording. The 'MacValve B' setting smoothed out the overall sound and brought a new dimension to the bass drum on one of the tracks. On another, the 'LA3, eh? 1-11 r @-14 thr' worked an amazing transformation on a guitar, piano, bass and drums recording - drawing all of the elements in the mix together in a very intimate way, while at the same time contriving to give each instrument its own 'space'. Recommended!

MC2000

The MC2000 is a multi-band version of the MacDSP CompressorBank plug-in, offered in two-, three-, and four-band configurations in both mono and stereo versions. As with the CompressorBank, the MC2000 compression channels each have the usual controls for make-up gain (Output), Threshold, Compression ratio, Attack and Release, along with additional controls for 'Bite' and 'Knee' characteristics. Like the CompressorBank, the MC2000 also offers pure peak detection, adaptive release, and auto attack/release.

All the MC2000 plug-ins have a master control section with a pair of output meters, a crossover frequency response graph, a compression curve graph, phase polarity and band-linking switches, and a pair of output level controls - plus the two, three or four compressor modules.

The MC2000 plug-in breaks the input signal into separate frequency bands using a multi-band crossover. The frequency responses of these separate bands are drawn in different colours in the Crossover Frequency Response Graph. The crossover filters in the MC2000 are taken directly from MacDSP's award-winning FilterBank design and are implemented as steep 24 dB/octave filters to minimize cross-talk between the compression bands. The crossover points can be adjusted from the text boxes just to the right of the Crossover Display, or by simply clicking and dragging the crossover points themselves on the graph. Holding the Command key while moving the crossover point allows fine control. The Compression Curve Graph shows the transfer characteristic, i.e. the graph of

*So named by McDSP as this software emulates the audio hardware LA-2A.

Input vs Output, as determined by the Threshold, Compression, and Knee controls of each compression band. Each band is drawn in the same colour as the corresponding band in the Crossover Display.

Figure 5.54 McDSP MC2000 MC4.

The input gain control ranges from –12 dB to +12 dB to cater for post-production facilities that work with audio at the –20 dB ('Dolby') level. There is no special 'saturation' handling at the input stage, so the user needs to check the input levels for each compression band to make sure none of the overload LEDs are lit. If any of the peak LEDs do light during playback, the only way to deal with this is to reduce the input gain until the LEDs do not light.

The MC2000 Output control ranges from –24 dB to +24 dB. The output stage follows the summing of the individual compression bands, and the master output meters show the level of the audio after being processed by this final gain stage.

Tip You can link the two output control faders on the stereo plug-ins so that they move as one. To unlink them, click on the small LED to the left of the lock icon. This glows yellow when the lock is on.

The polarity, or phase, of the signal can also be reversed at this final gain stage by enabling or disabling the 'Ø' LED. The phase of the signal is unaffected when the 'Ø' LED is off. The controls for any one of the compression bands can be made to control all the others using the Link popup selector. This lets you select which of the compression modules is to be the Master. The three different Time Constant circuit types (Auto, Type-1 or Type-2) can be selected by clicking the LED buttons to the left of each Attack control. The Auto control overrides the TC type selection and the attack and release controls. To help with setting up, each compression band can be solo'ed by clicking on the yellow Solo LED, or switched in or out by clicking on the orange LED in the upper right corner of each section.

The MC2000 plug-in presets were inspired by compressors such as the Urei 1176 LN, Teletronix LA2A, dbx 165, Neve 33609C, Avalon Designs 2044, Empirical Labs EL8 Distressor, and Altec Lansing 9473A. The MC2000 uses the compression algorithms from the CompressorBank plug-in to model this wide variety of vintage and contemporary compressors, but with one big difference - they are in multi-band configurations! Imagine a three-band stereo compressor with a dbx165 on the low band, the Neve 33609C on the mid band, and a Teletronix LA2A on the high band. Using the MC2000, these kinds of configurations can now be achieved with ease!

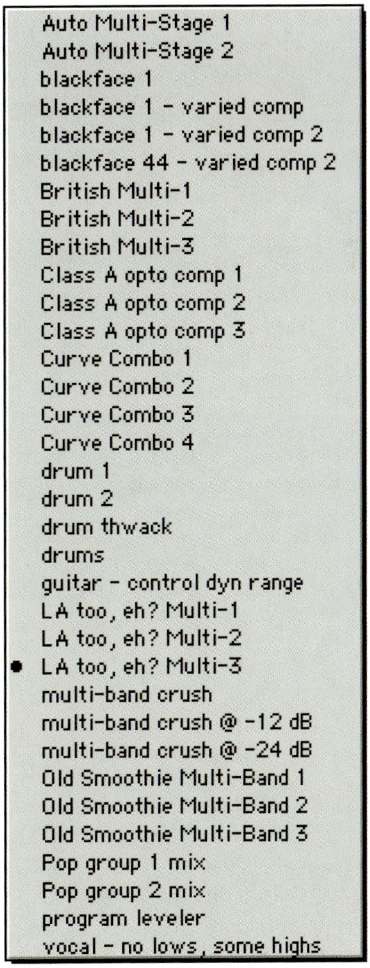

Auto Multi-Stage 1
Auto Multi-Stage 2
blackface 1
blackface 1 - varied comp
blackface 1 - varied comp 2
blackface 44 - varied comp 2
British Multi-1
British Multi-2
British Multi-3
Class A opto comp 1
Class A opto comp 2
Class A opto comp 3
Curve Combo 1
Curve Combo 2
Curve Combo 3
Curve Combo 4
drum 1
drum 2
drum thwack
drums
guitar - control dyn range
LA too, eh? Multi-1
LA too, eh? Multi-2
• LA too, eh? Multi-3
multi-band crush
multi-band crush @ -12 dB
multi-band crush @ -24 dB
Old Smoothie Multi-Band 1
Old Smoothie Multi-Band 2
Old Smoothie Multi-Band 3
Pop group 1 mix
Pop group 2 mix
program leveler
vocal - no lows, some highs

Figure 5.55 McDSP MC2000 presets.

AC1 Analog Channel Emulation

The AC1 plug-in is designed to emulate the characteristics of an analogue amplifier stage and can be used to model a wide range of vintage and contemporary analogue devices.

Clipping of the waveforms in digital audio systems causes distortion that always sounds extremely unpleasant. The AC1 Analog Channel TDM plug-in acts as a kind of pre-amplifier for Pro Tools TDM systems that allows even excessive amounts of gain without clipping. Instead of letting maximum signal peaks hit and distort at 0 dBFS, the AC1 provides gentle, continuous soft-limiting/compression (or saturation depending on the attack/release times) at a user-adjustable level. It can either be used to prevent digital clipping and smooth out signal transients, or as a creative tool to 'warm-up' digital tracks – closely emulating the sound of large analogue mixing consoles. A Drive control lets you define where the audio sits in the saturation region, i.e. how 'hot' the channel is, and you can adjust the Attack and Release times of the pre-amp's distortion characteristic.

Note A significant difference between the AC1 and a standard compressor is that the AC1 has no threshold control. The amplifier operation does not depend on input signals exceeding a threshold before acting. Instead, the amplifier circuit is 'always on' – and the transition from a linear region to a non-linear region of operation is totally continuous. For the user processing audio through the AC1, this equates to a process that can limit, compress or saturate high signal levels in a consistent and precise manner.

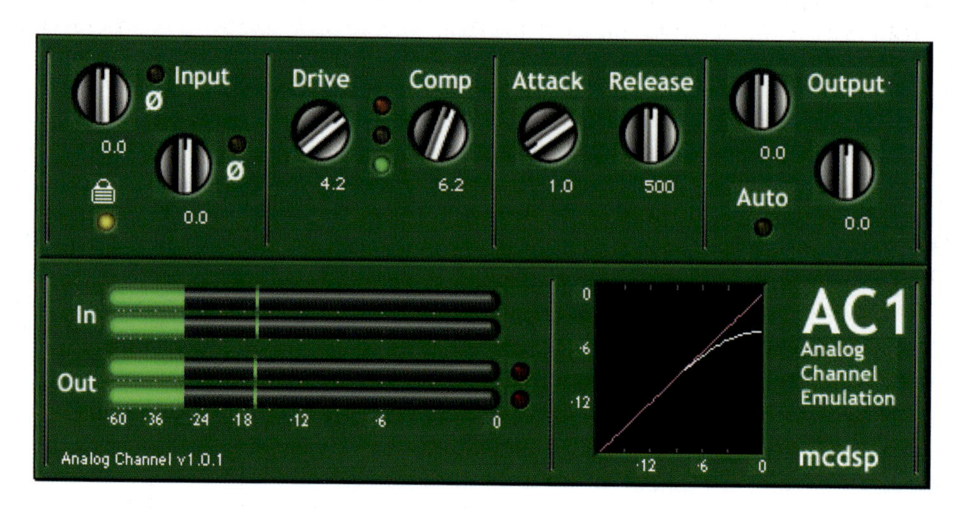

Figure 5.56 McDSP AC1 Analog Channel Emulation.

McDSP suggest that you should insert an AC1 on each of your Pro Tools channels to model the sound of an analogue console, then use an AC2 on the master output to model the sound of an analogue recorder! Three Pro Tools session templates are supplied with a mono AC1 inserted on every channel of a 24-track mixer and an AC2 placed on the stereo master fader. The AC1 is placed in the first insert of each mixer channel, with the AC2 on the master fader. With these setups, signal level should be set by the input control of the AC1 plug-in - with the Pro Tools mixer fader set for unity gain. The signal level can then be set without digital clipping, and the analogue amplifier characteristics will be imparted to the mixer as well.

As the advertising puts it - the most sought-after analogue consoles are just 'an insert away' using the McDSP AC1 Analog Channel plug-ins. Presets provided include various well-known large analogue mixing consoles and Class A amplifier designs.

AC2 Analog Tape Emulation

One of the complaints about digital audio systems such as Pro Tools is that they take away the flaws in the analogue audio systems that recording engineers had learnt to use creatively to 'weave their magic' while producing music. The AC2 plug-in provides Pro Tools users with a way to get back the typical sounds of analogue tape and analogue tape machines. The AC2 emulates most popular tape playback systems, and both modern and vintage tape formulations, at all the professional playback speeds. Controls are provided to emulate all the typical tape-machine parameters such as bias, tape speed, and IEC1/2 standards.

The three most common playback speeds are provided - 7.5, 15, and 30 inches per second (ips). When you change the playback speed in a real tape machine, this affects tape saturation, dynamic range and low frequency rolloff. Low frequency rolloff due to playback head wear and other factors and the deviations in frequency response due to the physics of the tape heads known as 'head bumps' have also been modelled and can be applied independently of tape speed. Broadly speaking,15 ips delivers better low-end response while 30 ips delivers better high frequency response.

All the popular playback heads have been modelled, including Studer, Otari, MCI, Ampex, Sony and Tascam types - referred to as Swiss, Japan-O, USA-M, USA-A, Japan-S and Japan-T in the AC2.

The AC2 plug-in has a Bias control and the tape playback response this produces is shown by the purple curve in the display. In general, under-biasing (bias < 0 dB) increases dynamic range and boosts high frequencies, depending on the tape speed and IEC standard used. Over-biasing (bias >0 dB) decreases dynamic range and reduces high frequencies - again, depending on the tape speed and IEC standard used. Over-biasing also increases the effect the tape

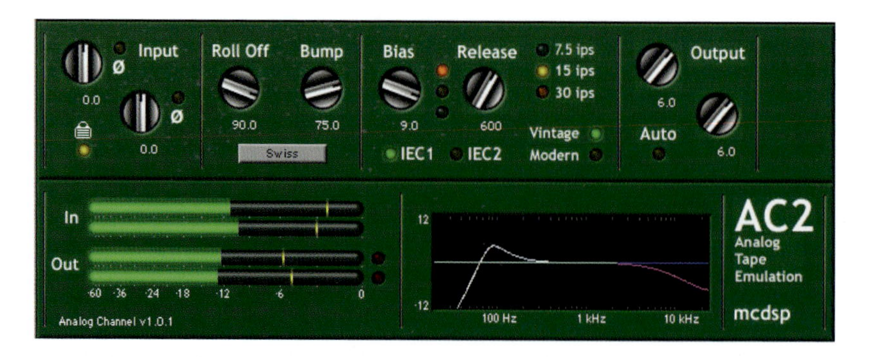

Figure 5.57 McDSP AC2 Analog Tape Emulation.

formulation character has on the audio and is considered desirable by most tape medium enthusiasts.

McDSP has modelled both 'vintage' and 'modern' tape formulation types. Modern tape formulations have greater dynamic range and offer improved linearity with less likelihood of saturation. Recordings made using these sound very close to the original material. Vintage tape formulations tend to have less dynamic range and introduce greater levels of distortion/saturation. These tapes, however, have 'character' that newer ones do not, and have been used with great success in recording audio.

The suggested maximum biasing level for the 'vintage' type (less dynamic range, more character) is between +6 and +9 dB. The suggested maximum biasing level for the modern type (less dynamic range, more character) is between +9 and +12 dB. The user can adjust the Bias control as desired, although some extreme over-biasing can occur if these maximums are exceeded.

The AC2 even allows the user to control the rate at which the system recovers from a tape saturation state. For each tape machine, a saturation state can be reached when the tape is over biased. The machine returns to a linear state once the input audio drops back to a nominal level, at a rate specific to each tape machine. To emulate this behaviour, the AC2 plug-in has a Release control that lets you set the length of time the plug-in takes to recover from the tape saturation state and return to the linear state of operation.

The International Electro-technical Commission (IEC) has developed two standard types of equalization for tape machines to compensate for deficiencies in the frequency response of the tape medium: IEC1, also known as IEC, CCIR and DIN, is mostly used in Europe, and IEC2, also known as NAB, is mostly used in the USA. Because these standards determine how the audio is recorded to and reproduced from tape, they affect the character of the tape saturation of the tape machine. To model this aspect, the AC2 includes IEC1 and IEC2 equalization, and the subsequent sonic affects are displayed in the Tape Playback Head and Tape Response Display.

The Tape Playback Head and Tape Response Display has a white curve to show the Playback Head response and a purple curve to show the Tape Response. The playback head response shows the effects of the low frequency roll-off and head bumps for the different head types. The purple curve shows the amount of tape saturation caused by the under/over biasing of the virtual tape machine, and how the biasing changes for different playback speeds and IEC equalization. Altering the Release control and using an extreme over-biased or under-biased setting is a good way to experiment (and audition) the tape saturation effect. The frequency scale is logarithmic (as are most frequency displays), with the 100, 1000 and 10 000 Hz frequencies labelled.

Antares

Antares Audio Technologies was founded in 1990 as Jupiter Systems by Dr Harold (Andy) Hildebrand. By 1993 Andy Hildebrand had developed the innovative Infinity Looping Tools software to allow the creation of seamless loops within Sound Designer II files. Hildebrand went on to develop three of the first professional plug-ins for Digidesign systems – the Multiband Dynamics Tools, the Jupiter Voice Processor and the SST (Spectral Shaping Tool). The market for these software products was initially fairly small, and Jupiter Systems fell by the wayside to be replaced by Antares. Then in 1997 Hildebrand came up with Auto-Tune. I recall speaking to him at the time and it was clear that they had a monster 'hit' on their hands! Auto-Tune was selling by the bucket-load! And then Cher had a hit with a vocal that had been deliberately altered in a cute way using Auto-Tune, and it sold even more – becoming the biggest selling plug-in of all time. In 1997, encouraged by this success, Antares decided to move into the hardware DSP effects processor market with the ATR-1, a rack-mount version of Auto-Tune. This was followed in late 1999 by Microphone Modeler, a plug-in that allows any reasonable quality microphone to sound like any of a wide variety of other microphones. This also looks likely to gain a fair degree of acceptance – especially now that the market for plug-ins has expanded so much. Again, this was followed in 2000 by a hardware version, the AMM-1. At the Audio Engineering Society Conference in September 2000, the Mic Modeler was honoured with the TEC Award as the year's Outstanding Achievement in Signal Processing Software.

Infinity Looping Tools

'Imagine taking a raw sample, and then, with just a few mouse clicks, creating a loop so perfect that you can't tell when it started or where it repeats. Imagine doing that with sounds that are impossible to loop smoothly using your current sample editor, like string orchestras, choirs, fat analogue synths, chorused guitars, and so on. Infinity lets you do that, quickly and easily.' That's how the manual manages to encapsulate what Infinity is all about.

'Infinity is not a replacement for your preferred sample editor. Instead it provides specialised tools for just one purpose - looping. Infinity can loop anything from orchestral instruments to synthesizers and you don't have to put up with rhythmic pulsing or lifeless sustains any longer. Infinity preserves the "liveness" of your samples because it has a looping tool appropriate for every type of sound.'

Real Time Loop Adjust lets you move the loop points during real-time playback, allowing you to locate the best loop points by ear.

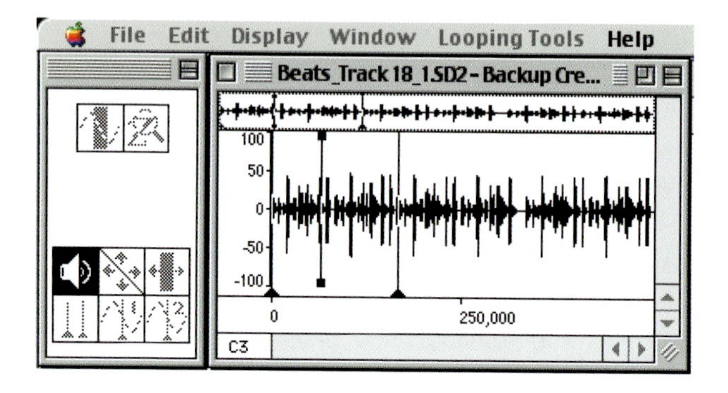

Figure 5.58 Infinity Looping Tools.

The SPR Looper automatically creates seamless loops in chorused and ensemble sounds while preserving the exact sound colour and stereo image of your sample. The resulting loop data can be automatically merged with the attack portion of your sample.

The Synthesis Looper allows you to control the frequency domain sidelobe energy of the loop, and selectively smooth out lumpy-sounding frequencies in chorused and ensemble sounds. The new sound is more stable and loop repetitions are less noticeable.

The Rotated Sums Looper allows you to randomly layer multiple copies of the loop data to 'homogenize' the sound while preserving your original loop

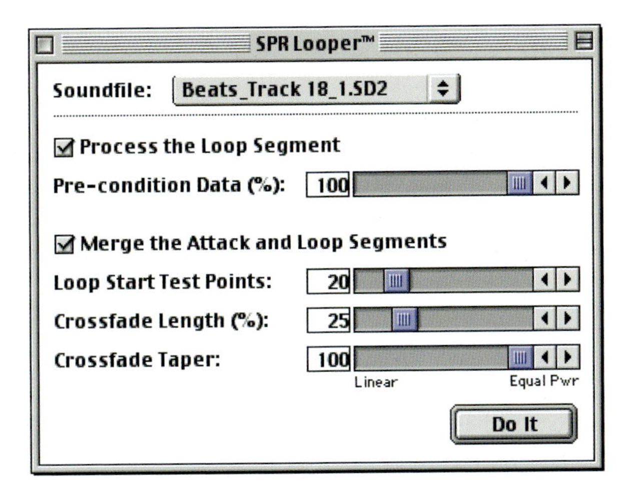

Figure 5.59 Infinity SPR Looper.

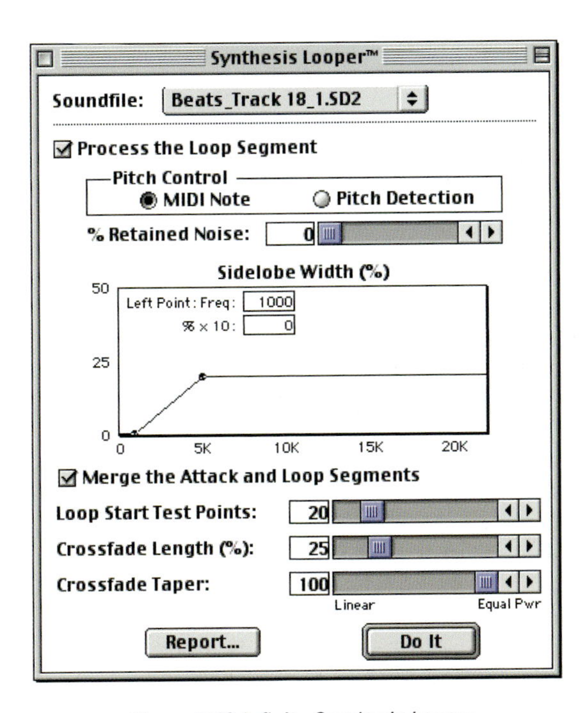

Figure 5.60 Infinity Synthesis Looper.

points. Perfect for improving those problem loops on CD-ROMs and for special effects.

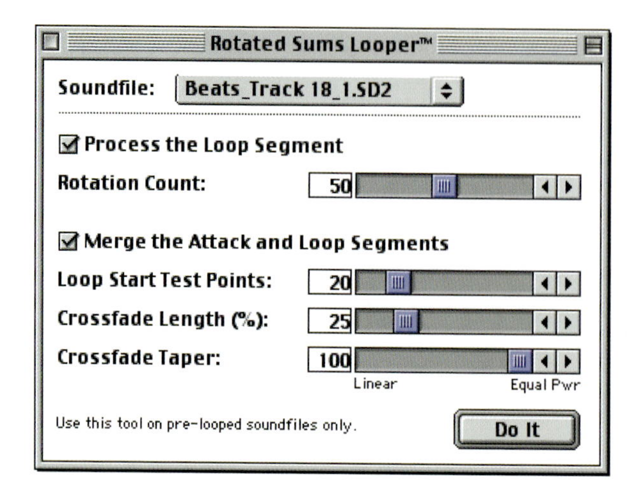

Figure 5.61 Infinity Rotated Sums Looper.

The Freeze Looper creates perfect loops in sounds like winds, brass and other solo sounds that have a clear harmonic series.

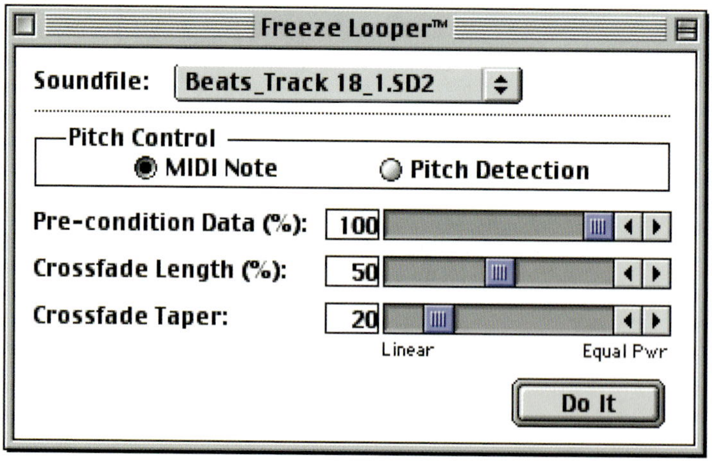

Figure 5.62 Infinity Freeze Looper.

The Crossfade Looper with Smart Auto-Scan is an enhanced crossfade tool with an intelligent loop point scanner that lets you find the best loop point automatically.

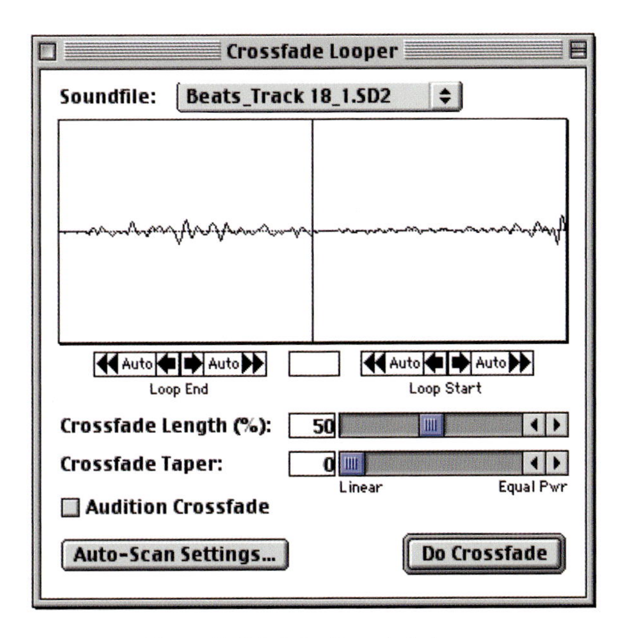

Figure 5.63 Infinity Crossfade Looper.

Infinity has many useful editing functions, including Cut, Copy, Paste, Clear Unselected, Clear, Mix, Invert, Reverse, Silence, Create/Delete Loop Points, Select Loop Points, Select All, Set MIDI Note and Auto Zero. It also has several special edit functions, including Real Time Loop Adjust, Variable Edit Blend Time and Taper, Switchable Window Modes, Enable/Disable Clipboard and Enable/Disable Undo.

Infinity is compatible with any Macintosh Sound Manager compatible hardware and reads and writes stereo and mono sound files in Sound Designer and AIFF file formats. It can communicate directly with Digidesign's SampleCell or SampleCell II through the SampleCell Editor software and includes Ensoniq's ASR-SCSI, to let you move samples back and forth between your Ensoniq EPS 16 plus or ASR 10 sampler and the Macintosh. Also included is a utility to transfer samples to Akai S3000 series samplers.

MDT Multiband Dynamics Tool

MDT - the Multiband Dynamics Tool - is a powerful DSP plug-in for Digidesign's TDM environment that allows you to shape the sound of your recordings using its unique graphic interface. And flexibility is the key word here. MDT can easily be configured to emulate just about any existing compressor, limiter or expander, downward expander or gate, 'tube' compander, or any combination of these. In multi-band mode, MDT can create de-essers, spectral enhancers and dynamic EQs as well as compressors and expanders that can suppress the effects of artefacts such as breathing and pumping.

MDT scans the audio data about to be processed for upcoming peaks while calculating the average peak level of the data it is currently working on, then uses both numbers to derive the final input level. MDT can look ahead in this way for up to 1028 samples to prevent signals that would cause clipping from getting through to the processor. If the input level is so high that clipping is about to occur, the Clip Sentry high-speed digital limiter 'kicks' into action to reduce the gain so that clipping will not take place.

In multi-band mode, MDT divides the energy of the signal into three or five equally spaced bands, with the peak level of each band displayed at the bottom of the In/Out grid. The Input Offset feature allows the relationship of each band to the I/O curve to be adjusted individually. Each band can be processed using a different section of the I/O curve and the input offset levels can be continuously adjusted so that the band levels are, on average, even with one another. This eliminates the major source of sonic coloration previously inherent in multi-band compressors. Zero attack and release times allow MDT to be used as a peak limiter.

MDT has five filter modes - Full Band, 3 Band, 5 Band, 3 Band AP and 5 Band AP. The filter mode menu controls the number of bands into which the signal is divided before processing. Multi-band modes divide the frequencies into equally spaced bands. The AP filter modes are optimized for minimum passband ripple for settings within ±12 dB of unity gain.

All the buttons and sliders, and the Settings Menu items, can be automated using Pro Tools, and MDT processes 16-bit or 24-bit sound files using an internal 56-bit accumulator to ensure the highest audio quality - in stereo or mono.

The user interface includes a graph of the transfer characteristic to show the effect of the processing, a peak level indicator, a settings popup menu, and controls for Attack, Release, Gain, delay and so forth. You can directly manipulate the transfer characteristic onscreen using the mouse.

This graph represents the relationship between the input level of the sound and the output level of the sound. Notice that the curve has a breakpoint at the threshold level. All standard compressors have a threshold level, above which compression takes place and below which the signal passes unaltered.

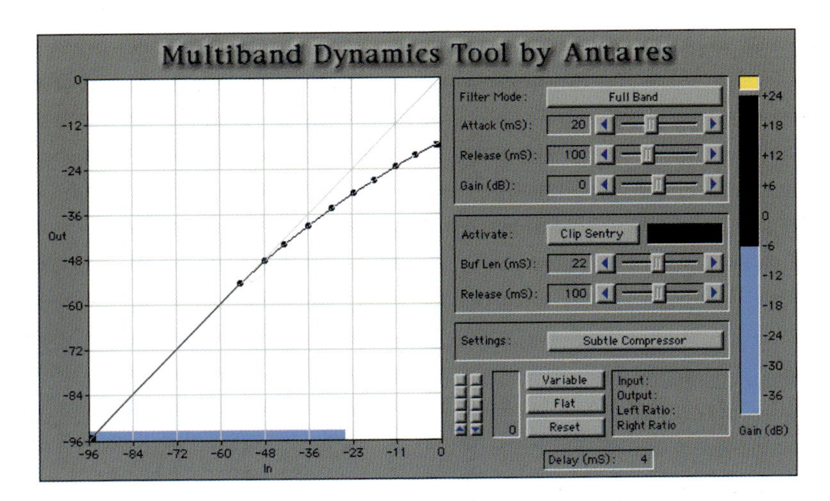

Figure 5.64 Multiband Dynamics Tool.

With conventional dynamics processors, knobs or sliders are used to control the threshold and ratio parameters. MDT takes a different approach in which you directly manipulate the graph to change the sound - so you get to see what you are doing as well as hear what you are doing. Also, it is much easier to set up multiple thresholds and their associated compression ratios by 'drawing' these on the graph. You can even see where the input level lies in relation to the graph by observing the blue bar that runs along the bottom of the graph - the input level meter.

In the example shown above, the input level is around –26, the threshold starts at around –56 with a soft knee and a low ratio. The actual gain reduction, looking at the difference between the straight line and the curve at the –26 dB input level, is just 6 or 7 dB.

The Settings popup lets you select any of the presets as a starting point to work from. So you get various 'soft knee' and other compressors, expanders, companders, gates, and so forth - and you can always save your own settings here.

'Soft knee' compressors are so-called because the change in ratio at the threshold point takes place gradually instead of switching instantly to the new ratio. Vintage tube compressors often produce this characteristic. Vintage designs also work as companders, where below the threshold the dynamic range is expanded and above the threshold the dynamic range is compressed - with the ratios changing gradually and continuously throughout the range. This design sounds very natural because only the peaks of the audio signal get compressed heavily, while most of the audio passes through at 1:1, assuming the average level is about halfway up

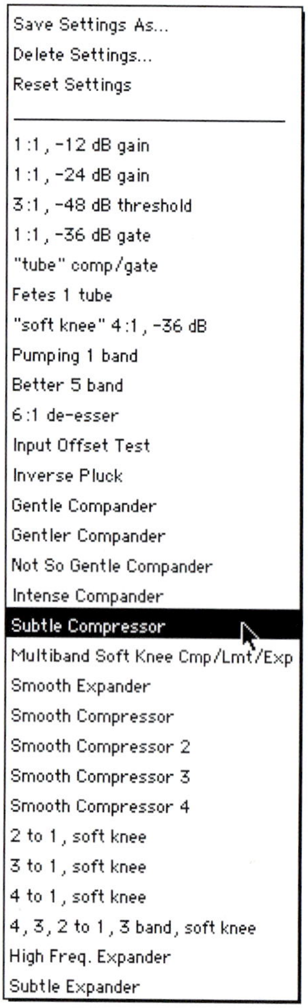

Save Settings As...
Delete Settings...
Reset Settings
1:1, -12 dB gain
1:1, -24 dB gain
3:1, -48 dB threshold
1:1, -36 dB gate
"tube" comp/gate
Fetes 1 tube
"soft knee" 4:1, -36 dB
Pumping 1 band
Better 5 band
6:1 de-esser
Input Offset Test
Inverse Pluck
Gentle Compander
Gentler Compander
Not So Gentle Compander
Intense Compander
Subtle Compressor
Multiband Soft Knee Cmp/Lmt/Exp
Smooth Expander
Smooth Compressor
Smooth Compressor 2
Smooth Compressor 3
Smooth Compressor 4
2 to 1, soft knee
3 to 1, soft knee
4 to 1, soft knee
4, 3, 2 to 1, 3 band, soft knee
High Freq. Expander
Subtle Expander

Figure 5.65 Multiband Dynamics Tool Settings.

the range. Quieter sounds get expanded downwards, making them even quieter as they get softer. This acts as a subtle kind of gating, getting rid of noise while maintaining a sense of wide dynamic range - yet still compressing the higher peaks in the material.

The manual contains an excellent tutorial section that guides you through the basics of the user interface first. Then it shows you, step by step, how to configure MDT as a single band compressor, limiter, gate, expander, soft knee compressor/gate and tube compander. The third tutorial section explains how to use MDT in multi-band mode as a compressor, de-esser, dynamic EQ or spectral enhancer. This tutorial, complete with example sound files, is one of the best tutorials on dynamics processing that I have come across.

Tip I can personally recommend MDT to anyone - student or working professional - who would like to gain a better understanding of how dynamics and multiband dynamics processors work.

All in all, MDT is a serious engineer's tool, which delivers excellent results. Applications include mastering, individual track processing, sound effects, sample editing and much more.

JVP Jupiter Voice Processor

The Jupiter Voice Processor TDM plug-in comprises a de-esser, a compressor with a downward expanding gate, a parametric EQ, and a multi-tap delay all controlled via one easy-to-use interface. It is particularly intended for vocal processing, but can be excellent for individual instruments, sound effects, or even complete mixes. The JVP can also be used very effectively for mastering stereo files, as its compressor/gate and EQ can be used simultaneously. The JVP processes up to 24-bit files using an internal 56-bit accumulator to deliver first-rate sound quality in stereo or mono. All the buttons and sliders, and the Settings

Menu items, can be automated using the standard Pro Tools automation and the JVP can be controlled using external control surfaces such as the JL Cooper CS-10 or the Mackie HUI.

To the left of the neatly designed control window there is an input level fader with an associated input level meter. Symmetrically, at the far right there is an output level fader with associated metering. The popup Setting menu at the bottom left lets you call up standard settings.

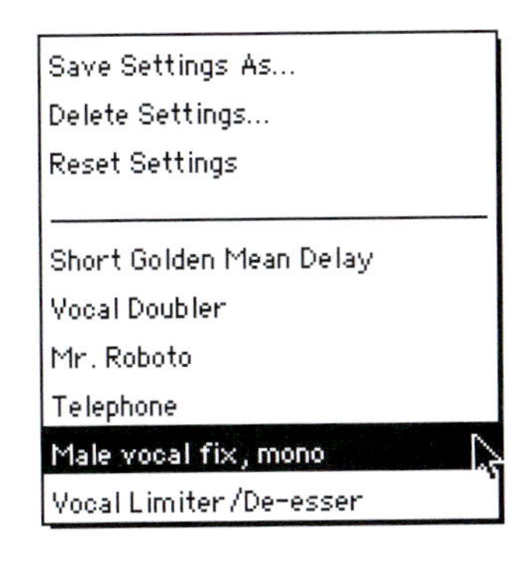

Figure 5.66 Jupiter Voice Processor settings.

Along the top, each section has a Bypass button and a Display button to let you switch the controls between the De-Esser, Comp/Gate, Parametric EQ and Delay FX sections.

The De-Esser has controls for Threshold, Frequency, Compression Ratio, Attack and Release Times, with an associated Gain Reduction meter.

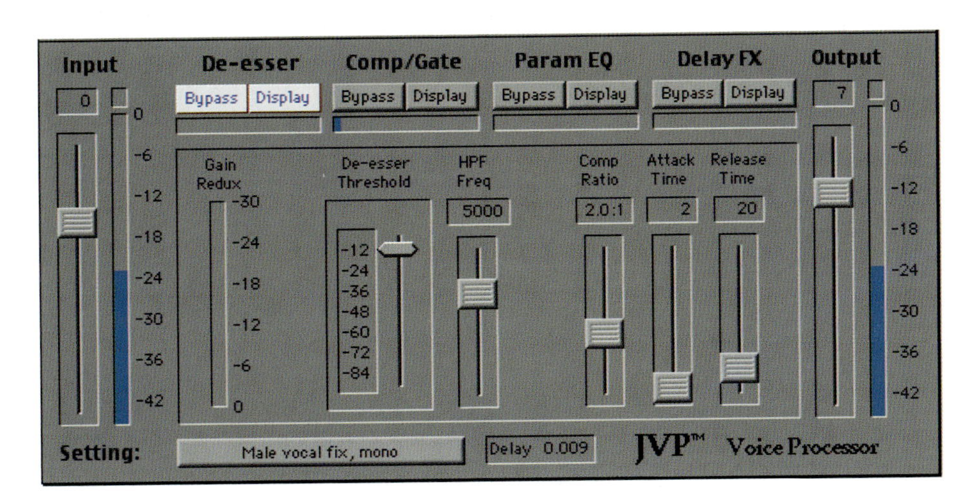

Figure 5.67 Jupiter Voice Processor De-Esser.

The Compressor/Gate has a variable knee and special compression algorithms to provide smooth, natural dynamic control. You get all the controls found on hardware compressors with the added benefits of 24-bit resolution plus a threshold meter to help you set gate and compressor thresholds.

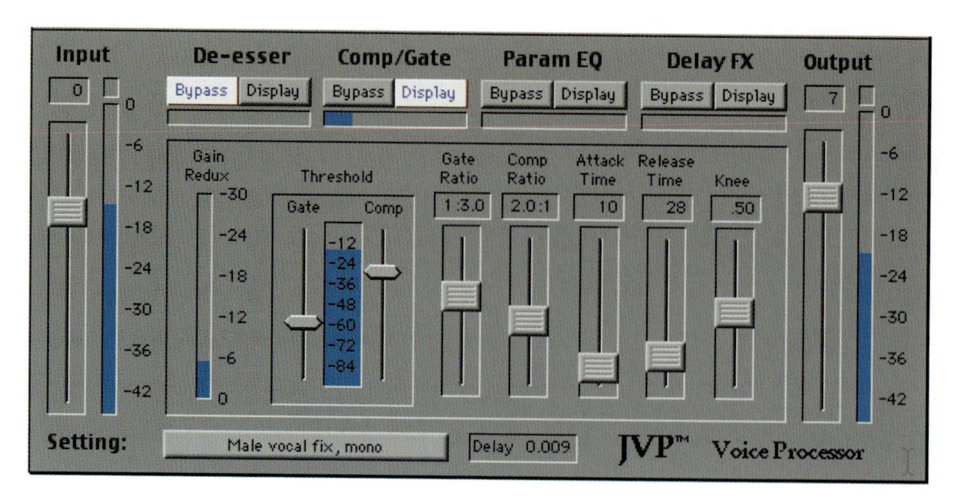

Figure 5.68 Jupiter Voice Processor Comp/Gate.

The Parametric EQ can handle 140 dB of dynamic range, and has three filter sections. Two of these can be configured as different filter types including Low Pass, Low Shelf, Peak/Notch, High Shelf and High Pass.

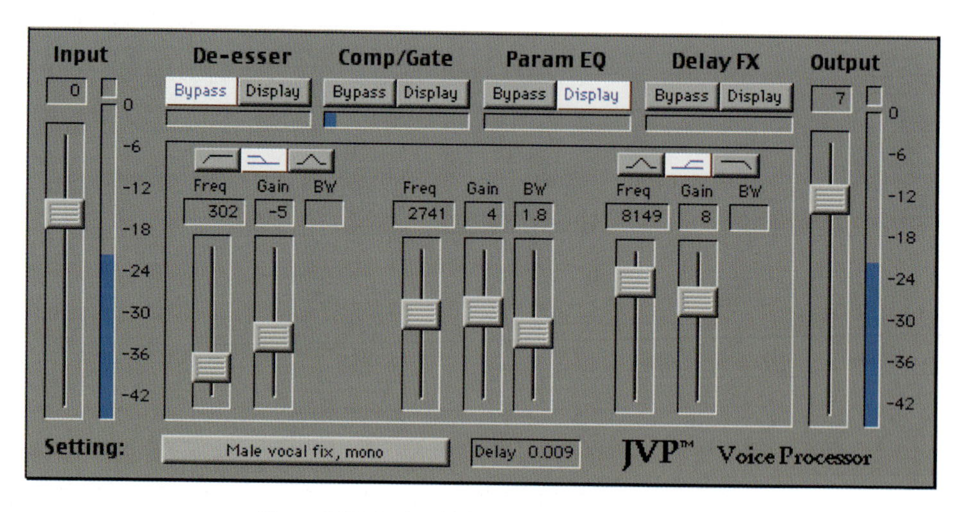

Figure 5.69 Jupiter Voice Processor Param EQ.

Finally, the delay section lets you create complex and rich sound textures. This is a multi-tap stereo delay with six taps each with feedback control that can be used to create very complex time-based effects. You can pan the outputs from the taps and you can use negative gain and adjustable comb filtering to create a wide variety of effects.

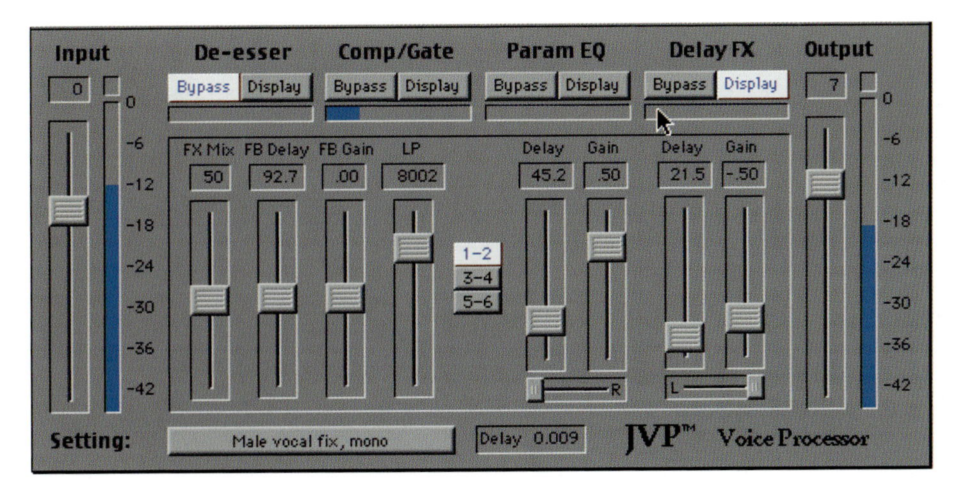

Figure 5.70 Jupiter Voice Processor Delay FX.

In action, the JVP sounds very 'smooth'. It has all the stuff you need for processing vocals and works well with instruments also. I tried the JVP with a jazz guitar solo and used the Male Vocal Fix preset to great effect. These settings 'flattered' the instrument, making it a little brighter and less muddy - yet without it sounding brash. And it didn't use up too much of my precious DSP resources. Recommended.

Antares Auto-Tune

Auto-Tune 1.0

Auto-Tune 1.0, first introduced in 1997, lets you fix out-of-tune vocals or monophonic instrumental material. It has two modes of operation - Automatic and Graphical. You can use the automatic mode to shift an entire recording to a well-centred pitch without any particular effort or expertise being required. You can compare the pitches in the recording to any scale you define, including microtonal and ethnic scales, and you can even remove poorly executed vibrato and add accurate vibrato if necessary.

If you have an accurate ear for pitch and you have the time and inclination, you can use the graphical mode to correct even the smallest deviation from the desired pitch. First you analyse the recording to produce a graphical representation of the pitches, then you draw in the corrected pitches you want to hear.

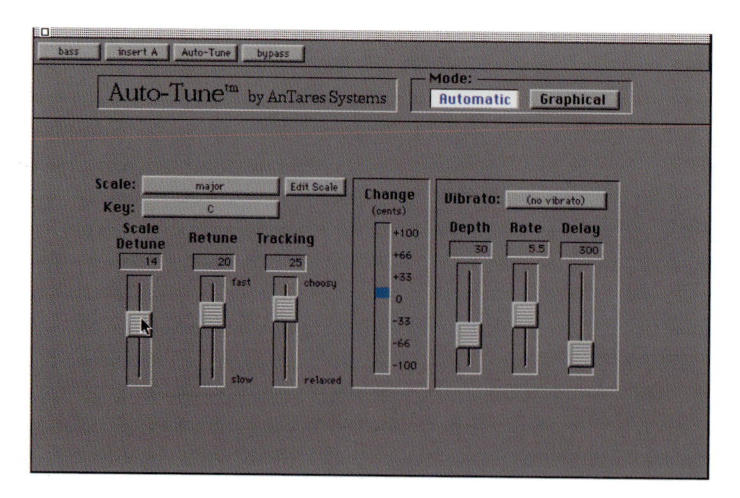

Figure 5.71 Antares Auto-Tune 1.0 Automatic mode.

I tried it on a reasonably well-performed vocal recording where the singer had sung just a little sharp or flat on certain notes. Automatic mode corrected most notes correctly, but failed with a couple. Then an Indian music producer brought me a vocal track sung using an Indian scale. This was 'all over the place' in terms of pitching – especially as the singer had been very nervous during the recording

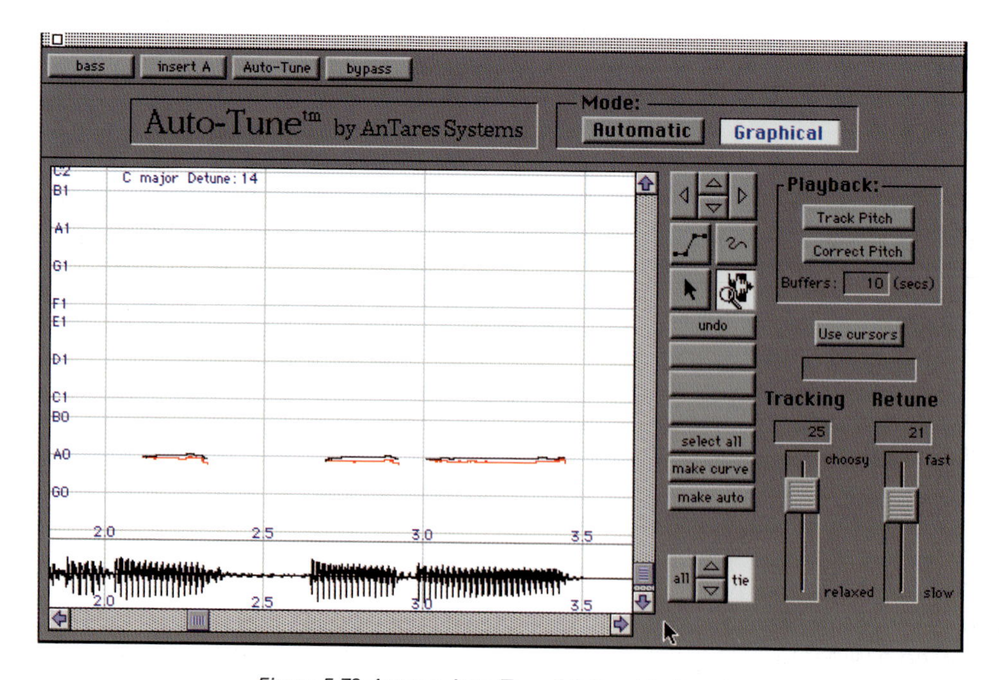

Figure 5.72 Antares Auto-Tune 1.0 Graphical mode.

session. There was no way that Automatic mode was going to get this correct, especially with the many notes that were connected by sliding pitch. The producer taught me the scale by ear, then I set about correcting each note, as necessary, in Graphical mode. Luckily, the song was not too long! The good news is that, using Graphical mode, even though it was painstaking work, I was able to make this performance 100% accurate in terms of pitching.

Auto-Tune 3

In 2001, Antares introduced Auto-Tune 3 with lots of new features and a snazzy new user interface. The pitch detection algorithm has been revised to improve the reliability of the pitch detection in both Automatic and Graphical modes. A number of 'Source Specific' pitch detection and correction algorithms have been added, including Soprano Voice, Alto/Tenor Voice, Low Male Voice, Instrument, and Bass Instrument. Matching the appropriate algorithm to the input results in even faster and more accurate pitch detection and correction. The Bass Instrument Mode has a lowest detectable frequency of 25 Hz – about one octave below the standard Auto-Tune. Low E on a bass guitar is about 41Hz, so Bass Mode lets you apply pitch correction to fretless bass lines as well as other low bass range instruments. You can set target pitches in real time via MIDI from a keyboard or sequencer track and the Make Scale From MIDI function allows you to simply play a line from a MIDI keyboard or sequencer and let Auto-Tune 3 construct a custom scale containing only those notes that appear in the line.

As with the original AutoTune, there are both Automatic and Graphical modes. As the manual explains: the automatic mode works by continuously tracking the pitch of an input sound and comparing it to a user-defined scale. The scale tone closest to the input is continuously identified. If the input pitch exactly matches the scale tone, no correction is applied. If the input pitch varies from the desired scale tone, an output pitch is generated which is closer to the scale tone than the input pitch. The exact amount of correction is controlled by the Retune Speed parameter.

The heart of Automatic mode pitch correction is the Scale. Auto-Tune 3 lets you choose from major, minor, chromatic or 26 historical, ethnic and microtonal scales. Individual scale notes can be bypassed, resulting in no pitch correction when the input is near those notes. Individual scale notes can also be removed, allowing a wider range of pitch correction for neighbouring pitches. The scale can be de-tuned, allowing pitch correction to any pitch centre.

Auto-Tune 3 can also apply a vibrato to the input sound. You can program the vibrato depth, vibrato rate and the onset delay of the vibrato. You can also choose the shape of the pitch variation in the vibrato (sine, square or sawtooth). By combining a fast Retune Speed setting with Auto-Tune 3's Vibrato settings, you can even remove a performer's own vibrato and replace it with Auto-Tune 3's programmed vibrato, all in real time. Also, unusual combinations of Vibrato Waveform, Rate and Depth settings can be used for some interesting special effects.

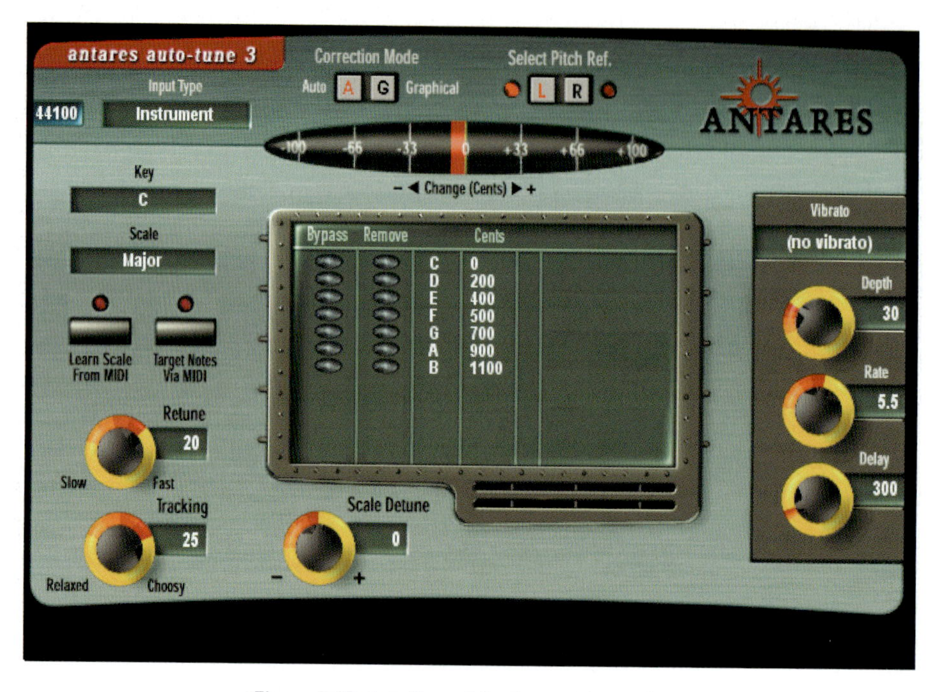

Figure 5.73 AutoTune 3 in Automatic mode.

The key feature of Graphical mode is the Pitch Graph display. On this display, the vertical axis represents pitch (with higher notes towards the top) while the horizontal axis represents time. The red curve represents the original pitch contour of the input track, while the desired target pitch or pitch contour is indicated in yellow.

Using Graphical mode is easy enough. Just select some audio in your host application, bring up Auto-Tune 3, set the buffer length to the number of seconds from the beginning to the end of the selection you want to tune, hit the Track Pitch button and play your audio. Auto-Tune analyses this and displays the pitches of the notes in red. Now you can either hit Make Curve, so that Auto-Tune automatically draws 'target pitch' curves in yellow for each note, or you can manually draw in target curves (using the line and curve drawing tools) for each note that needs correcting. You than have to manually move the target curves to the new correct pitches for each note that needs correcting - which can be a painstaking task if there are many notes. You may also need to adjust the Retune Speed and the Tracking parameters for best results. Once this is completed, all you need to do is to hit Correct Pitch and play back your audio to hear it corrected the way you have specified.

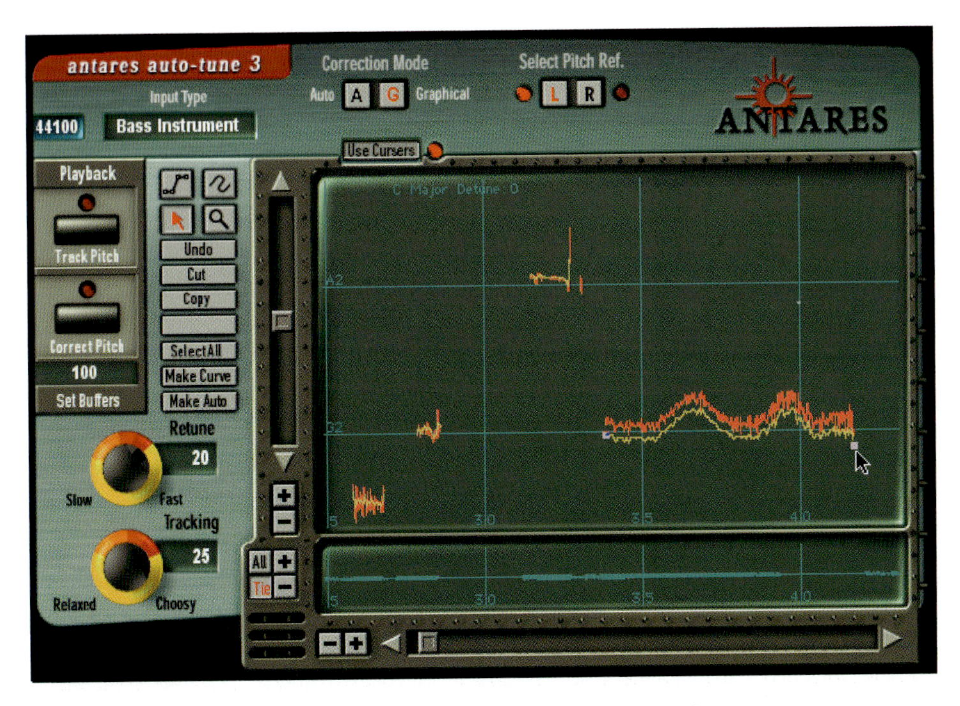

Figure 5.74 AutoTune 3 Bass in Graphical Mode.

Auto-Tune 3 is available in TDM, MAS, RTAS, Mac VST and DirectX versions and supports the new higher sample rates including 88.2 kHz and 96 kHz, with compatible host applications and audio hardware.

Auto-Tune can sometime appear to work miracles, but at other times it cannot. For example, the range of pitches extends up to C6, but if you encounter a higher pitch, this will be misinterpreted as sounding one octave lower. Also, Auto-Tune 3 will not detect pitch when the input waveform is aperiodic. So, for example, although Auto-Tune can correct a solo violin playing a sustained note off-pitch, it cannot correct a violin section playing this same note. This is because the violin section contains many notes that were all started at slightly different times and are at slightly different pitches - so the waveform does not repeat periodically - as does a single note from a single violin. A breathy voice or a noisy recording of anything will be aperiodic for similar reasons - the note will be mixed with a noise-type signal in these cases. Auto-Tune 3 can take this into account to help with pitch detection if you lower the Tracking control value, which relaxes the strict requirement for periodicity in the input signal.

Auto-Tune 3 does offer several advantages over the previous version, but I do feel that its user interface has become too 'busy' compared with the clarity of the original user interface. Nevertheless, it does the job - which is the most important thing!

Antares Microphone Modeler

Have you ever wished you could get your hands on a Coles 4038, RCA RCA44, Telefunken U-47 (Original Tube), or Neumann U 87 70th Anniversary Gold Edition? Or maybe you fancied a B&K 4007, Brauner VM1, Groove Tubes MD1, or Manley Labs Reference Gold - until you checked your bank account and found that your budget would not stretch? Well, now there's a solution to your woes - Antares Microphone Modeler.

Antares has been modelling microphone characteristics for several years now, and has built up a very comprehensive library of popular microphones - with options. Does the microphone have a low cut filter? What pickup pattern? Is the wind-screen on or off? Where is the microphone positioned? OK - you choose! You can use any reasonable quality microphone to record with and then make this sound like any of a wide variety of high-end studio microphones

The Microphone Modeler couldn't be much easier to use. You first select a Source Model corresponding to the microphone you have used to record your track. This strips off the characteristics of that microphone to leave a neutral signal to which you can then apply the characteristics of whichever modelled microphone you choose. A tube pre-amp model is also provided so that you can further process your audio using classic tube saturation distortion.

Using Microphone Modeler, you can now afford to record every track through your ideal choice of microphone, or use it on mixdown to change the sound already recorded. The Microphone Modeler is available as a plug-in for TDM, MAS, Mac VST, RTAS and DirectX, and whichever version you choose will cost you much less than even one of the sought-after vintage microphones in the library included with the software.

Figure 5.75 Antares Microphone Modeler.

The Source Mic section lets you specify the microphone and the settings that were (or will be) used to capture the input sound. The idea here is to remove the characteristics of the source microphone as completely as possible. Use the Low-Cut popup to select any filter you have used to undo the effect of this and use the Pattern popup selector to neutralize the effects of any particular pattern setting. If you know the distance between the original microphone and the sound source, you can set the Proximity value to this distance to remove any proximity effects.

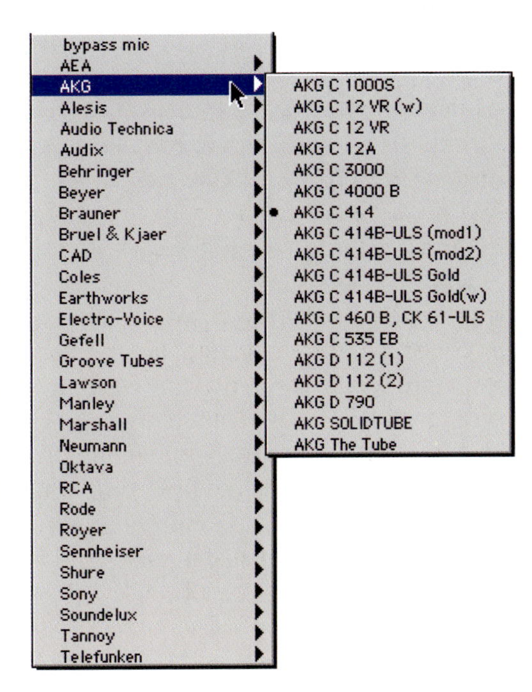

Figure 5.76 Microphone Modeler Source Mic popup menu.

With this all set up, you can now choose the microphone you want to use from the Modeled Mic popup.

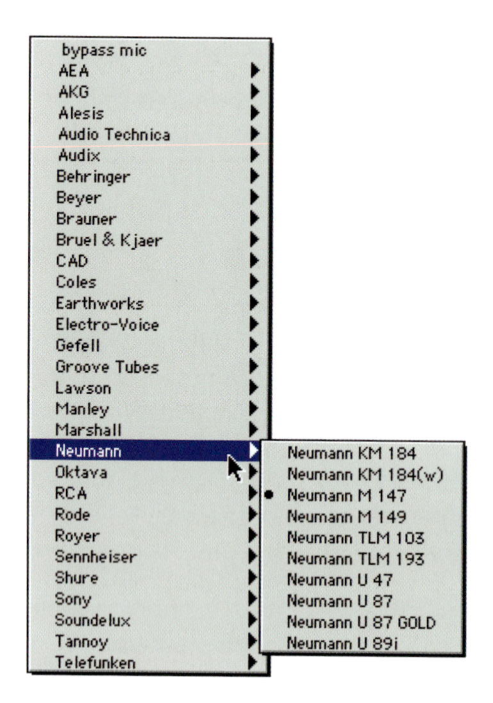

Figure 5.77 Microphone Modeler Modeled Microphone popup menu.

Use the Low Cut and Pattern popups to select any available filter or pickup pattern and use the Proximity knob to set the amount of proximity effect that will be applied to the new microphone you choose.

Now there may be occasions where you like the way one mic sounds in the bass but you prefer the way another sounds in the treble range. The Preserve Source controls lets you split your audio into its bass and treble ranges and process each separately to create a hybrid mic with the bass characteristics of one and the treble characteristics of another. Just click on either the Bass or Treble button - depending on whether you want to preserve the bass or treble characteristics of your source mic. So, for example, with the Bass button pressed, only the model's treble characteristics are applied to the signal. The net effect is that you get the source mic's bass characteristics and the modelled mic's treble characteristics. Finally, you can add some extra 'warmth' to the sound using the Tube Saturation section which models the typical distortion of a high-quality tube pre-amp.

Now most users will probably want to use Microphone Modeler to get as close as possible to the sound of a particular microphone - and Microphone Modeler certainly helps you to do that. But there are no rules to this game. You can always get creative and try any settings you like until you get a sound you like. And if it sounds good - it is good! That's the only consistent rule in this game!

Also, be realistic in your expectations. You are not really going to get world-class results from a £20 microphone, and even with a vintage mic you can get lousy results if you position it inappropriately. As the manual explains, 'If your audio is not well-recorded in the first place, the Microphone Modeler can to do very little to improve it.'

I got great results using an AKG C414 as my source mic, for example, although I didn't have the Neumann models that I chose to model available for A/B comparisons. I ended up taking something of the creative approach by trying several different modelled mics until I found a sound that I liked - like switching through the presets on a synthesizer! Then again - isn't that what engineers always do in studios when selecting microphones anyway?

Arboretum

Arboretum software offers five main software packages including Hyperprism, Realizer Pro, Ray Gun, Harmony and Ionizer. Hyperprism is a suite of plug-ins - everything from reverbs to Doppler effects. Realizer Pro lets you process audio files simply but effectively prior to MP3 encoding. Ray Gun provides effective noise reduction and can remove pops and clicks from vinyl transfers. Harmony lets you create harmonies and also allows you to correct faulty pitching. Ionizer does so many things that it is hard to categorize. It starts out as a sophisticated noise reduction device, then reinvents itself as a multi-band processor, then decides it can be a vocoder as well - and succeeds!

Hyperprism

Edgar Varese spoke about his 1923 composition Hyperprism, saying 'The universe of possible musical forms is as unlimited as the number of external forms of crystals.' Georges Jaroslaw, a French musician and composer, was inspired by these words to design Hyperprism for the Macintosh - a highly flexible real-time effects processor which lets you 'play' effects in real time.

Hyperprism is ideally suited to sound effects creation and will definitely be of interest to anyone working with sound for films and cartoons, or in more experimental areas of music. The intuitive graphic interface simply invites creativity, especially for those who feel less at home with numbers. Hyperprism offers interactive graphical control of effect parameters in real time using its unique blue window. You can control two (or more) parameters at once by dragging a point across the blue-coloured screen to set values according to the x and y positions within the screen. For example, left to right mouse-motion could change filter cut-off frequency while up and down motion simultaneously adjusts filter resonance. Recent versions allow you to control more than two parameters, so, for example, horizontal motion could control not only the cut-off frequency in a filter or the delay in a reverb but also diffusion, smoothing, brightness or whatever. If you just want to apply a static effect, simply click the mouse button once to set a particular point on the screen that will apply the effect parameters of your choosing. Drawing a path onscreen provides for more interesting moving changes to the sound.

Hyperprism can be thought of as a 'rack' of studio-quality effects processors. Each effect in the package can be selected from a menu or insert point in the host application, previewed and adjusted in real time and applied individually to any track, region or file. You can apply any number of effects to a single sound file and the software uses non-destructive processing, so your original sound files are always safe. You can also process 'live' sounds, playing these through the system and back out again with processing applied in real time.

Hyperprism comes in various different configurations to suit whichever computer or host software you are using. Hyperprism 2.5 is the stand-alone version of Hyperprism that runs on PowerMac computers. Amazingly, you can still get Hyperprism-68K. This is the classic stand-alone edition for older 68000-series Macintoshes with Audiomedia I/II, Sound Tools I/II or Pro Tools 442 systems. Hyperprism-TDM 2.5 is available for Digidesign Pro Tools|24 and Mix systems with TDM hardware. Hyperprism Plug-in Pack 2.5 contains Hyperprism-MMP (for Premiere compatible hosts), Hyperprism-DAS (for AudioSuite compatible hosts) and Hyperprism-VST (for VST compatible hosts). The Plug-in Pack also includes the Windows DirectX format. Hyperprism-DirectX 2.5 for Windows 95/98/ME/NT includes more than 30 different effects for the DirectX plug-in format for use with most of the popular Windows audio editing programs.

Hyperprism 2.5

Hyperprism 2.5 for Power Mac is the stand-alone version of Hyperprism with 42 different effects, parameter automation editing and 'live' multi-processing features.

Arboretum's HyperEngine real-time sound editing and signal processing application acts as the 'host' software environment for the Hyperprism plug-ins. HyperEngine works with the Apple Sound Manager and also has direct driver support for the Korg 1212 I/O card and Digidesign Audiomedia II and III sound cards.

Hyperprism has three different groups of controls: the HyperEngine monitoring and playback controls, the HyperEngine Document Reference windows, and the Hyperprism Blue Windows.

The Playback Window which appears on the left of your screen when HyperEngine starts up controls playback, monitoring and metering for Hyperprism.

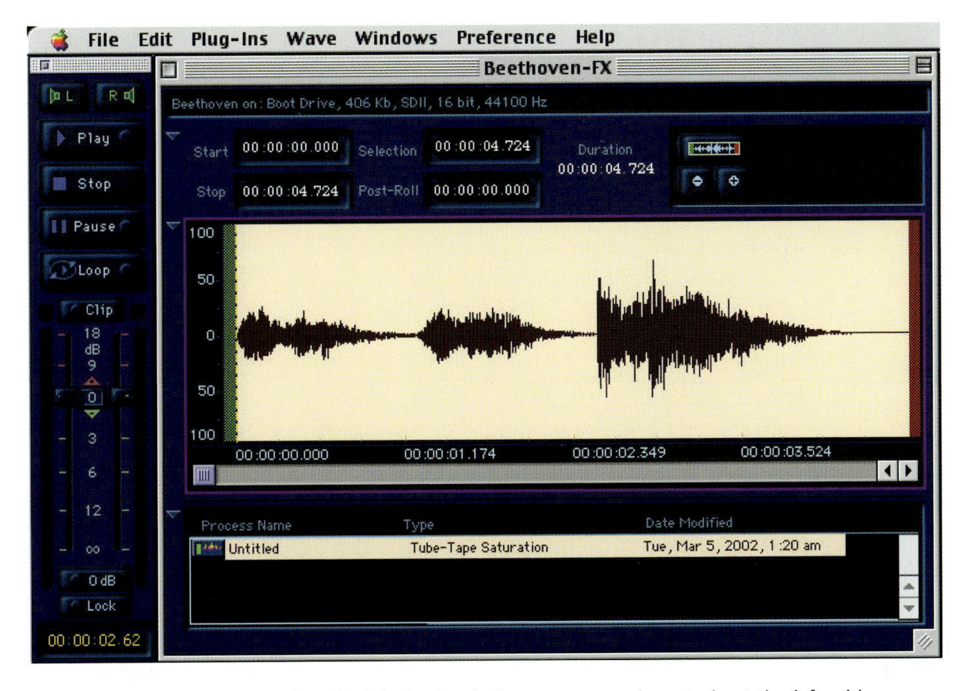

Figure 5.78 HyperEngine showing the Playback window meters and controls at the left with an open File Reference Document to the right.

There are two types of so-called Reference Documents that you can open in HyperEngine: the Play-Thru Document for recording audio or doing live, on-the-fly signal processing, and the File Document for working with sound files already on disk. The purpose of these Reference Documents is to keep track of all the edits and effect you apply to your original sound.

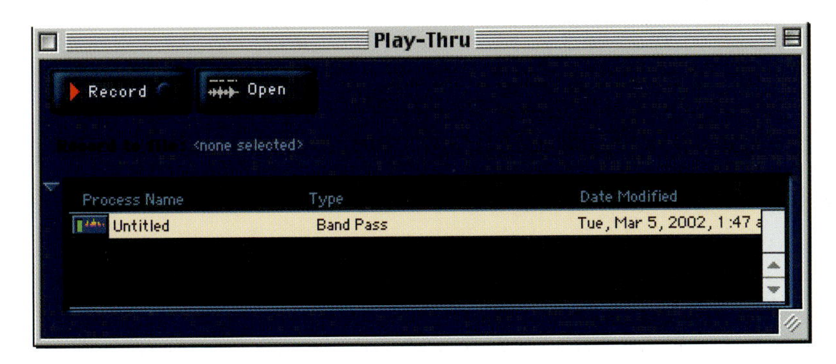

Figure 5.79 HyperEngine Play-Thru Document, for working with live audio signals. A Band-Pass Filter process has been inserted.

Both of these window types, the File Document window and the Play-Thru Document window, show a list of any Hyperprism processes that you have applied to the audio.

Tip HyperEngine's Play-Thru processing features allow you to run live audio into Hyperprism - from a tape deck, instrument or microphone - process the audio on-the-fly, and output it to a mixer, PA or recording device.

When you choose a Hyperprism process from HyperEngine's Plug-in menu, a Blue Window will open and the process will be listed in the File Document or Play-Thru Document's window. You can then use the Hyperprism Blue Window to alter the settings for each individual effect parameter.

Hyperprism's Blue Window is used to design audio effects. It was created with the goal of giving you gestural control over signal processing, allowing you to 'play' your processors as if they were musical instruments. Hyperprism allows you to connect any effect parameter to the Blue Window's horizontal (X) axis, or the vertical (Y) axis, so that you can change two or more parameters at once, by just moving the mouse around. You can also record and edit any Blue Window mouse movements, then, later on, you can apply these effects to a new sound file or to live signal processed using a Play-Thru Document.

Note The Blue Window Axis Switches, located to the extreme left of each parameter slider, allow you to map the parameter to the X or Y axis. Each switch has three possible states: Click once on any Axis switch to map its parameter slider to the horizontal axis, click twice for vertical mapping, or a third time to disconnect the parameter from the Blue Window.

Figure 5.80 Hyperprism's Blue Window interface, used to dynamically control effect parameters.

Hyperprism 2.5 has 42 different effects:

Filter processes

- Low Pass
- High Pass
- Band Pass
- Band Reject
- Low Shelf
- Hi Shelf
- Parametric EQ

Modulation processes

- Vocoder
- Frequency Shifter
- Phaser
- Flanger
- HyperPhaser
- Chorus
- Ring Modulator
- Tremolo
- Vibrato

Delay and reverb processes

- Single Delay
- Multi Delay
- Echo
- EchoTranz
- HyperVerb
- Room Reverb
- Hall Reverb
- Granulator

Stereo processes

- Auto Pan
- Quasi Stereo
- Doppler
- Stereo Dynamics
- More Stereo
- M-S Matrix

Miscellaneous processes

- Harmonic Exciter
- Bass Maximizer
- Tube/Tape Saturation
- Pitch Time Changer
- Pitch Changer
- Formant Pitch Shifter
- Vari-Speed
- Noise Gate
- Compressor
- Limiter
- Sonic Decimator
- Z-Morph

Hyperprism-TDM

Hyperprism-TDM 2.5 for Digidesign Pro Tools|24 and Mix systems with TDM hardware provides 23 real-time non-destructive effects each with Pro Tools parameter automation support.

Note There are not as many Hyperprism effects available for TDM as for other platforms, but AudioSuite versions are also available for non-real-time processing within the Pro Tools environment (see later).

The TDM version features the Hyperprism blue screen. Any of the sliders, or the input gain faders, can be controlled using the blue screen. To link any of these to the movements of the cross in the blue screen, simply click in the small blue box next to each slider. Now you can make gestural movements with your mouse or other controller to drag the cross around the blue screen – and in turn control whichever parameters you have linked to the cross. And you can use the standard Pro Tools automation features to record automation data for any of the Hyperprism plug-ins into any Pro Tools track and edit these graphically in Pro Tools.

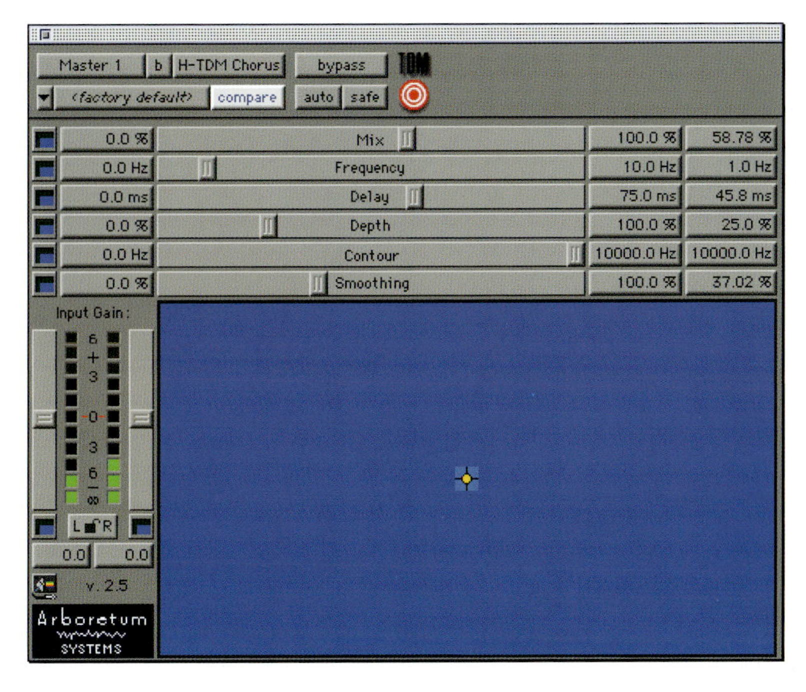

Figure 5.81 Hyperprism-TDM plug-in.

Hyperprism-TDM has 23 effects:

Filter processes

- Low Pass/High Pass Filter
- Band Pass/Band Reject Filter

Modulation processes

- Phaser
- Flanger
- Chorus
- Ring Modulator
- Tremolo
- Vibrato

Miscellaneous processes

- Bass Maximizer
- Harmonic Exciter
- Tube-Tape Saturation
- Pitch Changer
- Noise Gate
- Sonic Decimator

Stereo processes

- Balance
- Auto Pan
- Quasi Stereo
- Stereo Dynamics
- More Stereo
- M-S Matrix
- Hyper Stereo
- Doppler
- Hyper Pan

Hyperprism Plug-in Pack 2.5

Hyperprism Plug-in Pack 2.5 contains the MacOS Hyperprism formats: Hyperprism-MMP (for Premiere-compatible hosts), Hyperprism-DAS (for AudioSuite-compatible hosts) and Hyperprism-VST (for VST-compatible hosts). The Plug-in Pack also includes the Windows Direct-X format.

Hyperprism-MMP

Hyperprism-MMP plug-ins can be used from within any MacOS audio editing application that supports Adobe Premiere format plug-ins. These include Adobe Premiere, BIAS Deck and Peak, Digital Performer, Logic Audio, Metro 4, TurboMorph – and Opcode's Vision, Studio Vision and Studio Vision Pro.

The Hyperprism-MMP screen is made up of four main parts: The Blue Window at the centre, the Sliders at the top, the level and metering section at the lower left, and the playback, process and presets area at the upper left.

Figure 5.82 Hyperprism-MMP Premiere plug-in.

Hyperprism-MMP includes 38 effects:

Filter processes

- Low Pass Filter
- High Pass Filter
- Band Pass Filter
- Band Reject Filter
- Low Shelf Filter
- Hi Shelf Filter
- Parametric EQ

Modulation processes

- Vocoder
- Frequency Shifter
- Phaser
- Flanger
- HyperPhaser
- Chorus
- Ring Modulator
- Tremolo
- Vibrato

Miscellaneous processes

- Harmonic Exciter
- Bass Maximizer
- Tube/Tape Saturation
- Pitch Changer
- Formant Pitch Shifter
- Noise Gate
- Compressor
- Limiter
- Sonic Decimator

Delay and reverb processes

- Single Delay
- Multi Delay
- Echo
- HyperVerb
- Room Reverb
- Hall Reverb
- Granulator

Stereo processes

- Pan
- Auto Pan
- Quasi Stereo
- Stereo Dynamics
- More Stereo
- M-S Matrix

Hyperprism-DAS

Hyperprism is also available in AudioSuite format. AudioSuite plug-ins offer the same basic functionality as the TDM algorithms but because AudioSuite is not a real-time plug-in format, no Blue Window is included.

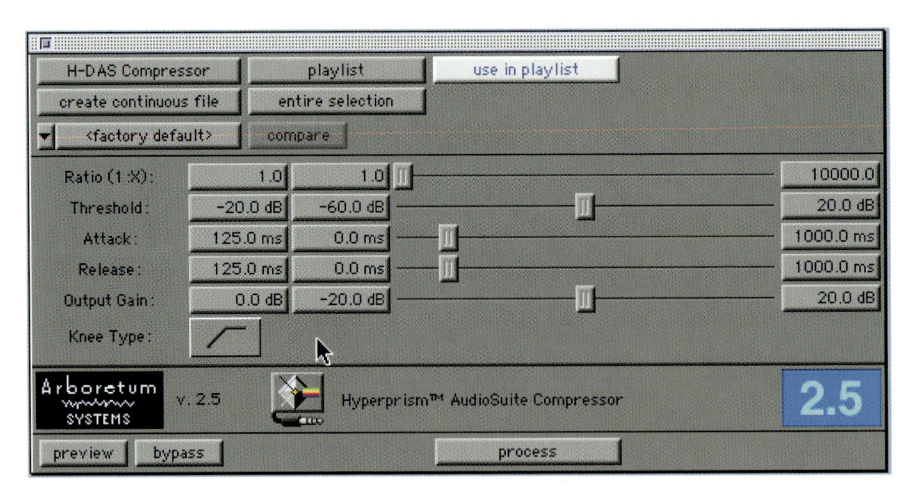

Figure 5.83 Hyperprism-DAS AudioSuite plug-in.

Hyperprism-DAS includes 35 different effect algorithms:

Filter processes
- Low Pass
- High Pass
- Band Pass
- Band Reject

Modulation processes
- Vocoder
- Frequency Shifter
- Phaser
- Flanger
- HyperPhaser
- Chorus
- Ring Modulator
- Tremolo
- Vibrato

Stereo processes
- Pan
- Auto Pan
- Quasi Stereo
- Stereo Dynamics
- More Stereo
- M-S Matrix

Delay and reverb processes
- Single Delay
- Multi Delay
- Echo
- HyperVerb
- Room Reverb
- Hall Reverb
- Granulator

Miscellaneous processes
- Harmonic Exciter
- Bass Maximizer
- Tube/Tape Saturation
- Pitch Changer
- Formant Pitch Shifter
- Noise Gate
- Compressor
- Limiter
- Sonic Decimator

Hyperprism-VST

With Premiere or AudioSuite plug-ins, you have to render each process directly to a file. The VST format works in real time so you can keep making changes to your effects as you build an arrangement, without committing to any effect until the final mix.

Once you have assigned effect parameters to mouse movements, simply click and drag the mouse within the Blue Window. When you find the 'spot' you like, you can process your sound with those particular effect settings. You may also use your host application's automation capabilities to record dynamic effect changes for any Hyperprism-VST effect. These will be played back in time with your music, allowing a much more spontaneous and creative use of effects than the traditional 'set-it-and-forget-it' behaviour of typical signal processors. Hyperprism-VST can be played much like a musical instrument, changing your effects right along with the music . . . and it remembers and recreates your every move!

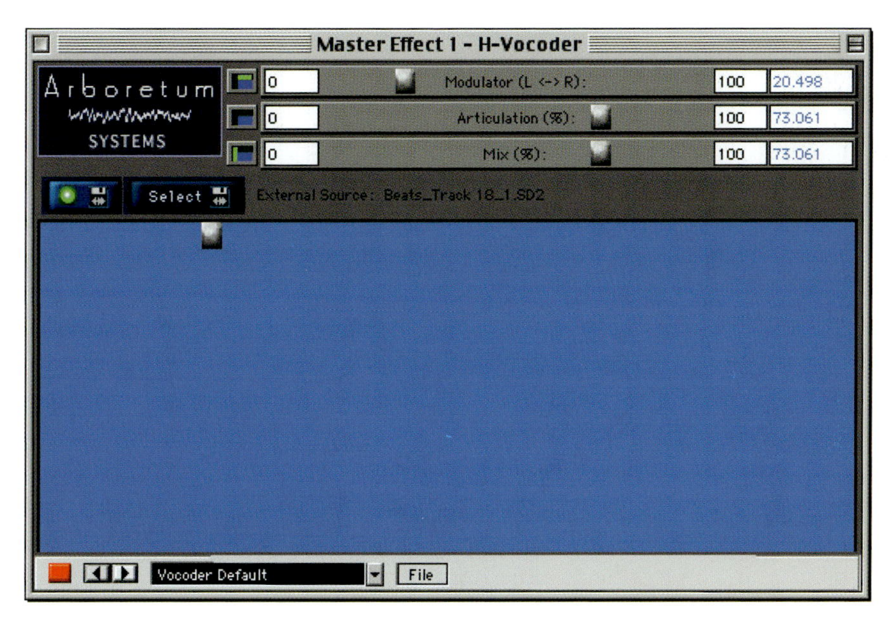

Figure 5.84 Hyperprism-VST.

Hyperprism-VST includes 37 different effect algorithms:

Filter processes
- Low Pass
- High Pass
- Band Pass
- Band Reject

Modulation processes
- Vocoder
- Frequency Shifter
- Phaser
- Flanger
- HyperPhaser
- Chorus
- Ring Modulator
- Tremolo
- Vibrato

Delay and reverb processes
- Single Delay
- Multi Delay
- Echo
- EchoTranz
- HyperVerb
- Room Reverb
- Hall Reverb
- Granulator

Stereo processes
- Pan
- Auto Pan
- Quasi Stereo
- Doppler
- Stereo Dynamics
- More Stereo
- M-S Matrix

Miscellaneous processes
- Harmonic Exciter
- Bass Maximizer
- Tube/Tape Saturation
- Pitch Changer
- Formant Pitch Shifter
- Noise Gate
- Compressor
- Limiter
- Sonic Decimator

Hyperprism VST effects

This section contains brief overviews of just about all the Hyperprism effects, showing the screens in Cubase VST as examples.

The Auto Pan is instantly effective and not only lets you control the depth of the effect, but also the position within the stereo field that the panning action takes place. With the Position control in the centre, you get the full left to right panning effect. When you move this to left or right, the panning will range from hard left to something less than hard right or vice versa.

Figure 5.85 Hyperprism Auto Pan.

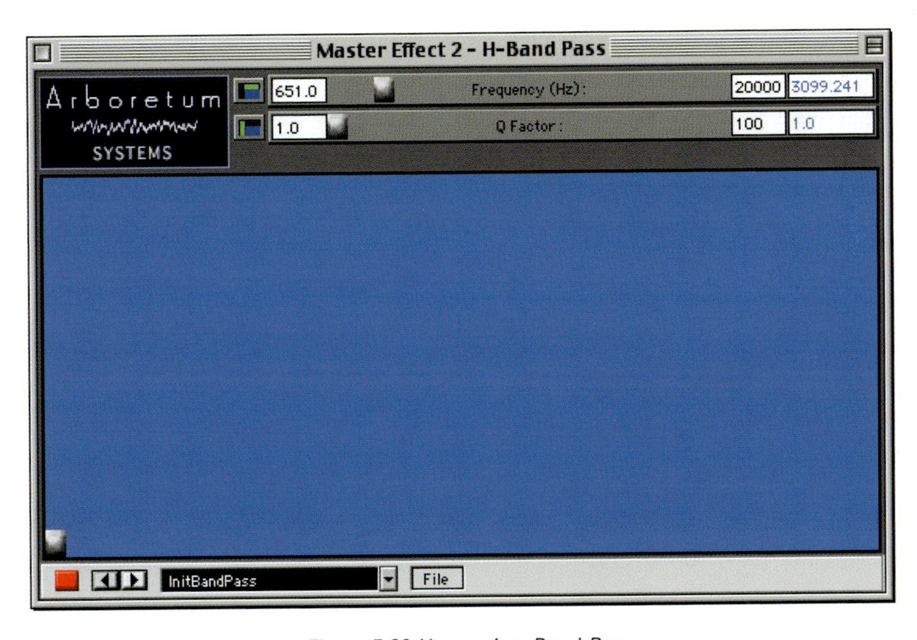

Figure 5.86 Hyperprism Band Pass.

The Band Pass filter lets you isolate a narrow band of frequencies from within your sound - a special effect that can be very useful for the sound designer. The Frequency control lets you set the centre frequency of this band, and the Q control lets you adjust the width of the band.

Figure 5.87 Hyperprism Band Reject.

Band Reject is the opposite of Band Pass in the sense that it rejects frequencies within the specified band rather than passing these through. The Frequency parameter lets you adjust the centre frequency of the rejected band, from 13 Hz to 20 kHz. The Q Factor lets you control the sharpness of the cut-off of the filter.

The Bass Maximizer effect adds harmonics into the sound to boost the impression of bass. Again, this can be much more effective than using simple EQ. This plug-in provides controls for a greater range of parameters than many of its competitors - helping to achieve superior results in many situations.

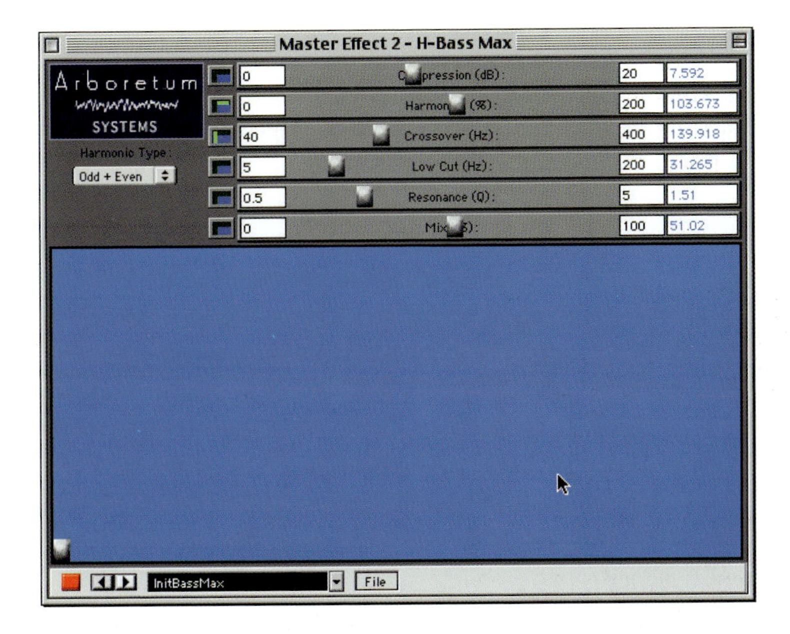

Figure 5.88 Hyperprism Bass Maximizer.

Figure 5.89 Hyperprism Chorus.

The Chorus effect has the usual controls for Depth and Mix along with controls for Frequency, Delay and Contour. The clarity of the interface, again, is one of its strengths, and, as with all these plug-ins, you can save your favourite settings for instant recall.

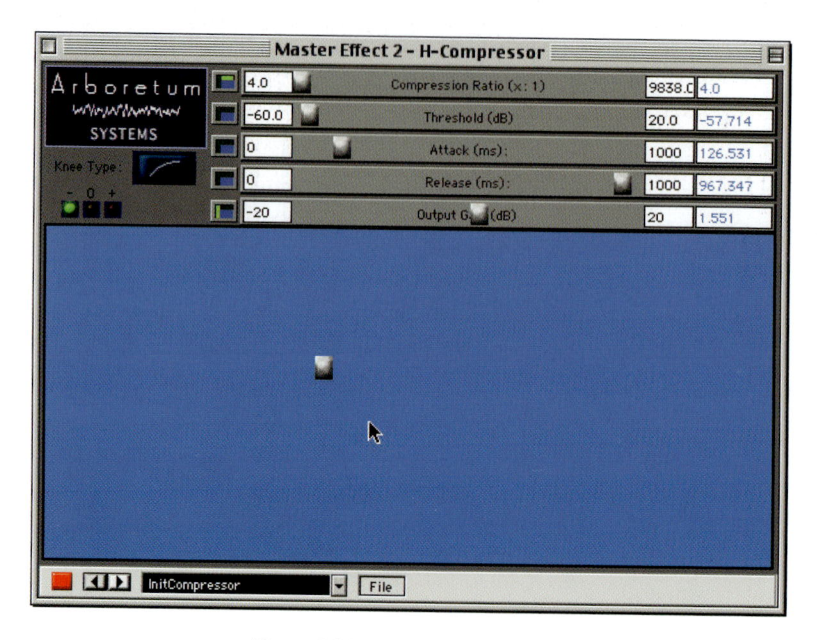

Figure 5.90 Hyperprism Compressor.

The Compressor has all the basic controls that you find on most compressors, but lacks the more sophisticated metering that most engineers find helpful when adjusting the settings. The simple metering LEDs show when your input signal is moving above and below the threshold. A green light indicates that the signal is below the threshold, so no processing is occurring. An orange light shows that the signal is in the soft knee region and incremental compression is occurring. A red light means that the signal is above the threshold, so full compression is being applied.

Note Metering lights are not included in Hyperprism-DAS for AudioSuite.

The Doppler effect lets you create the effect of the pitch changing as the source approaches and recedes from the listener. This is useful when you need to quickly set up a basic Doppler effect, but this plug-in lacks the sophistication that you will find in the Waves Pro-FX Doppler plug-in or the GRM Tools Doppler VST plug-in.

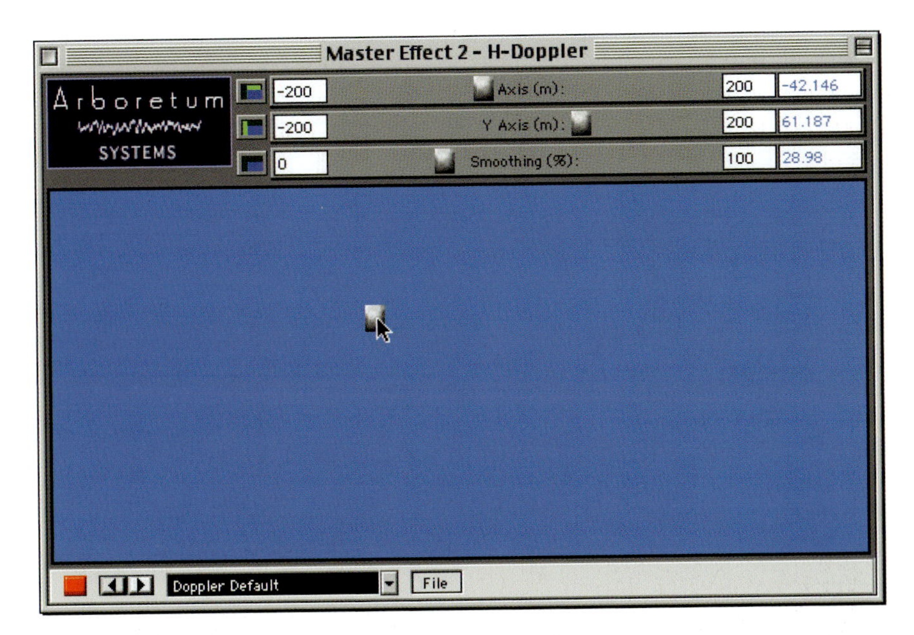

Figure 5.91 Hyperprism Doppler.

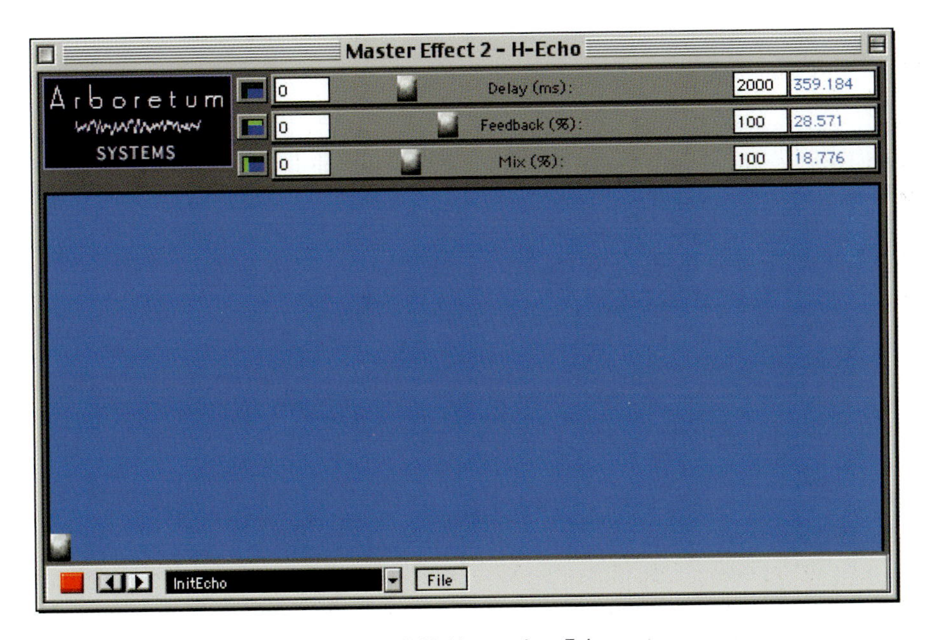

Figure 5.92 Hyperprism Echo.

Echo is always a useful effect to have at your fingertips, and the Hyperprism Echo plug-in lets you create the basic echo effect and apply feedback to cause repeats.

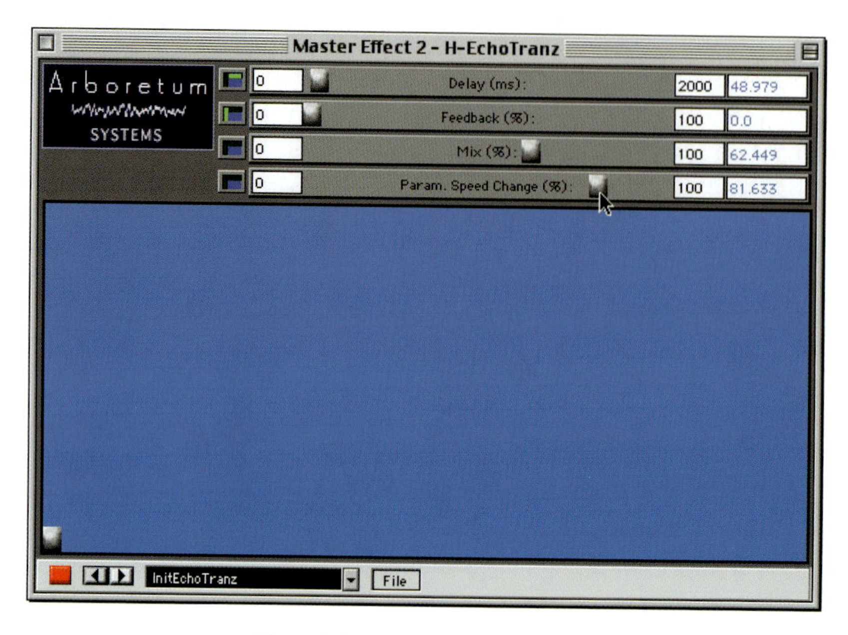

Figure 5.93 Hyperprism EchoTranz.

EchoTranz provides a variation on the basic Echo plug-in by simulating the speed change effect that takes place with an analogue tape-echo unit, such as the Maestro Echoplex, when you move the playback head. With the Echoplex, the sound seems to 'swing around' for a little while if you change the delay time while playing sound through the unit, before settling down at the new speed. This effect is emulated in EchoTranz using the Parameter Speed Change control.

The Flanger simulates the basic 'flanging' effect perfectly well, and provides all the usual controls for the effect, such as feedback and depth.

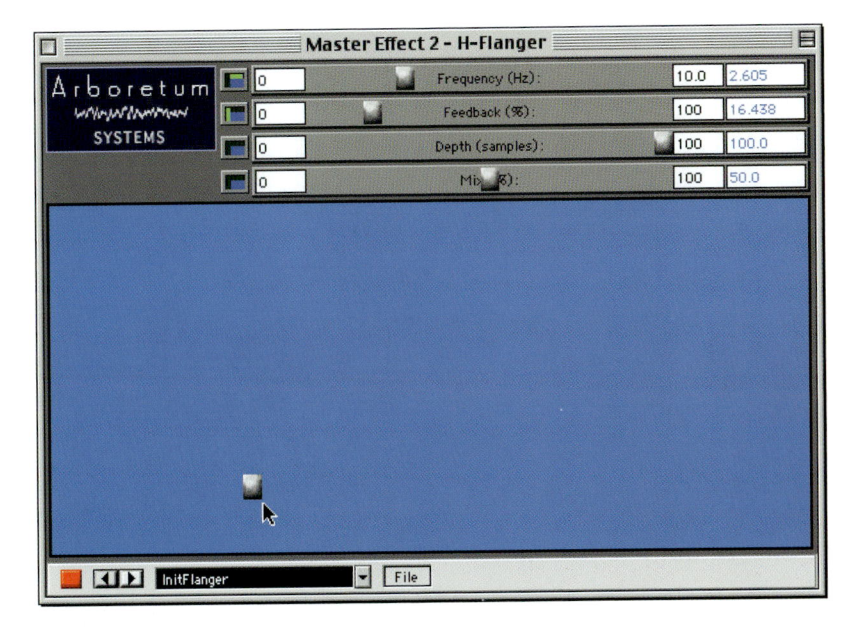

Figure 5.94 Hyperprism Flanger.

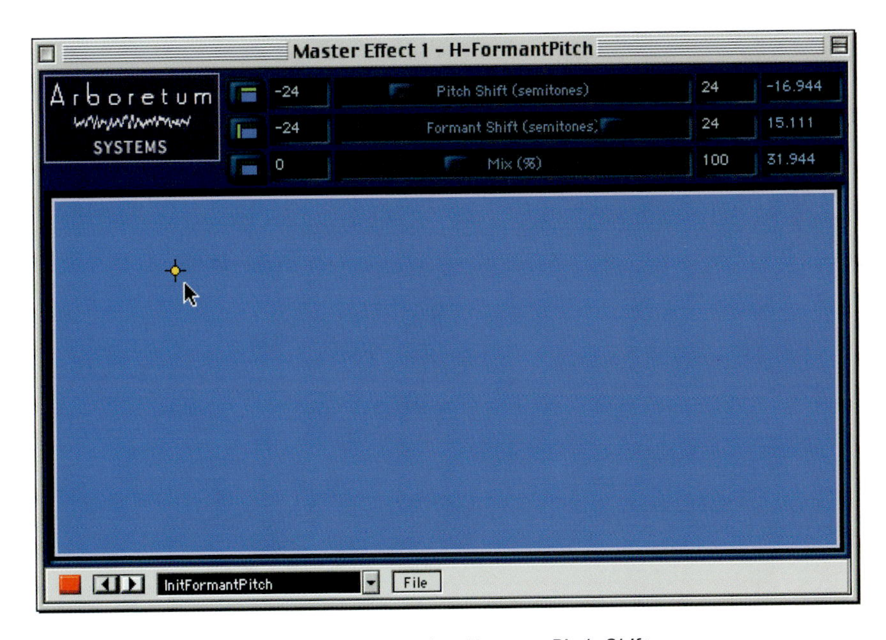

Figure 5.95 Hyperprism Formant Pitch Shifter.

The Formant Pitch Shifter plug-in gives you independent control of overall pitch and formant pitch so you can shift either of these while leaving the other unaltered – or you can adjust both. This plug-in is particularly useful for sound effects creation – great for creating 'exorcist'-type voices and suchlike. This is intended for use with vocals or melodic material and works over a four-octave range – two above and two below the original pitch. A fast computer is required for best results.

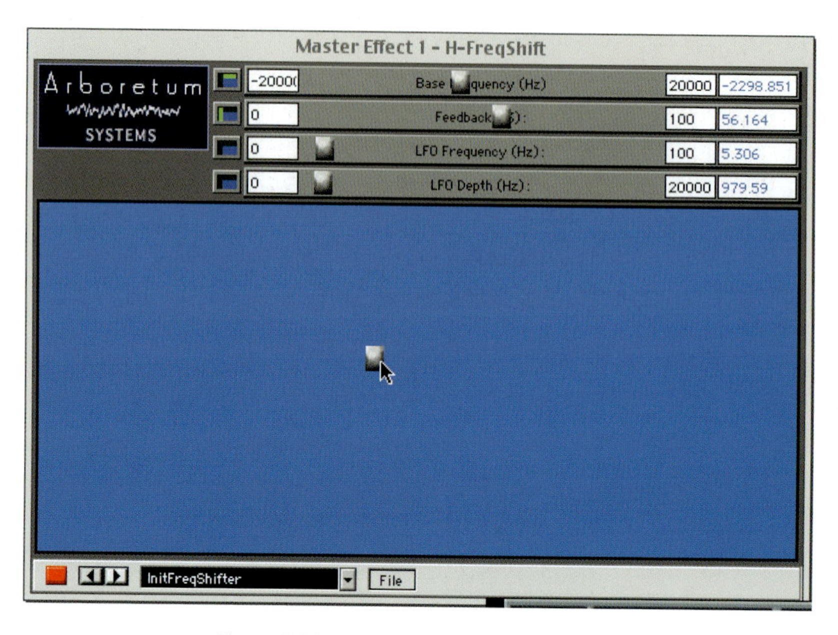

Figure 5.96 Hyperprism Frequency Shifter.

The Frequency Shifter plug-in is another for the creative sound effects designer. Try putting a vocal through this and change two or more parameters at the same time. This sounds like you are 'tuning' a radio through the different frequency bands with the vocal 'fighting' to come through!

For even more 'way out' effects, try the Granulator. This extracts part of the original audio and plays it back, rhythmically, at different pitches. Not easy to describe in words – this creates effects that you have to hear to appreciate.

Basically, the Granulator is a tempo-based multi-tap delay line with the number of taps ranging from 4 to 16. The spacing between taps is determined by the setting of the tempo slider and the note value selected. Each tap creates a 'grain' that has several parameters including length, attack and release. The length can be set with a slider, while the attack and release can be changed using popups. Grains

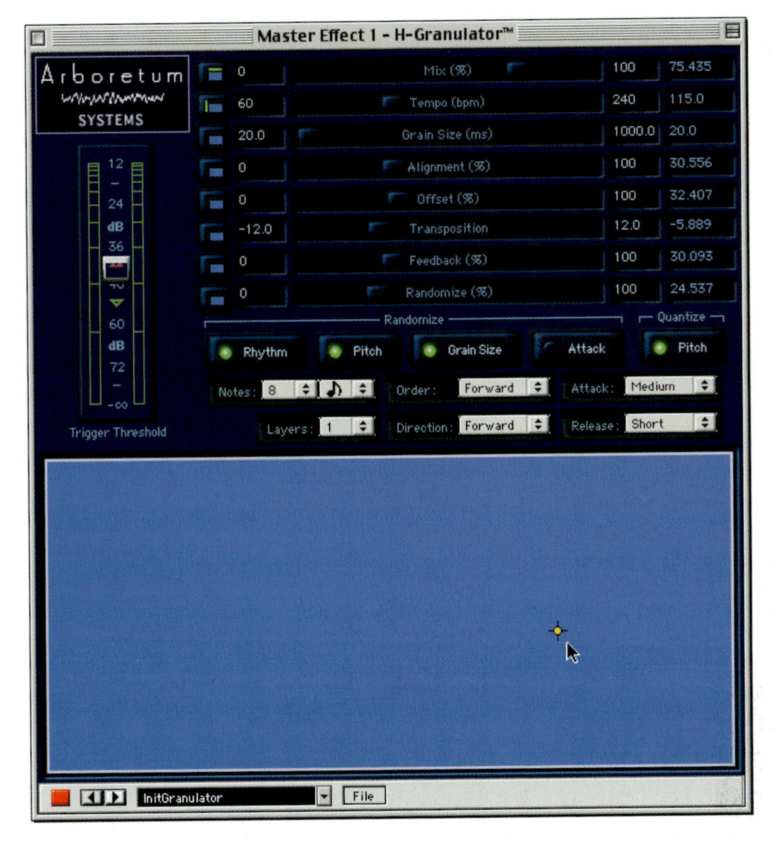

Figure 5.97 Hyperprism Granulator.

can also be transposed up to an octave up or down using the transpose slider and they can play forward or in reverse depending on the setting of the 'Direction' popup. The 'Alignment' slider allows further control over the positioning of the original and delayed signals, while 'Offset' controls the placement of the taps relative to the source signal. The delayed grains can then be re-injected into the effect using the 'Feedback' slider. A slider can be used to introduce even more variety by randomizing any combination of four parameters: 'Rhythm', 'Pitch', 'Grain Size', and 'Attack'.

Note The Granulator plug-in requires a large amount of memory.

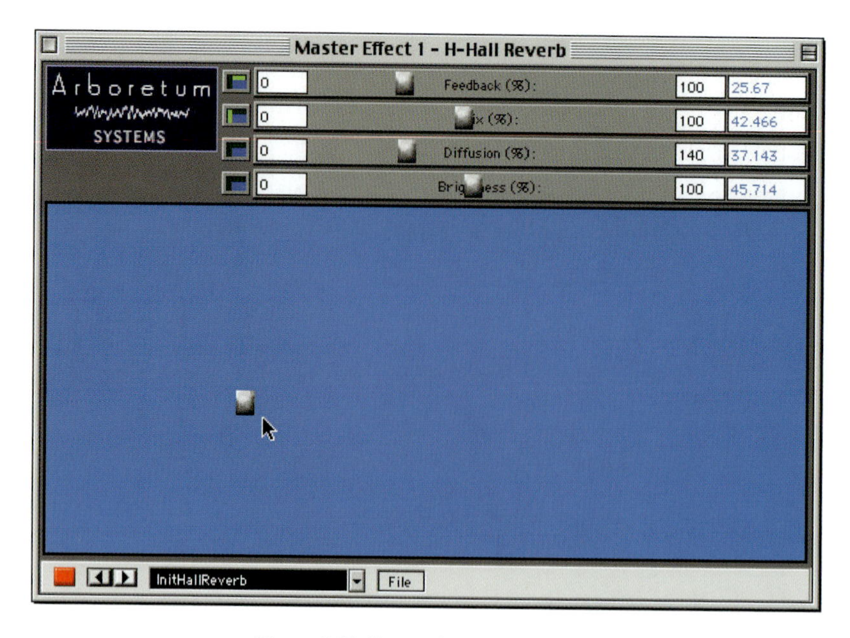

Figure 5.98 Hyperprism Hall Reverb.

The Hall Reverb provides a basic reverberation effect with Feedback, Diffusion and Brightness controls. Higher feedback settings lengthen the time it takes for the reverb to decay to silence and the brightness slider simulates the loss of high frequencies due to room reflections and atmospheric absorption. Diffusion controls the phase smearing of the reflected sound. Low diffusion settings result in more discrete delayed repeats, while higher diffusion settings smear out the attacks of each repeat - producing a smooth wash of sound, without identifiable reflected attacks.

The Harmonic Exciter lets you brighten up (or dull down) your sounds by affecting the harmonic content. This effect is particularly useful on acoustic guitar sounds, for example, providing a more effective alternative to using EQ.

A High Pass filter is always a useful tool for sound design, removing frequencies below the specified cut-off frequency. Having a variable Q, or filter bandwidth, makes this even more useful.

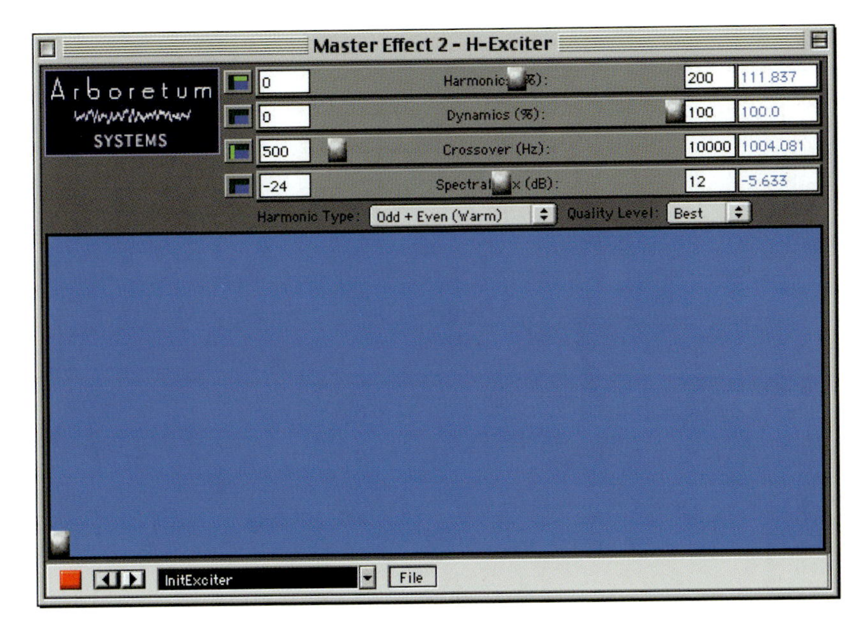

Figure 5.99 Hyperprism Exciter.

Figure 5.100 Hyperprism High Pass Filter.

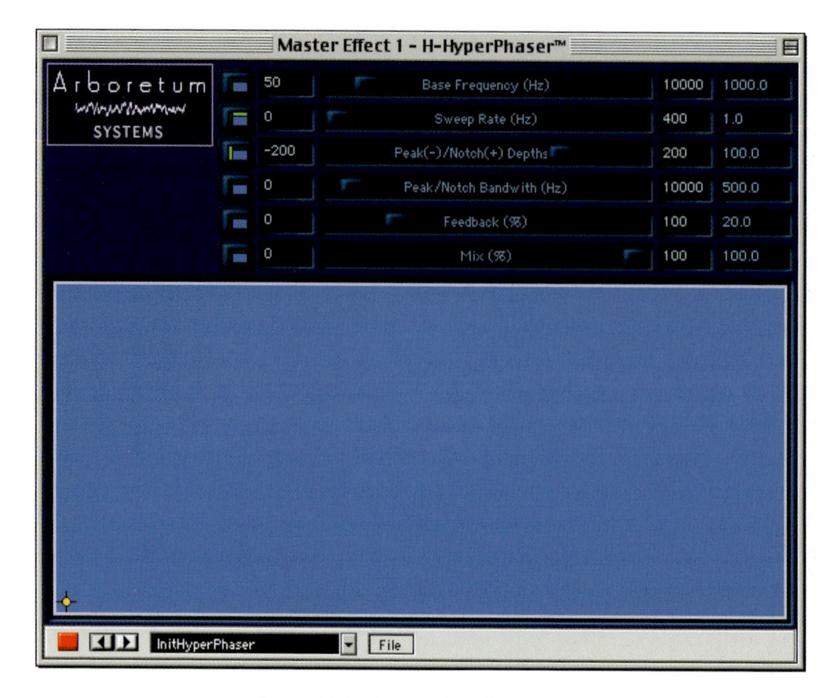

Figure 5.101 Hyperprism HyperPhaser.

The HyperPhaser is particularly impressive, producing deep, analogue sounding phasing and flanging. A wide variety of effects are possible, from slow filter sweeps to intense resonant phasing. The unique ability to create either spectral notches or peaks allows the user a great deal of control over the character of the sound. Using the Base Frequency slider you can set the centre frequency of the effect, while the Sweep Rate slider controls the rate at which the base frequency is modulated. Use the Peak/Notch Depth and the Peak/Notch Bandwidth sliders to control the depth and bandwidth of peaks (when the depth slider has negative values) or notches (when the depth slider has positive values). The Feedback slider adds a metallic ringing to the effect.

HyperVerb is an improved version of the standard Hyperprism reverb algorithm with a very smooth impulse response that sounds great on percussive tracks and doesn't colour the sound unnaturally. You get lots of parameter control and dense reflections without ringing or metallic sounding artefacts. You can use the Blue Window for gestural control over the effect parameters, but if you're after a natural effect it's usually best to just dial up the appropriate setting, save it as a preset, then treat your file with all the parameter settings kept constant.

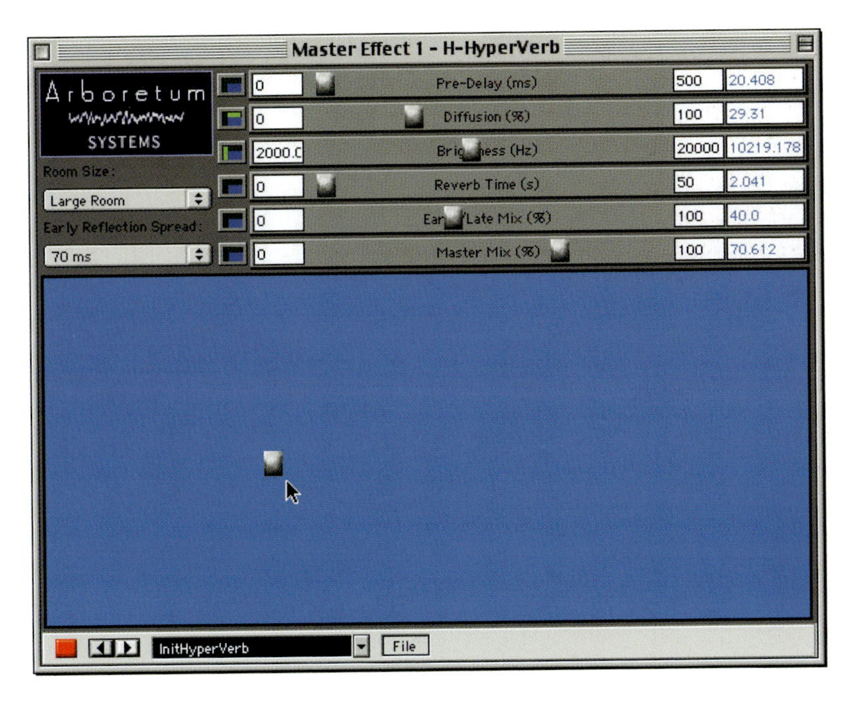

Figure 5.102 Hyperprism HyperVerb.

The PreDelay slider lets you set the delay time - from 0–500 ms - of the first echo after which reverberation begins. The Diffusion slider lets you control how smeared out the individual echoes are. High diffusion will wash the echoes together, creating a lush sound, while zero diffusion gives a bouncy, tight sound. The Brightness slider can be adjusted from 2–20 kHz to apply low pass filtering to the reverberation to create mellow or shimmering soundscapes. The Reverb Time slider is adjustable from 0–50 seconds. For your reference, large cathedrals have reverb times of about 7 seconds, and small rooms have reverb times under half a second. The Early/Late Reflection Mix slider controls the mix of early reflection to late reflections. Late reflections contain more of the washed reverberant sound, early reflections have more of the bouncy initial echoes. This control is called Early/Late in the DirectX (PC/Windows) version of Hyperprism. The Master Mix slider lets you control the balance between the original signal and the reverberation.

To the left of the slider controls, the Room Size popup lets you set the perceived size of the reverberation space - from small closet to large cathedral. Underneath this, the Early Reflection Spread popup selections range from 30–120 ms and control the distance between early reflection times, clustering these within the time you choose.

Note If you're processing a mono source HyperVerb gives you the option of having the output in either mono or stereo. Simply click on the Channel Switch to toggle between the two settings. If you're processing a stereo source your output will be stereo by default; the button will default to the stereo/stereo display and be deactivated from toggling.

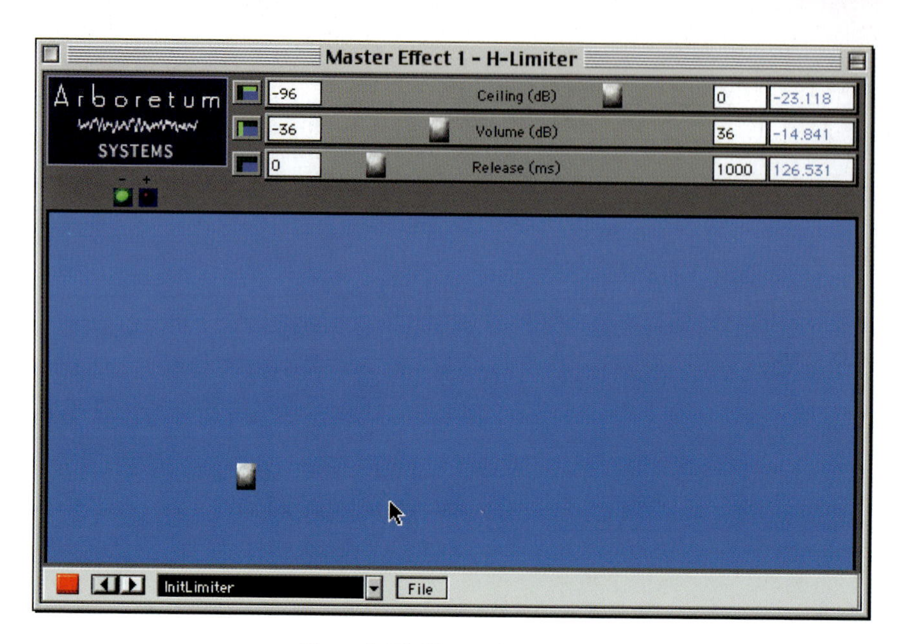

Figure 5.103 Hyperprism Limiter.

The Limiter is extremely simple to use and extremely effective! Adjust the Ceiling parameter to bring in the limiting action, tweak the Release time to suit the audio you are working with, and adjust the output level to suit your needs. It couldn't be easier!

The Low Pass filter progressively removes the higher frequencies as you lower the cut-off frequency. Adjusting the Q factor lets you 'tune in' on the harmonics within the sound.

The MS Matrix lets you decode M-S recordings to standard L-R stereo. M-S refers to either Mid-Side or Mono-Stereo. This classic stereo recording technique has the advantage that one can change the apparent width of the stereo image after the recording session, using a device called an M-S decoder. In the past, an M-S decoder took the form of an external analogue circuit. The M-S Matrix effect

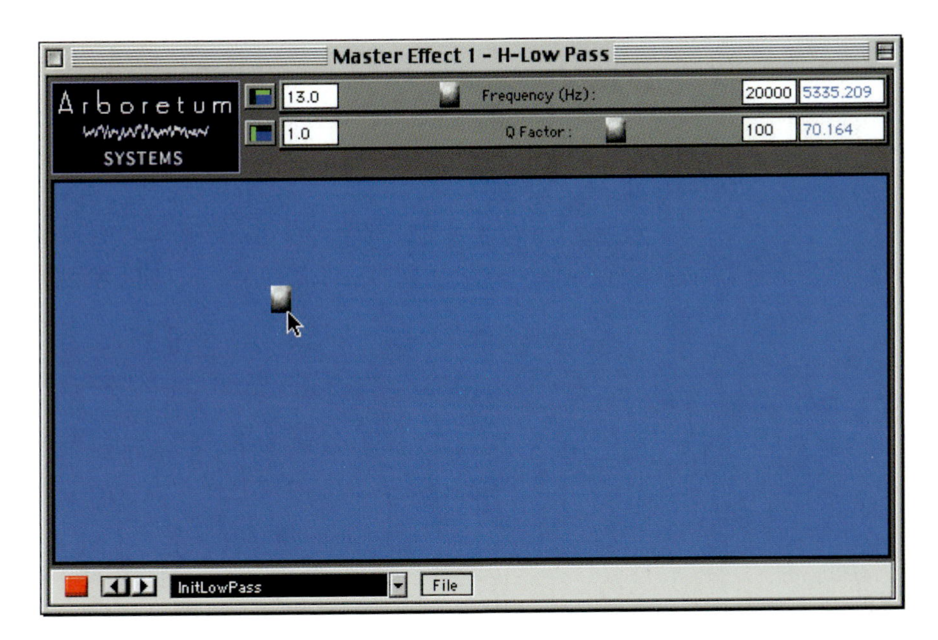

Figure 5.104 Hyperprism Low Pass Filter.

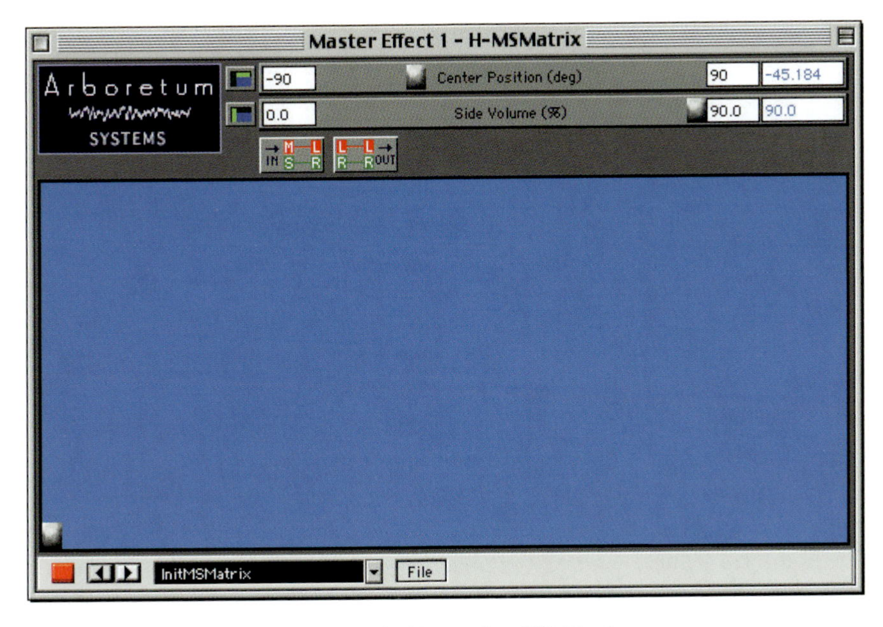

Figure 5.105 Hyperprism MS Matrix.

replaces this circuit, making experimentation with the M-S technique as easy as tracing a line in the Blue Window.

In order to perform M-S decoding you first need to make M-S encoded recordings. This requires two microphones. One of these microphones faces front, toward the audio source. Its polar pattern is usually a cardioid type. The other microphone is a bidirectional figure-eight type. The cardioid microphone captures the middle or mono part of the stereo image. The bidirectional microphone picks up the sides or discrete stereo part of the image. You record the output of these microphones on a two-channel recorder, with one channel called M and the other S, as opposed to the usual left (L) and right (R). If you have an M-S recording, you can connect the M output of your recorder to Channel 1 of your system, and connect the S output to Channel 2 to decode this to L-R stereo using Hyperprism.

Note Hyperprism-VST users should make sure you have your stereo tracks panned 100% left and right or you will get no effect with this process.

The Stereo Position parameter allows displacement of the stereo image in degrees [−90, +90], where the centre 0° position means no change. This can be useful in processing sound tracks recorded on location where the dialog is off-centre, and you want to restore it to the centre position. This parameter is labelled Center Position in Hyperprism for HyperEngine, Hyperprism-MMP and Hyperprism-DAS. The Stereo Position parameter maintains constant power at any position, so that at extreme left or right positions the level is boosted. If this causes distortion you can compensate by lowering the input level by 3 dB in the Settings window. The Side Volume parameter gives you control over the level of the stereo part of the signal. In M-S processing, the higher the stereo level, the wider the apparent stereo image. The default range is 0% to 150%, which corresponds to a range from −96 dB to +6 dB. This parameter's range is from 0 to 100% in Hyperprism for HyperEngine and Hyperprism-MMP. The Input button can be used to switch the input settings between 'Middle Left' and 'Side Left'. By default, the M-S Matrix assumes that the signal going into the Left input jack is the Mid part of the M-S signal. If you have the Side part of the M-S signal routed to your Left input, you can use this button to inform the M-S Matrix that this is the case. The Output button can correct for an out-of-phase microphone configuration. It determines the phase of the M source. M+S in Left is the normal case; the M input sums positively. Click on this button to switch output channels.

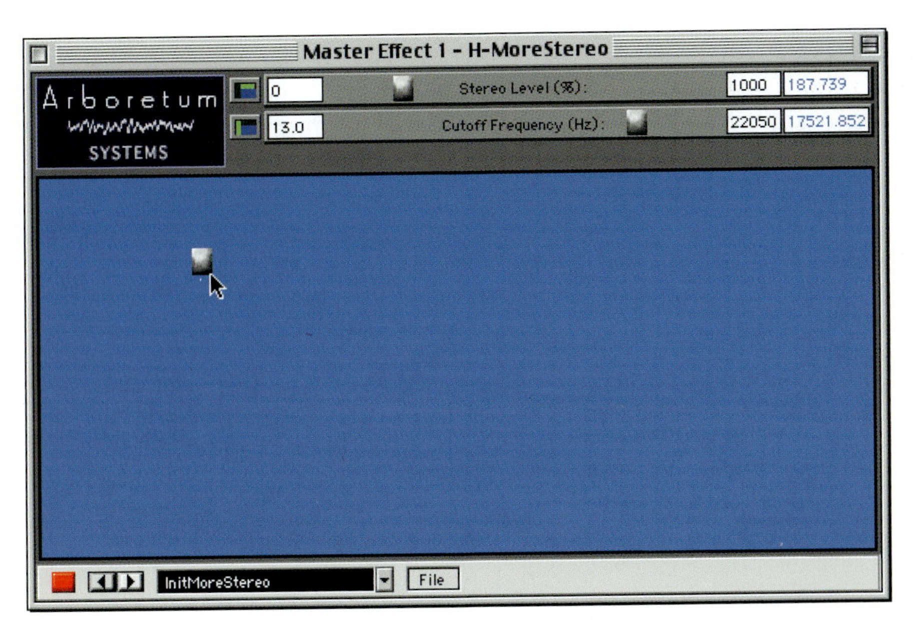

Figure 5.106 Hyperprism MoreStereo.

MoreStereo alters the stereo effect within a stereo file by widening or narrowing the spatial image, without changing the overall volume. It optionally applies a variable cut-off high pass filter to the signal, reducing potential phase-cancellation artefacts in the low-frequency content of the original sound file. When using a file with very different left channel and right channel material, a third image is created in the centre of the stereo field. When using a sound with similar material in both channels the stereo spread is enhanced. More Stereo has no effect on mono input signals.

Note Hyperprism-VST users should make sure you have your stereo tracks panned 100% left and right or you will get no effect with this process.

The Stereo Level slider controls the width of the stereo image. Setting this to zero means mono; 100 means normal stereo (no change to the input); 200 doubles the proportion of the discrete channel information in the output by attenuating the common or mono part of the signal; and 300 triples it. A horizontal value greater than 100 has the effect of widening the stereo image, pushing the sound to the extreme lateral poles. This parameter goes up to 300 in Hyperprism-MMP and up to 1000 in other versions.

The Low Cut parameter, called 'Cutoff Frequency' in Hyperprism-DAS and Hyperprism-VST, controls the cut-off frequency of the 12 dB per octave high pass filter. At the bottom of the Blue Window (a value of 0) the effect is switched off. As you raise the position of the tracer, a high pass filter attenuates low frequencies in the range up to 6000 Hz before they are injected into the widening algorithm. These frequencies will still be present in the final signal. By using the filter, you can control low-frequency artefacts at the expense of a slightly narrower stereo image. This ranges from 13 to 22 050 Hz in Hyperprism for HyperEngine and Hyperprism-VST.

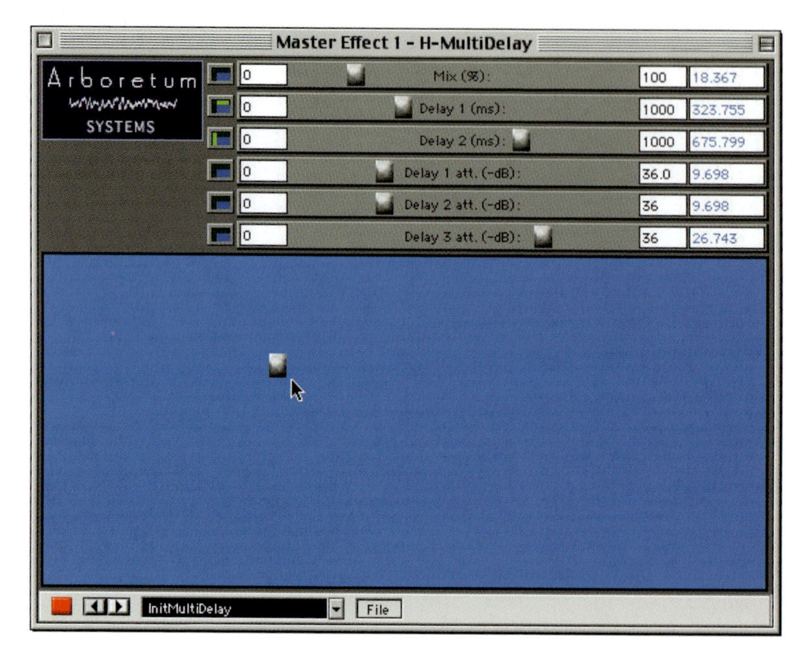

Figure 5.107 Hyperprism MultiDelay.

The MultiDelay delays the input signal three times. The user sets the first two and the program calculates the third Delay time by adding the times of Delays 1 and 2. To get an idea of the kind of effects you can create with this, load the example sound file supplied with the software called 'Gunshot'. This file contains a single shot fired by a hunting rifle. Loop playback and adjust the delay settings and you can make this sound like the gunfight at the OK Corral!

The Noise Gate has three simple controls. Set the Threshold to lie just above the level of any quiet sounds you don't want to hear in your audio, set the Attack time (labelled Open Time in Hyperprism-VST) so that the Gate opens smoothly, and set the Release time (labelled Close time in Hyperprism-VST) so that the gate closes smoothly.

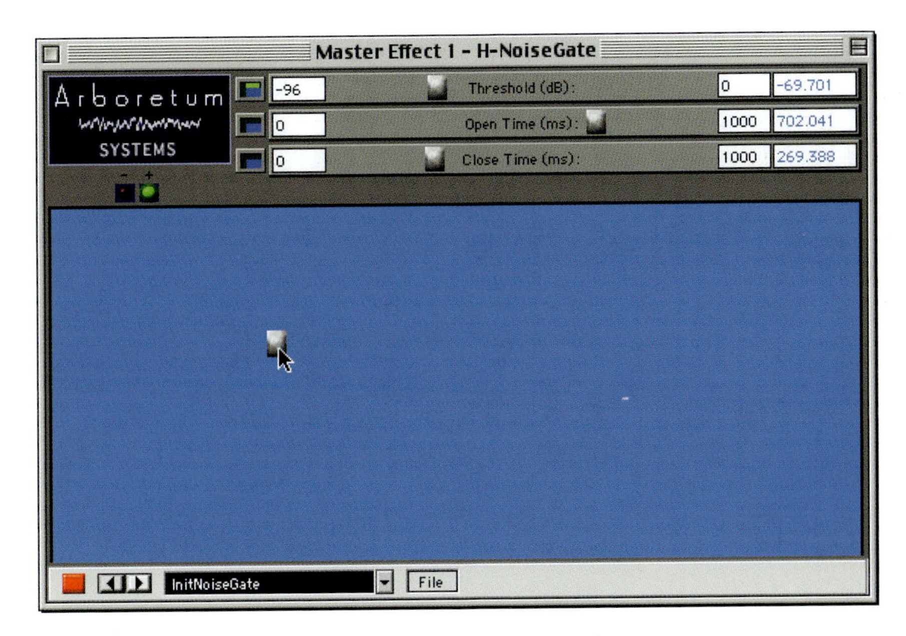

Figure 5.108 Hyperprism Noise Gate.

The Pan effect does just what it says – allowing you to pan a mono source between the left and right channels. A Gain control is also provided.

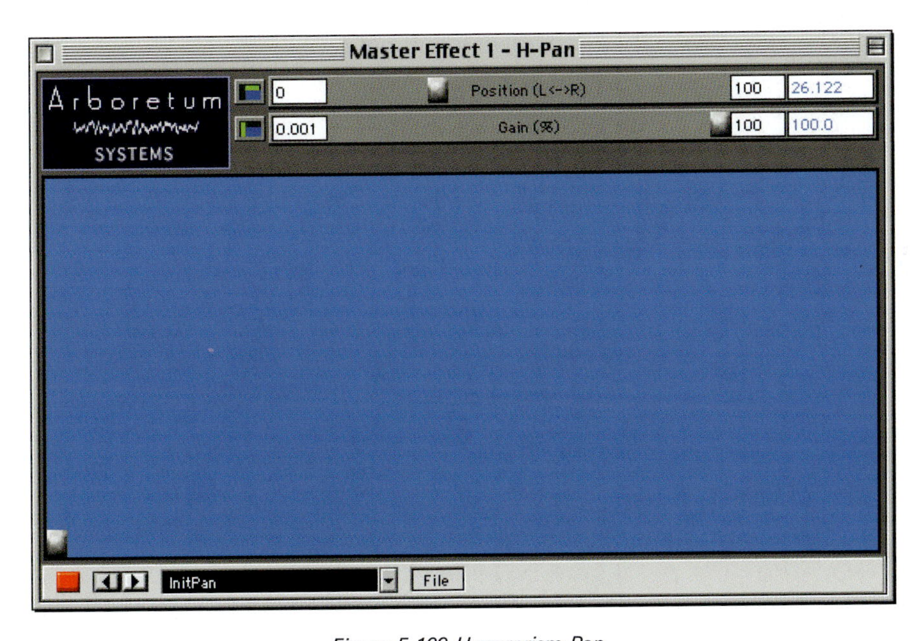

Figure 5.109 Hyperprism Pan.

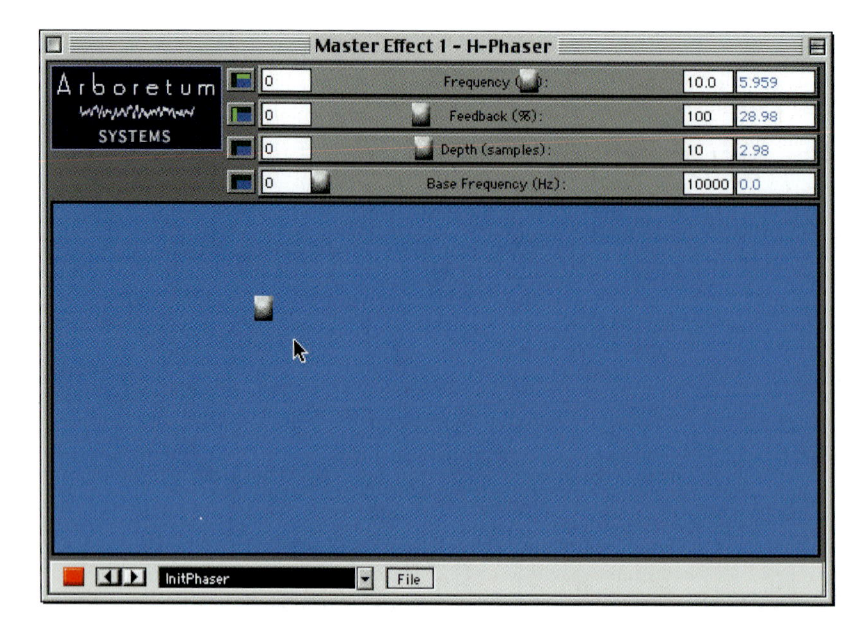

Figure 5.110 Hyperprism Phaser.

The Phaser produces an effect reminiscent of old phase shifter boxes from the late 1970s. Here, an all-pass filter introduces a frequency dependent delay, resulting in phase distortion. A wide range of results can be achieved by adjusting the amount of this distortion and using feedback to increase the intensity of the effect. Frequency is the number of delay cycles per second, from 0 to 10 Hz in Hyperprism-MMP, from 0 to 4 kHz in Hyperprism-DAS and Hyperprism-VST, 0 to 1000 Hz in Hyperprism for HyperEngine. Depth controls the intensity of the phase shift effect, with values ranging from 0 to 10 samples. The Feedback slider introduces regeneration, in which the output of the Phaser is mixed back at its input. A setting of 0% is no feedback, 100% will lead to runaway regeneration, as the full output of the effect is constantly being routed to its own input. The Base Frequency slider sets the lowest frequency of the original sound that will be processed by the Phaser algorithm. Higher settings will leave lower frequency components of the sound relatively unaffected. The range of this setting is from 0 to 10 000 Hz. The Phaser has the controls you need to set up a phasing effect, but you will typically want to automate this effect using your host environment's automation features – unless you are using HyperEngine. The static effect produced can be useful on occasion.

The Pitch Changer lets you mess with pitch to produce some first-rate effects. It also incorporates a 'granulator' effect.

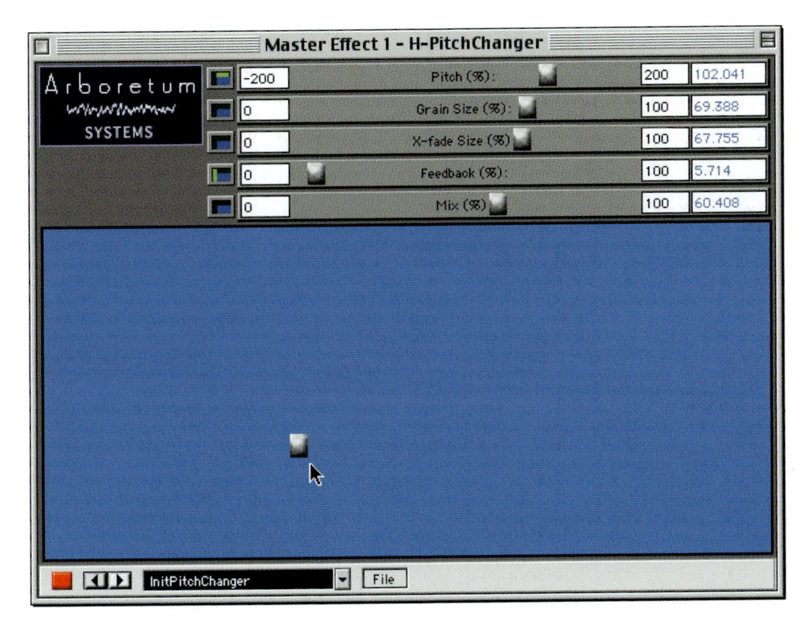

Figure 5.111 Hyperprism Pitch Changer.

Pitch can be varied from −200% to +200%. The way this works is that 0% represents a transposition towards extreme low frequencies and 200% transposes the input sound up an octave. Negative values of pitch-shift cause the pitch of the source signal to be shifted by the same amount as positive percentage values, but the playback direction of the individual sound fragments, or 'grains' are reversed. This can be especially interesting on speech or rhythmic sounds. For example, a pitch shift of −100% results in playing the sound at its original pitch, but broken into small, backwards chunks.

To stretch or shrink the time base of the original sound signal, the algorithm repeats (or skips) small segments, or 'grains' of the original sound. The length of these grains can be varied, resulting in a great difference in sound. The effect can range from echoing to 'granulation' to obliteration of the identity of the input signal. Small grain settings can lead to buzzing sounds, while long ones may be perceived as a doubling, or echo effect. The actual results depend heavily on the character of the original sound. Hyperprism works with a maximum grain size of 1024 samples; this parameter is controlled by a percentage value, ranging from 0 to 100%.

The X-Fade Size slider sets the cross-fade time, from 0% (no cross-fade) to 100% (where the cross-fade lasts the entire length of the grain. Short settings tend to produce audible artefacts at the boundaries set by the Grain size. X-Fades of zero will often click. Longer cross-fades will be smoother, but can lead to more chorusing or doubling sounds.

The Feedback slider introduces regeneration, where the output of the pitch time changer is mixed back at its input. A setting of 0% is no feedback, 100% will lead to runaway regeneration, as the full output of the effect is constantly being routed to its own input. Different feedback settings can result in anything from flanging or resonant effects to cascading showers of pitch-bending echoes as the mouse is moved.

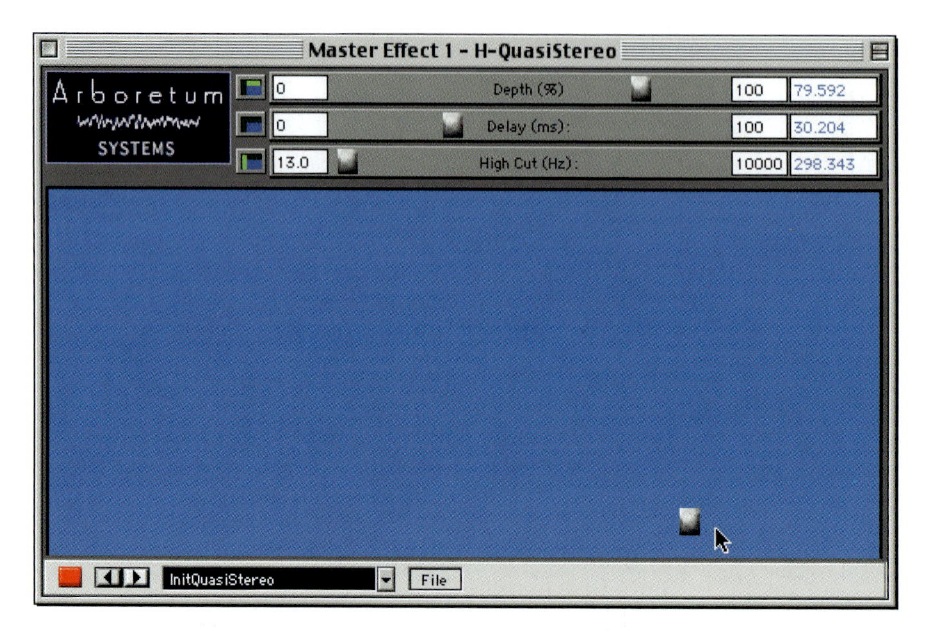

Figure 5.112 Hyperprism Quasi Stereo.

QuasiStereo instantly produces a stereo effect from a mono file or enhances the stereo effect with an existing stereo file.

It divides the mono signal path and applies complementary comb filters to each channel. These filters disperse their outputs to the left and right channels in such a way as to create the impression of a stereo image.

Note Hyperprism-VST users should make sure you have your stereo tracks panned 100% left and right or you will get no effect with this process.

The comb filtering used to create the pseudo-stereo image uses a short delay, 100 ms or less, which you can set using the Delay Time slider, and the Depth slider lets you control the width of the stereo image. The High Cut parameter in Hyperprism-VST, Hyperprism-MMP and Hyperprism-DAS is incorrectly labelled - it should be Low Cut, as it is with the other versions of this plug-in. This parameter sets the cut-off frequency for a 12 dB per octave high pass filter. At a value of zero the filter is switched off. As you raise the value, the high pass filter attenuates low frequencies in the range up to 6000 Hz (up to 10 kHz in Hyperprism-DAS) before they are injected into the Quasi Stereo algorithm. These frequencies will still be present in the final signal. By using the filter, you can reduce low-frequency artefacts at the expense of a slightly narrower stereo image.

Tip For an interesting spatial effect, apply Quasi Stereo to a two-channel stereo input signal. The effect acts on only the mono (common) part of the signal. The discrete stereo channel information is added to the processed pseudo stereo signals to form the output.

Figure 5.113 Hyperprism Ring Modulator.

The Ring Modulator can produce some very unusual distorted timbres, including metallic and bell-like tones. The input sound is used as a carrier signal. A second signal called a modulator frequency, is used to generate two frequencies that are the sum and the difference of the carrier and the modulator frequencies, each at half the amplitude value of the carrier.

The Frequency slider controls the modulator frequency in cycles per second. This ranges from 0 to 20 kHz in Hyperprism for HyperEngine, Hyperprism-VST and Hyperprism-MMP. In Hyperprism-DAS it ranges from 0 to 10 kHz, and is called 'Base Frequency'. A low frequency oscillator can be applied to the modulator, allowing you to create even richer sonic effects. The LFO Frequency slider controls the LFO frequency and ranges from 0 to 1000 Hz in Hyperprism for HyperEngine, Hyperprism-VST and Hyperprism-DAS, and from 0 to 40 Hz in Hyperprism-MMP. LFO Depth controls the amount of LFO applied to the modulator tone. The Mix or Depth control lets you adjust the direct/effect mix between the original sound and the Ring modulator's output. This parameter is called 'Mix' in Hyperprism-VST, Hyperprism-MMP and Hyperprism-DAS and 'Depth' in all other versions.

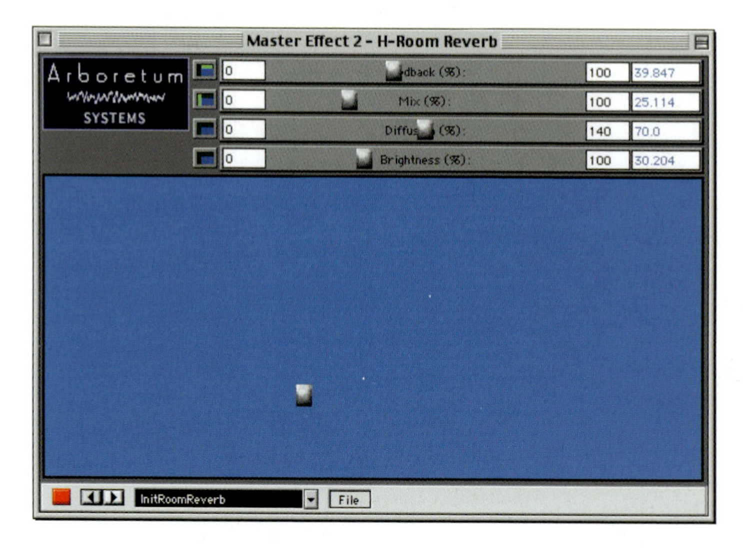

Figure 5.114 Hyperprism Room Reverb.

The Room Reverb gives you what it says in one easy-to-use window with just four main parameters that you can tweak – Feedback, Mix, Diffusion and Brightness. This is a very usable and attractive-sounding reverb. You can drag the button in the blue window to change any of these parameters at the same time – selecting which parameters and which direction by clicking in the boxes to the left of the sliding controls. For example, I set this up to control Feedback and Mix together – moving vertically to control the Mix and moving horizontally to control the Feedback.

The Single Delay gives you a single delay – what more can I say! You can set the delay time and adjust the balance between source and delay signals – and that's it!

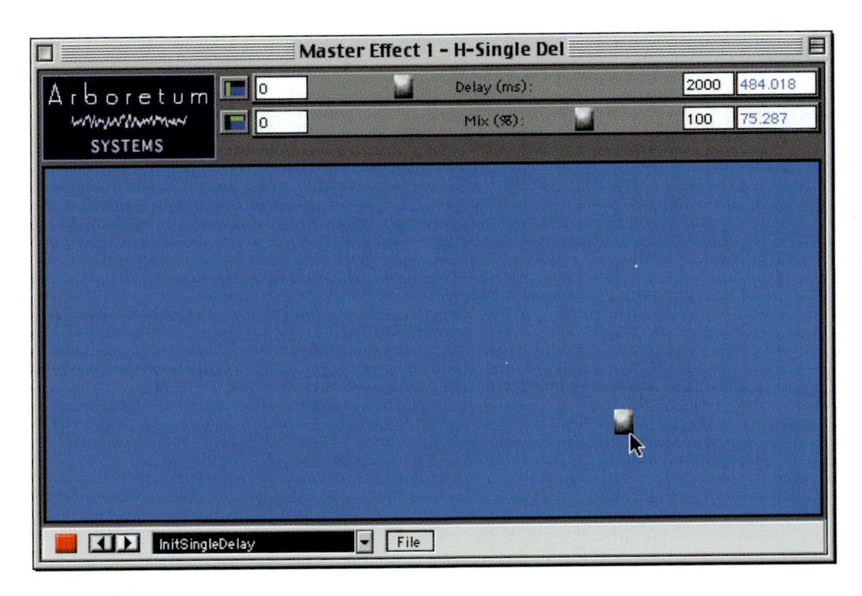

Figure 5.115 Hyperprism Single Delay.

The Sonic Decimator lets you reduce the bit depth and the sample rate, introducing various kinds of hiss and grunge into the sound. Useful if you want to emulate the sound of older digital audio equipment or if, perversely, you like to hear these types of artefacts in your sound.

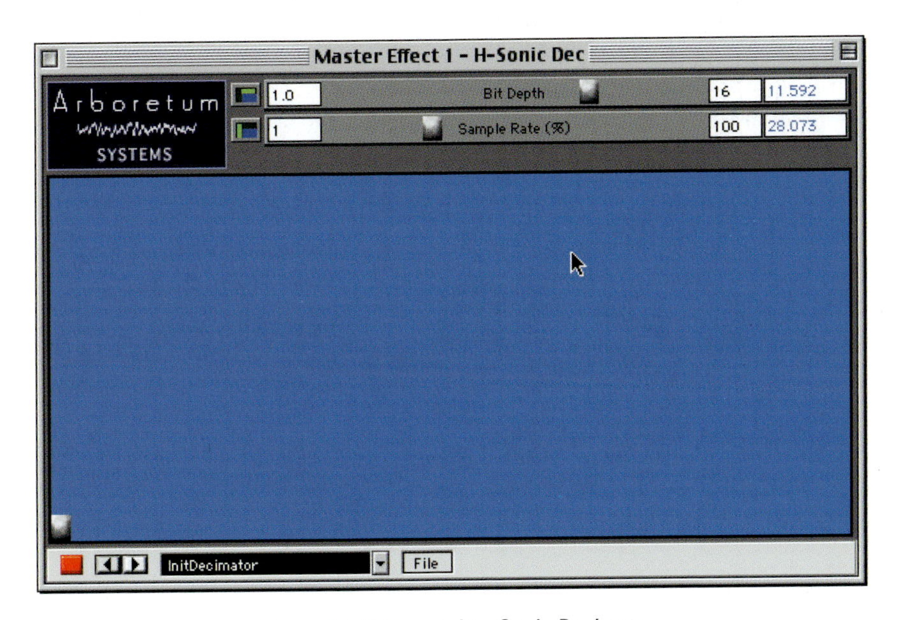

Figure 5.116 Hyperprism Sonic Decimator.

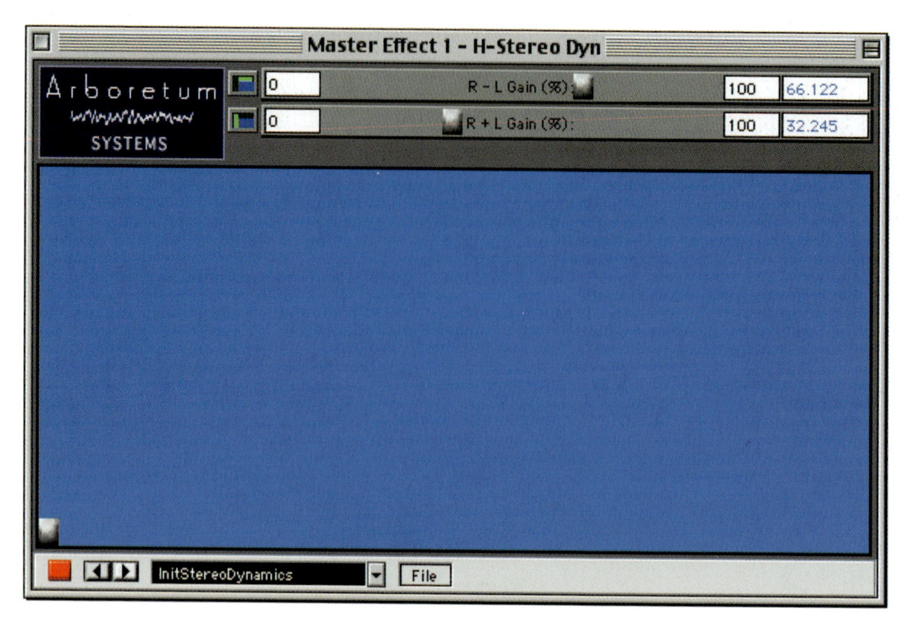

Figure 5.117 Hyperprism Stereo Dynamics.

Stereo Dynamics spatializes the input sound, positioning it in a virtual space situated in front of the listener. It moves the input sound from front to back and from left to right, depending on the position of the tracer in the Blue Window. This effect transforms a monaural input signal into a stereo output signal. Using the Process to New File . . . command creates a stereo sound file as the output. The 'Right minus Left Gain' parameter controls the proportion of one channel relative to the other. For example, moving the tracer to the right will increase the proportion of the right channel in the output. The 'Right plus Left Gain' parameter controls the overall volume level of both channels.

Note Hyperprism-VST users should make sure you have your stereo tracks panned 100% left and right or you will get no effect with this process.

The Tremolo plug-in does exactly what it says – producing a tremolo effect by varying the amplitude of the sound at a rate set by the Frequency control and by an amount set by the Depth control.

The Saturation plug-in offers two types of tube saturation and two types of tape saturation. Use these when you want to 'dirty' your sound.

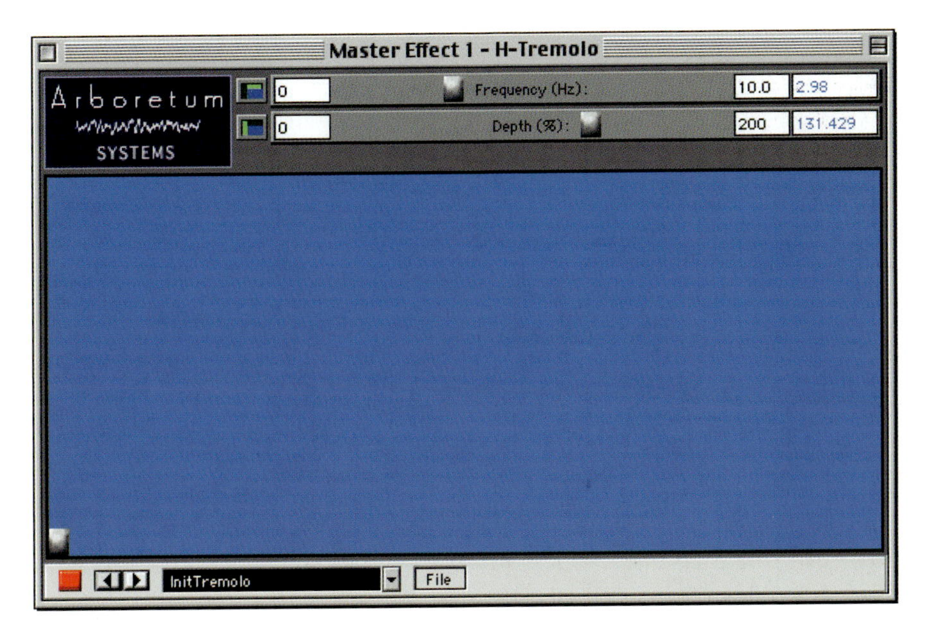

Figure 5.118 Hyperprism Tremolo.

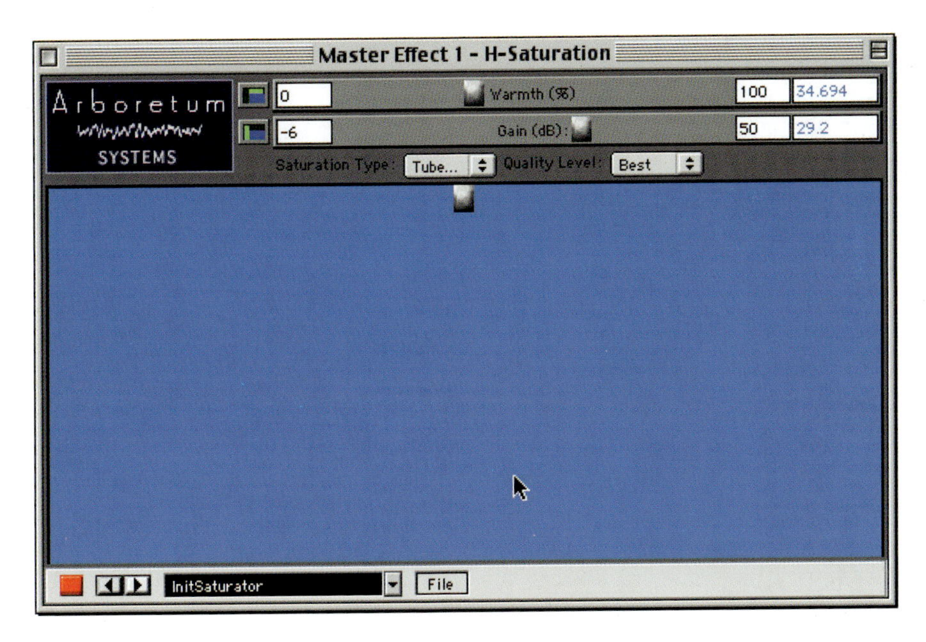

Figure 5.119 Hyperprism Saturation.

Figure 5.120 Hyperprism Vibrato.

The Vibrato plug-in does exactly what it says - producing a vibrato effect by varying the pitch of the sound at a rate set by the Frequency control and by an amount set by the Depth control.

Vocoding was first heard as a vocal effect on Wendy Carlos and Kraftwerk recordings and is still in widespread use on the latest films and hit records. You can use the vocoder to create talking guitars or merge synth pads with drum samples to create intense new loops.

The Vocoding process uses one signal to create a set of filters that are applied to a second signal. The signal from which the filters are set is called the modulator; the signal that gets filtered (and is heard) is called the carrier. The Vocoder looks at the modulator signal and adjusts its 26 filter bands to match the spectrum of the incoming signal. It then applies these filters to the carrier signal. Remember that the centre frequency of each of the 26 filters is fixed; the modulating signal is setting the gain for each of the filters. And either signal can be the modulator or carrier, to varying degrees.

The Modulator slider lets you control which sound is the source and which is the modulator. Zero % means the left channel modulates the right, or if using an external sound file, that the internal left channel modulates the external source.

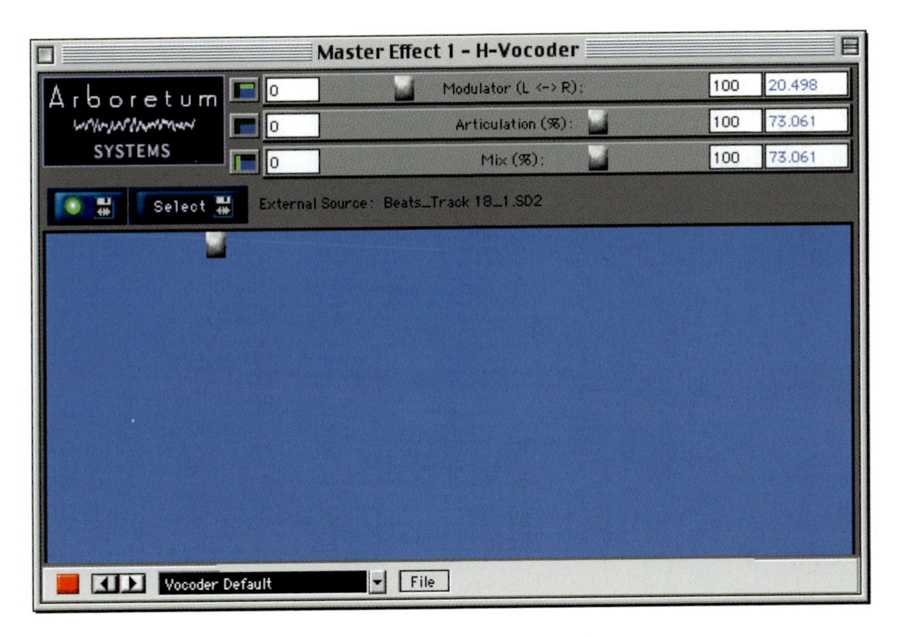

Figure 5.121 Hyperprism Vocoder.

When set to 100%, the right channel modulates the left, or the external sound modulates the internal source.

The Articulation slider controls how quickly changes in the modulation sound are applied to the source sound. When set at 100%, fast changes are applied, when at 0% no changes are applied. Turn the articulation down to smooth out the vocoded sound, or turn it up to make the sound sharper, and speech more intelligible.

With Hyperprism for HyperEngine or Hyperprism-MMP you can vocode mono sources, selecting any other sound file to be the modulator or source for vocoding. In Hyperprism for HyperEngine you can modulate any live signal with a sound file, or even vocode two live sources against each other by modulating the left channel against the right channel. In the AudioSuite, VST and DirectX editions you vocode the left channel against the right channel.

Note Hyperprism-VST users should make sure you have your stereo tracks panned 100% left and right or you will get no effect with this process.

Realizer Pro

Arboretum Realizer Pro is a cost-effective VST or Direct X plug-in that you can use to quickly prepare your music for MP3 or streaming formats for Internet delivery. The MP3 encoding process always causes a loss of quality and you can use Realizer Pro, based on Arboretum's Hyperprism effects, to compensate for this prior to encoding.

The Bass Maximizer brings back that full, round bass to compensate for the shortcomings of small computer speakers. It does this by synthesizing new partials from the existing low-end frequencies to create a psychoacoustic impression of more low-end than your speakers can actually produce. The two Mid-Range EQ sliders let you tweak the mid-range of the audio spectrum to bring out vocals or other sounds. The Harmonic Exciter can be used to restore the high-end sparkle and lustre that gets washed out by compression. The Harmonic Exciter, like the Bass Maximizer creates new harmonics, only in the higher registers. The result is clearer vocals, sharper percussion and more highs.

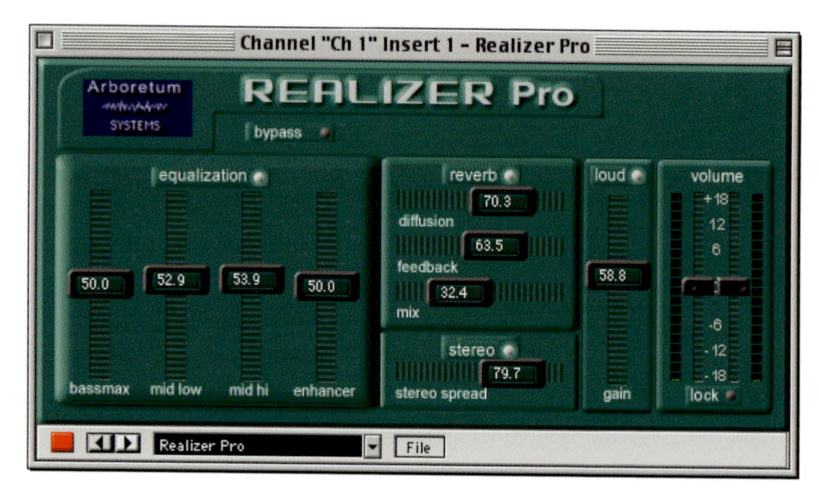

Figure 5.122 Arboretum Realizer Pro.

The Reverb section simulates room acoustics. It's simple, flexible and effective, with controls for diffusion, feedback and mix. Use this to add a touch of reverb if your sound is too 'dry'. The Stereo effect creates a dramatic widening of the stereo image, beyond the limits of your speakers, while the Loud effect solves the problem of uneven levels – bringing up the level of the quieter portions of the audio to match the loudest portions so everything sounds loud and clear. Finally, the Volume section lets you adjust the left and right output levels to compensate for the other effects.

Ray Gun

Ray Gun is Arboretum's low-cost vinyl and tape restoration utility for Mac OS and Windows. Ray Gun lets you clean up old vinyl and tape recordings by removing hum, hiss, pops, clicks and other unwanted audio, leaving your original sound intact. Designed especially for audio archiving and restoration, Ray Gun is perfect for removing the nastiest pops, clicks and scratches when converting vinyl collections to CD-R. You can also use Ray Gun for cleaning up noisy dialogue or telephone recordings, for eliminating camera sound and other noises from video or film soundtracks, and for filtering out line noise and earth loop hums. Combining sophisticated pop and click detection with complex spectral analysis, downward expansion, notch filtering and an intelligent 'search' function, Ray Gun is like a sci-fi device: small, fast, fun to use and packed with effective technology.

Using its 'Intelligent Search Function', Ray Gun finds and eliminates noise, all by itself. If necessary, you can fine-tune the processing using the Attenuation, Threshold and Sensitivity controls. The Noise Reduction section at the left of the plug-in's window uses fast spectral analysis, a downward expansion function and special 'search-and-destroy' noise-busting technology to reduce or eliminate broadband noise, including fan sound and tape hiss. To the right of this, the Pop removal section finds the spikes, clicks and pops in your file, and seamlessly removes them from your audio. To the right again, the Filters are designed specifically to remove rumble or 50 or 60 Hz hum. Finally, the Output section at the far right lets you boost the level to make up for any gain lost during the processing.

Ray Gun is dead simple to use. Just listen to your audio playing through Ray Gun as you turn on the Noise Reduction, the Pop/Click removal or the Hum/Rumble filtering. Start with Threshold, Attenuation and Sensitivity settings at zero. Adjust these settings to fine-tune the results, then apply the process to create a new noise-free sound file.

Tip Play live audio signals into your Macintosh and instantly zap the noise with Ray Gun, while routing the signal back out to your tape deck or other recording device.

Windows users can plug Ray Gun into any DirectX-compatible sound editing program. PowerMac users get AudioSuite, VST and Premiere plug-in format compatibility, allowing you to use Ray Gun with the full range of Mac audio

Tip You can also select the 'Don't Preserve Formant' option to use an alternate pitch-shifting algorithm that does not preserve the natural formant during pitch shifting. Although the results can be less realistic on large pitch shifts, this setting is quite useful for special effects. And for pitch correction tasks that require only small amounts of shifting, the 'Don't Preserve Formant' option can actually yield better results.

a Pitch Variation percentage to tell the pitch detector how near subsequent pitches in a single note need to be. With this information the pitch detector can spot wrong guesses and correct them even if the pitch is far from an expected value. The Pitch Sensitivity parameter helps the pitch detector to decide if a harmonic is the true pitch or if it is a super/subharmonic and the Correction Strength parameter determines how closely an out-of-tune part is brought into tune. At low settings out-of-tune parts are corrected more gently, so the singer's subtle variations and vibrato are preserved. The Global Formant and Pitch Shift controls are similar to those in the Mixer and Voices displays, but apply to all the harmony voices.

Harmony will generate up to eight individual harmony parts and will render the process to a mono or stereo SDII or AIFF file. Up to nine part arrangements can be made if you include the original input voice.

Harmony tutorial

Arboretum Harmony runs within Arboretum's HyperEngine and all basic file operations, such as opening, selecting and rendering, and all playback/record features are accessed through HyperEngine.

To get started, launch HyperEngine and open a sound file containing a recording of a single voice.

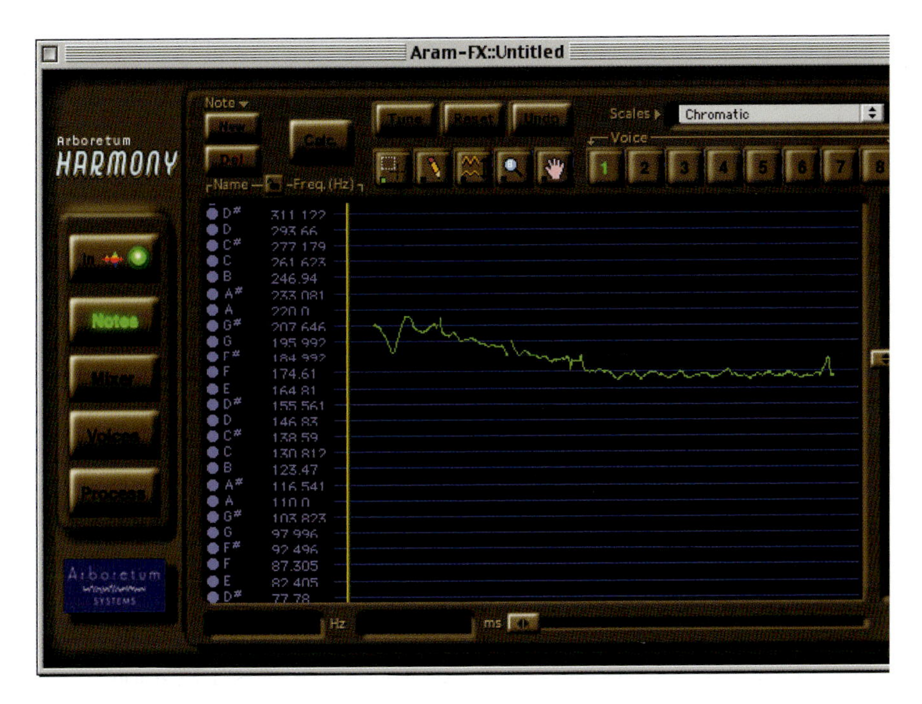

Figure 5.124 Arboretum Harmony Notes display.

When the audio appears onscreen, select a region for processing and then select Harmony from HyperEngine's plug-ins menu. The Harmony plug-in will launch with the Notes display showing by default.

Pitch Correction is an automated feature in Harmony and several settings are provided under the Process display to guide that operation. Pitch Correction relies on the Scale shown in the Notes display. The displayed Scale can be easily recalibrated. A selection of preset Scales are provided and custom user Scales can also be created.

The Notes display arranges pitch/frequency along the vertical axis and time along the horizontal axis. At the left side of the display there are calibration markings showing note names and frequencies. These are completely user-configurable; click and hold on any frequency number then mouse drag up or down to adjust the note frequency in Hz. Mouse click and type directly on the note name field to rename any note. You can transpose the scale relative to an edited frequency by holding down the shift key when you hit the tab key to enter a new frequency. To manually recalibrate any note, simply mouse-click on the dot beside the note name, and hold down the mouse while dragging the note up or down the scale. The Frequency display will update to reflect the current setting. The

Tools palette above the Notes Display contains a Selection, Pencil, Resize, Zoom and Hand tools. The Resize tool, for example, re-scales the selected pitch sequence around its pitch centre. You can use this to accentuate or limit vibrato. It can also be applied as an arranging tool for a complete re-mapping of the part's range.

You can correct the pitch of the notes using Harmony's Tune function. Click on the Calc button above the Notes display and after a brief calculation period you'll see a graph representing the pitch of your audio selection. This line shows the pitch of the input melody over time, and cannot be edited. If it's not already selected, click on the Voice button labelled '1' above the edit display to activate the first harmony, then type the keyboard shortcut Command-a to select the entire sequence. The pitch sequence for Voice 1 will become highlighted in gold in the edit display. Now when you click on the Tune button this will 'snap' the harmony voice into tune with the current scale. You can either hit the Enter key or the Space Bar to start playback and hear your results.

To balance the levels of the harmony voice or voices against that of the original, open the Mixer Window by clicking on the Mixer button at the left of the plug-in's window.

Figure 5.125 Arboretum Harmony Mixer window.

Nine mixer channels are provided. The original vocal line is in mixer channel '0', at the left, and the pitch-corrected version is on channel 1. Mute buttons let you turn each channel on and off as required and Solo buttons let you listen to one or more channels while setting up. Pan, Pitch and Formant controls are also provided for each channel. A pair of master output level controls are provided at the right of the window.

Editing the pitches in the Notes Display is easy enough once you get the hang of it. From the palette of controls just above the Notes display, choose the Selection Tool – the one immediately to the left of the Pencil Tool, so that its button lights up green. Drag the mouse into the Notes display, without clicking or holding the mouse button, and watch the Selection cursor change from a crosshair (+) to a pointing hand as it crosses over the gold highlighted pitch sequence. Now, using the hand cursor, click on any part of the yellow line and hold down the mouse button to drag the harmony voice to a new location along the scale, placing it, say, a fifth above the original line and audition your results.

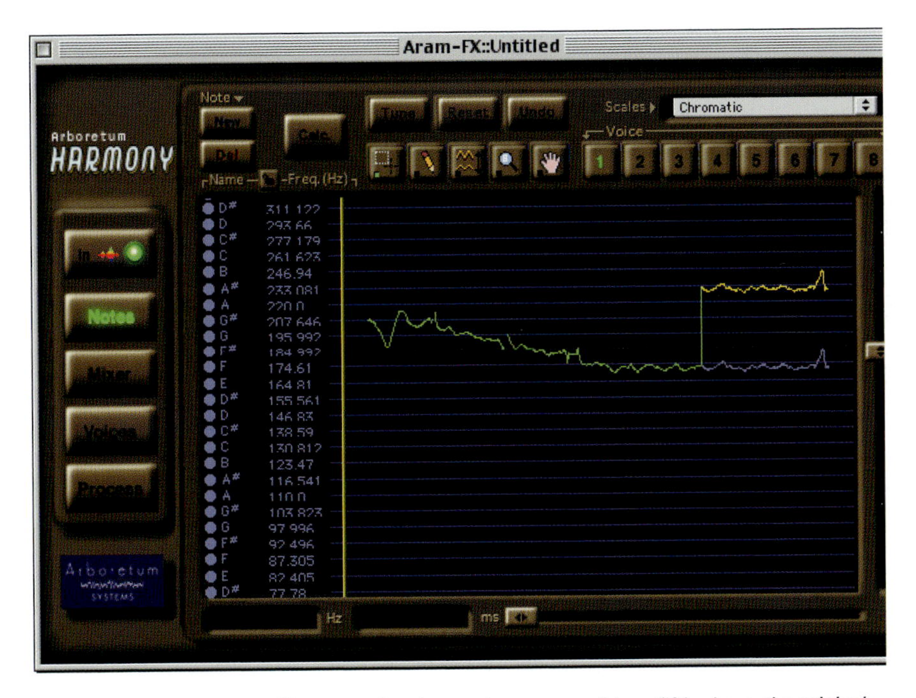

Figure 5.126 Arboretum Harmony showing a phrase moved to a fifth above the original.

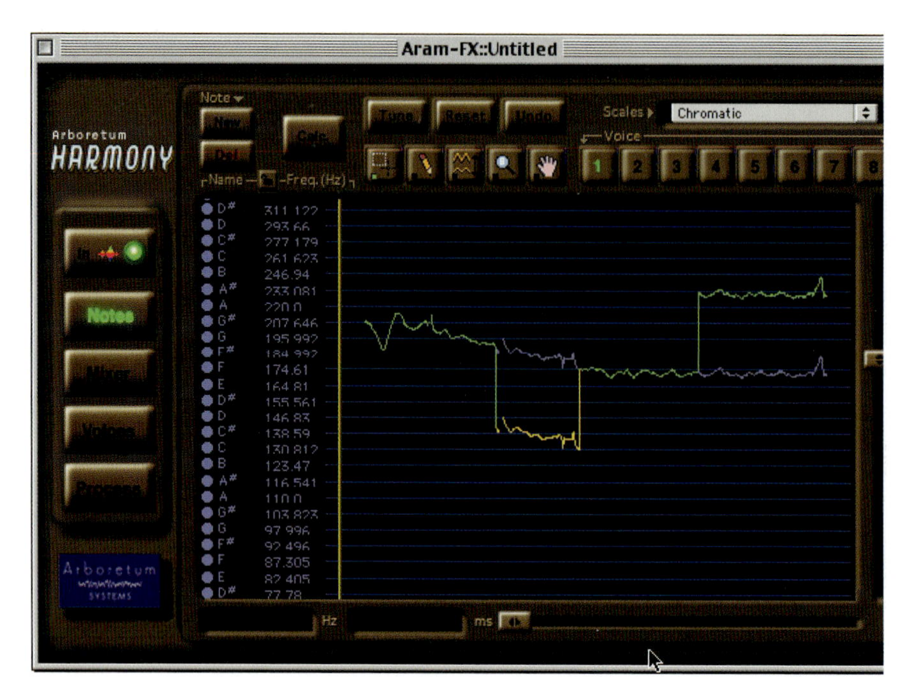

Figure 5.127 Arboretum Harmony showing a phrase moved to a fifth below the original.

Now click and drag across any part of the pitch sequence, highlighting the selected range in yellow, to work with another phrase. Move the mouse cursor over the selection so that it turns to a pointing hand. Hold down the mouse button and drag the phrase to a new pitch, perhaps a fifth below the original, and audition your results.

You can draw in new notes using the Pencil Tool. When you click on the Pencil Tool its green indicator light turns on and the mouse cursor changes into a pencil as you drag it into the edit display. Just click and drag with the mouse to draw new notes for the current voice. Play back the results and if you don't like what you hear, simply redraw the notes.

Of course, what Harmony is mostly used for is adding extra harmony voices. To add a voice, click on the next Voice button, which would be button 2 if you are

Tip If you mess up your edits and want to start over again, just type Command-a to select all and then click on the Reset button above the Tools Palette.

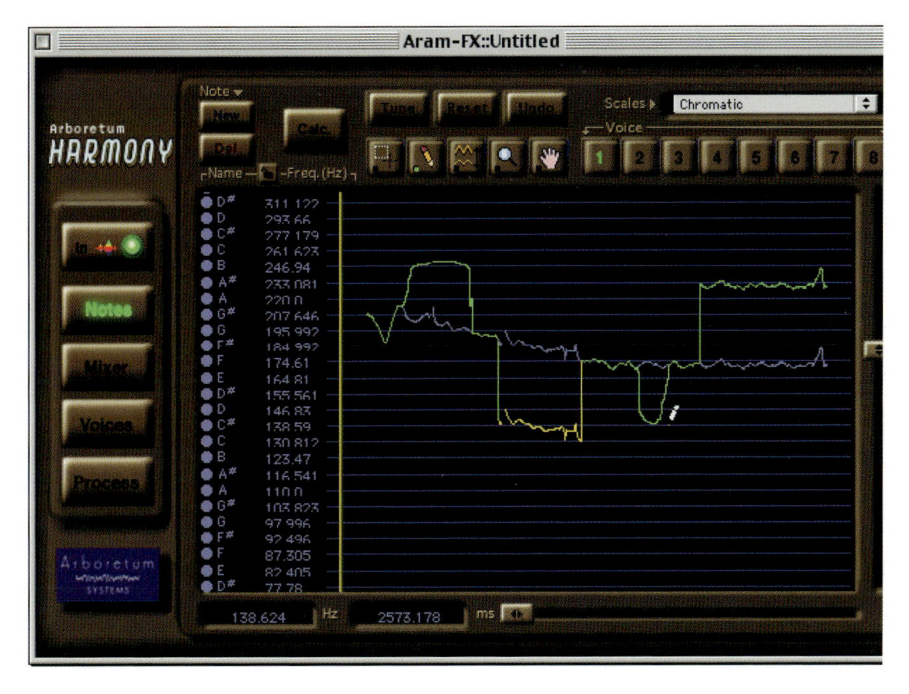

Figure 5.128 Arboretum Harmony showing new notes drawn for the current voice.

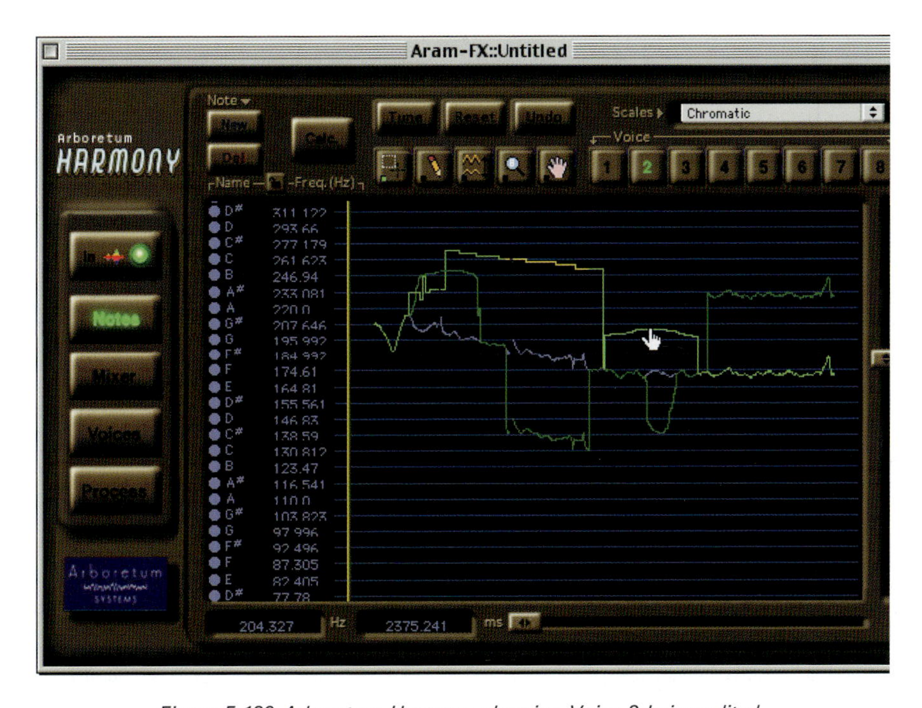

Figure 5.129 Arboretum Harmony showing Voice 2 being edited.

following this example. Now you can move a copy of the original pitches around to create a second harmony.

Tip Voices must be un-muted in the Mixer in order to be heard and edited in the Note Display, so don't forget that you will need to go to the Mixer Window and un-mute Channel 2 before you will hear this voice.

Note If playback begins cutting out or slows down after adding a new voice, you have exceeded your CPU's capacity and will need to scale back the number of real-time voices using the Mixer window to mute one or more harmony voices.

Once you're satisfied with your pitch arrangement, use the Mixer window to adjust the relative levels for each voice. To create a finished audio file, select Process to New File from the HyperEngine's File menu. A name file dialog will appear, followed by a brief wait as the harmonies are calculated into the new file.

To hear your newly processed file, type Command-n (New File Document) and select the audio file you've just created. The file will load into HyperEngine for playback.

Ionizer

Arboretum's Ionizer can be used to carry out a wide range of useful processing tasks, including noise reduction, frequency morphing, vocoding, graphic EQ, multi-band compression, expansion and EQ, frequency-dependent keying/ducking, and specialized sound design.

The Noise Reduction process analyses the noise in your signal, or uses the automatic noise-seeking feature, to remove hiss, hum, camera sound and other steady-state broadband noise. This works in real time and leaves the original music or sound intact. Ionizer's noise reduction includes special controls that eliminate the chirps, flanging and ringing artefacts associated with other noise reduction software.

The Frequency Morphing function allows you to apply the spectral characteristics of any sound to any other piece of audio. Analyse the mix of one any your favourite recordings at the touch of a button, then use the Ionizer Frequency Morph function to create filter curves that will give other recordings those same overall EQ characteristics. Or apply the sonic 'fingerprints' of instrumental samples, ambiences, sound effects or voices to any other audio source during sound design, mixing or mastering. Ideal for use in ADR work, the Frequency Morphing can make looped dialogue sound like it was recorded on set or on location!

Ionizer can also be used as a sophisticated 512-Band Graphic EQ in which the filters don't suffer from the phase distortion that occurs in most other analogue and digital equalizers. Ionizer incorporates a multi-band dynamics processor that offers fast spectral analysis and 512 bands of gated EQ per channel. It can be used for dynamic noise reduction, equalization, limiting, compression and expansion – and all at 32-bit floating-point internal precision. So, for instance, you could limit the low end, expand the mid-range and de-ess the high end all in a single pass. You can use the keying feature to map one signal's spectral profile to another signal in real time – to lock a bass line to a kick drum, for example. The keying feature also allows you to perform frequency dependent ducking for mixing voice with music. Amazingly, the Ionizer can also act as a 512-band vocoder. This is made possible by using the Keying option with the 512 filter bands. This vocoder sounds similar to a classic analogue vocoder and has the advantage of the much greater level of control that digital systems provide.

All in all, the potential for creative sound design is simply tremendous. You can use up to 512 bands of gated EQ to create extreme band pass filters to pull out individual harmonics, or create 'brick wall' filters to completely remove particular frequencies. And you can use the threshold settings to create incredible dynamic filters.

Ionizer tutorial

The user interface is fairly complex, but makes more sense after some explanation of the features.

Situated along the top part of the Ionizer window are three sliders that control the Correlation, Attack and Release settings for noise reduction applications. Careful setting of these controls will enable you to minimize artefacts such as chirping, flanging and squishing. Perhaps the best way to approach noise reduction is to analyse a 'noise-only' section of your recording first to get a 'noise signature'. Then the software knows exactly what to extract from the recording to remove the noise.

To obtain the noise signature, simply select a portion of the audio that contains only noise, such as an otherwise silent region at the beginning or the end of a recording. Use the Spectrum button, located at the far left underneath the control sliders, to calculate the spectral content of the selected portion of audio. After calculating the noise signature, Ionizer will display the Frequency Profile – a black curve indicating the gain at each EQ band in your original audio. Clicking on the Fit button 'tells' Ionizer to fit the Blue and Red curves to the noise signature.

Make certain that the black gain curve is visible by clicking the Black Curve display button, then click in the black circle of the Black Curve and drag it to about –24 dB. This will produce a significant reduction in the noise level of the signal.

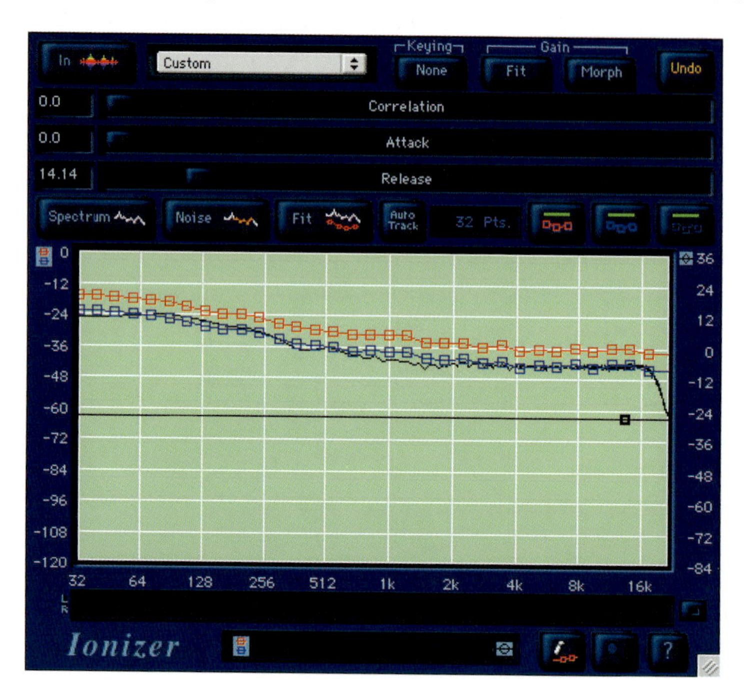

Figure 5.130 Arboretum Ionizer set up for –24 dB noise reduction.

> **Note** Gain reductions of 12 to 24 dB are often enough to push the noise floor below the perceptible level without inducing any noticeable artefacts.

Now let's explore Ionizer's automatic Noise calculation function. The Noise button performs the following steps: It looks at your audio file and makes an 'educated guess' about where the noise component of your signal is. Then it performs a spectral analysis of the presumed noise and displays the frequency profile. This eliminates several steps from the noise reduction process and is especially helpful when you are working with audio files that don't have enough 'silence' in them to get an accurate noise floor fingerprint with the Spectrum function. Beyond that, Noise works much like the Spectrum function described above, performing a spectral analysis on the derived noise signature, then drawing a black 'frequency profile' envelope on the grid. The frequency profile is a graphic representation of the noise's energy within each EQ band. It serves as a visual display of your noise's spectral content, and is a guide for the Fit tool.

Ionizer's Automatic Tracking feature can also be used during noise reduction to ride the noise level in your signal. When the Auto Track button is engaged, the Ionizer thresholds will be raised and lowered to match the changing amplitude of the noise signature, resulting in much smoother, more pleasing results.

The Keying function allows you to extract the spectral and dynamic content from one sound and use it to manipulate another sound. This mapping of spectral envelope characteristics from one sound to another can create a great number of unusual effects, including classic Vocoder sounds. The default mode for Keying is 'None.' In this mode, sound files are processed normally by Ionizer. The 'L to R' keying mode is used when you want to apply the spectral and dynamic content of the sound in the left channel to the sound in the right channel. The 'R to L' mode does the opposite - applying the spectral and dynamic content of the sound in the right channel to the sound in the left channel. Keying only works in stereo mode. The Keying 'source' sound must be present in one channel, and the Keying 'target' sound must be present in the other channel. This means that your final processed 'target' sound will be mono (with the same sound in both channels) since you can't process two channels at once here. When the 'Keying' option is activated and you are processing a stereo signal, Ionizer uses the gain calculations from one channel and applies them to the other channel.

Note By altering the Threshold and Gain curves and adjusting the Correlation, Attack and Release sliders, you can create a wide range of Vocoding effects.

The Morph button is the key to one of Ionizer's most powerful functions - Frequency Morphing, i.e. mapping the spectral profile of one piece of audio to any other, or, put very simply, 'EQ Theft'. Frequency Morphing at its most basic level is just a fancy way of adjusting the boost and cut on 512 different EQ bands so that one sound's EQ characteristics will mimic another's. For example, a mastering engineer could use Ionizer to analyse the EQ used on a particular track or album, then apply this to any of the mixes he is working on. A Sound Designer could make one voice or instrument mimic the frequency characteristics of another. Similarly, an ADR/Foley engineer could analyse the sound from the set or location, then apply those characteristics to looped dialogue or effects.

To use the Morph feature, you need to choose a 'source' sound that will be analysed, and a 'target' sound that will be morphed into the frequency profile of the source. Load the source sound and select the range you wish to use as the model for morphing, then open the Ionizer plug-in. Place Ionizer into Bypass then click on the Spectrum button to calculate the frequency profile of the source audio. If you are using AudioSuite, click on Spectrum, then Analyze. If you are using MAS, click on Spectrum, begin audio playback, then click Spectrum off at the end of the selection. Move the Red and Blue Curves to the bottom of the screen, with Blue above Red, so that Ionizer is in a graphic EQ configuration. Either save your current Ionizer settings as a Preset, or simply close Ionizer – the frequency profile of the source will be retained either way.

Now you can use this profile to process your 'target' file. Load your 'target' audio and select the entire range you wish to be processed. When you re-open the Ionizer the frequency profile of the source should still be displayed. Hit the Morph button. In AudioSuite, hit Morph, then Analyze and in MAS, begin playthrough, click on the Morph button, then click Morph off at end of selection. After a brief calculating period a new Black Gain Curve will appear which shows the amounts of boost and cut applied to Morph the target audio to the source's frequency profile to make the target sound like the source. Preview the result, switching Bypass in and out to compare the target's original sound with the new Morphed version. If you're happy with the result, hit the Process button. If you are working in HyperEngine, select Process to New File from the File menu, or save the settings as a preset. If you are working in MAS, select Bounce to Disk from the Audio menu or hit the Apply button if using Ionizer-MAS in offline mode from the Audio/Plug-Ins menu.

Other software is available that aims to provide this type of functionality. But because of the sheer number of bands used, Ionizer's high bit-depth internal precision, the phase-linear nature of the EQ filters and Arboretum's proprietary means for reducing artefacts, few, if any, other software solutions can match Ionizer's spectral remapping capabilities and sonic accuracy.

Compatibility

Ionizer works with any MacOS audio or video editing software that supports either the MOTU Audio System (MAS), Adobe Premiere or Digidesign AudioSuite audio plug-in formats. Host software includes MOTU Digital Performer, Bias Peak and Deck, Emagic Logic Audio, Cakewalk Metro, Gallery TurboMorph, Adobe Premiere, Opcode Vision, StudioVision and Studio Vision Pro. MAS compatibility means you can run in real time within Mark of the Unicorn Performer, Digital Performer and Audio Desk. AudioSuite compatibility allows Ionizer to be used with Pro Tools, Logic Audio, Avid Media Composer 7.0 and Avid Xpress 2.0.

Ionizer will also run as a stand-alone application on any Power Macintosh through Arboretum's HyperEngine shell. HyperEngine advantages include unlimited-length disk-based preview, the ability to run multiple Ionizer windows concurrently for complex multi-effects, and running Ionizer in play-through mode for processing external signals in real time, much like an outboard hardware effect device.

On the Windows platform, Ionizer-DX operates as a plug-in effect for Direct-X audio applications such as Sound Forge and Cool Edit Pro. Ionizer-DX works requires a Pentium Computer, 233MHZ or higher, Windows 95/98/NT/ME/2K and a Windows-compatible audio input/output card. Ionizer is compatible with the WAV audio file format, at 44.1 and 48 kHz sample rates.

Chapter summary

The Sony Oxford EQ plug-in with the George Massenburg option is an essential buy for all professional users, and I would strongly recommend getting Cool Stuff Labs Generator X - to make sure you always have access to test tones - and Metric Halo's SpectraFoo - to provide analysis and metering. Professional users will also value the MacDSP range of compressors, filters and analogue channel emulators. Apogee MasterTools will be useful if you plan on mastering using your Pro Tools system - but many professional users will pass their mixes to a professional mastering engineer who is more likely to be using Sonic or Sadie equipment.

For those of you with limited numbers of DSP cards installed with your TDM system, Metric Halo's Channel Strip makes a lot of sense as it contains all the most popular channel processing in one plug-in - conserving DSP resources. And if you need to do any time-stretching, you will quickly realize that you need to get either Speed or Pitch 'n Time to replace the standard Digidesign time stretching plug-in. Wave Mechanics Speed is best for changing the playback speed quickly while trying out ideas. Serato's Pitch 'n Time is not as 'instant' to use, but offers even higher quality processing. The Wave Mechanics UltraTools plug-ins are primarily aimed at music producers who will use Pure Pitch for creating harmonies; Pitch Doctor to correct out-of-tune vocals; and Sound Blender for pitch-shifting, delay, filtering, panning and modulation effects. For creative sound design, the GRM Tools have a lot to offer while the Native Instruments Spektral Delay is an interesting and welcome newcomer. The DUY plug-ins work well as sound modifiers to liven up even the most sterile digital recordings - while the ReDSPider gives you a varied selection of presets at an affordable price.

Infinity is still a unique tool for anyone making loops and involved in creative sound design. However, it could do with updating and being made available as a plug-in to make it easily accessible from within popular host software such as Pro Tools or BIAS Peak. MDT also remains unique on account of its innovative user-interface design and its excellent multi-band processing features. The JVP is a useful tool, whether for vocal or instrumental processing - and especially for Pro Tools TDM systems where it is very efficient in terms of DSP usage. Auto-Tune and Auto-Tune 3 have their rivals now, but have not really been bettered by any other products. My current favourite from the Antares range has to be Microphone Modeler. I always did want a Neumann U47 and could never afford one!

Hyperprism will be valued by the creative sound designer most, although it can make sense to buy this suite of plug-ins if you are on a budget - as you get just about every type of effect included. Realizer Pro, despite its name, is really a hobbyist-level tool, as is Ray Gun. Harmony is a viable alternative to Auto-Tune for tuning up vocals and offers some of the same functionality as Waves UltraPitch or Wave Mechanics UltraTools when it comes to creating harmonies. Ionizer, on the other hand, is a serious audio tool aimed at professional engineers that uses a unique and highly innovative user interface to provide everything from noise reduction to multi-band processing to vocoding in one plug-in.

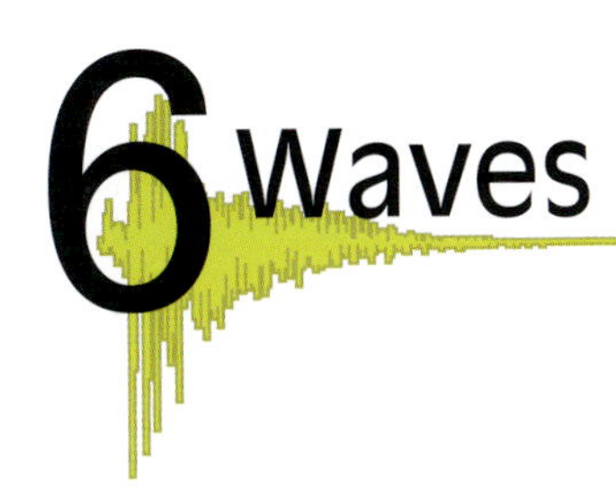

6 Waves

Waves plug-ins

Waves, based in Israel, was the first company to develop audio plug-ins - originally for Digidesign Sound Designer II software. Within a couple of years, development shifted to the Pro Tools TDM platform, and has since been widened to include all the other major formats, including Steinberg VST and MOTU MAS on the Mac and Direct X on the PC. This means that these plug-ins will work with Pro Tools TDM and LE, Digital Performer, Cubase VST and Logic Audio, Peak and Spark on the Mac, and with Pro Tools, Cubase VST, Logic Audio, and software from Sonic Foundry, Sek'D, CoolEdit Pro and CakeWalk on the PC.

Waves offer their plug-ins in various 'bundles' as well as offering them individually. For example, Gold, Platinum, Renaissance Collection 2 and Masters bundles are available for both Native (on the PC or Mac CPU) systems and Pro Tools HD TDM systems. The Renaissance Collection 2 bundle comprises the Renaissance VOX, Renaissance Bass and Renaissance DeEsser. The Masters bundle includes the award-winning L2 Ultramaximizer along with the Linear Phase Equalizer, Linear Phase Multiband, which both use Waves' latest zero phase distortion FIR filters to provide the level of transparency required for the most demanding mastering applications.

Waves Gold bundle

The Waves Gold bundle includes just about every audio signal-processing plug-in that you might need - with support for Pro Tools TDM, Realtime AudioSuite and AudioSuite, Steinberg VST and MOTU MAS - plus Direct-X for the PC. You can buy either TDM or Native versions of the bundle and 88.2/96 kHz support is now available for many of the plug-ins. The UK prices at the time of writing (summer

2002) for these bundles are £1950 + VAT for TDM Gold and £995 + VAT for Native Gold.

> **Note** A Digidesign MIX card is required for full functionality with the TDM version, although Pro Tools|24 and Pro Tools III cards are also supported. Check the Waves website at www.waves.com for details.

From essential tools for use in mastering, such as the Q10, L1, C1, S1, PS22 and PAZ through to special effects for sound designers in the Pro-FX range, such as Enigma, MondoMod, SuperTap, Doppler, UltraPitch and MetaFlanger – the Waves Gold bundle has all these and more. It includes every processor Waves make, such as the Renaissance Reverberator and C4 Multiband Parametric Processor, the TrueVerb, DeEsser and MaxxBass, and the excellent Renaissance series Compressor, EQ, DeEsser, Bass and Vox processors. Comprehensive libraries containing setup files for typical applications help to deliver speedy results for audio professionals – whether involved in mastering, remix, restoration, multimedia, film, web, games or any other audio work.

AudioTrack incorporates a four-band fully parametric EQ, compressor, and gate all in one space-saving window – so you can get started simply with this.

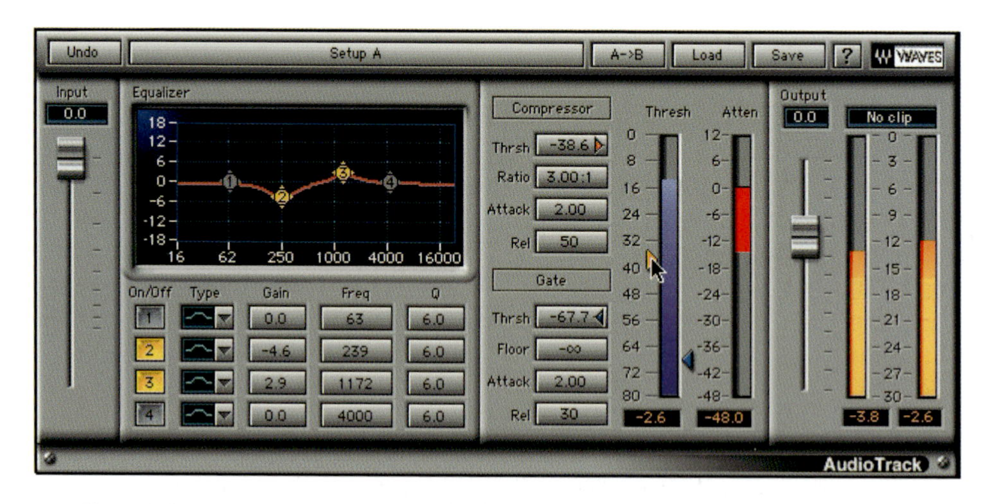

Figure 6.1 AudioTrack showing Threshold settings for the compressor and gate, and EQ.

When you want to get more serious, try the 10-band Q10 Paragraphic Equalizer. This has a Setup Library containing more than 200 equalizer types to suit anything from creative effects to precision mastering.

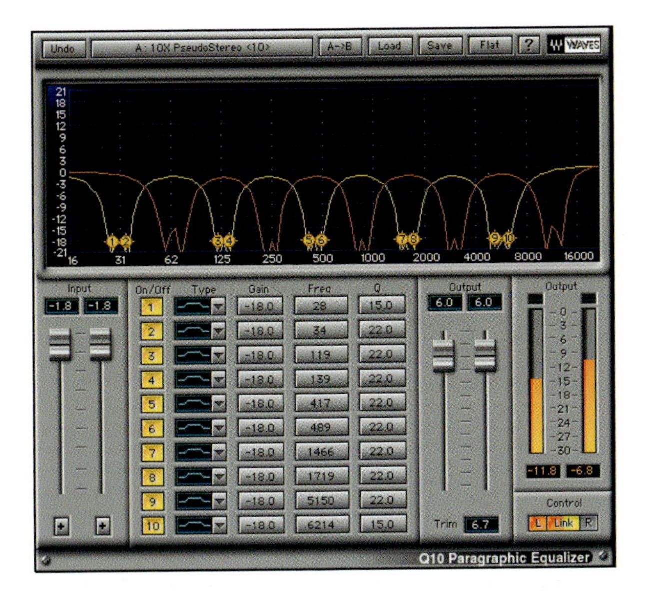

Figure 6.2 Q10 ParaGraphic Equalizer.

For dynamic control, the C1 Parametric Compander is much more flexible than a conventional compressor. It also has a Gate/Expander and a Filter section. The Filter section can be used for Sidechain EQ or with the unique bandsplit modes that allow you to compress just the bass while gating wideband, or to compress wideband while de-essing, for example.

Figure 6.3 C1 Compressor/Gate.

To raise the level of your audio to the maximum while avoiding distortion, use the L1 Ultramaximizer, Waves' most popular digital limiter, which now features 24-bit dithering and 48-bit, double-precision performance.

Figure 6.4 L1 UltraMaximizer: the 'full' mastering plug-in with limiter and all IDR options.

The S1 Stereo Imager includes a set of tools to readjust the stereo level-balance of a mix and has the ability to dramatically widen an existing stereo image. The stereo vector display, with its half-circle shape and soundstage graphic, lets you see and control the stereo soundstage of a file. The S1 can be used for mastering, to fix off-centre or unbalanced mixes, or for precision stereo image adjustment. It also incorporates an MS converter that converts left-right input into MS output or vice-versa.

There are four S1 plug-ins installed which you can choose to suit your purpose - the Standard S1 Shuffler, the 48-bit double precision S1 Shuffler, the S1 Stereo Imager and the S1 MS Matrix. The S1 Shuffler is the full plug-in with all the controls for Width, Rotation, Asymmetry, Shuffling, and so forth. The S1 48-bit Double Precision Shuffler is the high-quality mastering version of this plug-in - available for TDM only. The S1 Stereo Imager omits the Shuffler and Frequency controls, providing adjustments for Width, Rotation and Asymmetry only - to save processing power if Shuffling is not needed. The Stereo Imager is ideal for rotating (panning) stereo microphone recordings or sub-mixes into position within stereo mixes.

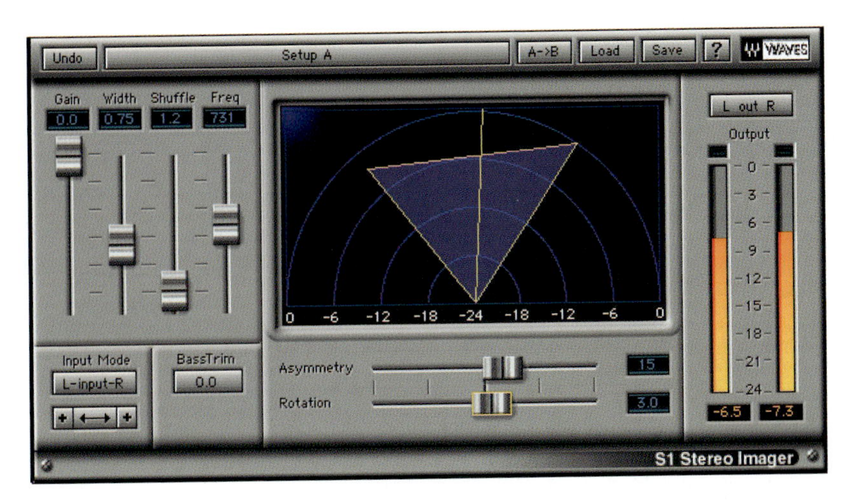

Figure 6.5 Waves S1 Shuffler.

The S1 MS Matrix is simply an MS converter matrix that converts left-right input into MS output or vice versa. It has no controls or metering, and is designed so that if the input does not clip, neither will the output.

Figure 6.6 Waves S1 MS Matrix.

Also for stereo, the PS22 StereoMaker is a set of tools for creating convincing stereo from mono source material as well as a set of processing tools for synthesizing a richer stereo effect from existing stereo material. StereoMaker improves on previous pseudo-stereo effects by leaving the tonal quality of the original source unaltered and by producing much less 'phasiness' and 'vagueness' in the sound. StereoMaker can be applied to mono sounds to produce a stereo spread effect. It also works with panned mono or true stereo inputs, spreading each individual sound source within the stereo input around its position in the original stereo soundstage to enhance the spread of each sound image within a panned mono or stereo mix.

Four PS22 plug-ins are provided - PS22-Spread, PS22-Spread (10), PS22-Split and PS22-Xsplit - along with four specialized delay plug-ins - DLA, DLA-X, DLA-XL and DLA-XLB, intended for use with these. All these plug-ins provide stereo-in/stereo-out operation and each produces a different kind of sound.

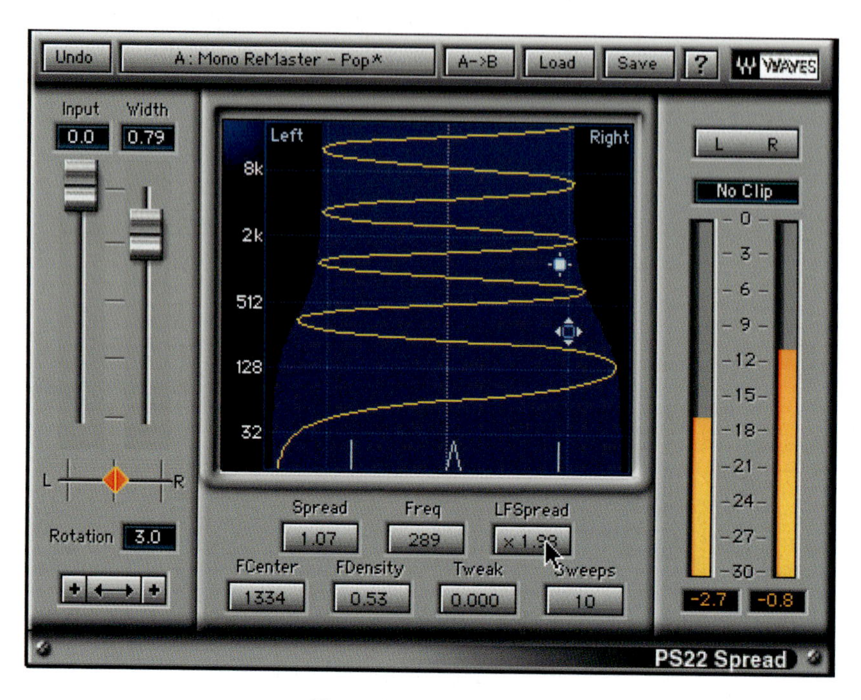

Figure 6.7 PS22 Spread.

To help you visualize what is happening with your audio, the PAZ Psychoacoustic Analyzer offers three measurement functions: a realtime frequency analyser, a Stereo Position display, and a set of peak/RMS Meters. All of these are essential tools for mastering and can also be used in a variety of creative situations. When you install PAZ, you get four versions of the plug-in to choose from. PAZ Analyzer has all the functions in one window, but if you only need to look at levels, just use the PAZ Meters plug-in. If you just want to analyse the frequency content of your audio, simply select the PAZ Frequency analyser or if you want to look for phase problems, use the PAZ Position plug-in. This way, you save on the amount of graphics processing your system has to do - although the DSP requirements are the same whichever plug-in you use.

Figure 6.8 PAZ Analyzer: The full plug-in with all three measurement functions – Frequency, Stereo Position, and Level.

The Pro-FX plug-ins are solidly aimed at creative applications. For example, Enigma – 20% phaser, 80% indescribable – is a new tool combining a very complex phaser/flanger with a reverb/feedback network with complex filtering, plus some modulation. Ideal for 'wacky' effects!

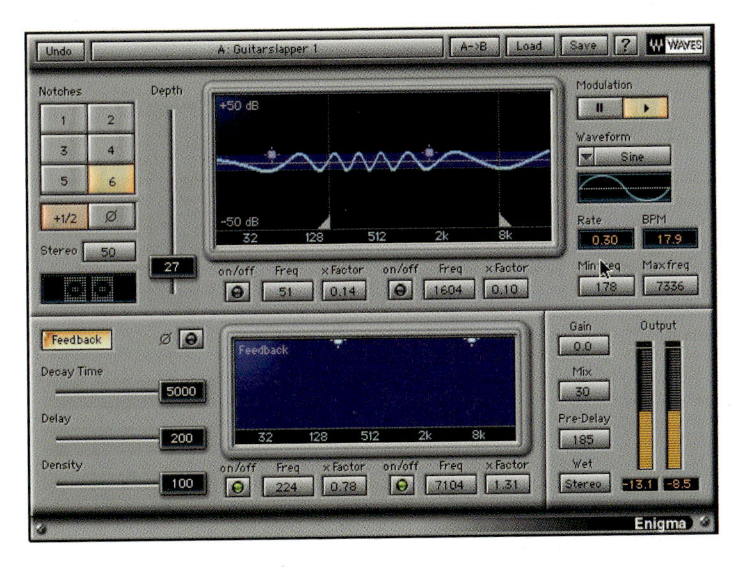

Figure 6.9 Enigma.

SuperTap is a multi-tap delay with up to six voices - each with up to six seconds of delay, independent gain, panning and filtering on each tap, plus modulation and feedback controls. You can use this to emulate just about any type of delay unit ever made.

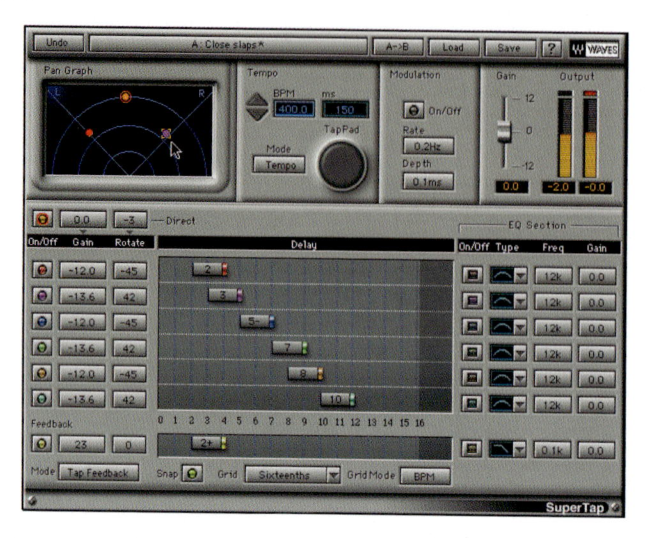

Figure 6.10 SuperTap screen showing delay and pan (polar) settings for a series of closely spaced slap delays.

MondoMod combines Amplitude Modulation, Frequency Modulation and auto-panning, enabling you to create effects ranging from subtle to bizarre. Preset names like 'Diffuser', 'Fried Metal' and 'See live alien!' should help you to get an impression of what sort of effects are possible. One of my favourites is the Full Rotor preset which simulates a 'Leslie'-type rotating-speaker system. Several other very useful presets are provided to let you simulate the various distortions that you get with analogue tapes and vinyl records.

Figure 6.11 MondoMod.

Doppler lets you create both realistic and 'larger-than-life' Doppler effects, simulating the pitch-shifting effects that occur in nature as sounds move towards and away from the listener.

Figure 6.12 Doppler.

MetaFlanger lets you create chorusing, phasing and flanging effects. It also lets you emulate just about every vintage flanger/phaser on the planet – such as the Mutron BiPhase, or the MXR 'grey stomp' pedal. The presets include simulations of classic flanging effects such as the jet plane swooshing effect used on the Small Faces hit single, Itchykoo Park. Rotating speaker effects include Slow and Fast Rotor speed settings and there are various useful phasing, doubling, chorusing and flanging effects.

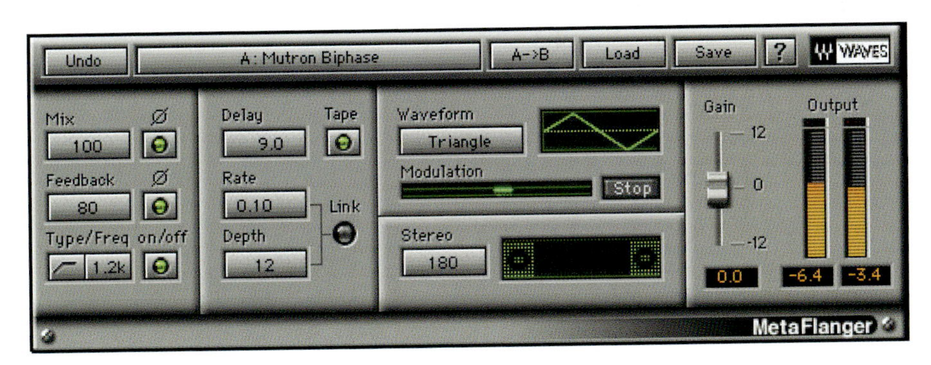

Figure 6.13 MetaFlanger.

My favourite from the Pro-FX bundle is UltraPitch. This is a harmonization tool that processes monophonic tracks such as vocals or single-note instruments to create up to six new 'voices'. You can produce extremely realistic effects using the stereo panning and delay on each output to create huge and thick stereo chorusing, doubling, parallel harmonies and lots more. Formant correction keeps the sound natural when you shift the pitch, and menu options are provided which optimize this for voice, brass, strings and so forth.

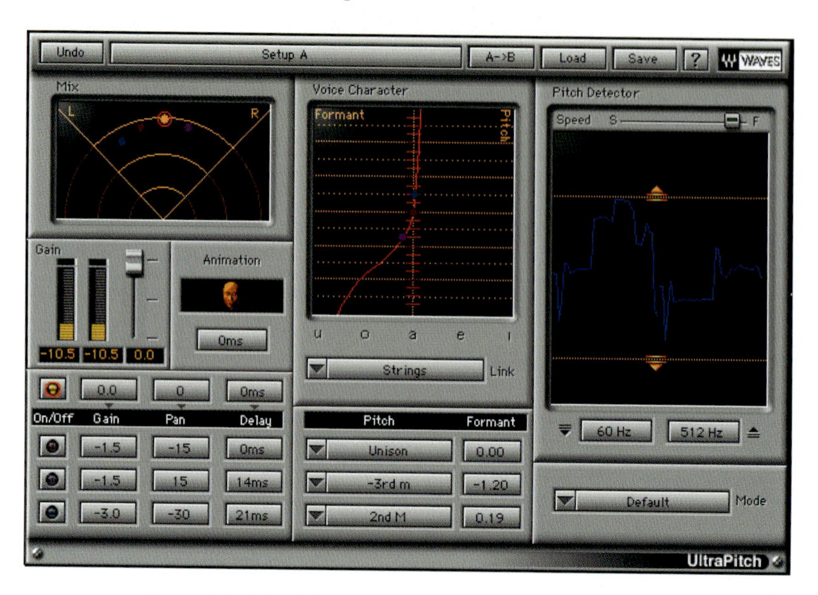

Figure 6.14 UltraPitch window showing three delay taps.

The Renaissance collection

The Renaissance series and most of the other Waves plug-ins are primarily intended for use while recording and mixing. I find myself using the Renaissance Compressor in preference to just about anything else these days. This features both Electro and Opto modes with ARC Auto-release and an L1 style limiter on the output. The Rennaissance Compressor, like most of the Renaissance series, has an appealingly simple user interface using sliding buttons to control Attack and Release, Threshold, Ratio and output makeup Gain, along with three selector buttons for Arc/Manual, Opto/Electro and Warm/Smooth. The input and output signal meters let you see at a glance where to set your threshold and makeup gain, while a large bar meter in the centre shows the signal attenuation.

The Renaissance Equalizer provides classic vintage EQ designs, such as Pultec-style resonant shelf filters, at double-precision 48-bit resolution dithered to 24-bit.

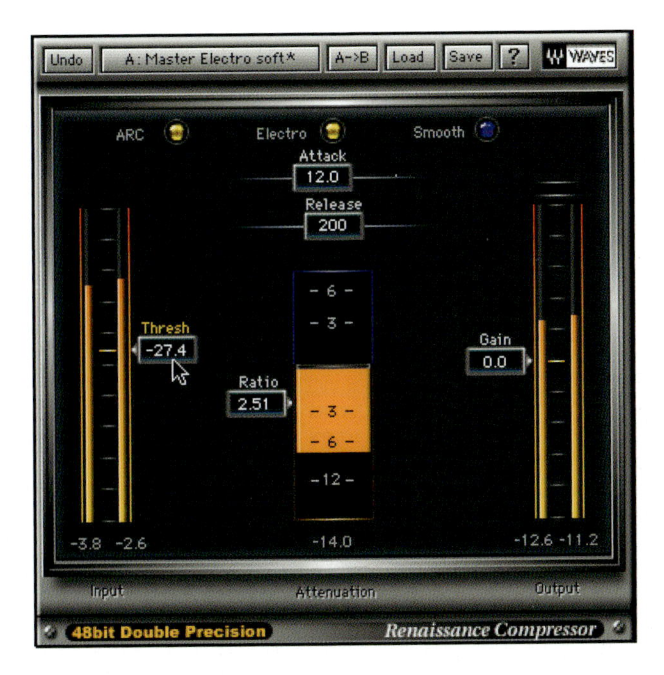

Figure 6.15 Renaissance Compressor.

You can choose six, four, or two-band versions of the plug-in, thus minimizing the processing power required by using the smaller versions whenever you need less bands of EQ. Stereo, dual-mono, and mono versions are available - although not necessarily on all platforms.

Figure 6.16 Renaissance Equalizer.

The Renaissance Vox combines compression and limiting and consolidates several controls into just two parameters. A single control alters the amount of compression and limiting, while providing automatic makeup gain. A second control lets you apply a downward expander to gently clean up the noise floor.

Figure 6.17 Renaissance Vox.

The TrueVerb Room Emulator offers some of the best room reverb emulations you will find, and has a unique distance control. The distance effect depends on the balance of Direct, Early Reflections, and Reverb – which you set using the single Distance control. TrueVerb is great at creating Early Reflections, so I often use it just for this effect and combine it with reverb or other effects from other plug-ins. My favourite presets are Drum Room, Studio A, Rock Guitar Room, Bijou Theatre and Tight NY Room. There are also plenty of large halls, cathedrals and so forth.

If you are looking for classic reverb effects, the new Renaissance Reverberator offers improved reverb 'tails' compared with TrueVerb – but without the Distance control. Waves took the early reflection system from TrueVerb as a starting point,

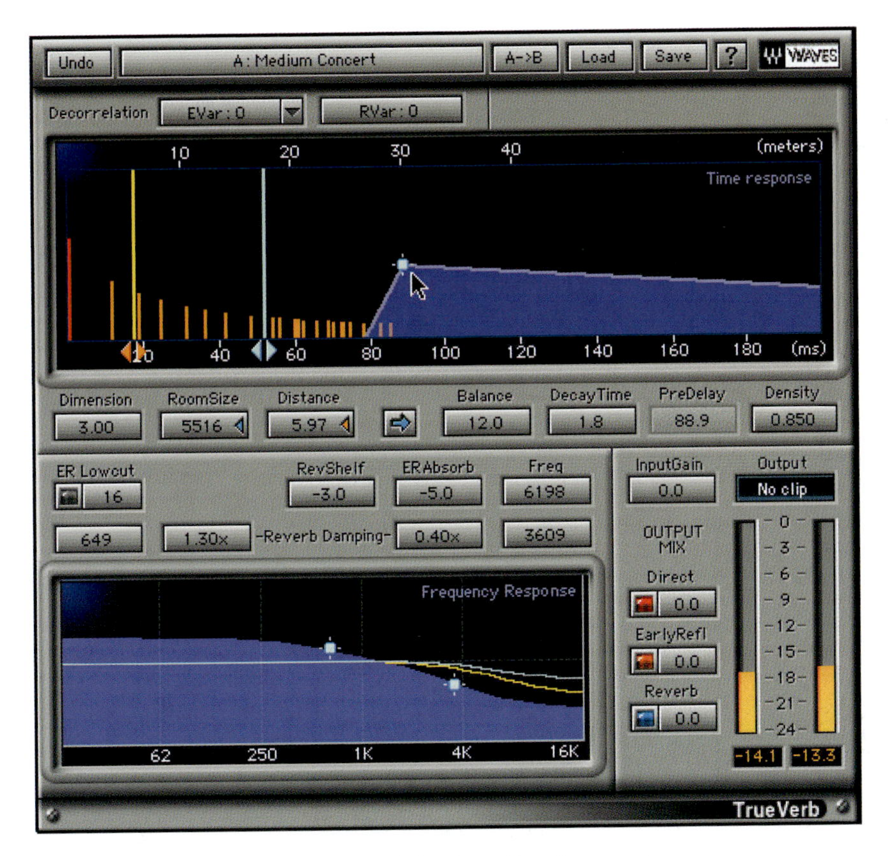

Figure 6.18 Waves TrueVerb.

improved upon this in many ways, created a new reverb tail, then created a new interface with lots of visual feedback and control.

Note Two plug-ins are actually supplied with Renaissance Reverb – the full Reverb plug-in and the Reverb Tail. The full Reverb TDM plug-in uses a chip from a MIX card, so it won't work on older Digidesign DSP Farm cards. The Reverb Tail plug-in will work with DSP Farms. This smaller Tail component does not have Early Reflections or any associated controls. It only has the Reverb 'Tail', and there are only a couple of reverb types to choose from – as the lack of early reflections limits the room or hall characteristics.

The standard Waves DeEsser is a precision high-frequency dynamic processor modelled after vintage gear for fast and easy de-essing/HF limiting. It has a simple, yet powerful user interface. The Factory Presets contain typical settings

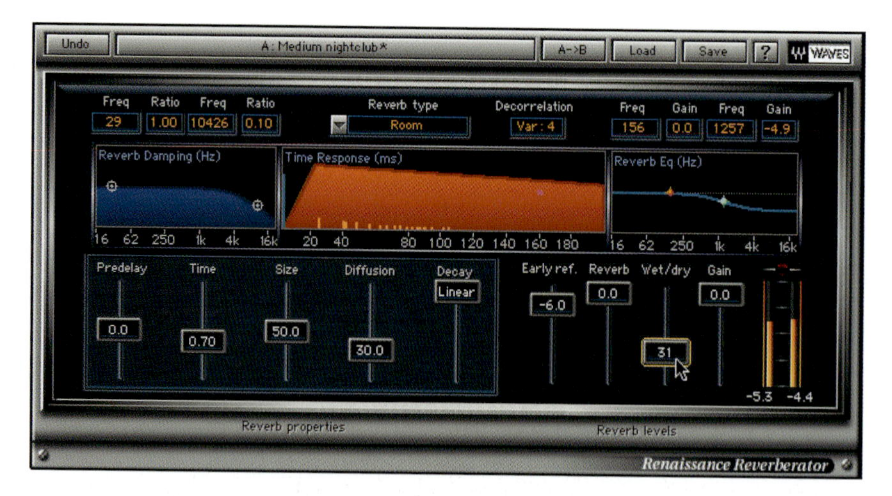

Figure 6.19 Renaissance Reverberator.

for Male Ess and Shh and for Female Ess and Shh sounds. You can also use the Shh settings for 'ch' and 'th' sounds, and for hard consonants such as 't', 'd', and 'k'. In a nutshell, the DeEsser does what it says and couldn't be much simpler to use.

Figure 6.20 DeEsser.

The Renaissance DeEsser has several new features that improve on this design. For example, it uses a phase-compensated crossover to eliminate phase modulation that would otherwise occur due to the amplitude modulation applied by the compressor. It also uses an Adaptive Threshold to dynamically adapt to the input signal in order to provide a more natural de-essing effect. And to make it even easier to use, it incorporates a graph to show exactly which frequencies are being de-essed.

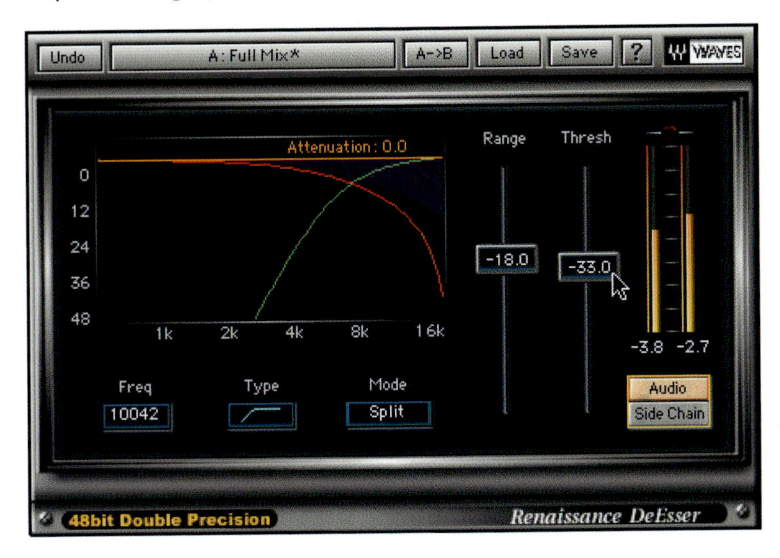

Figure 6.21 Renaissance DeEsser.

MaxxBass creates harmonics that you can add to any audio signal to improve the clarity of the bass frequencies. These harmonics trick the ear into perceiving low bass frequencies that may not actually be present in the output. This psychoacoustic illusion can be used for mixing and mastering to enhance the bass response for playback on any system. MaxxBass is a useful specialist tool, particularly for multimedia applications – and can be used as a general sound enhancer.

Figure 6.22 MaxxBass.

Renaissance Bass was developed from the original MaxxBass and offers significantly improved psychoacoustic bass performance, easier adjustment, and clip-free performance.

Figure 6.23 Renaissance Bass.

When your mixes are sounding more or less the way you want them, you can add the finishing touches using the C4 Multiband Parametric Processor. This offers four-band expansion, limiting, compression and EQ – ideal for 'finalizing' master mixes. The C4 is a state-of-the-art multi-band compressor that can be used to sort out problems with existing mixes, or can be used for a variety of mastering applications. For example, the Multiband Opto Mastering preset will give you reasonable compression and increased density of your mix. To enhance low-level signals (a great way to boost level without squashing dynamics), try the Upward Comp +5, or +3 version of the preset. This is great for adding level without losing punch. Now if you have a mix that has too much kick, the right amount of bass guitar, and needs a little 'cymbal control' and de-essing, you can use the BassComp/De-Esser preset to adjust the balance between the kick and the cymbals. And if you have to deal with a mix that has been seriously over-compressed you can try using the Uncompressor to restore the squashed dynamics using upward expansion.

Figure 6.24 C4 Multiband Compressor.

Native Masters bundle

The latest bundle of plug-ins from Waves, the Native Masters, includes the Linear Phase Multiband, the Linear Phase Equalizer, and the L2 Ultramaximizer plug-ins.

The Linear Phase Multiband is a more highly evolved version of Waves' C4 MultiBand Parametric Processor. Aimed squarely at specialist mastering engineers, this now has five (rather than four) discrete bands each with its own gain and dynamics controls for equalizing, compressing, expanding or limiting each band separately. Its Linear Phase crossovers provide a completely transparent signal path when the split is active, but idle. Of course, you can't avoid the processing delay, but there is absolutely no alteration of the sound until the thresholds you set are crossed. There are options for Automatic Makeup and gain Trim – and there is an Adaptive threshold. The accent here is on ease of setup and use. The user interface features Waves' unique DynamicLine display which shows the actual gain changes superimposed onto the frequency response graph. Despite the complexity, once you get used to it, the user interface works extremely well, delivering oodles of relevant information from a compact screen design. Multi-band signal processing is becoming increasingly popular as a tool for 'finalizing' mixes – although this is still best left in the experienced hands of a mastering engineer. The Linear Phase Multiband pushes the 'edge' of what is possible with multi-band processing - providing exciting new sonic dimensions for the more adventurous engineer.

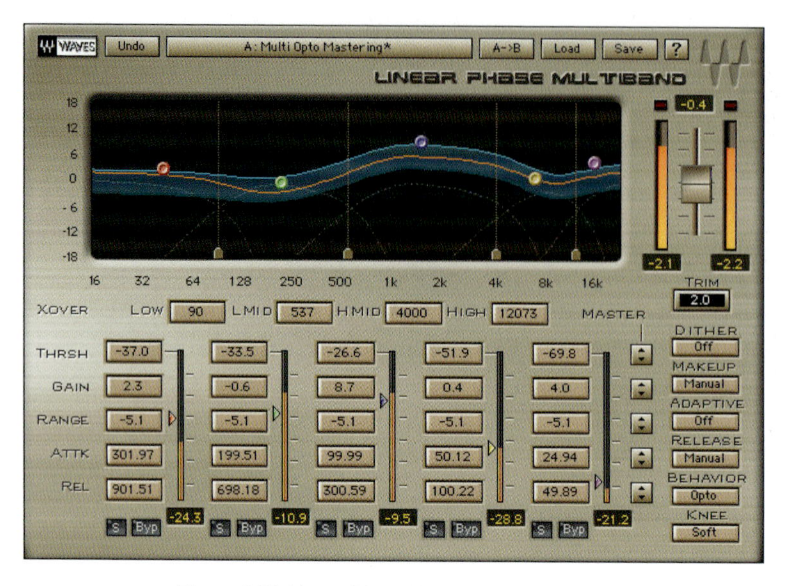

Figure 6.25 Linear Phase Multiband compressor.

Also intended primarily for use by Mastering engineers, the Linear Phase Equalizer provides extremely precise equalization with absolutely no phase shifting. Using Finite Impulse Response filters, it presents no quantization error and is 24-bit clean when idle. So what's so special about this then? Well, with normal EQ designs, different frequencies get different delays or phase shifts. These can blur transients - producing a duller sound. With the Linear Phase EQ all frequencies are delayed by the exact same amount - thus avoiding this problem and producing a more accurate sound. Again, this plug-in has been developed from previous Waves designs, such as the Q10 and Renaissance EQ 'paragraphic' equalizers, which provide both numerical and graphical control of the parameters. All the usual controls for gain, frequency and Q are provided and there are nine filter types, each offering two types of Shelf and Cut filters. A comprehensive library of presets provides settings for classic EQ types and for useful problem-solving tools.

It is much more memory - and calculation-intensive than any normal digital EQ - but it is truer to the source as it doesn't alter phase relations. We are all so accustomed to the sound of normal EQ with its attendant phase shift colorations that this EQ is almost certainly going to sound different to you. People who have heard it describe this EQ as more 'transparent' and say that it preserves the musical balance while still manipulating the harmonic spectrum very effectively.

You actually get two version of this plug-in – broadband and lowband. The normal full frequency range 'broadband' version has six EQ bands – one special low-frequency band and five general-purpose bands.

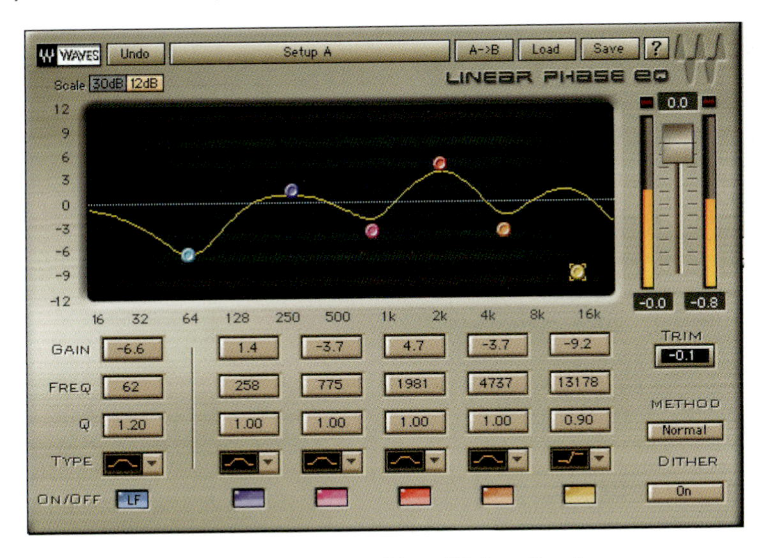

Figure 6.26 Linear Phase EQ broadband.

If you need more detailed low-frequency control you can use the three-band low frequency version instead.

Figure 6.27 Linear Phase EQ lowband.

Originally available just for TDM systems, the Native Masters L2 Ultramaximizer lets you set your digital audio signal to the maximum possible level by taming the peaks so that you can raise the overall level safely – without distortion.

Figure 6.28 L2 Ultramaximizer TDM showing original interface.

The L2 provides peak limiting and level maximizing for your final audio mixes, which you can also dither using Waves' proprietary IDR (Increased Digital Resolution) technology. This plug-in is actually a software version of Waves' own L2 hardware, so it features 48-bit double-precision processing, 9th-order noise-shaping in the IDR dithering system, and the same user-friendly Auto Release Control technology.

Note that the L2 Ultramaximizer's user interface has been restyled for the Native Masters Bundle – not particularly for the better in my opinion.

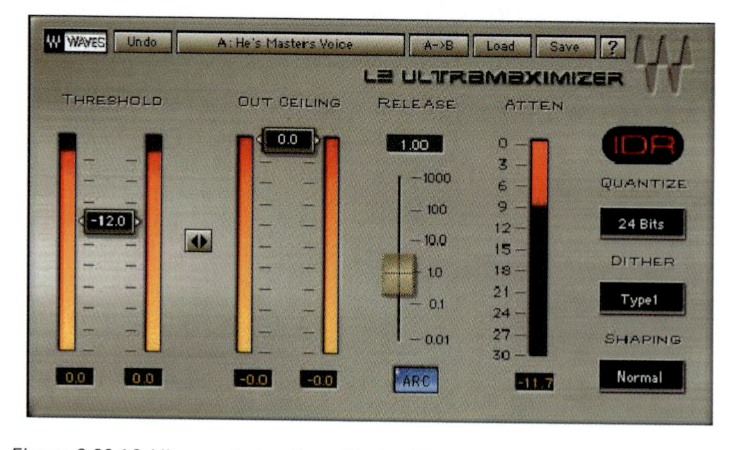

Figure 6.29 L2 Ultramaximizer from Native Masters Bundle with new interface.

At the time of writing, OSX versions were not yet available, but these Native Masters plug-ins will work with the latest Pro Tools HD and LE systems - or even with PT Free. Both VST and MAS versions are also included, so they will also work with Digital Performer 3.xx and all the major VST-compatible hosts including Cubase VST 5, NUENDO 1.5, Logic Audio 4.7, TC|Works SPARK 2.0, BIAS Peak 2.62, V-Box 1.01 and Deck 3, and Prosoniq SonicWORX 2.

Well, we are coming to the end of the list now, but before we wrap up, time to take a quick peek at the Waves Native Restoration tools. These are well designed, effective for all music restoration, post-production and forensic applications - and could not be much easier to use.

Native Restoration tools

X-Noise intelligently learns from a section of noise and then applies a broadband noise reduction to eliminate background noise from any source.

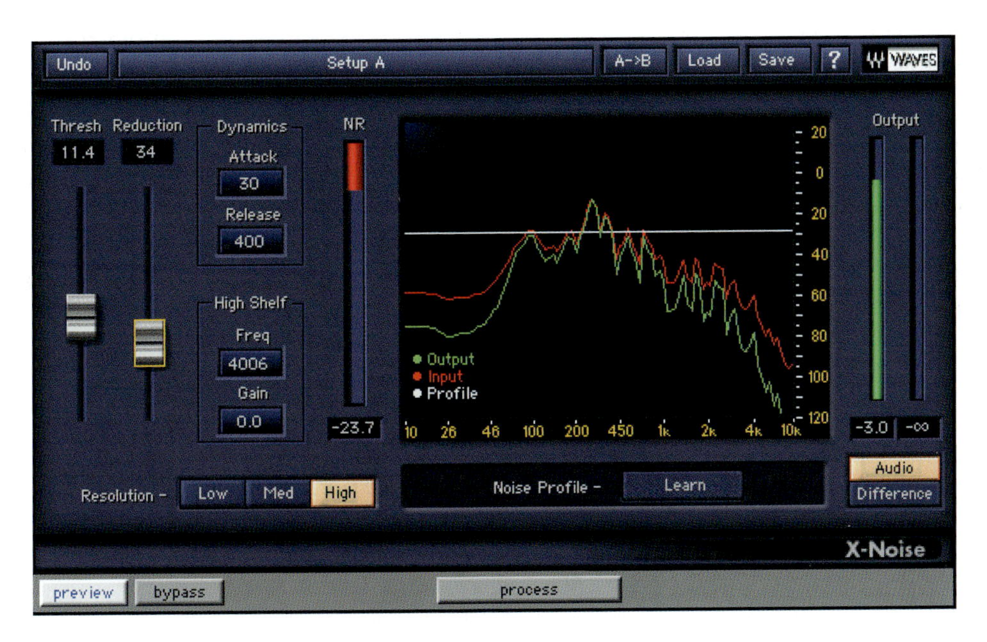

Figure 6.30 X-Noise.

X-Click effectively removes clicks from 78s and vinyl records as well as spikes arising from digital switching or crosstalk.

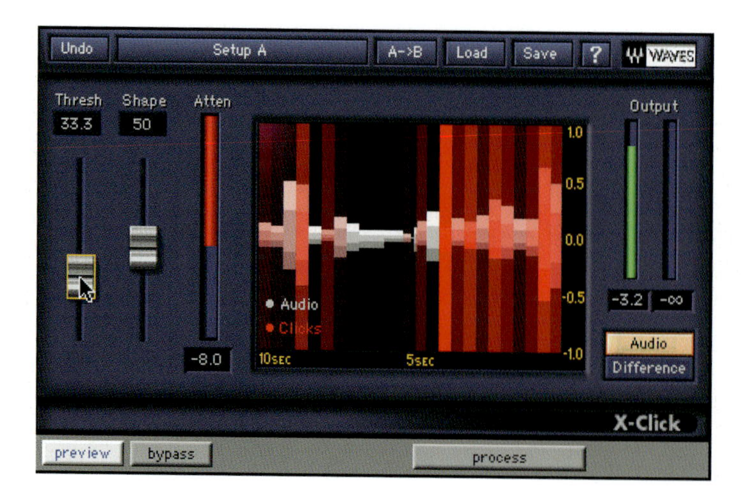

Figure 6.31 X-Click.

X-Crackle is the second stage in the restoration of old records – used to eliminate the crackles and surface noise left after X-Click.

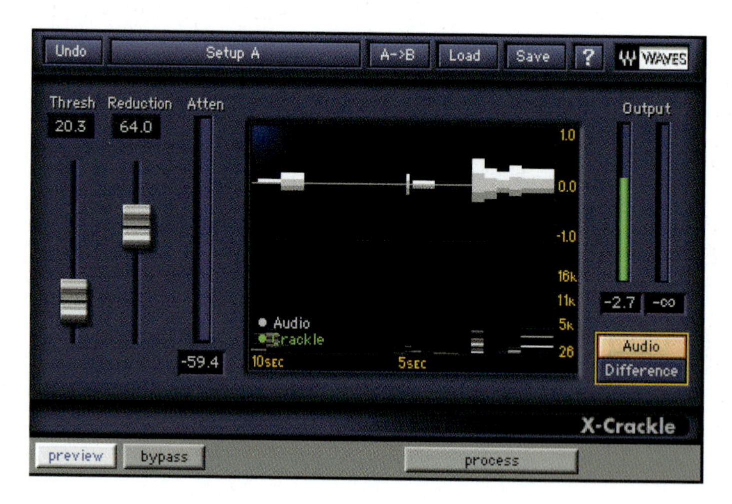

Figure 6.32 X-Crackle.

X-Hum attenuates pitched low frequency disturbances such as ground loop hum and its harmonics. It also reduces rumbles, microphone pops and DC-offset.

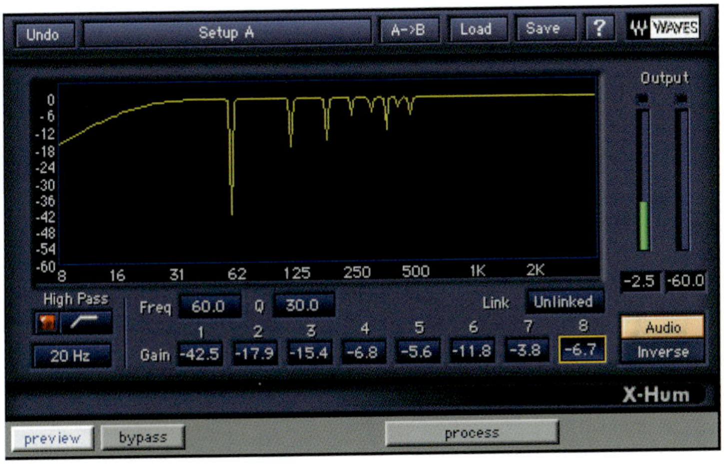

Figure 6.33 X-Hum.

Tip With X-Noise, X-Click and X-Crackle, the Output Monitor switches between the Audio signal (the audio processed by X-Noise) and the Difference signal (the noise type you are trying to remove. Listen carefully to the Difference signal to determine whether your settings removed parts of the audio signal in addition to the noise. The goal is to maximize noise reduction while capturing and degrading as little of the audio as possible. If you hear any of the original audio in the Difference signal, you can adjust the settings to achieve a better balance between noise reduction and signal loss/degradation with X-Noise or to hear whether the current parameters have removed any wanted audio transients with X-Click and X-Crackle. In the case of X-Hum, the second monitoring mode is called Inverse. The Inverse mode keeps the filters in the same position but boosts (instead of cuts) while applying appropriate gain reduction. This helps you to identify the unwanted hum frequencies, as you can hear these more clearly. You can then adjust the fundamental frequency of the notch filters to remove the hum more easily.

Finally, the smallest plug-in has to be IDR – which provides dithering from 24 bits down to 20 or 16 bits.

IDR is a noise-shaped dithering system that can be used in consecutive processes, and unlike some systems, is completely compatible with any other previously dithered files, such as some DAT machines might create using Sony SBM, or from Apogee A/D converters using UV22.

Figure 6.34 Waves IDR.

Chapter summary

Many of these plug-in types are available from other manufacturers, but no other manufacturer offers as complete and well developed a range as this. For the hobbyist or the casual user it is clearly overkill, but for audio professionals who have to have the best, there is no contest. They need the Waves Gold bundle. And for the person who has to have absolutely everything, the Platinum bundle combines the Gold bundle, the Renaissance Collection 2 and the Masters bundles!

Fig. 6.35 A typical Waves/Pro Tools TDM session.

7 Steinberg and Third-party VST Plug-ins

For sound design, Steinberg distribute a wide range of plug-ins developed by other companies, including the GRM Tools, Prosoniq Ambisone, Dynasone, Roomulator and VoxCiter – and the Orange Vocoder. The TL Audio EQ lets you simulate the audio 'signature' of this popular analogue audio hardware device. The Waldorf D-Pole plug-in provides the kind of filters that you find in Waldorf's popular analogue synthesizers, ideal for use with dance music. The Steinberg Voice Machine is similarly useful when working on pop arrangements to quickly create vocal harmonies and other effects. Guitarists will go for the excellent Hughes & Kettner Warp amplifier and speaker simulation. And Spectral Design's Magneto analogue tape simulator, SPL De-Esser and Q-Metric EQ are handy plug-ins to use within any mix. Once you have completed your mixes, the Mastering Edition suite which includes Loudness Maximizer, Spectralizer, FreeFilter, ME Compressor, ME Phasescope and ME Spectrograph, is ideal for CD mastering.

GRM Tools

GRM Tools is a set of real-time sound processing plug-ins for VST-compatible host applications such as Cubase VST, Wavelab and other programs. The GRM Tools plug-ins provide a palette of unusual effects particularly aimed at the sound designer working with electronic music or creating sound effects to picture. Each of these plug-ins feature various controls for their parameters plus sixteen memory buttons for the presets. You can make several different settings, store each of these, then click on any of the memory buttons to recall your settings. The current settings will then change automatically to the new set according to the time you have set (up to 30 seconds) using the slider at the right of the buttons. This way of working takes a little getting used to, but once you see how powerful it can be to either instantly switch or to slowly interpolate

Looking at the different plug-ins in more detail you will find that the BandPass plug-in actually combines a high-pass filter with a low-pass filter that can be used to create either a variable-width band-pass or band-reject filter. The cut-off frequencies can be individually set for the high-pass and low-pass filter using the HP and LP faders while the cut-off slope is fixed and set to a very high value of 560 dB/octave. A Mix fader lets you set the amount of unprocessed signal added to the BandPass output signal. HP and LP faders can be used to individually define the filter's high and low cut-off frequency settings. In the centre left of the BandPass window there is a rectangular frame with what looks like a small ball in the centre and if you grab this 'Super Handle' using the mouse you can change both the centre frequency and the bandwidth at the same time.

PitchAccum

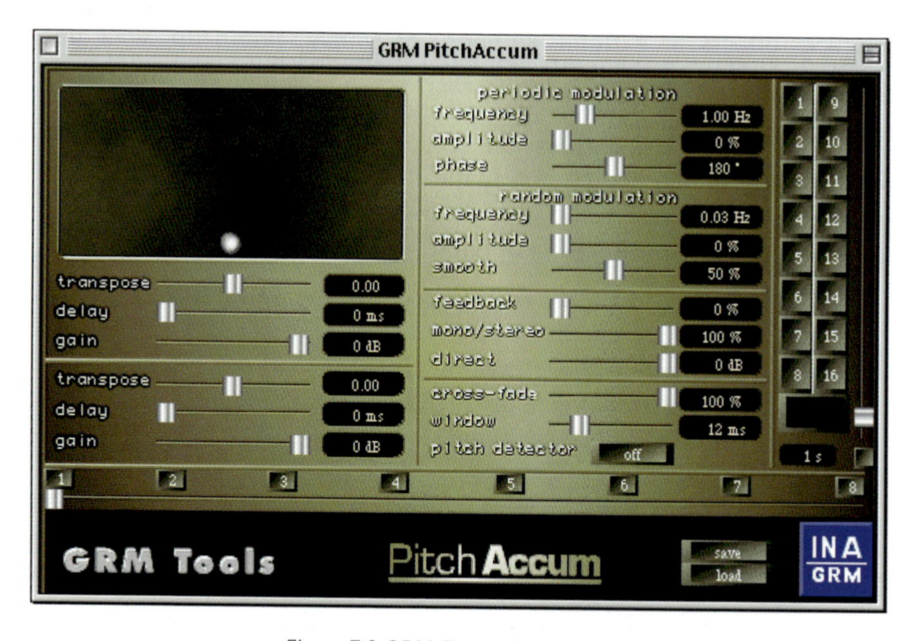

Figure 7.2 GRM Tools PitchAccum.

A 'Super Handle' display is also provided for the PitchAccum plug-in. This combines two transposers with a feedback delay device and in this case you can use the Super Handle to control both transpositions at the same time. You can delay each of the transposers and you can modulate the transposed signals by applying periodic or random modulations. Harmonizer effects and spectacular pitch-related creations that work particularly well on drums, percussion or voice can be generated in seconds. You can also create tremendous stereo delay effects for solo guitars or synthesizers.

Comb Filters

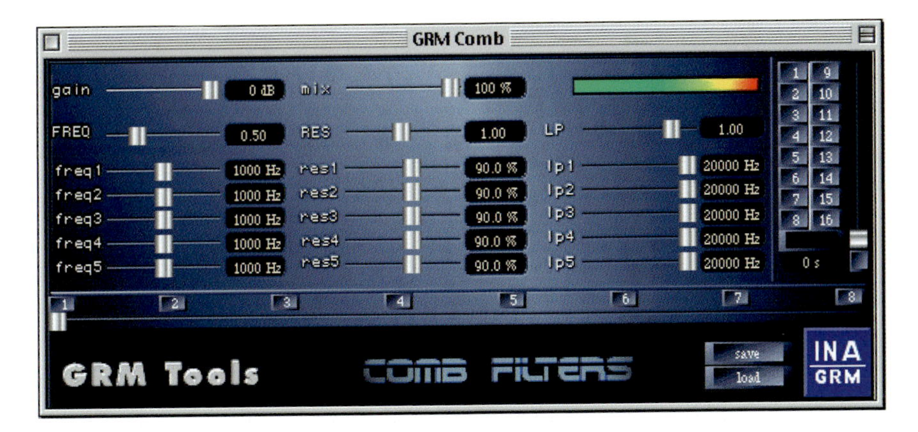

Figure 7.3 GRM Tools Comb Filters.

The Comb Filters plug-in features a set of five parallel filters with high resonance (Q) characteristics. Comb filters amplify signals at a given fundamental frequency and at all integer multiple (i.e. harmonic) frequencies. You can create great 'robot voice' effects using these filters at extreme settings, or simply 'fatten up' your vocal or instrumental sounds with more subtle settings.

Shuffling

Figure 7.4 GRM Tools Shuffling.

If you are looking for more radical-sounding effects, the Shuffling plug-in lets you have fragments of your audio source automatically moving around in the stereo field according to the envelope setting – while these fragments are shifted in pitch between the settings you make for initial and final pitch. You can also randomize the pitch settings. Again, a 'Super Handle' display is provided. These effects have to be heard to be understood, but I can assure you that you won't be disappointed if you are looking for unusual sounds!

GRM Tools Volume 2

Volume 2 of the GRM Tools includes a further set of four plug-ins – Delays, Freeze, Doppler and Reson. As with Volume 1, you can switch between different preset parameter settings in realtime or by defining a time frame within which a dynamic transition from one setting to another should take place. This transition time can be defined in seconds or as a musical note/bar value with VST 2.0-compatible host applications. You can also use the automation features in Cubase VST to record any changes you make to plug-in parameters.

Delays

The Delays plug-in packs a massive amount of effect-creating capability into its deceptively simple interface. You can actually create up to 128 delays and control the amplitude and timing of these to produce extremely complex delay patterns. The Amplitude Distribution fader lets you specify each delay's amplitude setting relative to the previous delays. A value of 1 means all delays get the same amplitude, values of less than 1 mean each successive delay gets a smaller amplitude, and an amplitude of, say, 2 gives a doubling of amplitude for each successive delay. Similarly, with the Delay Distribution fader, a value of 1 means all delays are evenly spaced; values smaller than 1 mean that successive delays are closer together; values higher than 1 mean each successive delay is longer than the previous delay, so they will be further apart. The Random delay fader lets you set the amount of time available for automatic random positioning of the individual echoes and the Variation Rate fader lets you set the speed at which this random variation takes place – i.e. how quickly the individual echoes are randomly moved to a new position.

The Feedback parameter lets you feed a percentage of the delayed signal back to the input to create repeats and the Mono/Stereo fader controls the distribution of delays across the two outputs. A value of zero creates a monophonic signal with the same signal on both channels, while a value of 100% causes delays to appear alternately on the left and right channels. The First parameter

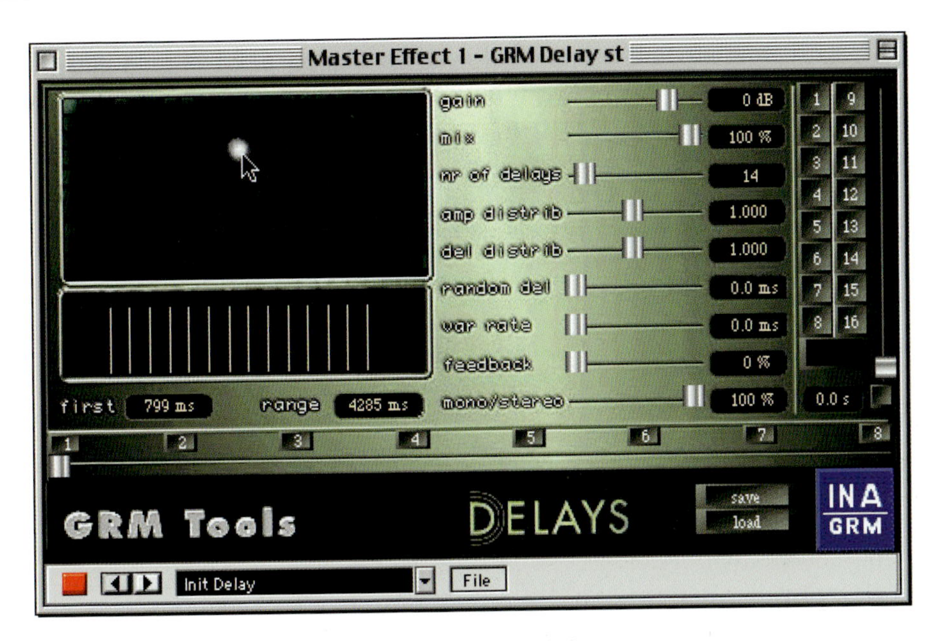

Figure 7.5 GRM Tools Delays.

sets where the first delay should occur, up to a maximum of 5.914 seconds, and the Range value specifies the time difference between the first and the last delay, up to a maximum of 5.944 seconds. You can switch these time values to musical values by Shift-clicking (Control-Shift-clicking on the PC) in the relevant value fields. The Delay Distribution parameter determines how the delays are distributed within this range. If you want instant gratification, just click on any of the buttons at the right hand side of the window to apply a preset delay configuration, and there's a fair chance that you will find something usable right away. If you like to 'roll your own', just start 'tweaking' the parameters by dragging the sliders or typing values. You can also grab the 'ball' in the window display and drag this around to vary the 'First' and 'Range' settings at the same time.

The delays produced are nothing like the standard delays you get from an Echoplex or even a Roland RE201. They are much more complex, with reverse sounds, reverberation and rhythmic effects and much, much more. You simply have to hear these to appreciate what this plug-in is capable of. Of course, when you add in all the possibilities for moving between preset settings and for other types of automation, you won't run out of things to try too quickly, that's for sure!

Dynasone

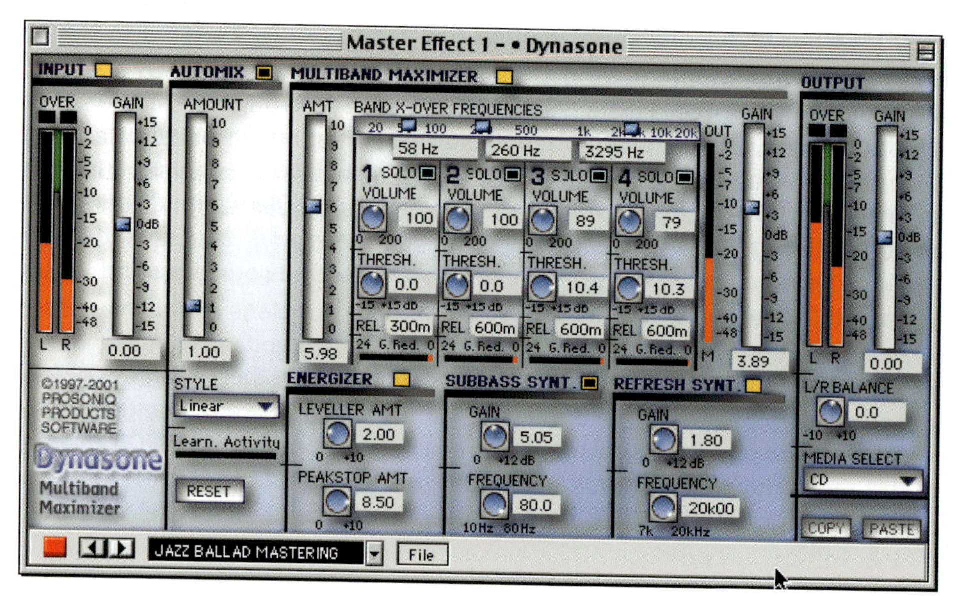

Figure 7.11 Prosoniq Dynasone.

When you have your mix, Dynasone is useful for processing this to make it transparent, add punch and tweak it to fit a specific medium. Its powerful multi-band dynamics compression and Automix functions will automatically balance the frequencies in your mix – making it sound transparent and giving it that special 'pro' touch. Different templates let you adjust the overall sound to match any musical style, from Classic to Dance, and you can choose an optimization scheme for your final release medium – such as 'Vinyl', 'Tape', 'CD, or 'Linear'.

The multi-band maximizer is essentially a dynamics compressor operating on different frequency bands. Typically, a compressor is used to compress the dynamic range of an instrument or voice to achieve equal loudness at any time (thus enhancing the intelligibility) by attenuating any signal above a certain threshold while leaving any signal below that threshold unaffected. One unfortunate side-effect of compression is to make background noise more prominent. Splitting your mix into different frequency bands and compressing these separately makes it easier to achieve equal loudness at every frequency. Overall, you get a higher output volume and a more pleasant sound – which is why multi-band compressors are often used by radio stations. Dynasone's Automix function uses a neural network structure to analyse your mix and compare it with recommended frequency response curves for different musical styles. You can use this feature to automatically balance your mix to sound 'good'. I found this especially useful where I needed to improve the overall sound of some of my older recordings that sounded much 'weaker' than my more recent work.

Other interesting features are the Subharmonic bass synthesizer, which generates artificial subsonic frequencies which can be used to great effect for dance and hip-hop music, and the Refresh function, which is a stereo high frequency exciter that can be used to restore and refresh older recordings by adding new spectral content.

Note Dynasone is intended to be used as a Master Effect. It needs to process and replace all incoming audio data with the processed version, so you should not use it in any send/return setup.

Roomulator

Figure 7.12 Prosoniq Roomulator.

Roomulator is actually a set of five reverb plug-ins offering Plate, Small and Medium Rooms and two Hall Reverbs. These all sound very natural, whether you are using the small room simulations or the dense hall reverbs with canyon-like decay times. You can use the Quality popup to set the quality of the reverb effect, bearing in mind that higher quality offers a finer reverb mesh but will also use up more processing power. The Roomsize slider lets you define the width of the simulated room and the Diffusion slider lets you define the density of the signal reflected by the simulated room, i.e. the roof, walls and floor. The Predelay slider lets you define the delay time between the unprocessed signal and the first reflections of the reverb. Higher settings are generally better, although the sound can become unpleasant at the highest settings. On the other hand, lower settings will produce a more indistinct signal. The section at the right of the plug-in window lets you control the Room Characteristics. Use your mouse to move the dot around this display to change the sound of the reverb effect from Warm to Colored to Cold to White.

Note It is best to set up the Roomulator reverb using a send/return path in your host software. To use the plug-in for a single channel, you should switch the channel mode to 'pre' and turn down the volume of that channel by setting the channel slider to its minimum. Otherwise you will get a mixture of the original and processed signal. Set the effects send knob to its maximum and turn on the effect in the VST effects rack. In order to make it work, you also have to set the input volume in the effects rack to any value greater than 0.

•π warp

The •π warp plug-in distorts the relationship of the partial frequencies within a sound to create novel and interesting textures and noises. You can use it on any input signal, although rich string sounds and drum loops probably create the most interesting textures and atmospheres. You get five parameters to play with - identified by unique symbols as they do not have a specific effect on every sound.

Parameter 1 controls the warping amount. Lower values will generally cause a more noticeable warping effect while higher values may sound 'grainy'. If you adjust this parameter carefully, you can match it to the basic tuning of a sound, which will yield especially beautiful effects for string textures.

Parameter 2 controls the output pitch. Higher values will generally yield a lower output pitch while very high values will cause interesting aliasing effects. Once you've found a good setting for parameter 1, try tweaking this parameter until you get satisfactory results.

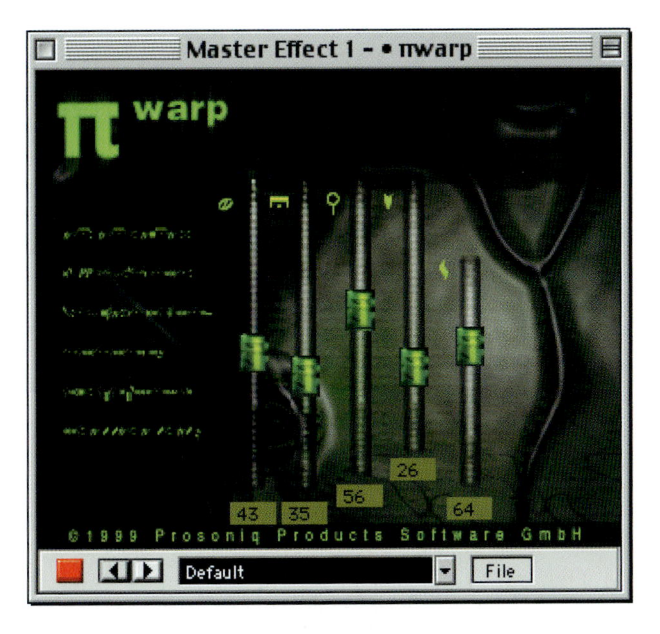

Figure 7.13 Prosoniq • π warp.

Parameter 3 controls a 6 dB/octave (one-pole) low pass filter. The output of the warping stage is post-filtered to smooth aliasing and to dampen the sound if it gets too harsh.

Parameter 4 controls the amount of post-warp reverb. A reverb effect is added to allow for dense atmospheres without sacrificing the scarce plug in rack slots. It also consumes very little processing power.

Parameter 5 controls the room size of the reverb effect. Higher values yield larger room sizes. This is a great little plug-in to use when you are searching for that 'impossible-to-describe' sound effect.

Orange Vocoder

The Orange Vocoder is a digital simulation of a 24-band high-quality analogue Vocoder and is used as an effects device for musical applications where the characteristics of one input signal, called the Modulator, are imposed onto those of another, referred to as the Carrier.

The Orange Vocoder has two banks of Band Pass filters - the Modulator and Carrier banks - which pass through a specified band of frequencies unchanged

Figure 7.14 Prosoniq Orange Vocoder.

and attenuate the rest. Using these filters, the frequency spectrum of the input signals is divided into 24 narrow bands of frequencies, producing a coarse representation of the spectral shape of each input signal. The level settings of the filters in the Modulator filter bank are then used to control the level settings of the filters in the Carrier filter bank - which is how the spectral shape of the Modulator is superimposed on the spectral shape of the Carrier signal. This creates a wide range of different-sounding results, especially if you use a broad-spectrum signal with a lot of sustain, such as strings, as the Carrier signal and a narrow-spectrum signal, such as speech, as the Modulator. In this case, the strings will appear to 'talk'.

You can use the four popup menus in the synthesizer section of the Orange Vocoder window to select a Modulation source for one of the oscillators. The Mixer at the top left of the window lets you balance the processed, Carrier and Modulator signals at the output of the Vocoder. The Graphic Equalizer to the right of this is used to define the frequency response curve according to which the Modulator signal modulates the Carrier. The reverb effect to the right of this can be used to add Reverb to the processed Carrier signal. The Control section has a few general controls such as the Filter Bank release switch that is used to shorten or lengthen the release time of the Vocoder effect. If you want to fatten the sound when using pad sounds, just try adding a little Carrier signal by raising the level of its fader in the Mixer. Or, using the Mixer faders and Reverb, you can create a mix where you hear the unprocessed Carrier signal in the foreground and the processed signal somewhere in the background.

The eight-voice Synthesizer unit can be used to generate an alternative Carrier signal for the Vocoder. This has 16 oscillators, two per voice, with ten basic waveforms and seven sampled sounds, an Oscillator hard sync switch and a Ring Modulator. You can choose between ten basic oscillator waveforms using the Wave popup menu or from one of seven sampled sounds. All the usual controls are provided to set the oscillator pitch and modulation source, depth and speed, filter type, distortion amount, low-pass filter cut-off frequency and resonance, oscillator tuning, and so forth. The Keyboard lets you control the pitch and number of notes in the Carrier signal and this can be switched to cover a range of six octaves.

Interestingly, many of the controls in Orange Vocoder can be automated using Cubase VST's automation. Using this feature you could change the chords you have defined in the Synthesizer, for example. Just copy and paste a complete set of parameters from one Orange Vocoder Program to the next. Then change the chord setting in the copied Program, play back your song and use Cubase's automation to switch from one Program to the next.

Steinberg SPL De-Esser

The SPL De-Esser user interface could hardly be simpler. You get an on/off switch for the Auto Threshold function, a pair of selector switches to select male or female voice types, an 'S'-reduction knob to control the amount of de-essing, and a simulated LED display to indicate the level by which the sibilant frequencies are being reduced – i.e. the amount of de-essing taking place. Choosing Male or Female adjusts the sibilant frequency settings and recognition parameters to the characteristic frequency ranges of the male or female voice.

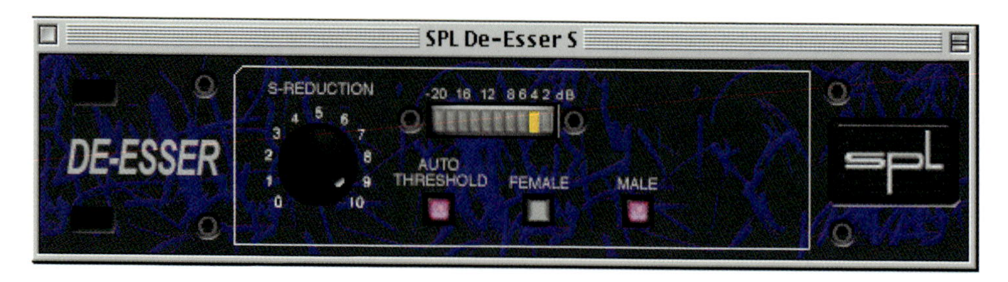

Figure 7.15 SPL De-Esser.

The Auto Threshold feature is ideal when recording less-experienced singers, for example, who often move around inadvertently in front of the microphone causing the microphone's output level to vary. In this case, when the microphone signal is too strong, too much de-essing takes place, while if the signal is too weak, no de-essing takes place. With Auto Threshold on, Threshold and Ratio settings are automatically and constantly adjusted to cope with changing input levels. Threshold is the level above which reduction of the level of the sibilant frequencies, i.e. the 'esses', takes place, so you set this to suit the level of the incoming signal from the microphone. Ratio is the ratio by which signals above the threshold are reduced - so, together, these two parameters control the amount of de-essing. This has to be one of the simplest plug-ins to use - it performs perfectly adequately on straightforward male and female vocals - and it should be ideal for novice users.

Steinberg VoiceMachine

Pitch-shifting effects have been widely available in audio software for the last ten years or so, but most of these have not worked very well with voices and certain instruments. The reason is simple: the human voice contains various fixed frequencies due to the natural resonances of the vocal tract, chest and head - known as 'formants' - which stay the same even though we may sing notes of different frequencies. So, although a basic pitch-shifting algorithm will change individually sung notes correctly, it changes these 'formant' frequencies at the same time - so the pitch-shifted voice sounds unnatural when treated in this way. VoiceMachine takes account of these formant frequencies, allowing you to adjust these voice resonances freely, changing the tone of the voice independently from the pitch of the individual notes. Various instruments, such as acoustic guitars, also have formant frequencies due to the resonances of the instrument body, so these will also benefit from processing via the VoiceMachine.

To take account of the 'latency' delay when processing signals through the VoiceMachine plug-ins, compensation is applied in Cubase VST when the processed track is played back - although not necessarily in other VST applications. With these other applications you can manually delay tracks for playback to compensate.

There are two plug-ins provided with the VoiceMachine, the Processor and the Generator. They can be used in mono, but work best in stereo - in which case you can adjust the stereo position of each voice separately to create 'wide' harmony vocals.

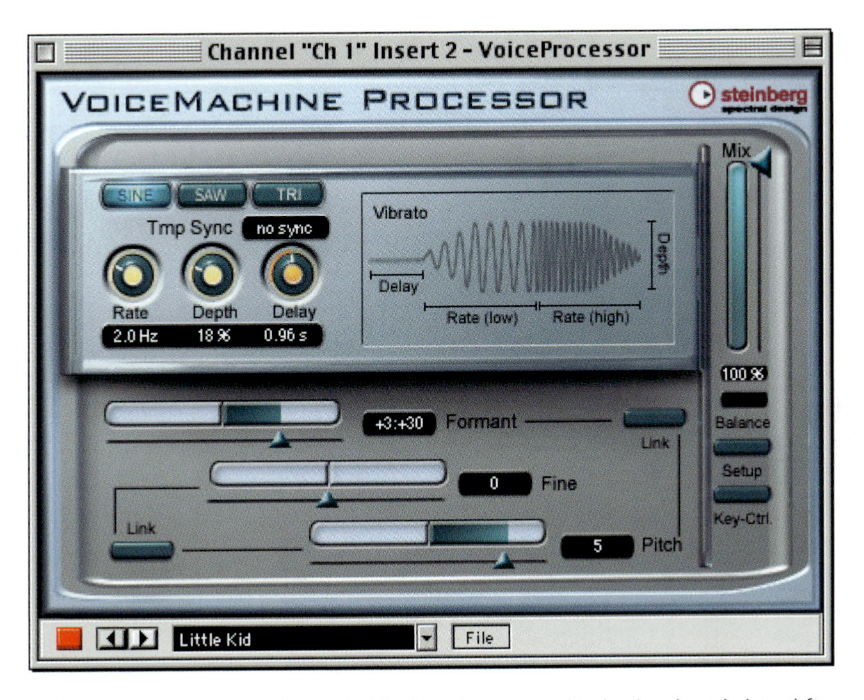

Figure 7.16 Steinberg Spectral Design VoiceMachine Processor plug-in showing pitch and formant controls and vibrato section.

The Processor lets you adjust the pitch and character of a single vocal line, leaving the timing and tempo unaffected. The Mix control lets you balance the original and processed sound and Balance affects the stereo balance of the processed sound when the plug-in is used as a stereo insert effect. The Pitch control lets you transpose vocals by an octave in either direction in semitone steps and fine control lets you adjust the pitch within a semitone, without affecting the formant frequencies. The Pitch and Fine Pitch controls can be linked to provide a continuously varying pitch control that can be more useful at times. The Formant parameter can be separately varied to change the formants without changing the pitch of the notes. Dragging the Formant slider to the left will make a female voice sound more masculine while dragging to the right will make the vocals sound brighter and 'smaller', creating childish- or chipmunk-sounding voices. You can also link the pitch and formant processing so that the processor works on both together if you want this. With all these types of processing, the timing and tempo of the vocals are unaffected - unlike when you speed up or slow down a tape-recorder, when both timing and pitch are affected. Vibrato controls are provided and vibrato tempo can be synchronized to a song tempo in your MIDI sequencer. You can also control the pitch of the processed sound when in Key Control mode by sending MIDI notes to the plug-in from your sequencer or MIDI keyboard.

Figure 7.17 Steinberg Spectral Design VoiceMachine Generator showing controls for each of the four sections, with keyboard underneath.

The Generator processes a single voice to produce four separate MIDI-controlled voices - each with independent formant, vibrato, level and balance settings. A keyboard is provided onscreen so you can conveniently check the effect while setting controls and a control is provided to 'humanize' the sound - introducing random delays between the different voices to simulate choral sounds.

Both plug-ins have nine presets to give you ideas for how to apply the effects. So you get 'Little Kid', 'Harmonic Duet' and 'Female Unison' in the Processor and 'Spare Choir', '4 Stranded Aliens' and 'Confused Children' with the Generator. The Processor's 'Harmonic Duet' produced very usable results with a cleanly recorded female vocal, but the other presets can best be described as special effects. This Processor is ideal for creating interesting vocal sounds for a film or maybe as a special effect on a record. The preset effects you get with the Generator all produce 'wacky' but interesting sounds - again, great for special effects for films and so forth. Then I tried changing the formants on the female voice without shifting the pitch. This time my dual 500 MHz G4 started 'objecting' and error messages appeared to say that 'a dropout' had occurred and recommending to decrease the processing load. The effect worked well enough when blended with the original voice - thickening the sound interestingly. However, the processed sound on its own was very 'muddy' sounding and not really usable other than as a radical effect.

MIDI control can be used with the Generator to create backing harmonies quickly while working on a song. Just create MIDI notes with the additional pitches you want to use and let these control the four available voices in the Generator. In practice, when I listened to the processed voices, although the pitch-shifting worked well, there were clearly audible 'warbling' and 'glitching' artefacts introduced into the sound. Careful balancing of the harmony vocals in relation to the main vocal produced a usable result at times - although in some instances this was impossible to achieve. As is often the case with digital processing of this kind, the 'cleaner' the source material, the more successful the processing will be.

VoiceMachine works with Cubase VST 4.1, Nuendo 1.5 or other VST-compatible host software and lets you easily achieve a range of interesting vocal effects. VoiceMachine is ideal for anyone looking for special vocal effects, and can be useful for creating harmonies to use in song arrangements. However, it requires a lot of processing power and the sounds it produces are not always very clear.

Hughes & Kettner Warp

The Hughes & Kettner Warp is an amazingly good VST plug-in, primarily designed for guitarists. To create this, Steinberg teamed up with Hughes & Kettner, the leading German brand for professional guitar amplifiers and makers of the zenTera digital guitar amp. Warp VST brings the Hughes & Kettner Dynamic Sector Modelling (DSM) Technology to users of VST-compatible host applications such as Cubase VST and Nuendo. The software draws on Hughes & Kettner's extensive experience with valve amplifier and loudspeaker cabinet design, teaming this with digital signal processing expertise from Spectral Design to produce one of the most realistic simulations of popular rock guitar amplifier and speaker combinations that I have heard.

The Warp VST features three amplifier models – Jazz Chorus Clean, Plexi and Warp – and three speaker cabinet models – a 12″ Combo and both 'British' and 'Greenback' 4 × 12″ cabinets.

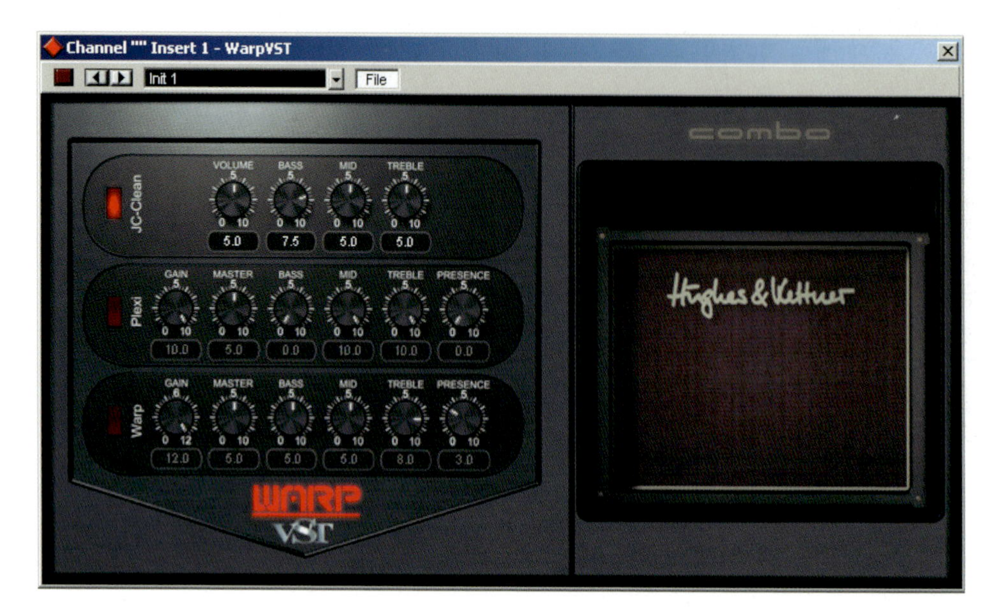

Figure 7.18 Warp JC-Clean with Combo speaker cabinet.

The Jazz Chorus amplifier, modelled on the popular Roland Jazz Chorus combo amps from the 70s, produces a big, warm, yet clean sound. Like the original, the controls provided are just the basic Bass, Mid, Treble and Volume.

The Plexi amplifier is modelled on a 'British' 50-watt 'head', such as the classic Marshall 50, so you get both Gain and Master level controls and there is an extra tone control for Presence that adds high treble.

The Warp amplifier is similar in character to the Rectified 'heads' made in California, such as the Mesa Boogie models.

Typically, you will use this plug-in as a channel insert – although you can use it as a send effect and route several instruments to the same plug-in if you like. When you record your guitar (or any other instrument), the audio goes directly onto your hard drive – it is only monitored through the insert plug-in. This means that you can change the amplifier and speaker settings any time you like after you have made your recordings. Of course, to 'fix' your sound, you will need to bounce your audio track to disk to create a new audio file – complete with the Warp processed sound – and then use this instead of the original clean recording. Once

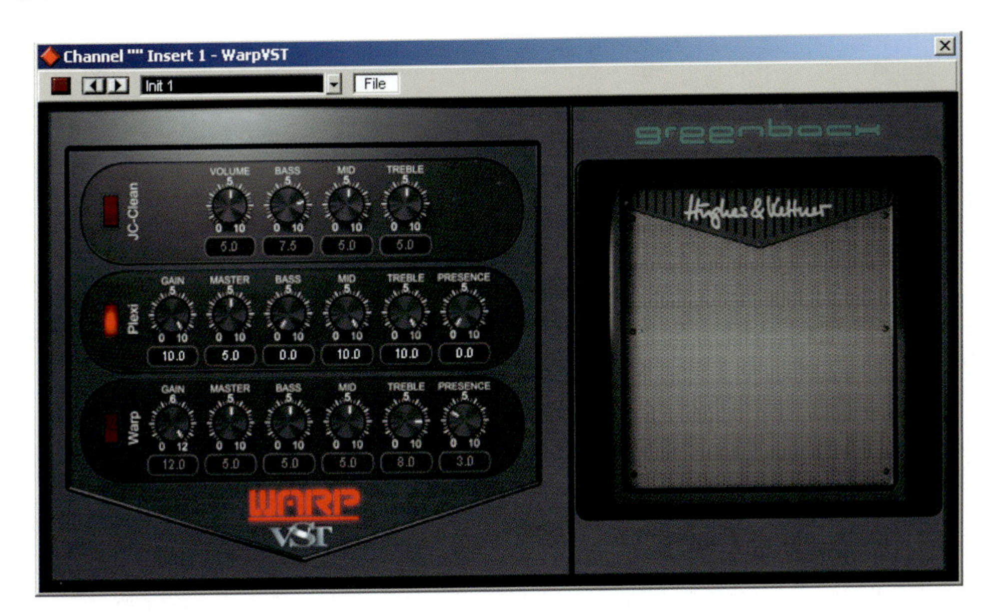

Figure 7.19 Warp Plexi with Greenback speaker cabinet.

you have gone through this process, you can remove the plug-in from the original guitar track, mute this track, and use your DSP power for something else. And if you change your mind, you can always re-process the original recording - as long as you keep this on your hard drive.

Figure 7.20 Warp Rectifier amp with British speaker cabinet.

Session guitarist Thomas Blug, who helped with the design of this plug-in, demonstrated the Warp plug-in for me at Steinberg's headquarters in Hamburg. Thomas played a '64 Stratocaster 'live' through the plug-in, performing a couple of stunning guitar instrumentals along with simple but effective MIDI backing tracks. His timing was absolutely impeccable, making it clear that there were absolutely no latency problems, and his Strat sounded just like it was plugged into a vintage amp and speaker setup with a great rock sound – similar to Steve Vai's. I asked Thomas how good he thought the system was and he told me that he was satisfied that they had got about 90% of the way towards delivering the sound of real amplifiers and speakers – not bad for software! Now I have previously encountered problems due to latency delays while over-dubbing through similar plug-ins, yet there was virtually no latency delay on this demonstration system. So why not? It turned out that the audio hardware used for this demo was Steinberg's own Nuendo AudioLink – which has a latency delay of just 2 or 3 milliseconds. Obviously, for similar results you need to use a similar card. The AudioLink cards are based on the popular RME Hammerfall cards, which I can highly recommend.

TL Audio EQ

Figure 7.21 TL Audio EQ-1.

EQ-1 is a plug-in version of TL Audio's highly rated Valve Equalizer. Use this on voice, strings or synthesizer sounds to warm up these recordings – or even put your whole mix through the plug-in to add some final 'colour'. The EQ-1 not only emulates all the features of the original TL Audio unit, it goes further by including a graphic display showing the frequency response curve. Just click on the grid showing the TLAudio EQ-1 logo to reveal this.

The rotary controls at the left and right of the plug-in window let you set the Input and Output Gain levels. If you raise the Input Gain and lower the Output Gain this will create a pleasing 'valve distortion' type of sound that you can use to enhance your recordings. Four filter bands are provided with musically appropriate preset parameter settings. Two of these bands cover the mid-frequency ranges and use fixed Q values. The other two bands are high and low shelving EQ filters with fixed slope characteristics. The controls for these filter bands are arranged in pairs – with switches for the filter cut-off or corner frequencies above and rotary faders for the boost or cut amount below and to the right of each of these. Low Shelving cut-off frequencies are at 60, 120, 250 and 500 Hz; Low Mid corner frequencies are at 250, 500, 1000 and 2200 Hz; High Mid corner frequencies are at 1500, 2200, 3600 and 5000 Hz; and High Shelving cut-off frequencies are at 2200, 5000, 8000 and 12 000 Hz. You can switch any of the filters in or out by clicking on the LF, LM, HM and HF buttons below the filter Gain controls and it makes sense to switch off any you are not using. (This way the CPU does not have to make any calculations for the unused filter.) There is just one more control to describe – the Channel Selector switch. You can use this to apply the EQ-1 filters either to the left or to the right, or to both channels. If you set this to L or R, the other channel will be set to Bypass and no processor power will be used for the bypassed channel.

Note If you want to process each channel of a stereo recording independently, you will have to insert two instances of the EQ-1 onto the stereo track, switch one to process the left channel and the other to process the right channel.

Waldorf D-Pole

The D-Pole Filter Module is, basically, the filter section from a typical analogue synthesizer and can be used to process any kind of input – from guitars to voices or whatever you like.

The Waldorf D-Pole filters are based on the filter algorithms used in Waldorf's Microwave II analogue filter section. D-Pole has an interesting amplifier section that includes an Overdrive control along with Volume, Panning and Delay Mix controls. The Overdrive control works well with guitars, for instance. A Sample

Figure 7.22 Waldorf D-Pole Filter Module.

Rate control lets you reduce the sample rate used to read the input signal – to allow for interesting digital 'grunge' effects – and a Ring Modulator section with its associated modulation controls lets you create typical ring modulation effects. The stereo delay section helps to create more spatial sounds and can also be used for Karplus/Strong synthesis. For example, if you use an extremely short Delay Time with relatively high Feedback and then damp the high or low frequencies the sound will become robot-like.

A sophisticated LFO section can be synchronized with the audio signal and provides a much wider frequency range than most low frequency oscillator designs. You can use the LFO's Pan Mod control to produce an automatic panning effect, or control the filter's cut-off frequency or the oscillator's frequency. A basic Envelope Follower section is also included. This lets you control the length of the envelope and the way the filter cut-off responds.

D-Pole can take tempo information from VST 2.0-compatible host software and use this to control its Delay Time and LFO Speed. When the Tempo parameter is set to 'on', this feature becomes active and the popup menus for these parameters will allow common note values to be selected. Also, when using VST 2.0 host software, all D-Pole's parameters can be controlled using MIDI Control Change messages from an external device. For example, you can use the Waldorf Microwave XT hardware synthesizer to directly control D-Pole using its controls.

D-Pole is best used either as a Channel Send effect or as a Master effect. If you want to use D-Pole on an individual mono instrument, open up the Channel Settings window in Cubase VST, insert D-Pole on one of the Sends and set the rotary level control to full on. Enable the Pre fader button to route the Pre fader signal directly to the effect and then set the channel volume fader in the Mixer window to zero – or press the Mute switch to mute the channel. This way you make sure that the original signal is only routed through the filter and not directly to the main outputs.

Figure 7.23 D-Pole used as a Channel Send effect in Cubase VST.

Describing the effects that D-Pole can produce is not easy to do in words, and this is reflected in the names Waldorf have given to the Presets provided with this plug-in. What does a 'Mertozoid' sound like, for instance? 'Thick Flanger' and 'Electro Funk' are as close as they get to being descriptive.

Note If you want to process a stereo signal without having to load two separate D-Pole plug-ins, then you should set D-Pole up as a Master Effect.

Nevertheless, if you listen to many popular dance music tracks from the various sub-genres, you will hear the kind of sounds that D-Pole makes. Of course, you could always go along to a music store and check out a Waldorf Microwave!

Multi Rhythm
Living Death
Run WMF, Run
Alien visiting Spain
Pass Stop
Electro Funk
Ring Wars
✓ Resonator Ride
AM radio
Thick Flanger
Imperial Code
Voice FX
slightly exited
Cutter
Muting
Mertozoid

Figure 7.24 D-Pole presets.

Steinberg Spectral Design Q-Metric EQ

If you are looking for a serious equalizer tool to 'sweeten' your audio recordings, then Steinberg's Q-Metric parametric equalizer deserves your full attention. Unlike many VST plug-ins that tend to have simplified feature

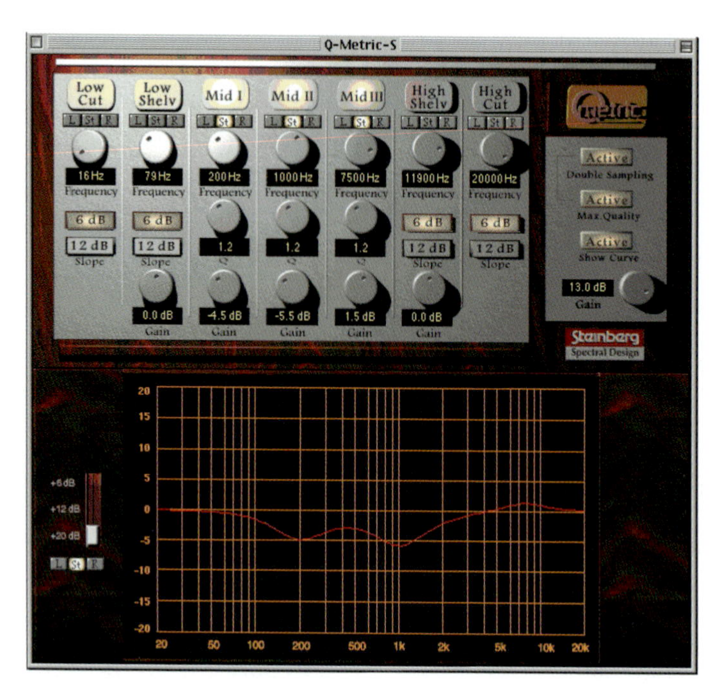

Figure 7.25 Steinberg Spectral Design Q-Metric EQ.

sets, Q-Metric includes all the features you need to 'zoom in' on frequencies of interest in your audio and adjust the response accordingly. The seven filter sections include Low Cut and Shelf, three mid-frequency filters, and High Shelf and Cut. These are arranged in vertical columns with an on/off switch at the top of each section and three switches below these to let you process left, right or both channels of the stereo input. Each filter has a frequency control that you can use to set centre, cut-off or stop frequency. Cut-off is the frequency at which a filter starts to act, so you set this to the frequency above or below which you want the filter to take effect for Low and High Cut or Shelf filters - which also have switches to let you select the steepness of the filter. Mid-frequency filters allow full control of their parameters - hence the name parametric - and here you set the centre frequency for the range of frequencies over which the filter acts. 'Q' controls let you set the bandwidth, i.e. the range of frequencies, which these filters control. Finally, all filters have gain controls to let you set the amount of filtering - apart from the Cut filters, which have a fixed amount. To sum up, Q-Metric EQ provides all the parameters you would expect to find on a high-end hardware device, is instantly usable and sounds great!

Steinberg Spectral Design Magneto

Figure 7.26 Steinberg Spectral Design Magneto.

Magneto is designed to simulate tape saturation and tape overdrive to add warmth, punch and brilliance to any sound. It works well with bass and guitar recordings – or on drum samples or loops. The meters look like the typical VU-type meters found on older equipment and these can be switched to show the level of the Input signal, an equivalent of the level recorded on the simulated Tape, or the Output level. A rotary control is provided to allow adjustments to be made to the level of the input signal to keep this within a sensible range – ideally as close to 0 dB without ever exceeding this. At the other end, a rotary control is provided for the Output level. Normally this should be left at 0 dB, because Magneto has an auto-gain function that aims to keep the output level as close to 0 dBfs as possible at high Drive settings. At very low drive settings you may need to increase the Output level. If you use very high HF Adjust settings you may need to reduce the Output level a little. The Drive parameter is the main control that you will use to set the simulated analogue tape 'recording level'. For example, a setting of 10 means that the 'tape' is being overdriven by 10 dB – producing the typical sound of tape saturation and causing compression of the sound. Of course, you should use this plug-in with material that has not already been heavily compressed or recorded onto analogue tape! The Tape Speed switch has two settings – 15 and 30 ips. There are slight differences in the character of the harmonics produced at these different tape speeds, although these can be difficult to hear with some material. You can also use the HF Adjust parameter to compensate for the loss of high frequencies that the overdrive effect produces. Unlike on a real tape recorder, this parameter can also be used to boost the high frequency content. To sum up: Magneto is a useful effect to have with your VST system, ideal for adding attractive colorations to the often-sterile sounds of digital recordings.

Steinberg Mastering Edition

If you want to get seriously involved in mastering your recordings and you are using VST-compatible software then you should definitely check out the Steinberg Mastering Edition which offers a suite of six plug-ins dedicated to finalizing and mastering audio files.

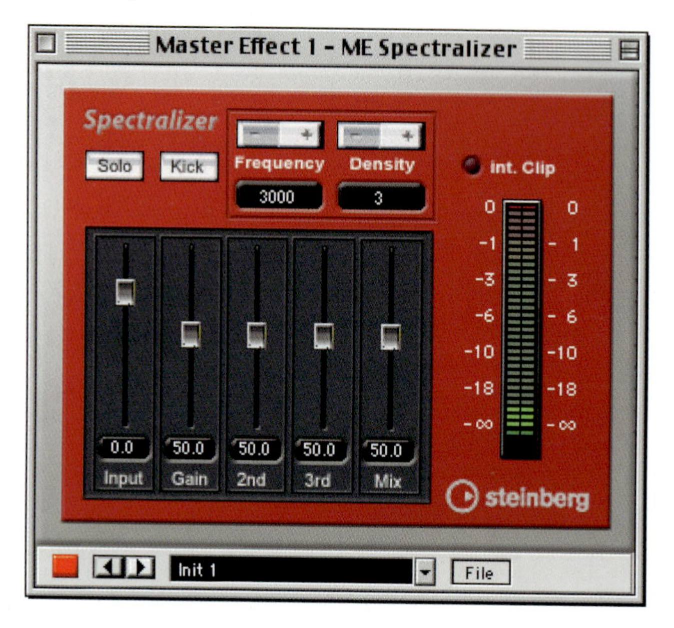

Figure 7.27 Steinberg Mastering Edition Spectralizer.

Let's look at the Spectralizer first. This is a 'sonic optimizer' which you can use to increase the width of the frequency spectrum in a recording, or to improve the clarity and transparency, or to make the recording 'warmer', more pleasing and more interesting to listen to. And how does it do this? Put simply, Spectralizer lets you boost the second and third harmonic frequencies in your audio and provides a 'Kick' switch that adds extra presence. One big advantage here is that, unlike analogue enhancers, no noise is added and the phase of the signal is not affected. The controls are mostly self-explanatory, but some deserve a little explanation. Kick adds extra harmonics to transients to give more attack or 'kick' to the sound. The Solo button lets you listen to what you are adding to the signal on its own. The Frequency control lets you set the high-pass filter which determines the frequencies above which harmonics are added. For example, if you set the

Frequency to 3000 Hz, the second harmonic generator adds harmonics from 6000 Hz upwards, and so forth.

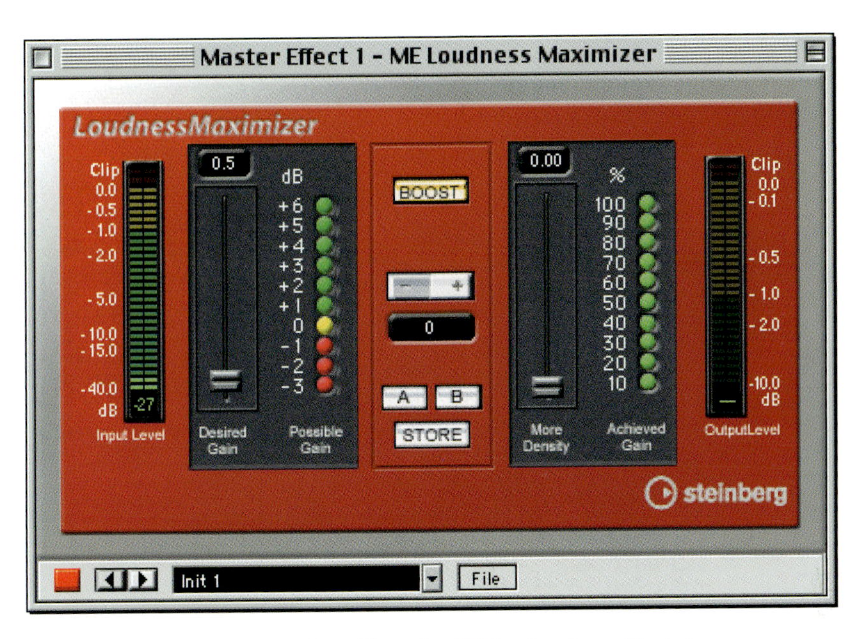

Figure 7.28 Steinberg Mastering Edition Loudness Maximizer.

Then you get the Loudness Maximizer. This raises the perceived loudness of the audio above the actual maximum amplitude without adding typical artefacts like 'pumping' or 'breathing'. So you can take music that is already normalized - with the loudest sections already at the maximum values - and make it louder still with a minimum of timbral changes. Its sophisticated algorithms adapt continuously to the audio material, making it extremely straightforward to use. The loudness is raised by increasing the density of the audio material and by limiting transients and simultaneously raising the general level. Use the Desired Gain slider to set the amount by which you wish to increase the loudness while keeping watch on the Possible Gain meter to make sure that this doesn't fall below 0 dB. The +/– buttons in the centre of the window let you adjust the response of the limiter for a harder or softer action. The Boost switch adds a further 2 dB of loudness if you feel you need this. The More Density control lets you adjust the balance between the compression and limiting action. The Achieved Gain shows what it says - the amount of gain by which you have boosted the audio. The Loudness Maximizer process does change the audio, but the trick is to shape the modifications so that they are either masked by the audio signal or perceived as musically appropriate - which, of course, will depend on the nature of the audio involved. This plug-in is ideal for digital mastering, preparation for broadcasting and other areas where you want maximum output and excellent sound quality.

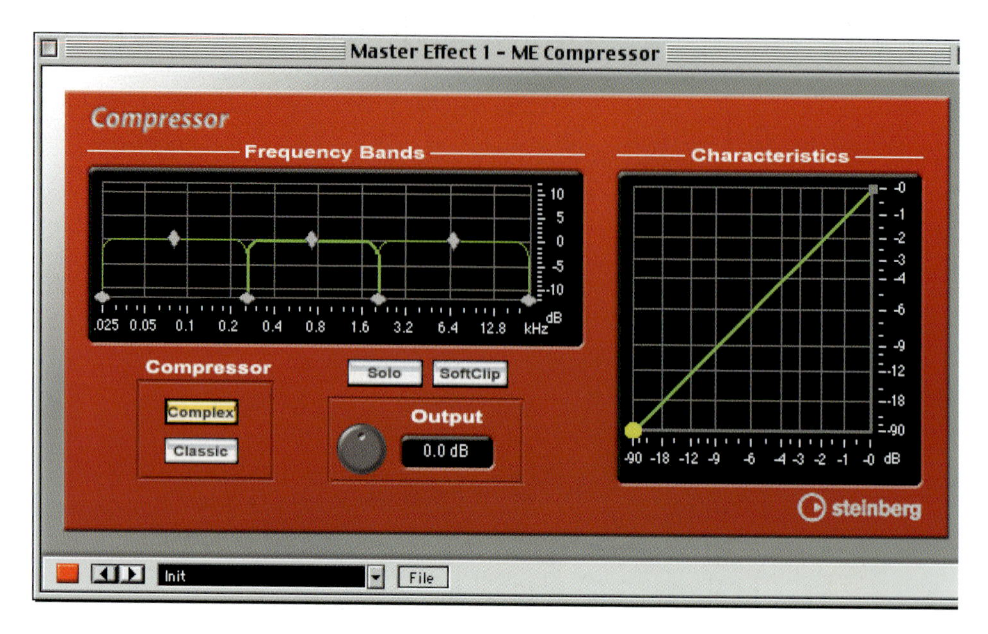

Figure 7.29 Steinberg Mastering Edition Compressor.

As you would expect, the suite includes a Compressor - in this case a multi-band compressor with five frequency bands, each with its own freely adjustable compressor characteristic. The Compressor also features graphic editing that makes the setting of compressor characteristics particularly easy and intuitive. The high degree of control provided makes this Compressor suitable for a variety of applications, ranging from single-instrument processing to mastering and finalizing complete mixes.

One of the most powerful and innovative plug-ins provided is the FreeFilter. This is a linear-phase real-time 1/3-octave equalizer with a number of features that go far beyond the norm for this type of device. In the Frequency monitor section of the window the input signal is displayed in green, the output signal in red and the filter curve in yellow - so critical frequencies can be easily detected and corrected. The Fader section directly below contains 30 faders that control the settings for the filter bank. If you select the Rubberband tool, you can also use the mouse to select a range of frequencies in the Frequency monitor which will now behave as a group, so all the faders within the group will move when you just move one of them, while maintaining their relative positions. Similar tools simplify creation of complex filter curves. A 'Learn' function is also provided which lets you analyse

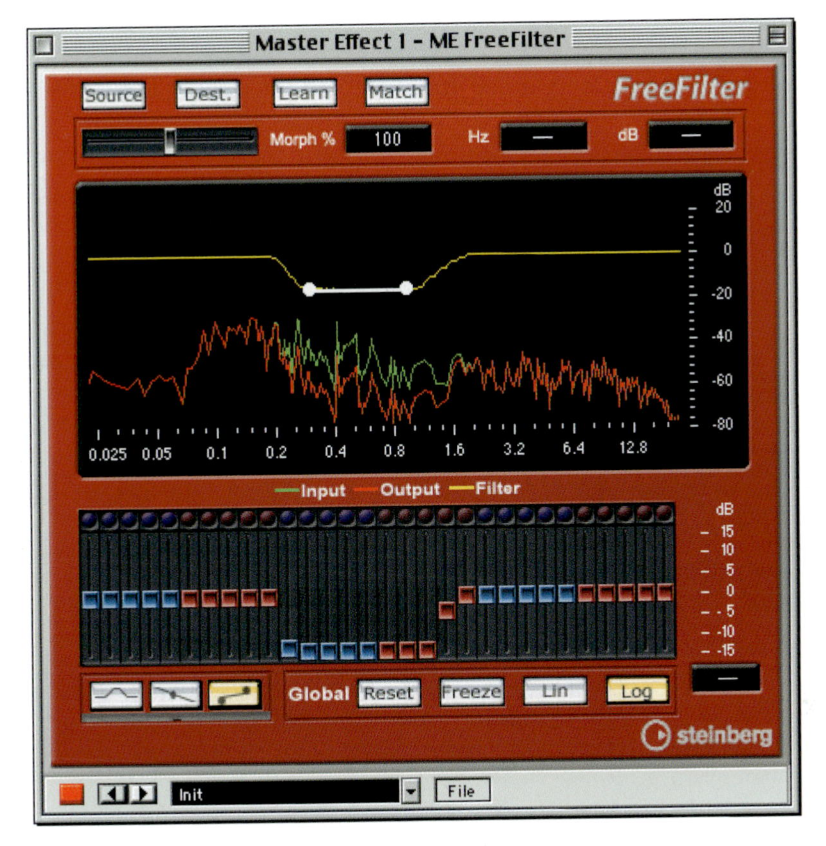

Figure 7.30 Steinberg Mastering Edition FreeFilter.

some source material, such as an existing track on a commercial CD. You can then apply the filter curve derived from this analysis to any audio file you like – to make your own mixes conform better to those on commercially released CDs, for example, or to help when producing compilation CDs.

The PhaseScope and SpectroGraph plug-ins complete the suite by providing a pair of serious analysis tools.

PhaseScope lets you examine the direction and mono compatibility of stereo sound material – combining the functions of a stereoscope and a correlation meter. The stereoscope helps you judge the direction, phase and relative levels of the stereo signal. The correlation meter indicates phase relationship and stereo image width. In addition to these optical controls, PhaseScope can also be switched to mono playback, which lets you verify mono compatibility by ear. The stereoscope is mainly used to control recordings created using the intensity-

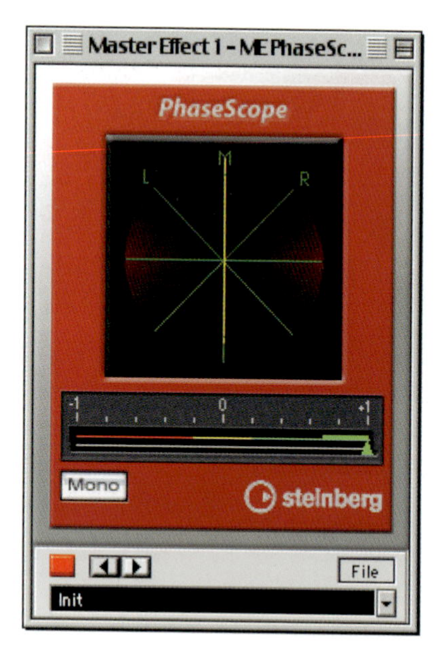

Figure 7.31 Steinberg Mastering Edition PhaseScope.

stereophony technique. Recordings that you make using the spaced-apart stereophony technique or mixed methods are by nature harder to interpret. Checking for mono compatibility is important for FM radio broadcasting and for vinyl record production, as well as for other radio and TV broadcasting – don't forget that there are still plenty of mono TV and radio sets 'out there'. The correlation meter can be used to examine the stereo signal in terms of phase position and stereo image width. The correlation indicates the degree of similarity of the signals in the left and right channels, or their current phase difference. The display will show you if one of the channels is phase reversed, or if one channel is missing, and will indicate whether a stereo signal is mono-compatible or not.

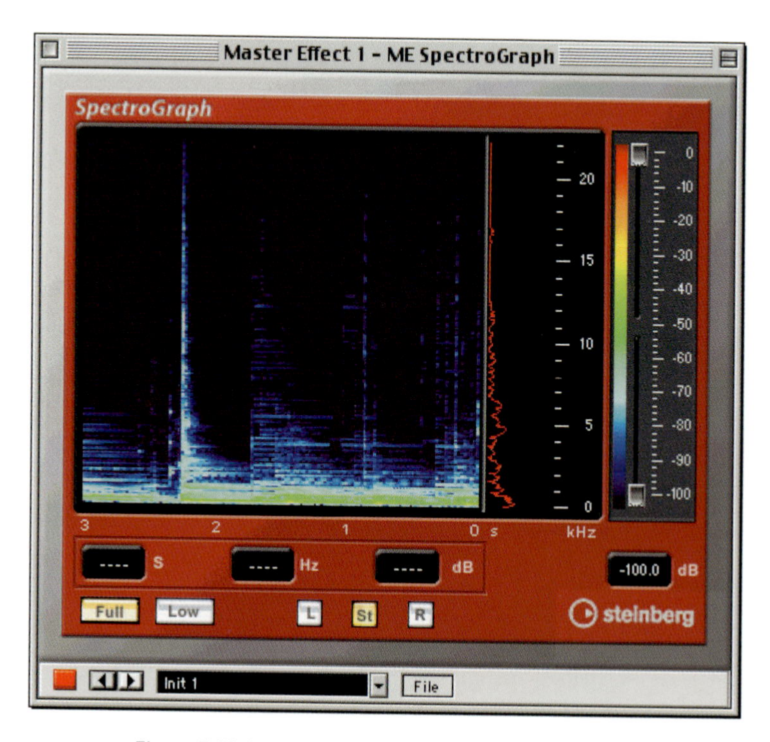

Figure 7.32 Steinberg Mastering Edition SpectroGraph.

SpectroGraph analyses frequencies and displays a spectral plot as a real-time sonogram, allowing you to check out even the smallest details in a recording. SpectroGraph can also be used for measuring purposes and for the detection of interfering frequencies, noise and DC offsets. Up to three seconds of audio can be viewed in the real-time Spectrum display with up to 64 different colours assigned to a level range from −100 to 0 dB. A separate Time Level display indicates the current level distribution across the frequency range. The displays can be switched to display just the lower frequencies between 0 and 1 kHz or to show just the left or right channel - although normally what you see is a composite of the two channels looking at the full frequency spectrum. I found this plug-in particularly useful when trying to identify the best EQ settings to remove the 'boominess' from some acoustic guitar recordings recently - and I would recommend using a spectrum analyser to help with any tricky EQing problems.

In a nutshell, the Steinberg Mastering Edition has all the tools you will need to make sure your mixes are truly ready for distribution or broadcast, sounding professional and with no lurking problems.

Chapter summary

There can be no doubt that Steinberg offers a solid range of signal processing plug-ins of interest to audio professionals. There is something for everyone here: the Hughes & Kettner Warp for guitarists, the GRM Tools for sound designers, the TL-Audio EQ and the Steinberg Voice Machine plug-ins for pop producers, the Prosoniq plug-ins and D-Pole if you are doing dance music and remixes, and the Spectral Design plug-ins (especially the Mastering Edition) for professional engineers.

Figure 8.1 Altiverb IR Pre-Processor.

Altiverb comes with a good selection of impulse responses taken from halls and rooms such as the Amsterdam Concertgebouw and the Arts & Sciences Building, along with those of several classic synthetic reverbs. To get started, you simply make your choice from the impulse response selector popup located at the centre of the Altiverb plug-in's window.

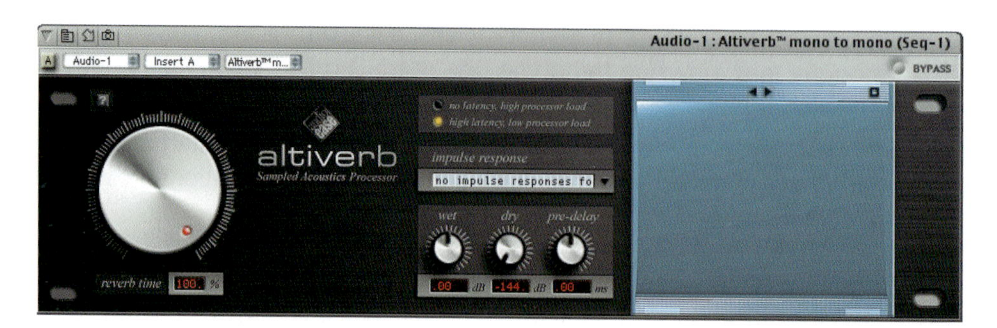

Figure 8.2 Altiverb plug-in.

The contents of this popup depend on the contents of the folder 'Altiverb impulse responses', and on which channel configuration you have inserted. For example, there are plenty of mono-to-stereo along with several stereo-to-four-channel impulse responses to choose from.

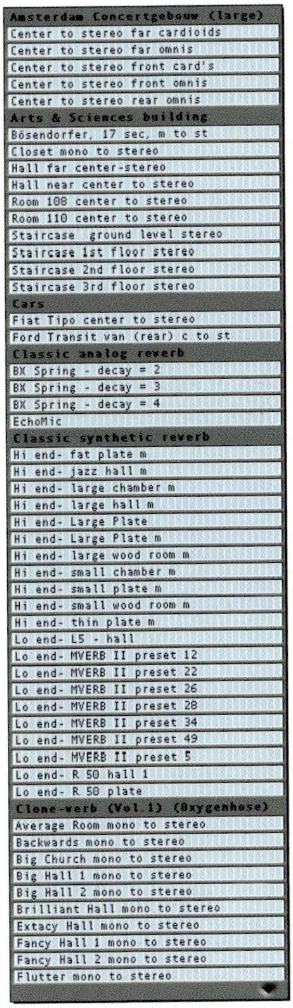

Figure 8.3 Altiverb impulse responses.

When you have selected an impulse response, a diagram indicating the basic setup and venue will appear in the plug-in's monitor window. Various interior and exterior photographs of the recording venue are usually provided as well. You can step through these using the forward and backward arrows at the top-centre of the monitor window.

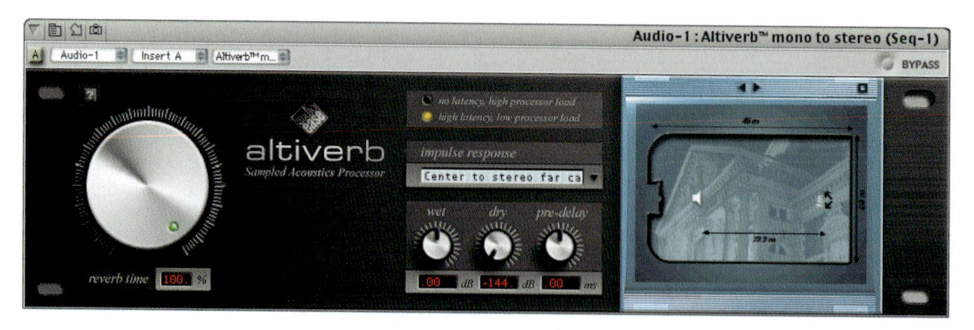

Figure 8.4 Altiverb plug-in with Amsterdam Concertgebouw IR selected, showing picture in monitor.

If you want to see a larger version of any of these pictures so you can see more detail, just double-click on the picture in the monitor window or click the small rectangle at the top right of the monitor window.

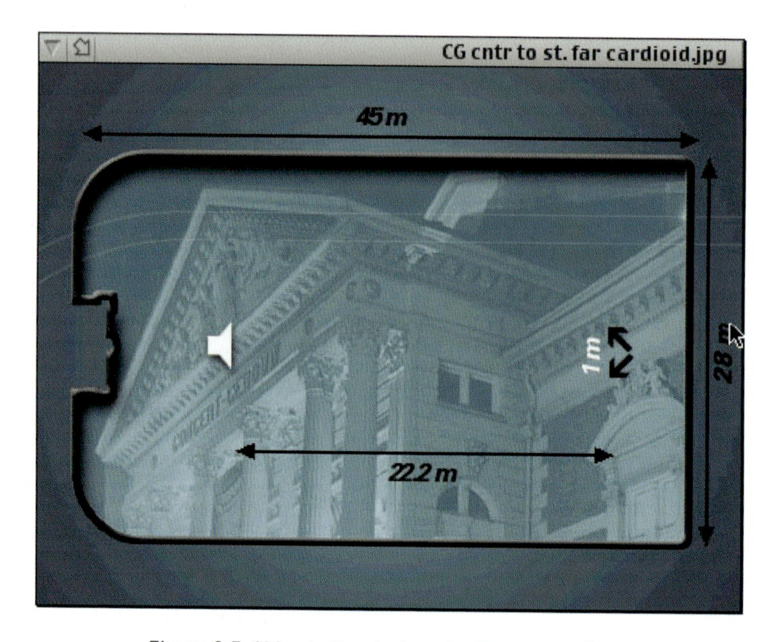

Figure 8.5 Altiverb Plug-in Impulse Response picture.

Most of the Audio Ease impulse responses are accompanied by a Recording Statistics picture. This provides information about the microphone setups and about where the speakers were positioned at the time that the impulse response was recorded.

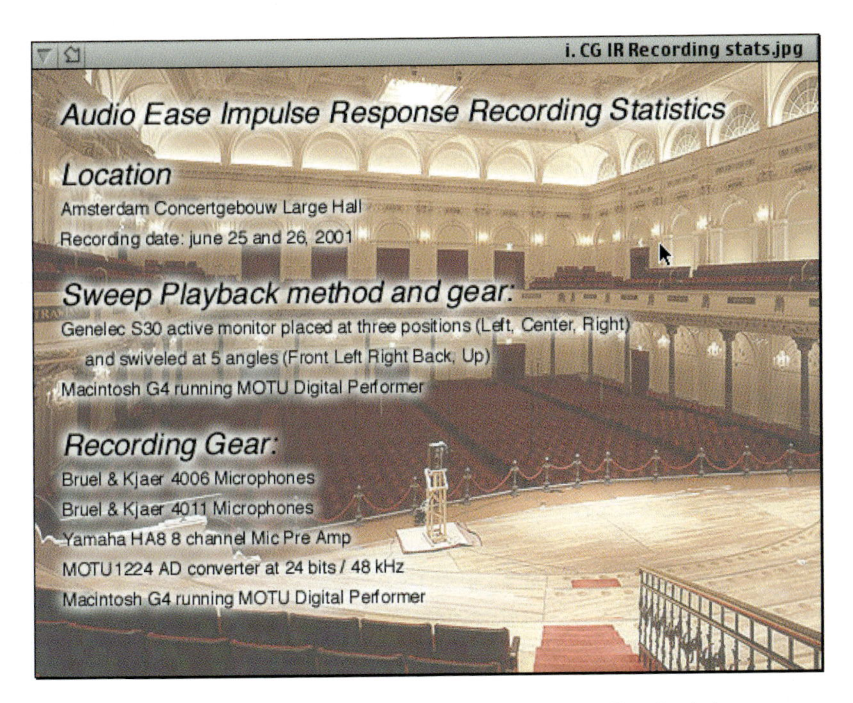

Figure 8.6 Altiverb Plug-in Impulse Response Recording Statistics.

The Impulse Response statistics screen shows the latency, the length of the impulse response, and the sample rate of the impulse response file.

Note When you use a different sample rate than the one displayed, Altiverb will convert the sample rate of the impulse response upon reading it from disk and use the new sample rate, not the one that it displays.

When you run your audio through Altiverb, the 'wet' output signal will sound as if the source sound was played back by the speakers and re-recorded with the microphones in this venue. To my ears, the Altiverb reverb sounds very much like you are in the real hall, with a smoothness that most digital reverb algorithms have yet to match. And Altiverb's user interface could not be much simpler to use – especially compared with some of the more advanced reverb processors that present you with dozens of parameters to tweak.

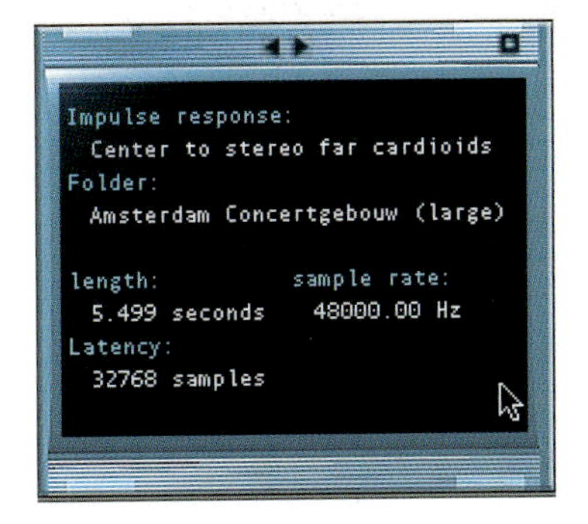

Figure 8.7 Altiverb Impulse Response statistics.

Two 'radio' buttons at the top centre of the window let you choose between 'no latency, high processor load' and 'high latency, low processor load'. The 'high latency' mode uses a minimum of processing power but can produce a delay of up to a second between input and output. Check the Impulse Response Statistics in the monitor window to see the exact delay. In 'no latency' mode, there is no delay between input and output of Altiverb, but at the expense of increased processing power demand.

At the lower centre of the window, there are three rotary controls - one each for Wet, Dry and Pre-Delay, again with associated numerical readouts. A completely wet signal will provide the most realistic acoustic effect, but you may wish to mix in some of the dry signal when creating special effects. Pre-delay is a delay that you can add between the dry sound and the wet sound. When the knob is at 12 o'clock, there is no delay between the two. When you turn it to the right, you will delay the wet sound so that the dry sound will become less 'attached' to the reverb. When you turn the knob to the left, the dry sound will be delayed, possibly placing it in or behind the early reflection pattern of the wet sound. This can create a better stereo image and more realistic positioning of the dry sound.

At the far left of the plug-in window, there is one large rotary control for Reverb Time with an associated numerical readout. The default is 100% of the time recorded in the original impulse response, and you can use this control to shorten the reverb tail if you like.

Note It is not possible to lengthen the tail, you can only shorten it.

Technically, the software applies an exponentially decaying envelope to the impulse response as you reduce the reverb time. When you turn the reverb time knob, Altiverb needs to calculate the new decay shape. While it does that it flashes a red light in the knob's indicator. Once the indicator has turned green again the reverb time re-calculation is completed. Again, you need the fastest computer you can get for best results here.

The interesting thing about Altiverb is that it is not restricted to creating reverb - Altiverb can convolve the input audio with any impulse response. So you could take impulse responses from microphones, loudspeakers, outboard gear such as equalizers, the body of a guitar, human vowel formant filters, or whatever you like. Some of these would work best using the mono-to-mono Altiverb. You can also combine impulse responses. So, for example, you could merge two impulse responses to create the sound of a particular loudspeaker in a particular room. And although Altiverb is primarily intended to be used to convolve impulse responses with the input sound, there is nothing to prevent you experimenting by convolving any sound you like with the input. Try using a guitar strum or white noise and see what happens - you might like the effect!

Audio Ease Rocket Science Bundle

There are three plug-ins in the Rocket Science Bundle - Orbit, Follo and Roger. These are all currently exclusive to MAS-compatible host software (e.g. Digital Performer).

They can all respond to Parameter Automation data as well as MIDI data playing back from Digital Performer - allowing comprehensive automation.

Orbit - Psycho Acoustic Flight Path Simulator

Audio Ease sub-titles the name Orbit as a 'Psycho Acoustic Flight Path Simulator'. Between the two names, you should have a fair idea of what Orbit does. The demo file supplied with Orbit is a 'sound fantasy' with lots of spatial movement creating interest for the listener. Immediately you can hear what Orbit is all about - moving sounds around in space and providing early reflections to support the illusion.

The localization cues that the listener gets from Orbit are generated by calculating the reflections of the sound source from the walls of the outer room. You can choose the number of reflections that will be calculated using the buttons underneath the room display. The higher the setting, the longer your reflection pattern and the more processor power Orbit will need. On the other hand, if you choose zero, then only the direct sound will reach the speakers in the simulated inner room. In this case, the acoustic characteristics of the outer room will not be imposed on your sound although the positioning effect will still be heard.

You can use Orbit simply as an automated panning device, or you can use it as a sophisticated flight path simulator that offers control over the position, speed and location of the source, the size of the virtual space and the absorption of high frequencies by the walls. Orbit can easily be used to emulate the Doppler effect, for example, so that the pitch of the source will shift up as the sound travels towards you and down as it moves away.

Figure 8.8 Orbit - Psycho Acoustic Flight Path Simulator.

Orbit's user interface features a large graphic display showing the localization parameters. The small light blue inner square in the display is the room you are in. You can drag this up and down in the outer room and its size can be set using the Speaker Span parameter. The tiled surface represents the outer room and this can be adjusted with the Room Size parameter. Each dark or light blue tile in the outer room represents a metre.

The small red ball in the display represents the sound source and this may remain stationary or move around in the display, according to the motion type selected. The motion selector popup at the left of the window lets you select the source's motion type and you can choose between five settings. You can have a static source or one that follows a path set by clicks in the outer room. The source can also follow an adjustable oval, travelling left to right or right to left - or it can just fly about randomly if you prefer.

Figure 8.9 Orbit Motion Selector popup.

When you select the first motion type, the static source, the red ball will just stay wherever you click the mouse in the outer room. This requires less processing power than the four moving types, so you may be able to select more reflections.

If you select the second motion type, the red ball will move at a pace set by the Speed control to wherever you click in the outer room.

With the next two motion types, the red ball will follow the trail set by an adjustable white oval, in the clockwise or counter-clockwise direction, at a pace set by the speed knob. You can change the horizontal and vertical dimensions of the oval by dragging any of its white squares. You can change the location of the circle by clicking inside the circle and dragging it. You can also set the circle so narrow that it becomes a vertical or horizontal line.

With the fifth motion type, the red ball moves randomly within the oval, again at a pace set by the Speed knob.

You can set the maximum speed of the source in metres per second using the Speed knob. You can hold option (alt) down while dragging for extra precision, or you can type in a value. The High Damp knob lets you set the frequency response of the walls of the outer room. For example, wooden walls reflect less high frequencies than concrete, so to simulate these you should apply some High Damping. When High Damp is set to zero, reflections will sound as bright as the direct sound.

Note Source localization works best if you set the Speaker Span value to the distance between the speakers that you are actually listening to. If you set this value too high, you may get a 'hole in the middle' and if you set it too low the stereo image may become too narrow. Click in the Speaker Span field and drag to left and right to adjust the setting. Hold down option ((alt) while dragging for more precision. In the graphic display you will see the light blue inner room size change accordingly. The minimum setting of 0.2 metre is meant for use on headphones.

The Speed and High Damp knobs can be automated, and the first two motion modes allow clicks in the window to be automated as x- and y-coordinate parameters.

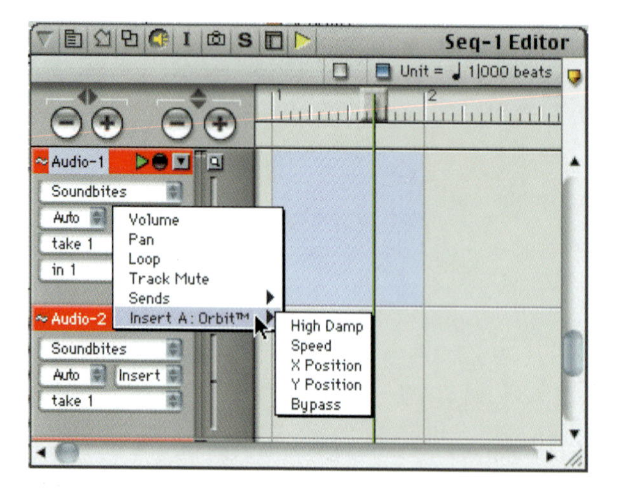

Figure 8.10 Orbit automation parameters in Performer.

Note Orbit is mainly used as a mono-to-stereo plug-in, although there is a stereo-to-stereo version that mixes the left and right inputs before they are sent to Orbit.

Tip To simulate the small movements a real musician would make, you can use the random motion type and confine the player within a small circle.

Bear in mind that moving sources require more processing power than static ones and that when you increase the number of reflections the processor load goes up.

Tip Suppose you have four close-mic'ed (or MIDI'ed) instruments on separate tracks of Performer. Insert an Orbit in every channel, set the number of reflections to 0 and the motion static. You can now push every instrument to its own location, creating a much more realistic panorama than you could have with plain panning. If you then want to add reverb, bear in mind that Orbit calculates the early reflections for your sound sources, so you should disable the early reflections in your reverb processor.

Orbit is a bit like a Doppler Effect plug-in 'on steroids'! It's a 'snap' to use and you can create some great effects with very little effort. And after you spend even a little time working with the automation parameters, you will find it easy to get sold on the idea of featuring subtle movements of various elements within your mixes as well as creating the more extreme effects.

Follo – Energy Driven Band Booster

Follo automatically sweeps a filter between preset frequency limits to produce an output that changes in response to the input or to control information.

Technically, Follo works by moving the peak frequency of a resonating bandpass filter according to the level of the incoming audio – and according to any incoming MIDI modulation wheel or breath control. So, when the volume goes up, the filter goes up and when the volume goes down the filter goes down. It's a bit like those guitar stomp-pedals that automatically sweep through harmonics from low to high depending on how hard you play, or like the filter section on an analogue synthesizer that can be set up to produce a similar effect.

> **Tip** Try inserting Performer's Echo plug-in before Follo in a channel. When the input sound stops, Echo with feedback makes it fade away in 'bumps'. And Follo will follow!

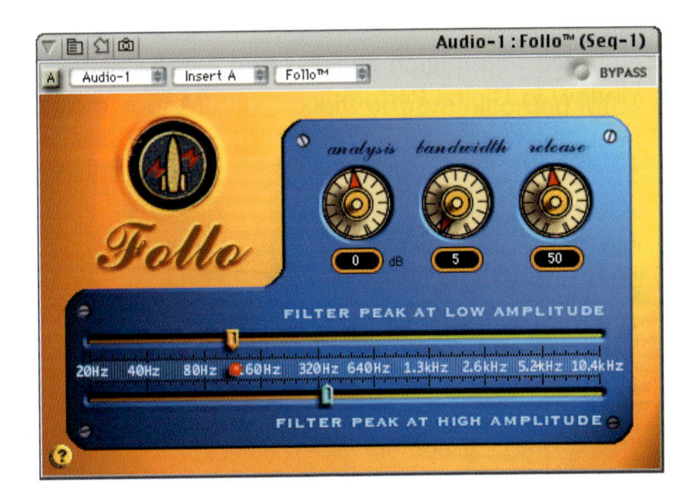

Figure 8.11 Follo – Energy Driven Band Booster.

In Amplify mode, if you drag any fader up or down, all the faders move up and down at the same time – without disturbing the relationship between these. So the shape of any EQ curve you have created will not change – you will simply hear an overall gain change.

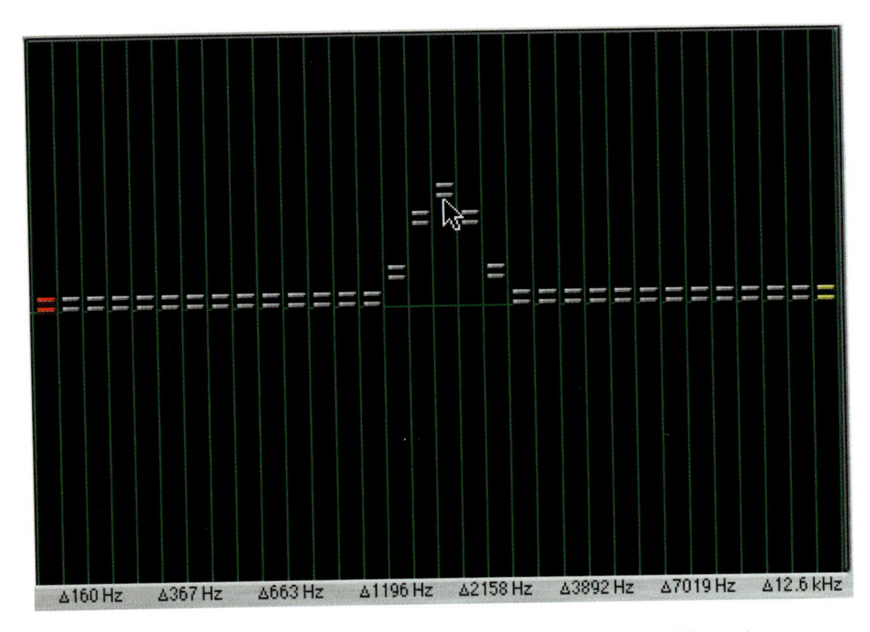

Figure 8.26 Periscope Equalizer Display before using Amplify mode.

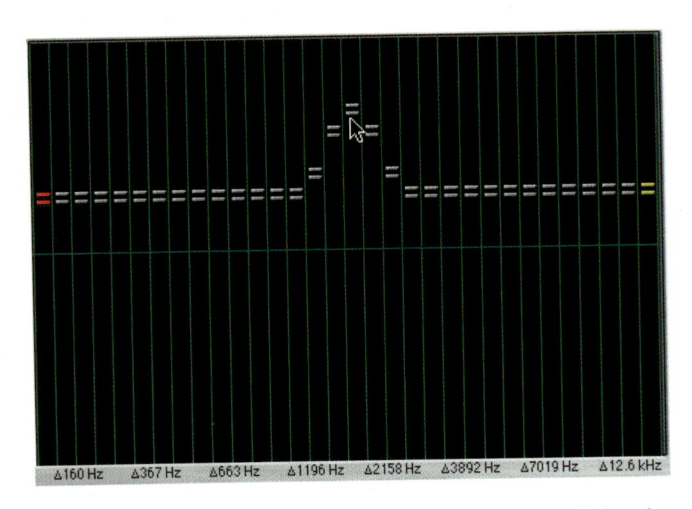

Figure 8.27 Periscope Equalizer Display in Amplify mode with one fader being dragged upwards.

In Exaggerate mode, pushing one fader up or down will cause all the faders to move up or down, but in this case the shape of any EQ curve you have created will change: pushing a fader up will exaggerate the curve, while pulling a fader down will tend to flatten it. So, for example, if you pull down a fader to half its previous height, all the faders will be drawn back to half their previous heights.

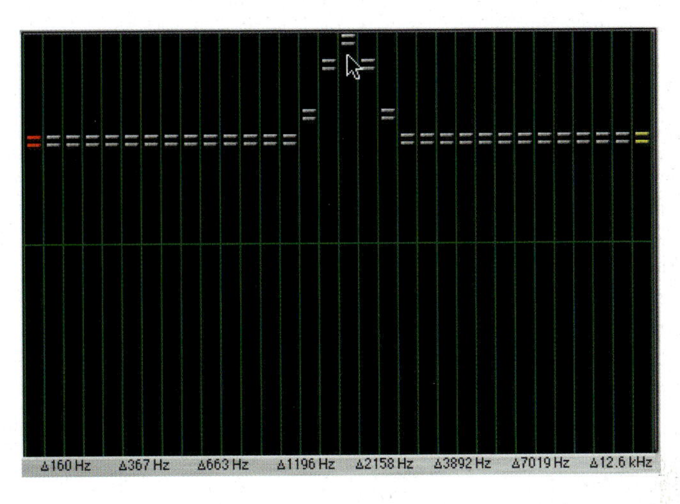

Figure 8.28 PeriScope Equalizer Display in Exaggerate mode with one fader dragged further upwards, moving the overall gain upwards while exaggerating the distances between the faders.

As you might imagine, working all this out takes some doing, so you will need to use the fastest computer you can get for best results. For example, when Periscope is run on a G4 it performs the bulk of its calculations using the Altivec processor, which makes it five times faster on a G4 than on a G3 with the same clock speed. And when it runs on a G3 it takes about 12% of a 400 MHz G3 processor's attention when you have set the buffer size to 2048 bytes in the hardware setup. This goes up to over 20% when you run with a 512-byte buffer.

Now Periscope's filters are said to be 'phase-correct' - whatever that means. The words sound good anyway. I suppose they mean that Periscope will not adversely affect the phase relationships within your audio signals. If so, that can't be a bad thing, but the 'proof of the pudding' is in the listening for me. What I can say is that, as far as my ears are concerned, Periscope is able to do what it says on the packaging without messing up your sounds - as long as you make sensible choices. The filters can boost each band by up to 36 dB and cut each band by up to 144 dB - so the potential for major destruction of your audio is plain to see. Just use extreme settings in crazy places and you could completely ruin your sound. On the other hand, Periscope can almost certainly 'reach the parts that no others can' when it comes to focusing on particular areas of interest within the input frequencies and manipulating these correctively or creatively.

The really amazing thing is that you can always see the real-time frequency analysis displayed behind your faders so that you can apply your EQ exactly where it is needed; PeriScope takes a lot of the guesswork out of EQing! Also, PeriScope is just about the most precise EQ you will find. You can 'crank up' the low end of your mix at the mastering stage while all the other frequencies stay in phase. And you can filter out (remember, up to 144 dB attenuation) anything under 30 Hz, 'brick wall', without hurting the rest of your mix. So you can find that TV beep at 15 kHz in your recording and take it out with just a 10 Hz band of frequencies around it - leaving everything else untouched. Truly a revolution!

RiverRun - Real-time Granular Synthesizer

RiverRun is a so-called 'granular' synthesizer. It takes small 'particles' of your audio, called 'grains', applies an envelope generator and a pitch transposition and then scatters these grains across the output channels. You can create immense choirs from a single voice or chop a guitar strum into many distinct stages, and using Rhythmic mode you can have random samples locked to Digital Performer's tempo while they are simultaneously transposed to pre-selected chord pitches. The demo file shows this off to great effect by playing some speech, sampling some of this, then applying automation to various RiverRun parameters to create various effects.

Figure 8.29 RiverRun - Real-time Granular Synthesizer.

Here's how it works: you insert the plug-in onto one of Performer's tracks, play the audio, listen for a section that you find interesting, then hit the large red and white Record button in the plug-in's window. The audio appears in the input display near the top of the plug-in's window, entering from the right and travelling towards the left of the display. When you have sampled enough, just hit Record again to stop the process – 'freezing' the audio you have sampled in the input display.

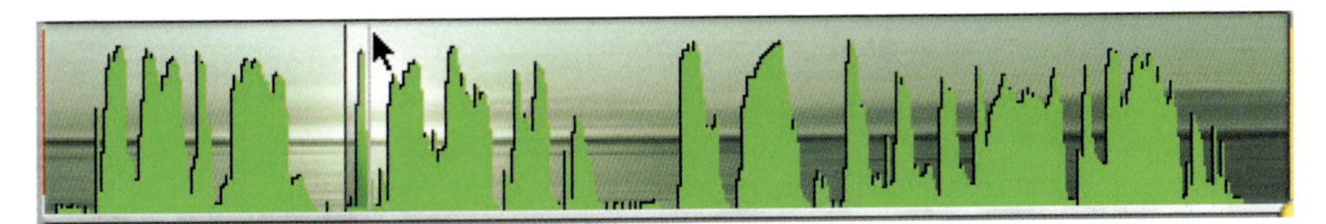

Figure 8.30 RiverRun input screen with arrow pointing to the 'Grain Glass'.

The so-called Grain Glass in the input screen can be manually moved by dragging it in the display to cover any section of interest within the sampled audio. The audio 'grains' will then be taken from this section. The position of the Grain Glass can also be controlled using the 'Walk' slider underneath the input screen. If you move the red 'thumb' to the right, or if you set a positive value numerically using the '+' button, the Grain Glass will 'walk' across the screen from left to right at a speed proportional to the distance by which you moved the red 'thumb' or the numerical value you set. Similarly, if you move the red 'thumb' to the left or set a negative numerical value using the '–' button, the Grain Glass will 'walk' across the screen to the left. You can stop the walk at any time by clicking on the zero position in the Walk slider. The 'walk' is constrained by the positions to which you have set the red and yellow 'needles' in the display. When the Grain Glass has completed its walk to left or right, it starts again from the opposite end and repeats the walk continually.

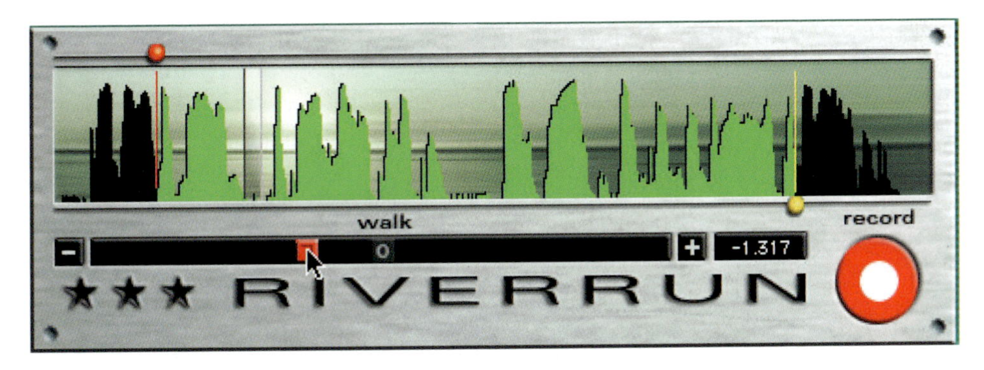

Figure 8.31 RiverRun input screen showing the Grain Glass walking to the left between the yellow and red needles.

Underneath the input section, there is a popup selector for Shape and a slider for mono/stereo. Shape is an envelope that is superimposed on every grain taken from the input screen. Here you can choose more flowing or more punchy envelopes depending on the effect you are trying to achieve. The Mono/Stereo slider lets you pan the output from mono at one extreme to all the way across the stereo at the other extreme. When set to mono, all the output grains are sent to the exact centre of the output grain stream. In this mode the distinctly audible pitch that results from repetition of the grains becomes most obvious. When set to stereo, every grain ends up at a randomly determined position between the far left and right of the stereo output panorama. Slowly moving the fader from mono to stereo produces the effect of 'opening up' the grain stream – an interesting effect that can be applied using parameter automation.

To the right of these controls you will find a pair of buttons marked Load and Save. You can use the Load button to load Sound Designer II (SDII) files into RiverRun's input screen and use the Save button to save the contents of the input screen in SDII format.

Note When you quit Digital Performer, or close a song, RiverRun automatically saves the contents of its input screen as a Sound Designer II file in a folder called 'RiverRun Scratch *f*'. This folder 'lives' inside the main Digital Performer folder. When you next open this song, RiverRun recalls this file and opens it into the input window again – just the way you left it.

In the lower left control section, there are two modes of operation that you can choose from, using a pair of selection buttons – Flowing or Rhythmic. The Flowing mode is designed to create dense textures, while the Rhythmic mode allows you to create beat patterns locked to Performer's tempo. When Rhythmic is selected, grey squares appear in the Grain Speed Slider, and the red 'thumb' jumps to the beginning of each of the alternate grey and black fields in turn as you slide it along. The black fields represent even divisions of Performer's tempo such as whole, half, or 1/4 notes all the way down to 1/128 notes. The grey fields represent the triplets in between these note values.

Three sliders let you control grain speed, add randomization to the grain speed, and adjust the grain length. The Grain Speed slider lets you choose the number of grains that will be taken from the input screen. Together with the Grain Length slider setting, this parameter determines the density of the output grain stream.

Note If you set both grain speed and grain length to their upper limits, the output grain stream could be up to 1000 layers deep which may be too much for your processor to handle.

If the Grain Speed Random slider is set to its minimum value, the grains appear at regular time intervals in the output grain stream and you can often hear the repetition as a steady tone, the pitch of which is set by Grain Speed. However, if you set Grain Speed Random to a positive value, the grains appear at random time intervals in the output grain stream. The output texture loses the grain pitch and becomes much more chaotic and organic. And in Rhythmic mode, randomizing the grain timing can lead to total anarchy!

Using the Grain Length slider you can adjust the size of the Grain Glass, and, in turn, the size of the grains that you hear. At the far left the grains are just a few milliseconds in length while at the far right the grain lengths are set to the maximum of half a second.

Figure 8.32 RiverRun input screen showing the Grain Glass set to its largest size.

At the bottom right of the plug-in's window you will find the Pitch control section. Once a grain has been taken from the input screen and provided with an envelope from the Shape popup, the pitch control section is used to apply a pitch transposition to the grain that depends on the Pitch Grid and Pitch settings. This pitch can then be fine-tuned using the '+' and '−' buttons at the bottom of this section before being passed through to the output.

If you set the Pitch Grid to the 'no grid' Seamless setting, the Pitch parameter lets you adjust the grain pitch to any semitone value in the audible range using the Pitch knob, as you would expect.

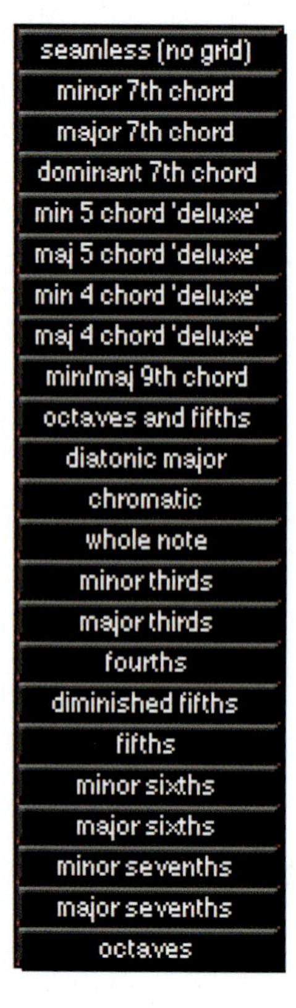

Figure 8.33 RiverRun Pitch Grid settings.

The rest of the Pitch Grid settings randomly apply different pitches to the output grains depending on the chosen grid and on the setting of the Random knob.

Tip To make it easier to hear this effect, choose a clear sound such as a piano note using the Grain Glass and halt the 'walk' while you are setting the effect up.

The figure below shows how the Pitch Grid and Random settings actually dictate RiverRun's choice of pitch:

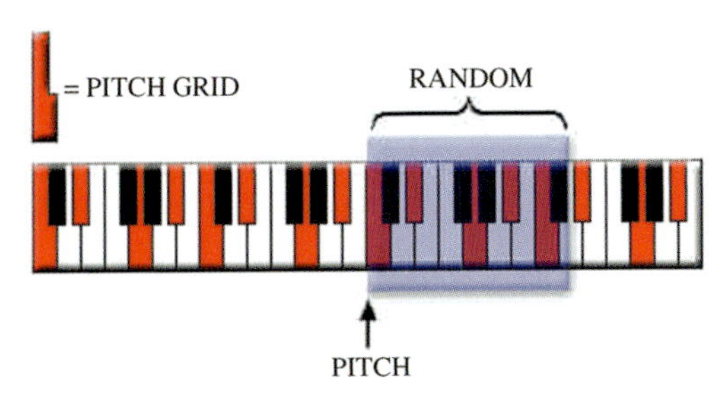

Figure 8.34 RiverRun 'C Minor 7th' Pitch Grid parameters.

For example, if you choose 'C Minor 7th' as your Pitch Grid, the grid will identify the notes of the keyboard that make up this chord – shown in red in the accompanying screenshot. As you increase the Pitch knob, the pitch will jump to each of these notes in turn. And with the Random control at its minimum setting, this is what you would hear at the output. As you increase the Random control, the area shown in blue in the accompanying screenshot grows in size to encompass a number of higher pitches from the chord. And here's the difference: the pitches for the output grains are then randomly picked from the red keys within the blue area. Oh, and, finally, you can adjust the chosen pitch using the Fine Tune controls.

All RiverRun's parameters can be automated using Digital Performer's automation controls.

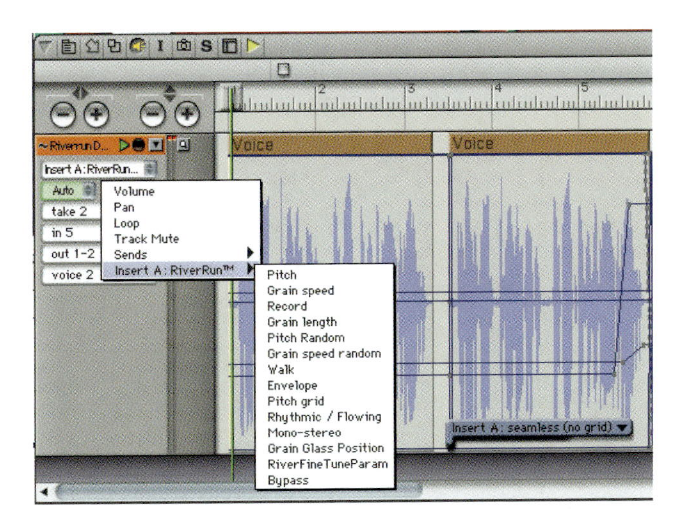

Figure 8.35 Digital Performer Editor window showing RiverRun automation parameters.

The Demo file supplied with the software manages to use 13 of the 14 available automation parameters to produce some very creative effects with a sampled spoken voice.

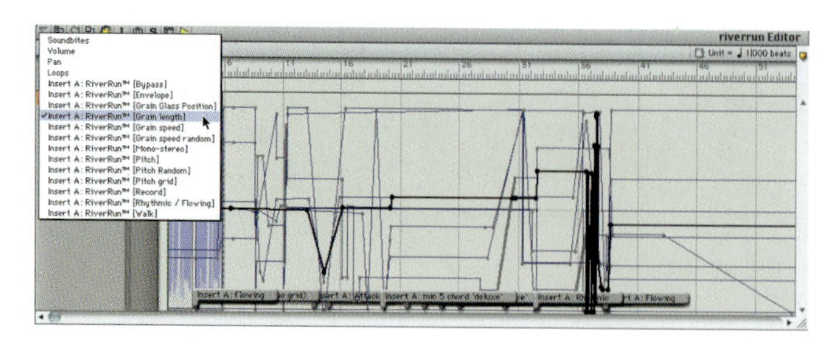

Figure 8.36 Digital Performer Editor window showing RiverRun Demo parameter automation.

For the creative sound designer, RiverRun is a 'must-have' tool. The control it allows over the different aspects of the sound and the way you can apply automation and edit this visually in Digital Performer is nothing short of astonishing! I'm impressed!

Audio Ease VST Wrapper

The VST Wrapper for MAS enables users of Mark of the Unicorn Digital Performer, Performer 6 and Audio Desk to run VST plug-ins and VST instruments as if they were native MAS plug-ins. This is a neat trick, as there are many highly desirable VST plug-ins that are not available for MAS – but now MAS users can install these using VST Wrapper. Take Steinberg's Halion sampler, for instance. VST Wrapper lets you install this plug-in and route its ten outputs to ten of Digital Performer's mix buses. You can load or save VST standard presets and banks and when you re-open a song that uses VST plug-ins, these will open up loaded with the correct samples and settings.

If you are using a lot of VST plug-ins, you will need to increase Digital Performer's memory partition. Also, the more plug-ins there are in your VST plug-ins folder, the longer it takes MAS to launch. Don't forget that you can always remove any plug-ins that you are not currently working with from the VST folder and put them in an adjacent folder called Plug-ins (Disabled), for example.

Each VST plug-in has the ability to maintain an internal list of presets. Some, but not all, plug-in manufacturers store useful things in here that you can use as they are, or modify to your heart's content using the plug-in's controls. You can access the internal bank of presets by using the leftmost popup menu in the VST Wrapper control bar.

The rightmost popup menu in the VST Wrapper control bar is the Options menu. Menu items include: Load Preset, Save Preset, Load Bank, Save Bank and Output Settings. The Load/Save Preset options save the plug-in settings as they currently are to a file, or set all settings to values taken from a previously saved file. This file contains one single set of parameters. The bank files are different in that they contain as many presets as the plug-in has slots for, so the whole list of presets internal to the VST plug-in is saved on 'Save Bank' and replaced when a 'Load Bank' occurs. The format of the files is compatible with the VST specifications, so you can load factory presets or swap bank and preset files with users of other VST hosts.

The number of presets in a bank is particular to the plug-in: a plug-in can have as many as its developer thought necessary. For instance, a VST instrument usually has 128 preset slots that can be chosen via MIDI program changes as well as via the preset list described above. Bank files always contain that particular number of presets - it is not possible to save or load partial banks. The preset and preset bank files are stored in a standard VST format, so they can be exchanged between different VST host applications.

The first two outputs of a VST plug-in will always be routed to the track's outputs. If you want to use more, and these are available from the plug-in, you can use the 'Output Settings' item available in the Options popup. The Output Settings dialog, as shown below for Steinberg's HALion sampler, lets you route optional plug-in outputs to MAS buses. Its use is very simple: the left column shows the names of the extra outputs, the second column is a list of buses to route to. The up and down arrows scroll the buses past the outputs, effectively assigning that output to that particular bus.

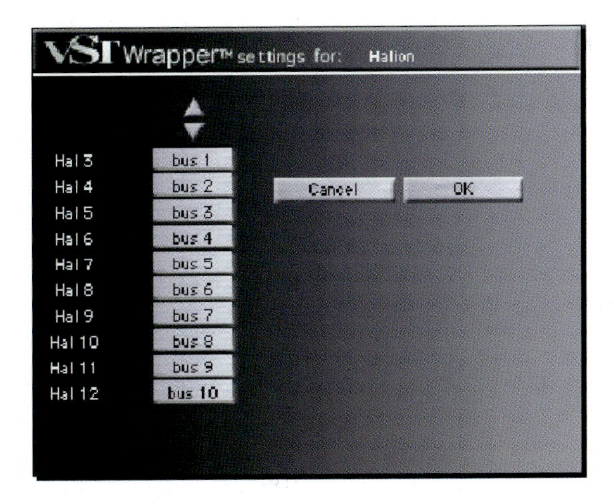

Figure 8.37 VST Wrapper Output Settings dialog.

To be able to use a VST instrument or effect as a MIDI destination, you have to enable 'Interapplication MIDI' in FreeMidi Setup's preferences. This allows plug-ins to receive MIDI data from the host applications. MIDI reception for plug-ins only works with Digital Performer version 2.6 and up. When using FreeMidi in 'OMS mode', a plug-in should automatically be available as a MIDI device. Your Performer MIDI destination popups will display all the VST synths that you have activated in your channel inserts. It will identify a particular instrument by its name, audio channel and a character identifying its insert slot. If appropriate, it also shows the choice of MIDI channel, as some VST Instruments are multi-timbral.

Tip Because Digital Performer does not transmit MIDI data during bounce operations, you cannot bounce your mix to disk along with the audio from any VST instruments in Digital Performer - as the MIDI tracks will not cause audio to appear from these instruments during the bounce process. One way to get around this is to record the output of any VST instruments to disk first, so that they can be played back as audio tracks within Digital Performer and subsequently included in any bounce operations. First, assign the outputs of any tracks containing VST instruments to a bus pair. Then create a stereo track that listens to that same bus pair, record enable the new track, and press Record. This will generate a soundbite on the new track containing a mix of the audio outputs from all the VST instruments.

Note The times between MIDI events that VST Wrapper V3 passes to the VST instrument are accurate to within 4 ms. However a constant delay is introduced. This means that between the onset of a MIDI note on a MIDI track and the recorded output of an instrument on an audio track, there will be a delay. This delay can be up to three MAS buffer sizes long and is the same for every MIDI event. Setting the MAS buffer size low, such as to 128, will give you a smaller (but still constant) delay. The MAS buffer size setting can be found in Digital Performer's 'Configure Hardware' dialog. Please note that some VST instruments do not support accurate MIDI timing internally. While the VST Wrapper may pass the proper timing information to the instrument, the instrument may ignore it and interpret the MIDI event at the start of each audio buffer instead.

VST Wrapper provides an elegant solution for anyone wishing to use VST plug-ins within Digital Performer. In practice, although it works well with many VST plug-ins, it is possible to encounter problems with others. This is not necessarily the fault of VST Wrapper, though. Often the plug-ins themselves are not so well written, in terms of the computer code used. So although they may work fine with Cubase VST, they may not work so well with other VST software such as VST Wrapper. Nevertheless, the situation is constantly changing with upgrades appearing both for VST Wrapper and for the individual plug-ins; you just have to keep abreast of things as best you are able.

DUY EverPack

DUY offer the 'EverPack' software bundle with versions of their plug-ins that work with MAS, VST, Premiere, RTAS and Audiosuite plug-in formats. The EverPack suite has versions of DaD Valve, DUY Shape, Max DUY and DUY Wide (reviewed in the TDM chapter), but not DaD Tape, DSPider, ReDSPider or SynthSpider. EverPack also has one plug-in that is not available for TDM - a natural sounding reverb called Z-Room.

Z-Room

Z-Room offers natural-sounding reverb algorithms together with a friendly and intuitive user interface and several unique features. For example, Z-Room is the only reverb that allows you to control the time base of the reverb's algorithm. And the exclusive Rehearsal Mode allows you to set the reverb parameters while listening to the real impulse response of the reverb. You can choose between several Early Reflections responses (reverse, gate, hall, room, stage, etc.) and all these presets can be modified with the Stretch and Pre-delay parameters.

Z-Room is divided into six control sections - Input, Diffusion, Color Control, Early Reflections, Mass Control and the Mixer - with a narrow strip of controls called the Utility bar at the top of the window.

Figure 8.38 DUY Z-Room.

Let's look at the Utility bar first. At the far left you will find the Patch Manager popup which has the word 'None' written on it until you choose a patch – in which case the patch name appears here. Click on this part, and a popup menu will appear to let you select a patch.

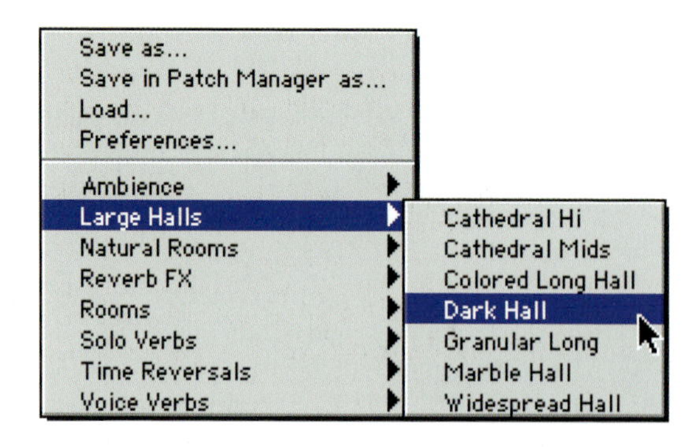

Figure 8.39 DUY Z-Room Patch Manager popup showing presets.

The patches are organized into families such as 'Ambience, 'Large Halls' and so forth. Each one of these contains several presets and you can also create new presets to add to the list. Save and Load commands are provided here, along with a Preferences option. Select this to open the Patches Preferences dialog:

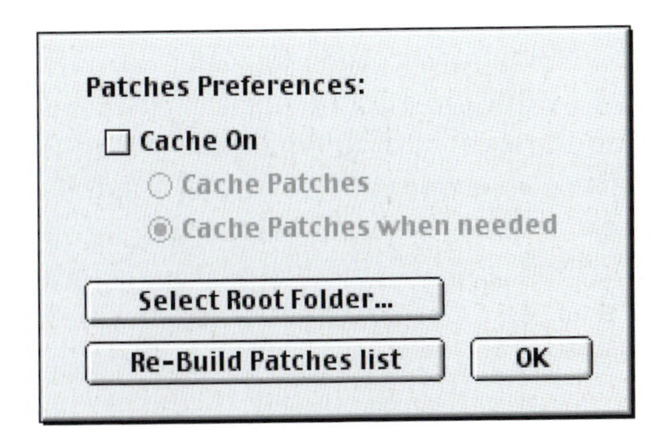

Figure 8.40 DUY Z-Room Patch Manager Preferences.

The 'Cache On' option lets you choose whether to store all the available patches in memory or not. If you want them in memory, you can choose to have them all available permanently or only the ones you need at the time. The 'Select Root Folder' options lets you choose the folder that you would like Z-Room to read the presets from in order to show them in the Patch Manager pop-up menu.

Note Once you've selected a folder, you must activate the 'Re-build Patches List' command and then press OK. If you don't click the 'Rebuild' button, it will not regenerate the list of presets after you've pressed OK.

To the right of the Patch Manager popup, you will find buttons to Copy and Paste patches, and a pair of Undo/Redo buttons. These let you revert the parameters of the process to the previous state before the last edit. Up to 10 edits can be undone/redone.

The Copy/Paste commands allow you to copy the settings from one preset and paste them into another. Four memory locations marked A, B, C and D can be used to store temporary settings and in-between changes and options while you are working on your patch before you finally save it as a preset. To store data in A, B, C or D, simply click on the arrow at the left of each letter and to recall, click on the letter itself. The last two icons in the utility bar let you activate the 'Balloons Help' and 'Talking Help' – very useful while you are learning how to use Z-Room.

The Input section includes controls for Input Gain, Low Pass filter, and Pulse Frequency; a Rehearsal switch and a popup operation Mode selector. In Economy Mode, Z-Room frees up some processing time from your processor. Deluxe Mode uses more processing time to ensure quality results - and Earlies Mode just uses the Early Reflections algorithm. The Pulse Frequency slider changes the frequency of the pulses used to test the parameters of the reverb, controlled by the Rehearsal button. This slider only has effect when the Rehearsal switch is on. The display shows the time between one pulse and the next in seconds. The Rehearsal Switch lets you evaluate the quality of the reverb patch that you are working with. When you turn the Rehearsal switch on, Z-Room enters the Test Mode. A series of very short pulses is produced to let you hear the response of the reverb. The pulses are repeated at a rate set by the Pulse Frequency slider, while the gain is controlled by the Input Gain slider.

The Diffusion section lets you process the sound before it enters the reverb stage. The Base time slider lets you set the seed time used to calculate the series of tap delays that diffuse the sound. The cross-decay (x Decay) parameter controls the time differences between the diffuser's internal blocks. The Depth slider lets you set the amount of Diffusion of the sound. Values near zero produce no diffusion, but will create an additional delay proportional to the current Base Time and x Decay. The number of iterations of the calculations used inside the Diffuser section can be set using the Level slider - 1, 2, 3, 4, 5 or 6. Lower levels produce less diffusion and require less processing time. Higher levels produce greater diffusion and require more processing power.

Figure 8.41 DUY Z-Room Geometry menu.

The Color control section lets you split the frequencies into low, middle and high bands and set different reverb decay times in the three frequency bands. A pair of controls let you adjust the crossover points from low to mid and mid to high to define the three ranges. Three vertical sliders are provided to control the decay times of the three frequency bands.

The Early Reflections section lets you define the room type by setting parameters for Pre-Delay, Stretch and Geometry. The Stretch parameter controls the size of the Room at the Early Reflections stage. The popup Geometry menu allows you to select the type and shape of the Room to be modelled by the Early Reflections response. Twelve different room models are provided.

The Mass Control section provides controls for the main internal parameters of the Reverb processor - Pre-Delay, x Decay and Base Time. The Base Time slider sets the seed time used to calculate the series of delay lines used to create the reverberation. The x Decay parameter controls the percentage time differences between the reverb's internal blocks. Twelve different room models can be selected in the 'Geometry' box included in the Early Reflections section.

The Mixer section has independent controls for Direct gain, Early reflections Gain and Reverb Gain. Direct Gain sets the final Gain of the unprocessed original sound – displayed in gray on the Scope at the bottom of Z-Room's interface. Setting this value to its minimum mutes the original sound. Early Gain sets the final Gain of the Early Reflections section – displayed in blue on the Scope at the bottom of Z-Room's interface. Setting Early Gain to its minimum disables the Early Reflection section, thus freeing processing time. Reverb Gain sets the final Gain of the reverberant sound – displayed in red on the Scope at the bottom of Z-Room's interface. Setting this value to its minimum mutes the reverb sound.

There are two 'plasma' meters provided – one each for the Input and Output signals. The output meter also has a 'clip' indicator situated just above the meter. If this lights red, it means that there has been a signal overflow during the processing calculations, so the input gain probably needs to be lowered.

At the bottom of the Z-Room window you will find the Scope Output Display, which lets you visualize the evolution of the output signal in the time domain. This display shows the three component signals that make up the final output – with the Direct signal in grey, the Early Reflections in Blue and the reverb signal in red.

OK – that's it for the technical stuff. Now, how does it sound? Well, I tried Z-Room on some mixes for a Beatles-like pop group from Liverpool and was very pleased with the results. On one song I found a natural sounding preset that was perfect for the harmony vocals, while on another, a more artificial-sounding preset produced a very similar effect to the echo chamber often used on Elvis Presley's vocals. And the selection of presets provided is sufficiently varied that you should be able to find at least a starting point for whatever reverb effect you are trying to create. Highly recommended!

Chapter summary

There is no doubt about it for me – Mark of the Unicorn have made some great alliances with third-party developers to provide a set of interesting and creative plug-ins for MAS-compatible software (such as Digital Performer), several of

When you are using Cubase VST, the audio outputs from the Model E are automatically routed to the VST Channel Mixer where you will find four stereo channels named ModE for each Model E plug-in you are using. Using the Output and Panorama controls at the bottom of the Model E plug-in's window, you can assign the audio outputs from the 16 multi-timbral channels within the Model-E plug-in to any of these eight audio channels. This all works in a similar way when you are using other host applications, although the detail may differ a little.

The Model E comes with a number of sound Banks that contain up to 128 preset Programs designed by well-known sound designers. Each Program represents a complete set of parameters for one of the 16 channels available in each Model E. You can load or save single Programs or Program Banks using the File popup menu in the VST Instruments Rack in Cubase or Nuendo, then use the Program selector arrow buttons in the lower section of the plug-in window to choose any Program. Clicking on the Program number display brings up a popup menu that you can use to choose these if you prefer.

For anyone who appreciates the Mini-Moog, the Model E will prove to be a joy to use. Just think how many times you have wished that a particular Moog sound was polyphonic - or that you could save and recall your patches! And if you have never used this kind of synthesizer before, you may well be more than pleasantly surprised when you hear the range of sounds it can produce.

Steinberg LM-4 drum-machine

The LM-4 is a software simulation of a drum-machine - offering many of the features that you will find on instruments of this type. You can play the LM-4 polyphonically in real time or from a sequencer track via MIDI - just like a normal drum-machine but with the added benefit of total integration into your MIDI + Audio sequencing environment. This means you can process the audio using plug-ins, balance and mix the audio outputs, and convert into an audio track when you are ready.

The LM-4 supports both 16- and 24-bit resolution and each instance of the plug-in that you are using has six virtual outputs into your host software's mixer - arranged as one stereo pair and four individual mono outputs. A library of sounds is included and you can use your own samples if you wish. There are 18 'pads' that each play a sound when clicked on. In the upper part of the plug-in's window, there are two faders associated with each of these pads to control volume and tuning. At the top right there is an overall Volume control, with a Velocity control underneath this that can be used to set the sensitivity of the LM-4 to

Figure 9.5 LM-4 drum-machine.

incoming MIDI velocity data. At the bottom right of the window, the Output and Panorama controls let you assign individual positions within the stereo panorama and one of the six separate outputs to the currently selected pad. The default setting assigns all the pads to the stereo output pair. There are ten 16-bit drum sets installed as standard – with another ten 24-bit sets available as an optional install. Most of the standard sets conform to the General MIDI specification, and include types such as Jazz, Soul, Heavy, Real and Electro. To use the 24-bit sets you will need lots of RAM – as much as 128 Mb or even more.

Note All the LM-4 parameters can be automated in real time. When you change any parameters they transmit MIDI System Exclusive data that can be recorded into your sequencer then played back to automate the parameter.

Tip Make sure that your sequencer is not filtering out MIDI System Exclusive data when you want to record LM-4 automation data. In Cubase VST, for example, the default setting is to filter out SysEx data.

The LM-4 provides a first-rate set of basic drum sounds to suit popular music styles and is perfectly straightforward to use. The 'icing on the cake' is that you can use your own drum samples and make up custom kits to suit whatever you are working on.

TC|Works Mercury-1

TC|Works Mercury-1 VST Instrument is a monophonic 'analogue'-sounding synthesizer capable of creating fat bass sounds, leads and classic synth effects.

Figure 9.6 Mercury-1 synthesizer.

Mercury-1 is loosely modelled on the Mini-Moog synthesizer, although there are several points of departure from this classic design. For example, there are two main Oscillators, each with switchable waveforms, filter and amplifier sections, two envelope generators, a mixer section and a Glide (Portamento) control. There is also a dedicated LFO section, a Pulse Width Modulation control for Oscillator 2, Oscillator Sync and Ring Modulation switches, and various other controls.

Mercury-1 is actually four-way multi-timbral as well. This means that it is effectively four monophonic synth modules in one plug-in. You can either use four different sounds, each with their own MIDI channel and output settings, or you can layer up all four to produce one huge sound. The controls to set this up are sited just above the main synthesizer controls. Here you can set a Key Range for each of the four synths to respond to, a Transposition if you want them to play back in different octaves, a Velocity Range to which each will respond, a Pan position for each, and an Output assignment to any of the four available outputs.

Note Mercury-1 transmits MIDI controller information whenever you change a parameter and you can use this to automate these parameters.

Mercury-1 is a useful instrument to have in your collection and the sound FX presets are particularly good. Steinberg's Model E offers a lot more in many ways, though, with its polyphony and greater multi-timbral capabilities.

TC|Works Spark Modular

TC|Works also have a modular synthesizer simulation – Spark Modular – which has a keyboard module, a sequencer module, an oscillator module, a filter module, and an amplifier/envelope module. Spark Modular was developed originally to work within Spark's Matrix Window. TC|Works Spark and Spark XL applications are primarily intended for mono or stereo soundfile editing and include a selection of useful signal-processing plug-ins.

Interestingly, Spark's Matrix Window can itself be loaded into VST-2.0 compatible host software as a VST Instrument called the 'Spark FX Machine Instrument'. This Matrix can have up to ten parallel streams of plug-ins, with up to five plug-ins within each stream, so you can put together a Spark Modular synthesizer by inserting several of the modules into one or more of these streams in the FX Machine matrix. You can even combine Spark modules with other VST instruments such as the Pro-52, or with signal-processing plug-ins.

To help while setting up sounds in Spark Modular, a 'virtual' MIDI keyboard module is available that you can click on to play notes. This would typically be inserted first in the FX Machine matrix. You could follow this with the Sequencer module. This is a simulation of the old monophonic sequencers or arpeggiators that were used with modular analogue synthesizers. Several basic sequence patterns are provided as presets (see Fig. 9.9). You can, of course, use several of these within the FX Machine matrix to produce different patterns, and the sequencer modules can be synchronized to MIDI.

Figure 9.7 Spark Modular Synthesizer open in Spark XL.

A typical analogue synthesizer has a Voltage Controlled Oscillator followed by a Voltage controlled Filter and finally a Voltage Controlled Amplifier. Spark Modular is basically set up in a similar way – with the Oscillator, Filter and Amplifier sections inserted into the Matrix in that order.

The monophonic dual Oscillator module gives you all the classic waveforms, including Pulse Width Modulation, an additional Sub-Oscillator, Oscillator Sync, Ring Modulation and an LFO. A useful set of presets let you get started right away (see Fig. 9.10), and you can save and load your own as you develop these.

The Filter module has a switchable 12, 18 or 24 dB low-pass filter with Resonance, an additional high-pass filter, its own envelope, key follow, single and multi trigger. Similarly, the Amplifier module has all the classic controls

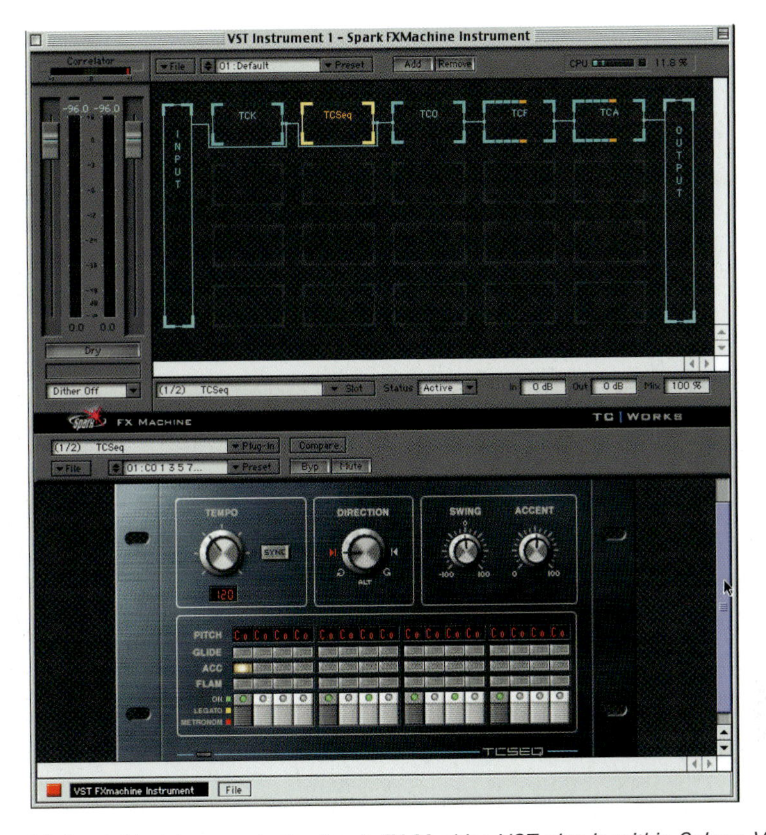

Figure 9.8 Spark Modular open in the Spark FX Machine VST plug-in within Cubase VST.

1: C0 1 3 5 7...
2: C0-C1 Alter.
3: CMaj. Arp 1 up
4: CMaj. Arp 2 up
5: CMaj. Arp 1 down
6: CMaj. Arp 2 down
7: CMaj. Arp 1 up/Down
8: CMaj. Arp 2 up/Down
9: 1/16 T
10: Swing HH
11: All OFF C0
12: TC303 Lead

Figure 9.9 Spark Modular Sequencer presets.

1: Default
2: Fat Squares
3: Ring Mod Lead
4: Sawtooth Sync
5: PWM
6: Multi octaves
7: -------------
8: Synth Bass
9: Softstrings
10: Voice
11: Bass Drum
12: Snare Drum
13: Hihat
14: Bell
15: TC303 Lead
16: Space FX
17: Echo
18: Musical Box

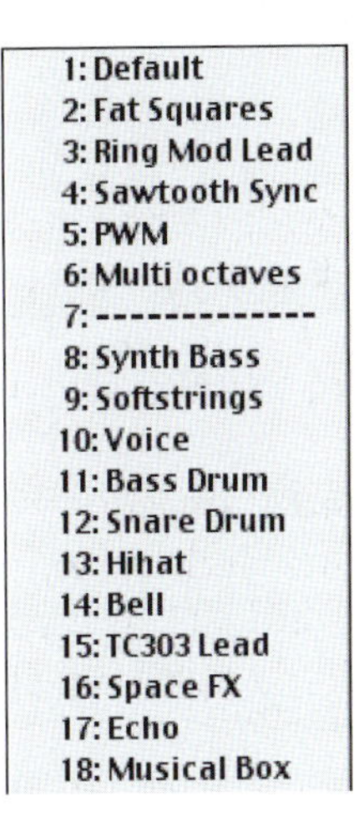

Figure 9.10 Spark Modular Oscillator presets.

including an ADSR envelope control section, a Drive control to add analogue-sounding distortion, and a single/multi trigger switch. Single will trigger the envelope once and won't trigger it again if one key is held and other keys are then pressed. Multi triggers the envelope every time a key is pressed - even when another key is still being held.

Spark Modular is best used within the Spark FX Machine Instrument with this running within a suitable MIDI + Audio sequencing software environment, such as Cubase VST. The presets are OK, but there are not too many of these, so you will want to be developing your own sounds as soon as possible. The neat thing is that you can combine other VST Instruments and plug-ins inside the Spark FX Machine Instrument matrix to build all manner of interesting and complex sounds.

Waldorf PPG Wave 2.V

The PPG Wave 2.V is a software version of the legendary PPG Wave 2.3 synthesizer which first became available in 1983.

Figure 9.11 Waldorf PPG Wave 2.V.

This was one of the first digital synthesizers and its most amazing feature was its ability to sweep through 64 waveforms stored in a 'wavetable'. This resulted in a sound that continually changed its character and enabled soft transitions between similar-sounding waveforms or extreme jumps between totally different sound spectra, giving this synthesizer the ability to create absolutely stunning and unique sounds. The PPG synthesizers quickly reached cult status and were used on many music productions of their time. I remember hiring one to use on various recording sessions myself around this time. I was particularly impressed by the outrageous clavinet-type sounds and the many interesting bell sounds and harmonic sweeps.

The new PPG Wave 2.V is a VST Instrument - which means it runs within Cubase VST, rather like a processing plug-in, and its audio outputs appear directly in the VST Channel mixer, where additional effects and EQ can be applied. Cubase MIDI tracks are routed directly into the input of the 2.V - or you can route your MIDI keyboard through Cubase to play it 'live'. A 61-note keyboard simulation is also included. You can hide this to save screen space, but it is quite useful for testing sounds using mouse clicks.

The software sounds very close to the original and even includes a simulation of the 'aliasing' noise that was present in the original. Differences between the old and the new? Well, the original basic sequencer has been omitted (as Cubase is so much more powerful), but the original Arpeggiator is included. And the 2.V has some great new user-interface features - for instance, easy and intuitive graphic editing of envelopes, generators and Filter Cutoff.

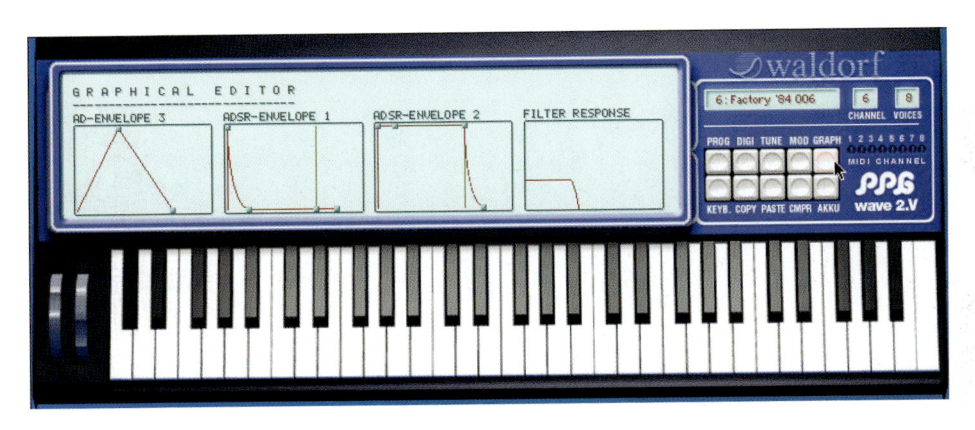

Figure 9.12 Waldorf PPG Wave 2.V Graphical Editor.

Also, while the old Wave had up to 64 voices of polyphony, the number available in the new one depends on the available processor power. The software comes with the original Wave 2.3 'patches' plus several sets of newly programmed patches - many of which sound better than the originals.

Up to eight units can be used at the same time, depending on CPU power. Latency (the delay between playing via a MIDI keyboard and hearing the sound) can be a problem unless you are using one of the newer low-latency cards such as the Nuendo Audiolink 96 or RME Hammerfall models.

Waldorf Attack 1.0

If you fancy some electronic percussion sounds for Cubase then you could do worse than to try the Waldorf Attack VST Instrument plug-in. Attack can emulate the sounds of classic analogue drum synthesizer sounds such as the TR808, TR909 and Simmons SDS-5 bass and snare drums, 808 and 909 side stick, 909 claps, 808 cowbell and CR78 hi-hats, plus a sampled cymbal sound. Toms and congas, shakers and maracas, claves and woodblock can also be simulated with excellent results. Complete simulations of these classic kits are provided - or you can 'roll your own' versions using Attack's powerful synthesis controls.

Figure 9.13 Attack Edit Window showing controls.

A five-octave keyboard is provided in the Attack Edit window so you can audition any sound at the click of a mouse. You can also control any of the functions of the Attack using MIDI controller data from an external unit with knobs and faders, or from a MIDI master keyboard.

Two oscillator sections are provided, each with a choice of waveforms, tuning and envelope controls. You can also use the 'Crack' controls to create handclap sounds using an amplitude modulation technique. Two envelopes are provided, each with the usual attack, decay, sustain and release controls, and a mixer lets you blend the sound of the oscillators and Crack modulator before feeding these to the comprehensive filter section. A delay section lets you create modulation effects and the Amplifier section provides eight audio outputs which are routed via the channel mixer in Cubase so you can add EQ or effects to the drum sounds.

Figure 9.14 VST Channel Mixer showing Attack outputs routed into eight mixer channels.

Each Attack drum kit can have up to 24 different sounds and 30 preset kits are provided which will appeal strongly to lovers of electronic-sounding music. These kits are grouped into Banks of 16 complete kits and two banks are included on the CD-ROM to get you started.

Installing and using Attack in Cubase is very straightforward. The software comes on CD-ROM and doesn't appear to use any copy-protection scheme – making installation extremely simple. Once installed in the VST plug-ins folder, you can use the VST Instruments window in Cubase to select Attack from the list of installed virtual instruments on your system. A click on the Edit button in this window brings up the edit window for Attack. Now you can select up to 16 Attack instruments as MIDI track destinations in Cubase's Arrange window, each on separate MIDI channels.

I compared the simulated TR808 kit with a real TR808 and found the Attack simulations fairly disappointing. On the other hand, the CR78 simulation was excellent and several of the electronic kits sounded very usable.

Attack's wide choice of simulated percussion synthesizer kits makes it easy to use 'straight out of the box'. The simulations of some of the classic sounds it claims to

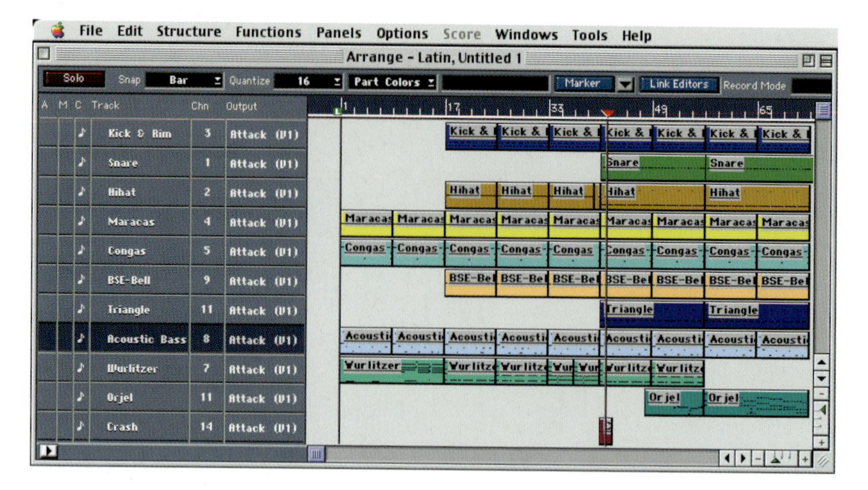

Figure 9.15 VST Arrange window showing Attack used as the output for the various MIDI tracks.

emulate are not very accurate though. Nevertheless, at the asking price this one is worth going for if you are into electronic-sounding percussion.

Native Instruments Battery

Native Instruments Battery percussion sampler, previously available as a VST Instrument, is now available as part of the Native Instruments Studio Collection Pro Tools Edition, which also includes the Pro-52 and B4 software instruments.

Native Instruments was the first manufacturer to offer virtual instruments for RTAS as well as Digidesign's newly developed 'Host TDM' (HTDM) plug-in formats. The RTAS versions use the host computer's CPU for processing and work with Pro Tools LE and Pro Tools Free systems. HTDM plug-ins are a hybrid of TDM and RTAS technologies. They can be used as TDM plug-ins, but, like RTAS plug-ins, the processing is carried out in the host computer instead of on the DSP chips in the TDM hardware. One SRAM chip on the Pro Tools card is used to handle the data transfer between the host and the TDM bus for up to 16 different stereo HTDM plug-ins.

Battery features 32-bit internal resolution and comes with a comprehensive library of 30 sets of percussion sounds. It can be used simply as a sample playback unit, or you can manipulate the samples using velocity layers, individual tunings, volume and pitch envelopes, apply bit reduction and waveshaping and various modulations. Battery can read samples from Akai, SF2, LM4, AIFF, SDII, WAV

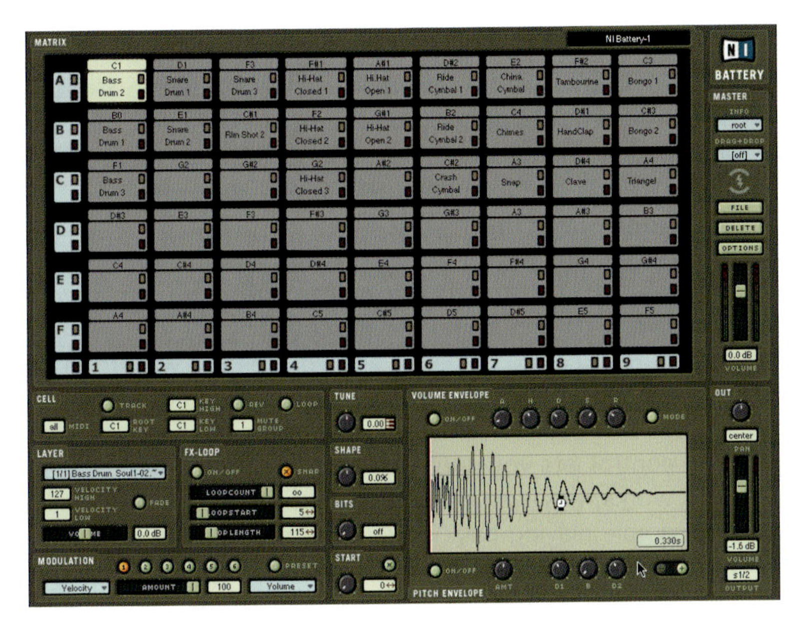

Figure 9.16 Battery.

and MAP sources with any bit resolution from 8 to 32 bits – so you can load in sounds from more or less any common source.

To get started, you can load a kit by clicking on the 'File' button in the Master section at the right of the plug-in's window. This brings up the standard 'Open File' dialog box which you can use to navigate through the folders on your connected disk drives until you find a Battery '.kit' file or another compatible file type, such as an LM4 file from Steinberg's popular VST drum-machine. Once you have loaded a kit you will see the names of the individual drum sounds written in the different cells in the matrix of 'pads'. Click on any pad to hear the sound play back. You can drag and drop samples and parameters between these cells, and underneath the cell matrix you can set various parameters for each cell, such as the MIDI note or note range that will play any particular cell.

Below the Cell parameters, the Layer section lets you define velocity layers for the samples contained in the current cell, set individual volumes and adjust crossfades for each layer. These layering capabilities really help to enhance the realism of your drum programming and couldn't be much easier to use. Another great feature is the FX-Loop section. This lets you loop any part of a sample as many times as you like – extending Battery's capabilities far beyond that of a one-shot sample playback unit. Again, this is easy to set up by dragging the sliders for Loop Count, Loop Start and Length – and if you hit the Snap button the Loop Start and

Note The stand-alone version works with the Apple Sound Manager or with ASIO, DirectConnect or MAS.

Programming is as easy as can be expected with such a complex synthesizer. You use the Navigator window to open and close the different programming windows and you can keep this open most of the time so that you are just a click away from the window you want to work with. There are too many windows to keep them all open all of the time, and most of these are not resizable, so you will have to get used to moving them around your available screen-space.

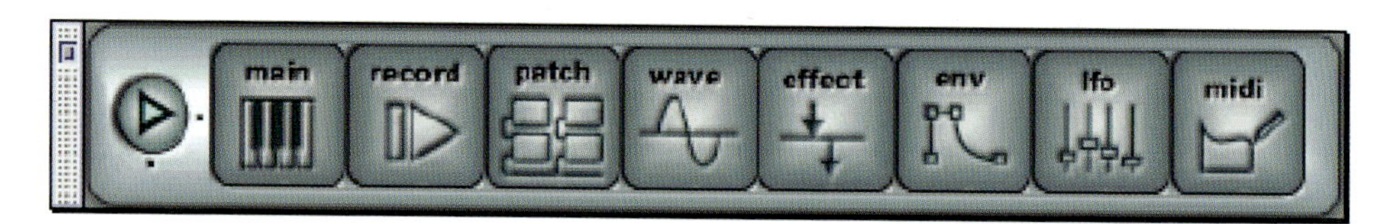

Figure 9.24 Absynth Navigator.

The Main window lets you access the stored sounds and audition these by clicking on a simulated keyboard. You are presented with sets of 16 sounds at a time, with a total of 128 available in each bank, and you can switch to any set of 16 by clicking on the numbered buttons above the patch names.

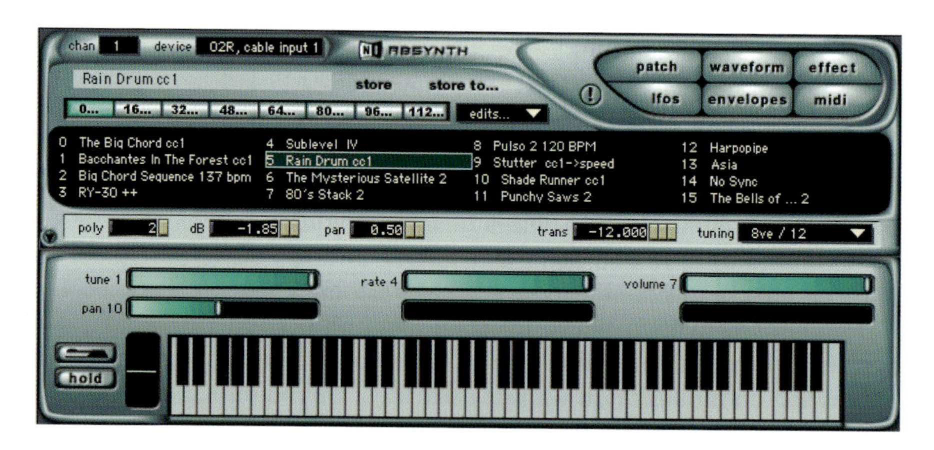

Figure 9.25 Absynth Main window.

If you find a sound you like and want to record some of this as an audio file, you can use Absynth's Record window to conveniently capture this – with options to save as an interleaved file or separate left and right files for stereo.

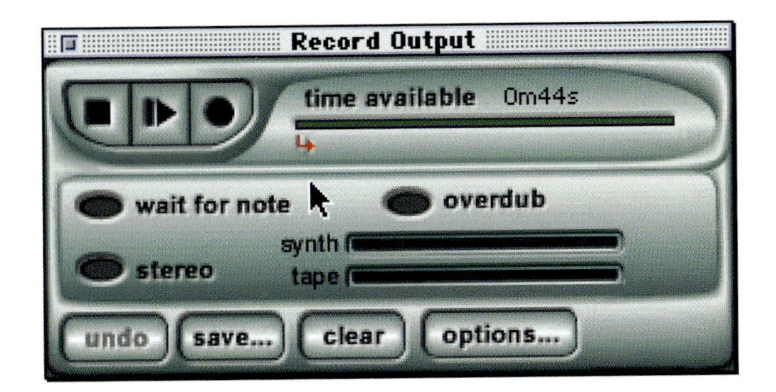

Figure 9.26 Absynth Record window.

You can use the Patch window to create the basic sound – or to modify existing patches. Here you can turn DSP modules on and off and modify the parameters of each module. The modules are arranged into three independent channels with three columns – each containing Oscil, Filter and Mod modules. The outputs from these three channels are mixed together and fed into the master channel. Level sliders underneath each column let you set the output level of each channel. The master channel consists of Waveshape, Filter and Effect modules arranged as a row of three modules at the bottom of the Patch window.

The Waveform window lets you edit waveforms that you have created using the wave popups in the Patch window's Oscillator modules – or you can edit the factory presets. There are two editing modes – waveform and spectrum. Waveform editing is the normal time domain waveform editing method that is found in most waveform editors. For example, you can draw directly in the waveform or you can apply processing to normalize, reverse or filter the audio. You can also perform frequency modulation on the waveform or 'fractalize' it to create bright, diffuse, organic-sounding waves. Alternatively, in the Spectrum mode, you can edit the first 64 harmonics of the waveform in the frequency domain.

Figure 9.27 Absynth Patch window.

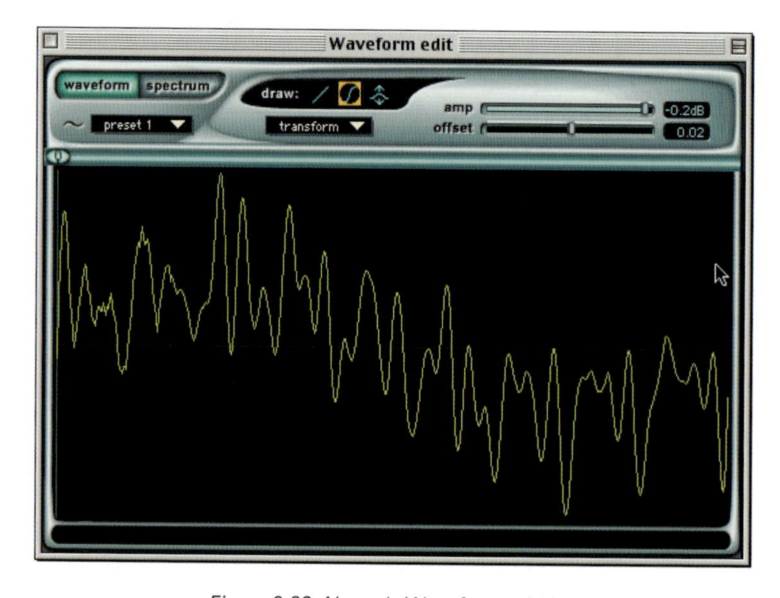

Figure 9.28 Absynth Waveform window.

The Effect edit window has three modes. Multicomb mode gives you up to six independent delay lines with feedback and a low-pass filter in the feedback loop. Delay time can be modulated by the LFOs or by a MIDI controller.

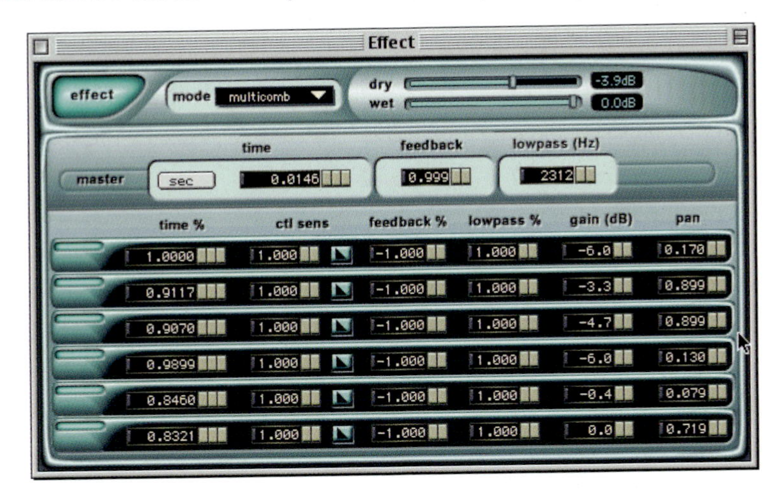

Figure 9.29 Absynth Effect window in Multicomb mode.

Pipe mode can be used to create great rotary speaker effects, flanging or pitch-shifting.

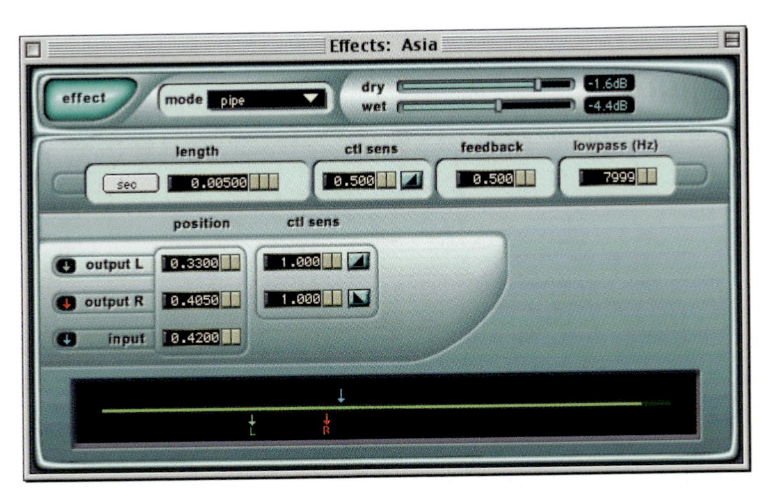

Figure 9.30 Absynth Effect window in Pipe mode.

Multitap mode provides a single delay line with three output taps. The delay time ranges from one sample to one second and the first tap has a feedback control. The three taps can be modulated by the LFOs or by a MIDI controller.

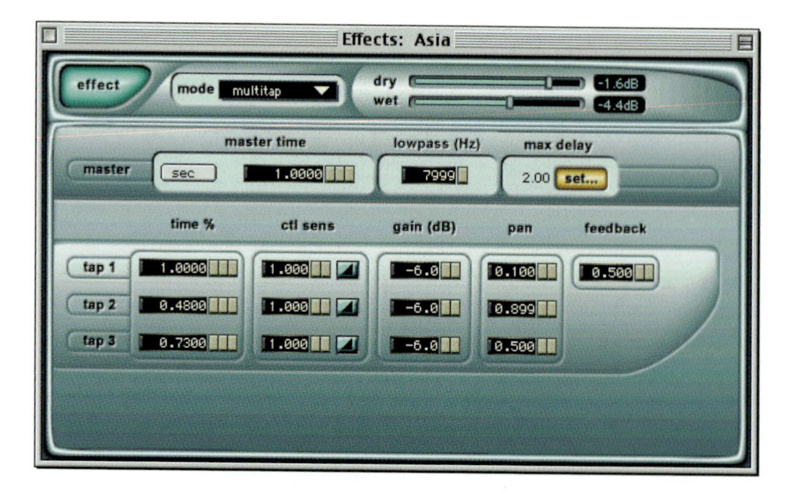

Figure 9.31 Absynth Effect window in Multitap mode.

Absynth uses breakpoint envelope generators to vary parameters in time. Each envelope can have up to 68 breakpoints – providing an incredible amount of flexibility.

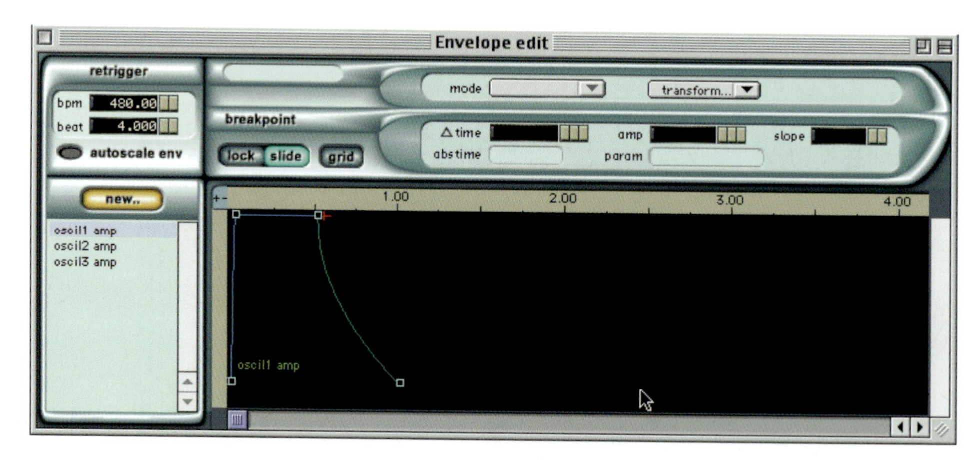

Figure 9.32 Absynth Envelope window.

Absynth has three LFOs, accessible via the LFO window. Each LFO can modulate several parameters at the same time and each parameter can also be modulated by all three LFOs simultaneously. Controls at the top of the window allow a waveform to be created for use as an LFO and a sample-and-hold mode is provided. The Depth control section lets you set the depth of the modulation effect for the various parameters that can be modulated, such as pitch, amplitude or filter frequency. The Controller section lets you assign which MIDI controllers you

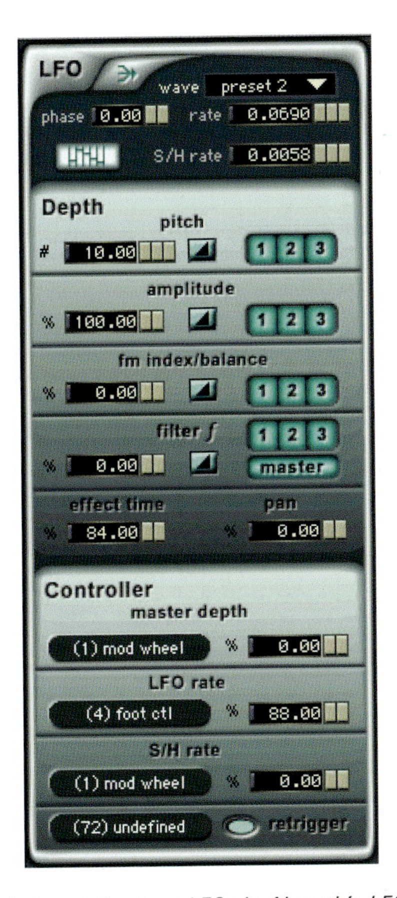

Figure 9.33 One of the three LFOs in Absynth's LFO window.

want to use to modulate the Master Depth, LFO Rate and Sample & Hold Rate parameters of the LFO module.

The MIDI window has pages for MIDI controllers, Velocity, and Note Scaling. In this window, the MIDI controllers page lets you set how the preset responds to continuous controllers, pitch bend and aftertouch.

On the Velocity page you can set the velocity depth independently for Oscil Amplitude, FM Index/Balance, and Filter Frequency for each of the three channels and for the master channel. The Note Scaling page lets you scale the value of a parameter according to the note that is played. So, for example, you can use this to increase the frequency of a filter as higher notes are played.

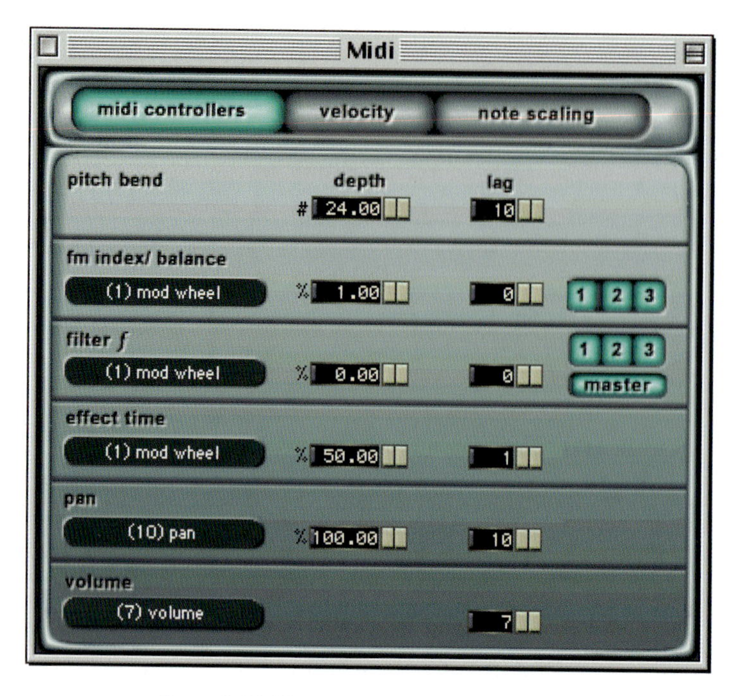

Figure 9.34 Absynth MIDI Controllers window.

Absynth with Pro Tools

Direct Connect allows you to use Absynth with Pro Tools. Launch Pro Tools first, then launch Absynth and make sure that its audio system is set to use DirectConnect. In Pro Tools, to listen to Absynth, you need to create a new Aux track and insert the Absynth DirectConnect TDM plug-in.

Note To record the audio from Absynth into Pro Tools, you need to bus the output from the Aux track across to a normal Audio track.

Absynth as a VST Instrument in Cubase

To use Absynth as a VST Instrument in Cubase, insert this into the VST Instruments window in the usual way.

When you hit the Edit button, Absynth's main Edit window appears in Cubase and you can select patches and carry out basic adjustments here.

Figure 9.35 Absynth VST Instrument.

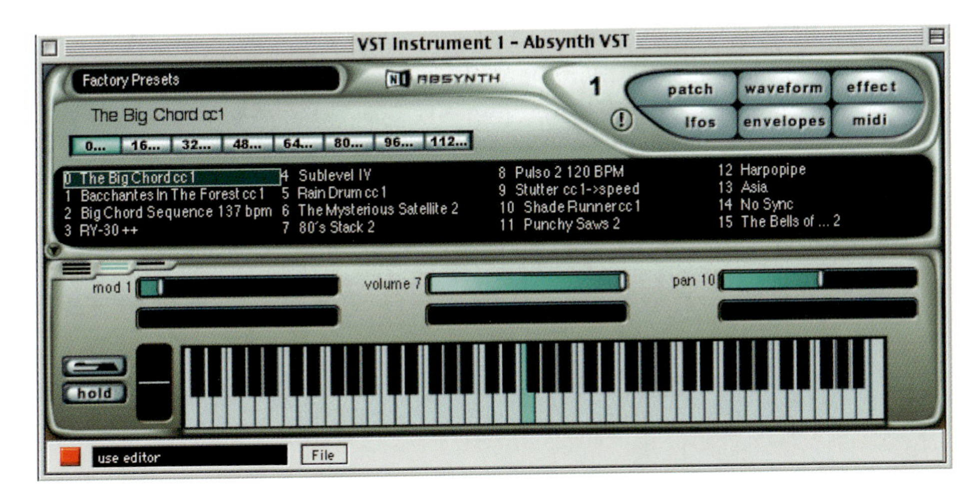

Figure 9.36 Absynth main Edit window in Cubase.

At the same time as you insert Absynth as a VST Instrument, the Absynth Engine (very similar to the stand-alone program) is launched in the background. If you need to do any detailed editing of the waveforms or envelopes or whatever, you can easily switch to this using the Applications menu in the top right-hand corner of your Mac's computer screen.

Figure 9.37 Mac Applications menu showing Absynth Engine and Cubase VST.

Alternatively, a click on any of the Edit buttons (Waveform, Effect, Envelopes, etc.) in the main VST Instrument Edit window in Cubase will bring the Absynth Engine to the front, with the selected editing window open for you. The Absynth Engine actually supports up to eight instances of the plug-in and it will display which of the eight instances is active and let you select which instance to edit. When you have made your edits, just click on the navigator button marked 'Host' to return to your host software – in this case Cubase VST.

As soon as you insert Absynth into Cubase, a pair of channels is automatically added to the VST Channel Mixer to take the audio outputs from Absynth into the VST mixer.

Absynth is also automatically added as a destination in the list of output devices in Cubase's Arrange window.

Figure 9.38 Absynth audio outputs Vstx 1 and Vstx 2 in VST Channel Mixer.

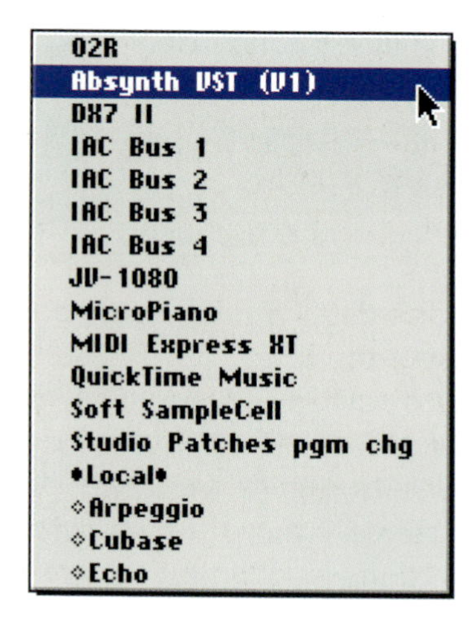

Figure 9.39 Absynth VST showing up as an output device in the Output popup menu in Cubase's Arrange window.

Native Instruments FM7

The Native Instruments FM7 is based on the design of the DX7 synthesizer first produced by Yamaha in 1983. This became the biggest-selling model of its era, on account of its relatively realistic simulations of acoustic instruments. Now available as a plug-in, the FM7 has 32-bit resolution, so it sounds much 'cleaner' than the 'hissy'-sounding lower-resolution audio produced by the original hardware. The FM7 also provides much more flexible programming, more extensive modulation and effects, a filter module, and multiple waveforms. There is an 'easy' programming page to complement the original more complex programming pages and Polyphony has been upgraded to 64 voices rather than the 16- or 32-voice polyphony of the original DX range. The FM7 can also be used as a plug-in to process incoming audio signals.

The FM7 provides you with just one 'single' FM synthesizer - it is not multi-timbral, and doesn't let you recreate the multi-timbral setups of the dual DX synthesizers such as the DX1, DX5, DX7/TX7 combination and DX7II, or of the eight-way multi-timbral TX816, TX802 and TX81Z models.

The latest Yamaha DX synthesizer is the DX200 Desktop Controller Synthesizer and this is recommended for use as a hardware MIDI controller for the FM7. When you realize just how many faders and switches there are on the FM7, you will definitely appreciate the benefits of being able to control at least some of these from a dedicated hardware controller. Interestingly for owners of original Yamaha FM hardware, the stand-alone version of the FM7 can receive voice data from any of the classic Yamaha DX-Series synthesizers - including the DX7, TX7, TX216, TX816, DX11, DX21, DX27, DX100, TX81Z, TX802 and DX7-II - although not from the TG77, SY77 or SY99.

Voice data can also be imported from a raw SysEx file or from a standard MIDI file using the Import SysEx button on the LIB page in the FM7 Editor. The data is automatically converted to the FM7's own advanced format. Banks of 32 sounds are stored in the FM7's memory in whichever bank of 32 memory locations is currently visible on the LIB page - 1-32, 33-64, 65-96 or 97-128. Single sounds are received into the FM7's edit buffer, so to keep these sounds you need to store from the edit buffer into one of the FM7's memory locations.

Note The FM7 cannot receive MIDI SysEx data when it is working as a plug-in, so you need to use the stand-alone version for data transfer.

A click on the FM7 logo at the top left of the plug-in window opens the About window. To close the window click again on the FM7 logo. In the About window a click on the FM7 logo opens the browser with the FM7 product page on the NI homepage. A click on the NI logo at the top right of the window opens the webpage, where you can find the latest update of FM7. In between these logos you will find displays for the Preset number and name, the spectrum and the waveform.

Immediately below these displays there is a row of buttons, resembling the membrane-type switches on the original DX7. As with the original hardware, these buttons let you select between the different programming 'pages'. The first eight pages let you control the eight FM 'operators'. The next six buttons open pages for the pitch, LFO and modulation parameters, the Master controls, the Easy Edit controls and the Preset Library.

The Library page lets you access the preset sounds. When you switch to this page, you will see 32 presets, labelled 1-32, which you can recall by clicking on the preset name or by using the up and down arrows on your computer keyboard. Another three sets of 32 presets each are directly accessible from this page. Initially, Bank 1 of the FM7 sound library is loaded and this contains four sets of 32 sounds - for a total of 128. You can load in any of the classic Yamaha DX7 voices by clicking on LOAD and choosing any of the files in the Presets or Presets Backup folders.

Figure 9.40 FM7 plug-in showing the Library page with the first bank of 32 sounds visible.

The original DX7 had six operators and the FM7 extends this to eight, labelled A to F, X and Z. Each of these operators has its own page where you can program its parameters.

Note These operators can be arranged in various configurations or algorithms to produce different types of sounds. There are 64 of these FM algorithms stored in the file 'FMMacros.fm7' in the FM7 software folder. In the Macros folder there is an HTML-document called 'FM Algorithms.htm' which contains a schematic overview of the first 59 of these, which are equivalent to the classic six-operator FM algorithms. The labels and the ordering depends on the number of carriers and the number of modulation paths.

You can view the current algorithm from any Operator page by clicking on the Matrix button above the top right of the Envelope display. This button name then changes to 'Envelope' - and a click on this resets the display to show the envelope again.

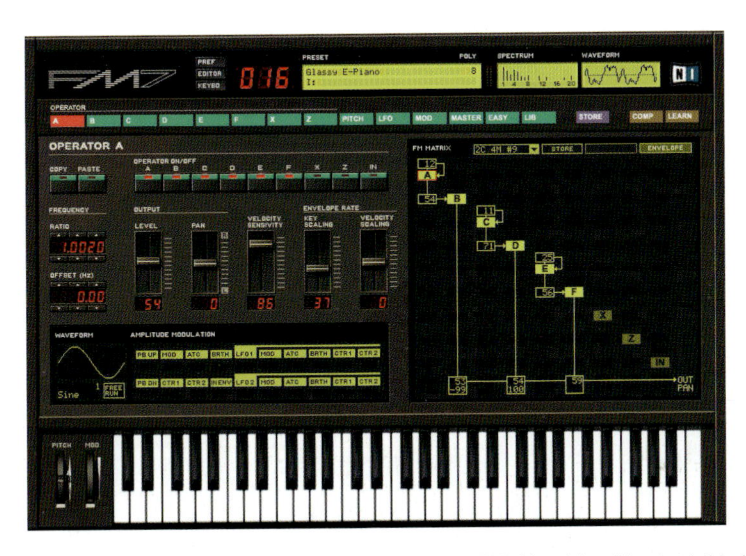

Figure 9.41 FM7 plug-in showing an Operator page with the FM Algorithm Matrix visible in the right-hand section of the window.

Each of the operators A to F is identical in features, and similar to the original DX7. A row of switches, looking like those on the front panel of the original DX7, lets you disable or enable any group or operators or a single operator while you are setting these up. This way you can focus on the particular components of the sound you are working on. All the operator controls, which will be familiar to DX7-owners, can be found here, including the frequency ratio controls, operator output level, key scaling and so forth. A graphical representation of the key scaling is provided and above this there is a sophisticated envelope control section into which you can insert breakpoints.

Figure 9.45 FM7 plug-in showing the Pitch controls.

The matrix is set up with columns of modulation sources and rows of modulation targets. At each junction, you can vary the amount of modulation going to the target destination by clicking on the display and dragging the mouse up or down to set the numerical value.

Figure 9.46 FM7 plug-in showing the LFO controls.

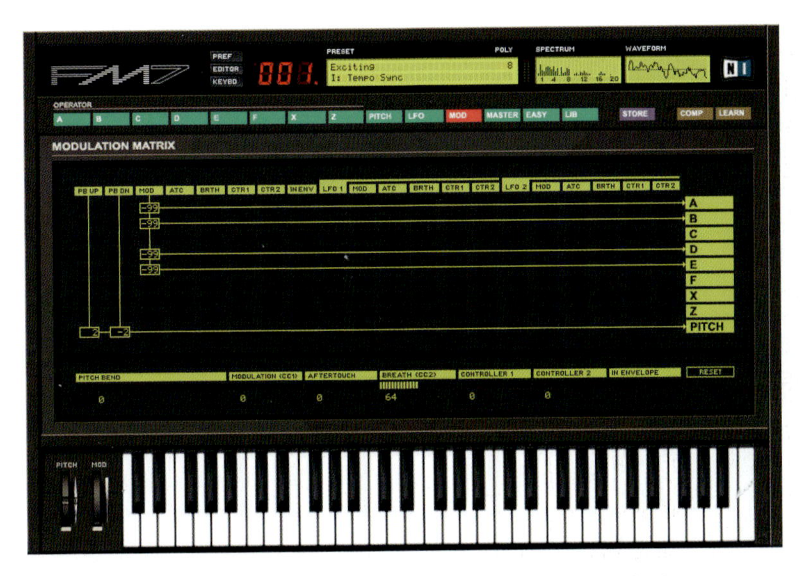

Figure 9.47 FM7 plug-in showing the Modulation Matrix.

The Master page lets you adjust the various parameters that affect the whole instrument. For example, here you can set the output volume or the input level of any signals that you feed into the FM7 when using it as a processor. You can set the polyphony - i.e. the number of voices - and you can set the master pitch for the whole instrument. The Effect section is controlled from this page, and this includes a four-tap stereo delay line whose output can be filtered. Modulation controls allow flanging, chorusing and other modulation effects to be created, and a Tempo Sync feature allows the delay to be synchronized to any incoming MIDI clock tempo.

Yamaha's FM synthesizers were always notoriously difficult to program, as they were the first to break away from the relatively easy-to-understand subtractive analogue synthesis model. Talk of 'operators' rather than 'oscillators' alone was enough to frighten many people away! Although the original FM programming is available in the FM7, an Easy Edit page is also provided. Here you can adjust straightforward-sounding parameters such as Brightness and set the envelopes using typical Attack, Decay, Sustain and Release controls.

At the bottom of each window there is a 60-note keyboard that you can click on with your mouse to play. This is useful while setting up sounds, but, ultimately, you will need to connect a suitable MIDI keyboard or other MIDI controller. To the left of the keyboard you will find a Mod and Pitch wheel. These actually work when you drag them with the mouse - although, again, you will normally control these using external hardware.

The Envelope/Filter page features two envelope generators, a filter section and an amplifier section.

Figure 10.4 The Envelope/Filter page.

The Modulation/Tune page lets you apply several modulators and also contains two LFOs, tuning controls and a Voice Grouping section.

Figure 10.5 The Modulation/Tune page.

The Options page lets you import external file formats and make memory settings.

Figure 10.6 The Options page.

The Channel/Program page shows the settings for HALion's 16 MIDI channels and its 12 'virtual' outputs.

Figure 10.7 The Channel/Program page.

So, lots of controls, but these are all laid out very clearly and logically – HALion is very easy to use.

HALion clearly has advantages over rivals such as the Emagic EXS24 or IK Multimedia SampleTank on account of its ability to play back samples direct from hard disk. And HALion's third-party sample file format support is the best available. A first-rate software sample for VST-compatible applications. Needs lots of RAM and a fast processor for best results. Works with Cubase VST 5.0 or higher, Nuendo 1.5 or higher or other Virtual Instrument compatible host software.

Emagic Xtreme Sampler EXS24

Emagic's software sample for Logic Audio, the EXS24, was virtually guaranteed to be a hit with Logic users from the outset. It integrates beautifully with Logic Audio, as you would expect, and has a good selection of factory sounds to get you started.

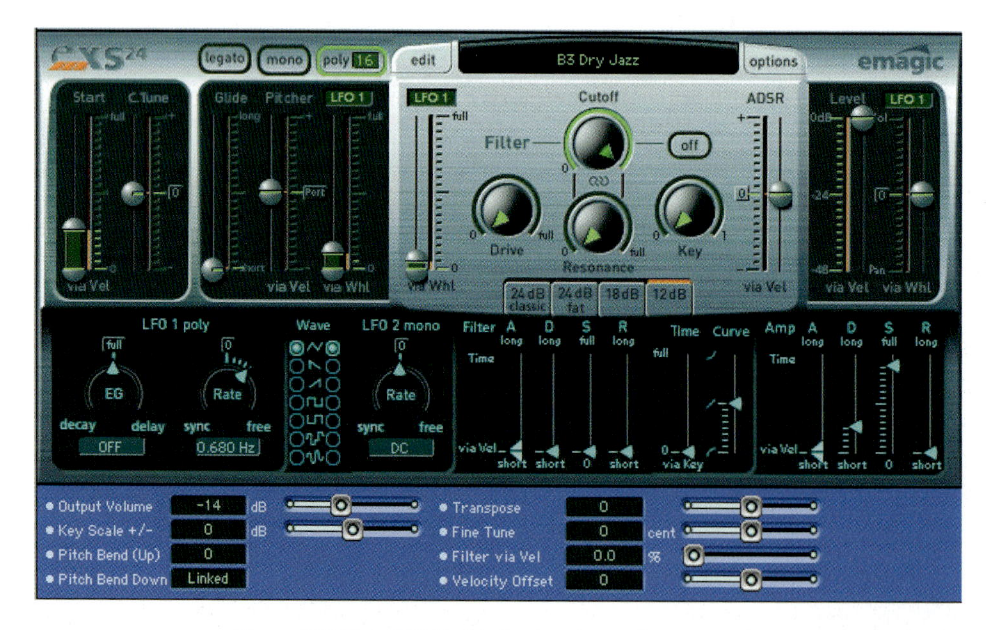

Figure 10.8 The EXS24 plug-in showing various controls.

Most of the controls you need to access frequently are presented in the main window, and you can load your sounds by clicking and holding on the name that appears at the top of the plug-in's window - B3 Dry Jazz in the example shown. The factory sound set is packed with useful stuff like drum kits, Moog basses, sampled guitars and synth pads.

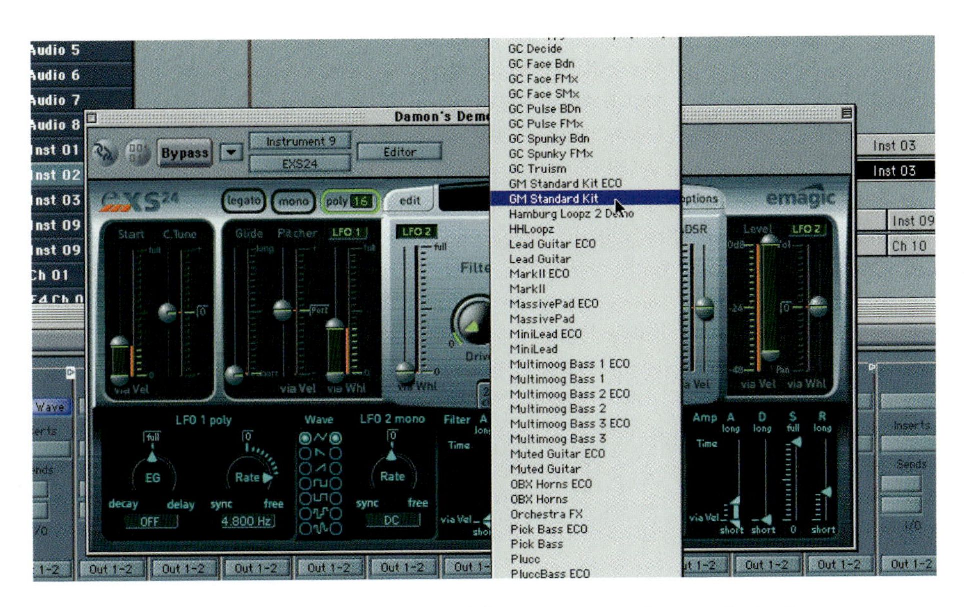

Figure 10.9 EXS24 showing part of the popup menu for sound selection.

You can open up to 16 EXS24's as Audio Instrument objects in Logic Audio, if you have enough RAM. I allocated 128 Mb to Logic Audio, for example.

Each instance of the EXS24 offers up to 64 mono or stereo voices. How many you can actually play at once depends on the CPU speed and availability. The EXS24 also provides high-quality digital playback quality at up to 24-bit and 96 kHz. The

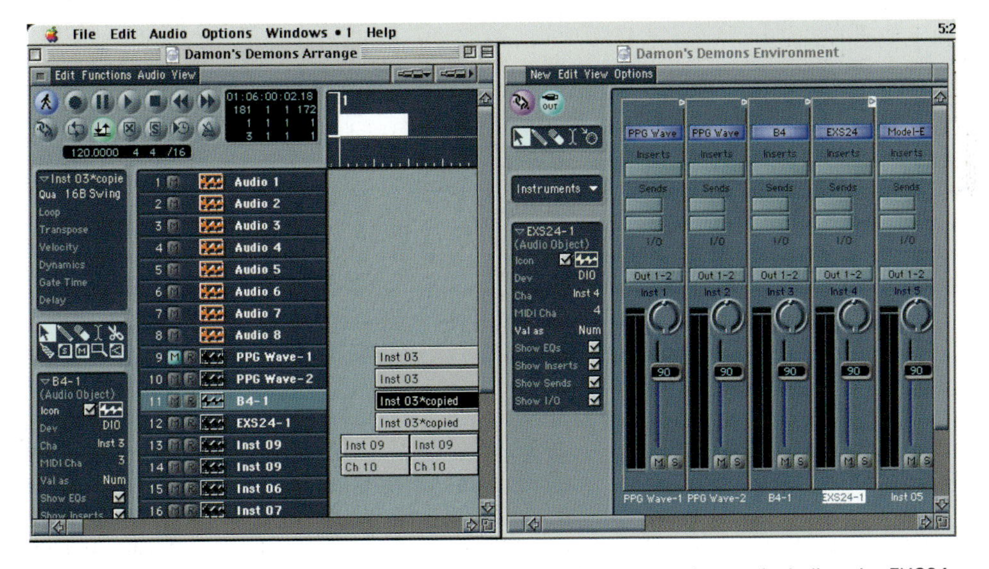

Figure 10.10 Logic Audio screens showing several virtual instruments in use, including the EXS24.

audio quality you will achieve in practice will depend on the quality of your D/A converters when playing back and your A/D converters when recording new samples. The user interface is clear and straightforward – in contrast to many hardware samplers which are far more 'fiddly' to use.

The supplied sample library is fine, but does not really compete with the many libraries available from third-parties in various formats. Fortunately, the EXS24 can read and convert Akai format samples using the Akai Convert feature in the Instrument Editor. Support for other formats is promised.

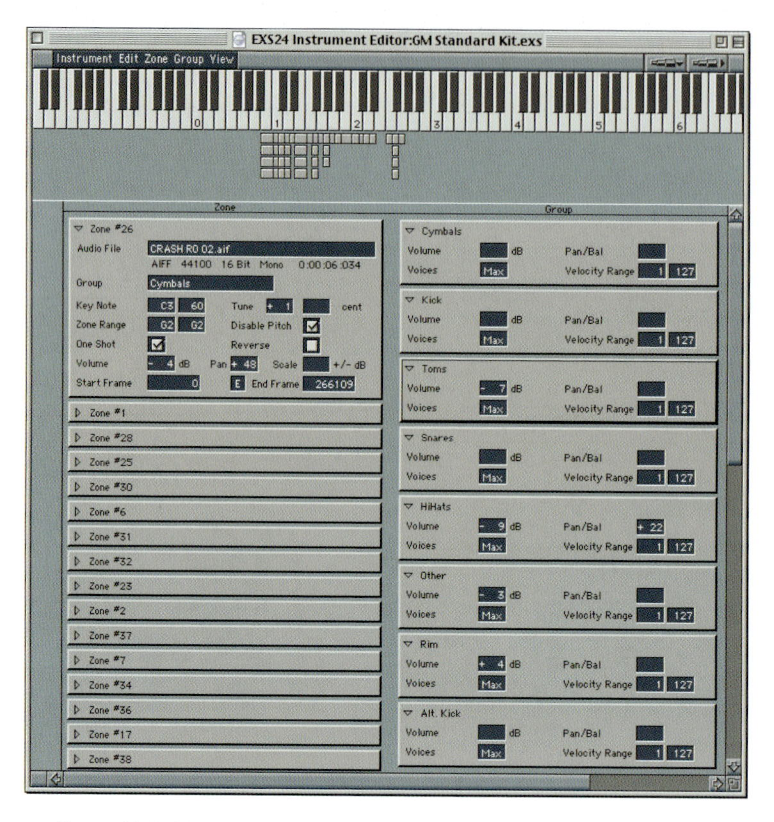

Figure 10.11 The EXS24 Instrument Editor showing Zones and Groups.

You can load in your own samples using the EXS24 Instrument Editor. Here you arrange your samples into Zones (keygroups) and combine these into Groups so you can velocity-switch or layer sounds.

The plug-in window available for each instance of the EXS24 gives you access to all of the EXS24's synthesizer parameters. Here you can alter filter settings and envelopes, offset pitches of samples, set up portamento effects and so forth. You can change this window to a Controls view using a flip menu in the upper window area. The Controls view lists all the parameters and lets you adjust them using sliders or by typing numbers - or, in some cases, using popup menus.

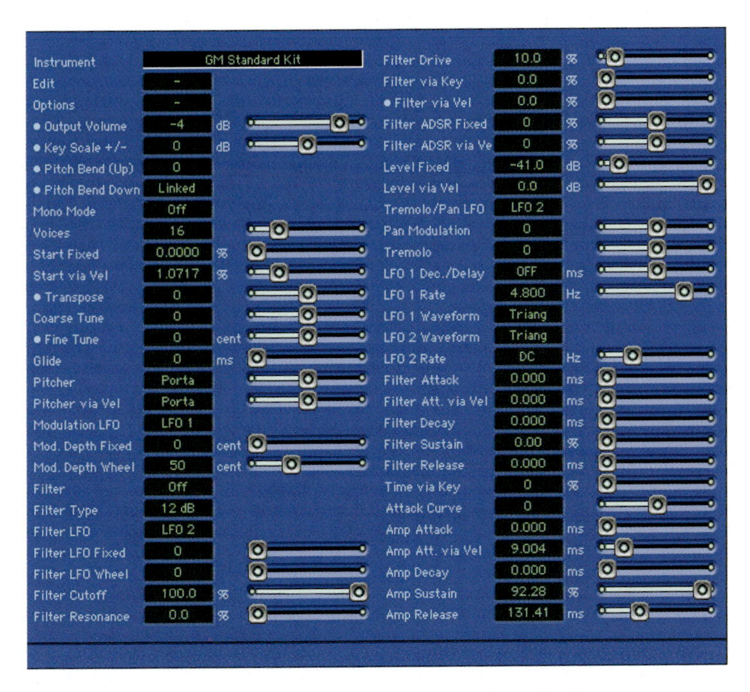

Figure 10.12 The EXS24 Controls display showing editable parameters.

To achieve best results with the EXS24 you will not only need the fastest dual-processor CPU, but also lots, and I mean lots, of RAM. To get the maximum number of voices to play, for example, you need to load your samples in 32-bit floating format into RAM - and 24-bit 96 kHz samples (if you use these) are going to use up even more RAM.

So, the EXS24 allows you to play music and sound effects samples digitally at the highest quality from your computer using Emagic software. Compared with the cost of buying a hardware sampler with similar capabilities, the EXS24 looks like a bargain. However, don't forget the cost of the computer, audio card and converters that you need to run the software. You could easily spend £3k-£4k, or more, for the very highest-quality kit - which is comparable with the price of a high-end hardware sample with all the 'bits'.

Emagic System Bridge (ESB) and Xtreme Sampler 24 (EXS24) TDM plug-ins

If you are a Pro Tools TDM user, then you will want to be using the EXS24 TDM with the ESB TDM plug-ins for Logic Audio. The EXS24 TDM runs on TDM hardware so you can plug this into Logic Audio by inserting it, say, on an Aux track and playing this back via the DAE software link to the Pro Tools TDM hardware. The System Bridge is a DirectConnect TDM plug-in that accepts audio routed from the Instrument channels in Logic Audio's mixer environment and feeds this into the Pro Tools TDM hardware. Using these plug-ins, you can stream up to 32 extra channels of audio into your Pro Tools TDM mixer from Logic Audio. So, for example, you could have lots of instances of the EXS24 sampler all playing audio via your Pro Tools TDM hardware. Or you could use these channels to play software instrument plug-ins through Pro Tools TDM hardware.

First you set up a selection of Audio Instrument mixing channels in your Logic Audio Environment and set these to use Direct TDM output. The ESB plug-in then lets you route the output signals from Logic's Direct TDM (DTDM) mixer to TDM Aux channels. You can also use the ESB TDM plug-in to route audio from Emagic's EVP88, ES1 and ES2 instruments or from VST 2.0 instruments into the TDM mixer.

Tip Remember that in Logic Audio you can always use combinations of hardware outputs – with some mixer channels set up using an ASIO driver, for example, while other mixer channels use Digidesign TDM hardware – or the DTDM outputs to the ESB plug-ins.

The outputs of these Audio Instrument channels have to be set up to play through one or more ESB DirectConnect plug-ins inserted into Aux channels in the Logic Audio Environment. In turn, these Aux channels are set up to use the DAE to connect to the Pro Tools TDM hardware.

With this all configured, the output channels of the 'native' Direct TDM mixer are made available to the TDM hardware via the ESB plug-in, which you insert as the first plug-in in any Aux channel of the TDM mixer.

Be aware that in the ESB plug-in window you also need to select whichever output channel or output channel pair of the Direct TDM mixer you are using

Figure 10.13 Logic Audio Instrument channels set to DTDM with EXS24 plug-ins in use. Logic and/or VST plug-ins can be inserted for additional processing.

Figure 10.14 Logic Audio Aux channels set to output via the DAE – note the popup selector showing available hardware. ESB plug-in inserted on one Aux channel.

Figure 10.15 ESB DirectConnect plug-in window showing output pair 1-2 selected.

in the Logic Instrument channel as the corresponding DirectConnect output channel or channel pair in the ESB DirectConnect plug-in window. In the example shown, this is output pair 1-2.

The really neat thing is that the EXS24 TDM plug-in lets you work with up to 32 EXS24 samplers and you can process the audio outputs from these using a combination of Logic, VST and TDM plug-ins.

The ESB TDM plug-in lets you route up to eight audio instrument channels of the native Logic Audio mixer into the Pro Tools hardware TDM mixer. With a virtual instrument or sampler, such as the EXS24, present in the topmost insert slot in the Instrument mixing channel, you can insert Logic or VST plug-ins into subsequent slots to process the audio using your favourite Logic or VST effects. The audio is then fed into the TDM mixer via the ESB plug-in or plug-ins inserted into one or more Aux channels - where you can add further processing using TDM plug-ins. This feature is not available for Cubase VST or for Digital Performer - giving Logic Audio something of an advantage in this respect.

The EXS24 TDM plug-in's latency performance is not as good as that of dedicated hardware samplers; its main advantage is its integration directly into your music production software-environment. And when you are working at your computer sequences, it gives you control right at your fingertips plus a relatively large screen for editing parameters. As with all these plug-ins, the faster your CPU and the more RAM you have available to allocate to the applications, the better the results.

IK Multimedia SampleTank XL

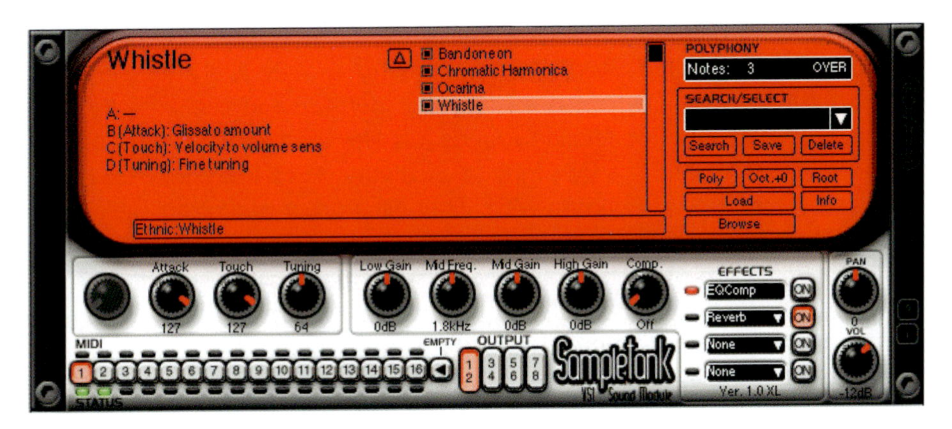

Figure 10.16 SampleTank.

Some of the biggest complaints I have heard from users of software samplers is that they are very fiddly to use and set up and that it can be awkward to browse and load samples quickly. Also, the quality of the factory samples provided can often be questionable. A VST plug-in that addresses these issues is IK Multimedia's SampleTank. This offers 32-bit quality playback with up to 128-note polyphony - depending on the CPU. The sample library in the XL version comes on 4 CD-ROMs and includes more than 450 sounds, encompassing loops for house and techno, a good selection of electronic and acoustic drums, percussion, ethnic instruments, synths, guitars, basses, strings, brass, woodwinds and keyboards - just about everything you might need to put music together. And if this isn't enough, you can import Akai S1000/3000 sound banks and convert these using the separate SampleTank Converter application.

SampleTank also features up to four effects that can be inserted on each sound and all the effects parameters are MIDI-controllable. There are 20 effects provided, including: Compressor, Equalizer, Reverb, Ambience, Reverb Delay, Delay, Filter, Wah-Wah, Chorus, AM and FM Modulation, Flanger, Autopan, Tremolo, Rotary Speaker, Lo-Fi, Distortion, Phonograph and Slicer.

The interface is very basic, which helps with the speed of use, and it has everything you need in one compact window. In Cubase VST, instruments can be routed via four separate stereo outputs, although in Logic Audio only stereo outputs are supported. I checked SampleTank out using both of these and I was stunned at how fast the sounds load - just double-click on any sample and there it is, ready for you to play instantly!

So what are the sounds like? Well, the TR808 samples are the best I have heard yet - and I have a real TR808 to compare with! The three acoustic drum kits all have long-decaying cymbals, full-sounding toms and very usable sounding bass and snare drums - and the Studio kit does sound exactly like a kit played in a small, dead studio! The DX piano sounds just perfect and the acoustic grand piano is very usable. The Pop Violins sounds just like the cheap string synthesizer which I used in the 70s and the B3 organ sounds very much like a B3 - and, again, I have a real Hammond to compare with. There is a great percussion selection as well, with ethnic instruments such as the Darbuka and Dumbeck, triangle, tambourine, shakers - and great congas. The best thing about these is that you get several different types of slaps and hits, unlike most samples which just provide two or three main sounds. In short, the sound library is excellent.

SampleTank is extremely easy to use, the samples all sound great, and a wide range of useful effects is provided. This is one plug-in that I can unreservedly recommend. SampleTank works with Cubase VST 4.1.1 or later, Logic Audio 4.5 or later, or any other sequencer compatible with VST Instruments. A limitation with versions 4.x of Logic Audio is that only two outputs are available from SampleTank. SampleTank is also available for MAS and for Pro Tools HTDM and RTAS.

Note SampleTank is the first plug-in sampler instrument that has been developed for RTAS - although you can be sure that others will follow.

BitHeadz Unity DS-1 version 2.1

The Unity DS-1 was one of the first software samplers developed for the Mac. Over 300 Mb of sounds are provided to get you started and the Unity DS-1 can read Akai, SampleCell, Sound Designer II, AIFF, Wave, and other formats - providing excellent compatibility with existing sample libraries. Version 2.1 has support for multiprocessor Macs with optimization for the G4 along with many other enhancements, including the ability to work as an RTAS plug-in within Pro Tools or as a VST plug-in in applications that support this. It also works with ASIO cards, ReWire, MAS 2.0, DirectIO and DirectConnect, providing excellent integration with Cubase VST, Logic Audio, Digital Performer and Pro Tools. Once the audio from the synthesizer is brought into the sequencer, you can apply effects, record the audio and automate the mixdown. Record to disk 'live' or from a sequencer or play the Unity DS-1 'live' or from your sequencer. This software is flexible - and it works great on a laptop! And using a MIDI controller such as the Phat.Boy you can even control the synth parameters in real time. On the 'down' side, installation is rather 'fiddly', with several extra files needed to configure for OMS and for the different sequencers - so plan on setting some time aside to get everything set up correctly.

Figure 10.17 Unity DS-1 Keyboard.

The Unity DS-1 Keyboard lets you conveniently check that everything is working OK and once everything is working properly, you can get instant gratification by simply opening the Unity DS-1 MIDI Processor to play sounds from your external MIDI controller.

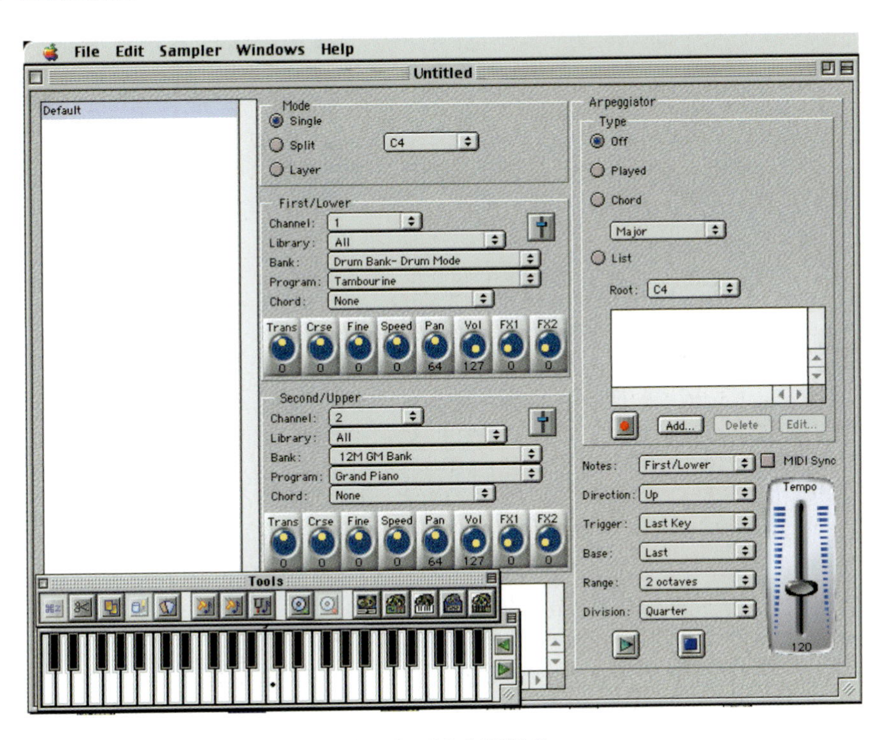

Figure 10.18 Unity DS-1 MIDI Processor.

Here, you can access the installed sounds, set up splits or layers, and use the arpeggiator. You can also play sounds using the Unity DS-1 Mixer, which also lets you monitor audio activity on all the channels and control the global effects processors.

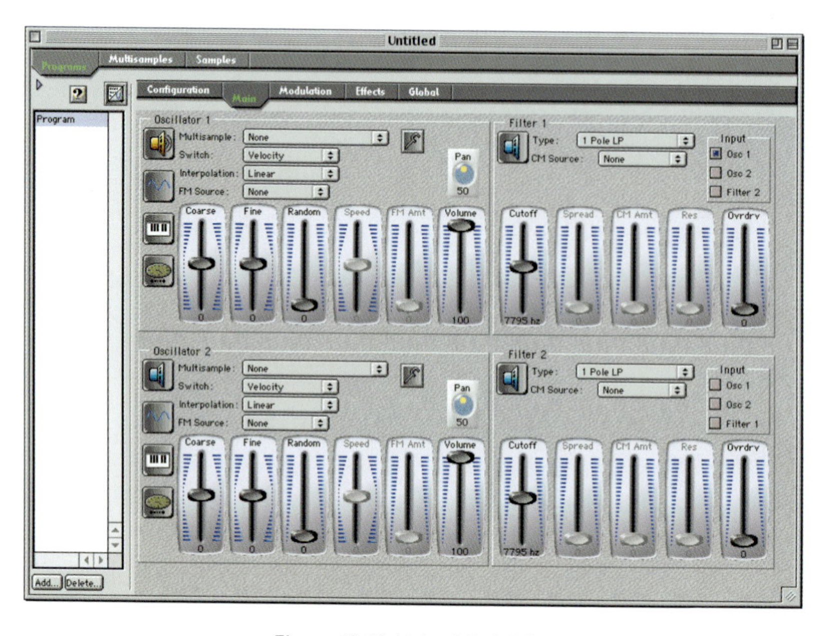

Figure 10.19 Unity DS-1 Editor

When you need to modify the factory sounds or create sounds from scratch, use the Unity DS-1 Editor. This provides two oscillators and two filters that you can use to play back and modify a pair of samples. You can apply EQ, delay, chorus and distortion effects using insert effects and two global effects let you add overall reverbs and delays. The Unity DS-1 has up to 256 stereo voice polyphony, depending on how fast your CPU is, and sample memory is allocated from available RAM – so you need lots for best results. The Unity DS-1 Engine supports up to 96 kHz sample rate and has 32-bit internal processing. Audio is saved as 8-, 16-, or 24-bit files.

The verdict

The Unity DS-1 offers integration with a wider range of software than most of its competitors, such as the Emagic EXS24, and has comprehensive editing, processing and mixing features. On the downside, the sheer number of windows and the way the software is split into several separate applications makes installation something of a pain, and operation somewhat less than instant. Nevertheless, a very usable sample library is provided and the Unity DS-1 has excellent support for other sample library formats. Technically, it performs well and sounds good - and the price is fair.

Chapter summary

If you are looking for a software sampler to use with Logic Audio, the Emagic EXS24 is probably the best choice - along with the ESB TDM plug-in if you have Pro Tools hardware. Cubase VST and Nuendo users can choose between Halion or SampleTank. Pro Tools and Digital Performer users can go for BitHeadz Unity DS-1, which will also work as a VST plug-in, or for SampleTank which now works with all Pro Tools versions.

11 BitHeadz

OK, so now you are wondering why BitHeadz has a chapter of its own! Well, just as I was putting the final touches to the book, having covered the AS-1 and DS-1 briefly in the chapters on software samplers and synthesizers, along came the brand new Unity Session package for me to evaluate. When I realized that Session incorporates a whole suite of plug-ins and that BitHeadz have developed their catalogue to include various sampling libraries, looping software and so forth, I felt that it was worth including a chapter all about BitHeadz - and I left the stuff I wrote about the original AS-1 and DS-1 for anyone still interested in these.

The latest BitHeadz catalogue includes the Unity (formerly the Retro) AS-1 analogue synthesizer and the Unity DS-1 sampling synthesizer, the Unity Player sample playback software, and Unity Session - which incorporates all of these and more. BitHeadz also offer Phrazer loop composition software, Voodoo software drum machine, Black & Whites virtual piano module, Steve Reid's Global Percussion virtual percussion library, Tubes, Tines, and Transistors synthesizer sample library, Tempo Tantrum breakbeats, Tubes, Tines, and Transistors vintage sample library, Pop Drums loop library and Harry Sharpe Guitars loop library.

Unity Session 3.0

Unity Session is a whole set of plug-ins - eleven in number - that you can run from within a wide range of host environments. You get software sampling, analogue synthesis, physical modelling, MIDI effects, and audio effects - all from within your favourite sequencer. You can import and playback most popular sample formats using Unity Session. Stand-alone operation is also provided for, using the Unity Editor, Player and Mixer applications.

The various samplers and synthesizers are implemented as plug-ins that use the basic Unity 'engine' for playback. These include the original Unity DS-1 sampler and Retro AS-1 analogue synthesizer, plus the new SP-1, SampleCell, GigaSampler, SoundFonts, and DLS sample playback modules, and four physical modelling synthesizers - the CL-1, FL-1, BW-1, and HS-1 which model the clarinet, flute, bowed string and hammered string, respectively. These different synthesizer plug-ins create their sounds in different ways. The DS-1, SP-1, SampleCell, Giga, SoundFonts and DLS all use samples, while the AS-1 and the physical models CL-1, FL-1, BW-1 and HS-1 create their sounds using mathematical algorithms. With the sample-based plug-ins, samples are used to build multi-samples, then these multi-samples are used by program oscillators to build the sounds.

The AS-1 is a very capable analogue synthesizer with three stereo oscillators, a ring modulator, portamento, frequency modulation, multiple stereo filters with 16 different filter types, and a flexible modulation matrix. The DS-1 is a stereo sample playback unit with two stereo oscillators, two stereo filters with 16 different filter types, portamento, and a modulation matrix. The oscillators play multi-samples, not just single samples, and they can stretch samples to different tempos without changing their pitch. The SP-1 is similar to the DS-1 but is a single oscillator sample playback synthesizer. Like the DS-1, it plays multi-samples, not samples. The Gigasampler, SampleCell and SoundFonts plug-ins each use a single oscillator algorithm similar to the SP-1. These plug-ins recognize the different file types and convert them into Unity format so that Unity Session can play them back.

The last four plug-ins all use physical modelling techniques to simulate various instruments. The CL-1 models the clarinet and the FL-1 models the flute. You can even hook up a breath controller to the breath input of the mouthpiece and dial in the type of reed (or mouthpiece in the case of the flute) you want - choosing the characteristics from your own custom table. The CL-1 and FL-1 also model the bore (or body) of the instrument and an output filter - creating the sound mathematically, like the AS-1. Similarly, the BW-1 models a bowed string and the HS-1 models a hammered string - like a picked guitar or hammered piano string.

The plug-ins don't stop with just the instruments though. Unity Session has plug-ins for everything. For example, Unity Session has a set of seven MIDI plug-ins that lets you process any MIDI that you are sending to a Unity synthesizer to create arpeggiated chords, splits, layers and other effects. A set of 23 audio plug-ins is also provided to let you add effects such as EQ, delay or reverb using the 'Unity Editor' and 'Unity Mixer' applications. Unity works with ASIO, ReWire, MAS 2.1, VST, DirectIO, RTAS and DirectConnect. These interfaces provide excellent integration with products such as Digital Performer, Pro Tools, Logic

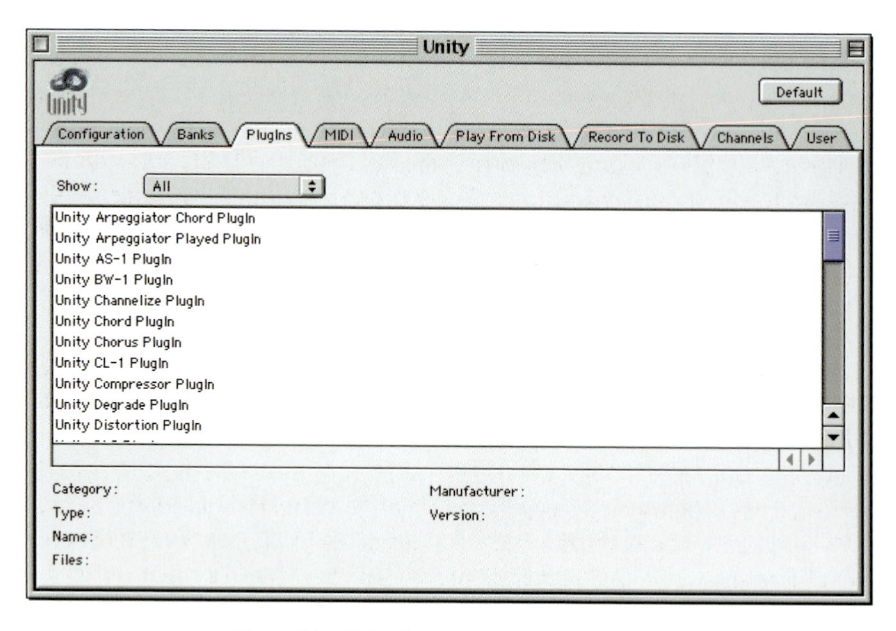

Figure 11.1 Unity Control Panel plug-ins page.

Audio and Cubase VST. Unity Session can also read a wide range of file formats that can be accessed simply by dropping them into Unity's Banks folder. These include GigaSampler, Akai, Roland, SampleCell, Sound Designer, AIFF, WAV, SoundFonts 2.0 and DLS formats – and audio CD-ROMs.

Installation is fairly straightforward; Unity uses the challenge/response method of copy protection. Once installed, you need to set up the MIDI to work with, say, OMS, and set up the audio to work with, say, DirectIO using the Unity Control Panel. Then open the Unity Mixer, the 'main' Unity application. This has a 'master' section for the main stereo outputs, fed by 16 input channels, each of which can be used to play back a different Unity instrument. You can insert two MIDI effects and two audio effects plug-ins onto each of the 16 channels, then mix the 16 channels down to a single audio output that can be routed to your ASIO or DirectIO sound card. In the 'master' section, there are two master MIDI effects, two master Send effects, and two Global effects that can be applied to the main outputs for the entire synthesizer. The watchword here is 'flexibility'.

Unity Session has been improved in many ways compared with the original AS-1 and DS-1. For instance, the Unity VST plug-in now includes a VST editor with library, bank and program menus so you can select from any of your installed Unity sounds. You can also limit the number of Unity banks so that the VST host software's popup menus aren't too unwieldy. And you don't have to switch manually to plug-in mode to use Unity with your sequencer - Unity automatically switches when you launch the sequencer. Another improvement over the original plug-ins is that the FreeMIDI driver functionality has been absorbed by the Unity MAS 2.1 plug-in. When Digital Performer is launched, the Unity Session server is launched in plug-in mode and automatically creates a Free MIDI inter-application communication port for Digital Performer to send MIDI to. These improvements go a fair way toward answering my criticisms of the original AS-1 and DS-1, which were very fiddly to set up and use.

Three CD-ROM libraries are provided: 'Black & Whites CD-ROM', 'Pop Drums CD-ROM', and 'Orchestral Strings CD-ROM'. The BitHeadz Osmosis conversion utility which runs on Mac or PC is also supplied with the package. This lets you convert Akai S1000/S3000 and Roland 760/770 formatted disks to Unity DS-1 or Digidesign SampleCell formats. Osmosis can even convert Roland S760 and S770 CD-ROMs and removable media such as Zip disks. You can also save the disk directory to a text file. This is extremely useful for cataloguing disks, or for finding specific samples on disks. All of the Akai partitions and volumes, Roland volumes and performances, and programs and samples, with their lengths, are listed. So now you can use all your old Akai and Roland disks with Unity Session or SampleCell. And once you've converted the samples to AIFF, Sound Designer II, or WAVE files, almost every Macintosh audio application will have access to your new sound library!

The AS-1 synthesizer has lots of great patches to get you started, and now you can explore the new modelling synthesizers as well. Most people are going to be impressed by the range of sample playback options. If it's a sample - Unity can almost certainly play it back. This means you can use most of your existing sample libraries, and the amount of time you will spend converting or configuring them for use with Unity Session is kept to a minimum. The libraries of sounds supplied with Unity Session are very good as well, with first-rate pianos on the Black & Whites disk, a good selection of loops and kits on the Pop Drums disk and some very convincing string samples on the Orchestral Strings disk. I also checked out the BitHeadz Discrete Drums sample library for Unity and was particularly impressed with the quality of the samples and general versatility of this set. Discrete Drums is based on the 24-bit Discrete Drums library produced by Rick DiFonzo and it sounds great! All these drums were digitally recorded using a Pro Tools Mix 24 through a Neve 8232 mixer. The recordings were made at Hum Depot in Nashville in a large wood-panelled room. The drum sets each have three different banks: 'Dry' is just the natural drum set; 'Room' is just the room tracks by themselves; and 'Mix' contains links to both Dry and Room banks and also has several preset mixes of the room level. Using Unity, you can mix between the

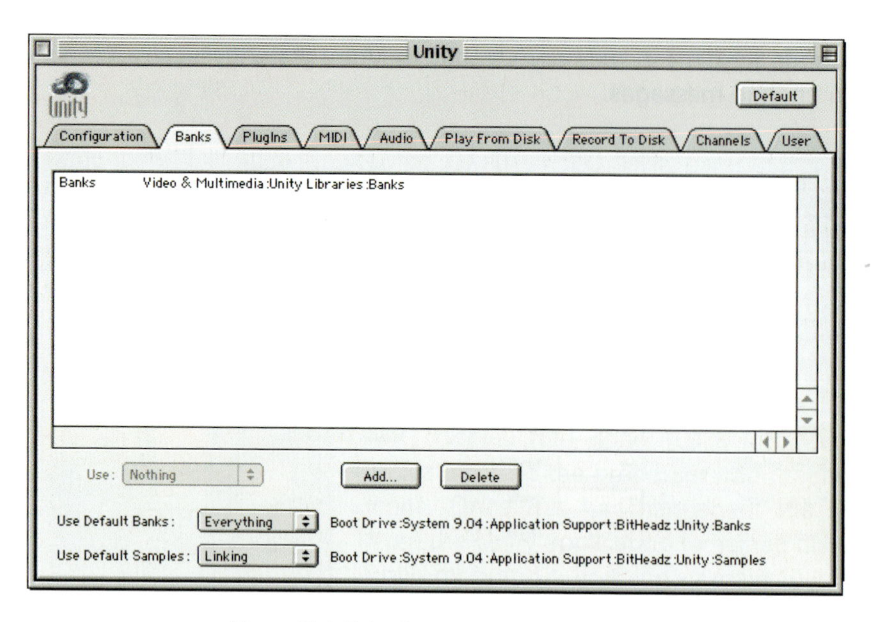

Figure 11.4 Unity Control Panel Banks page.

By default, the Unity installers place the Banks and the Unity plug-ins in a folder called 'Unity' inside a folder called 'BitHeadz', inside a folder called 'Application Support' inside the active System folder on your hard drive. If, like mine, your primary hard disk drive partition is getting full, you can simply copy these banks to another drive partition, or to a different drive, then delete them from the original hard drive. In this case, you will need to let Unity know where you have put them. Switch to the 'Banks' page in the Unity Control Panel and click on the 'Add' button. This brings up a dialog box that you can use to locate and select the new Banks folder.

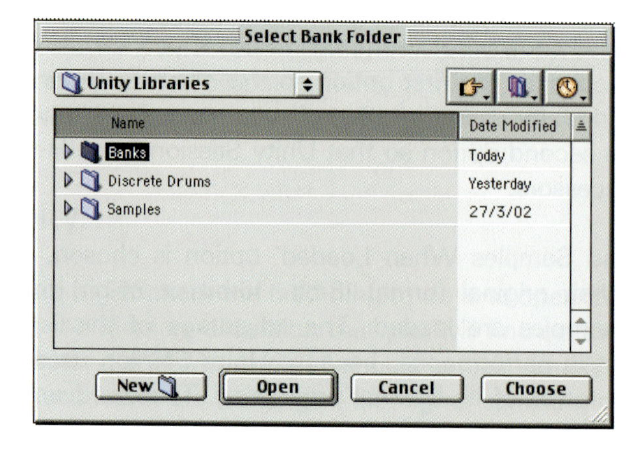

Figure 11.5 Unity Select Bank Folder dialog.

Now, Unity will load the banks from the new location you have specified, as well as from the default location in the Application Support folder.

The next thing to do is to route MIDI from your sequencer into Unity Session using the OMS or MAS driver. The Unity Session installer automatically installs the Unity OMS Driver and the Unity MAS 2.1 plug-in for you, but you must install and set up OMS or FreeMIDI yourself. Usefully, any application that supports OMS or FreeMIDI names, such as Cubase, Logic, Performer, or Pro Tools, will automatically get all of Unity's installed bank and program names.

Note When Unity Session works with OS 10.1, it includes the 'Unity CoreMIDI Input' utility for getting MIDI into Unity Session for standalone operation, and the 'Unity CoreMIDI Driver' for accessing Unity Session from a third-party sequencer.

So, after installing Unity Session, you need to set up your OMS configuration, if this is not already done. Even if you have an existing OMS setup, you will need to update your current setup document to include the Unity application. To do this, choose 'Midi Cards & Interfaces' from the OMS Setup application's Studio menu, and click on the Update Setup button.

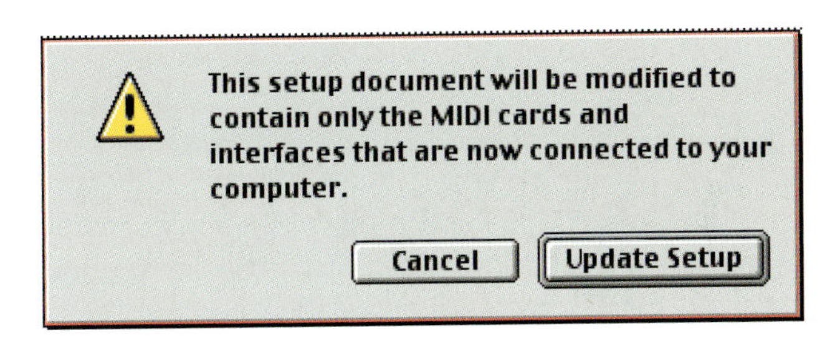

Figure 11.6 OMS Setup Update.

An icon for Unity will appear in your OMS Setup document.

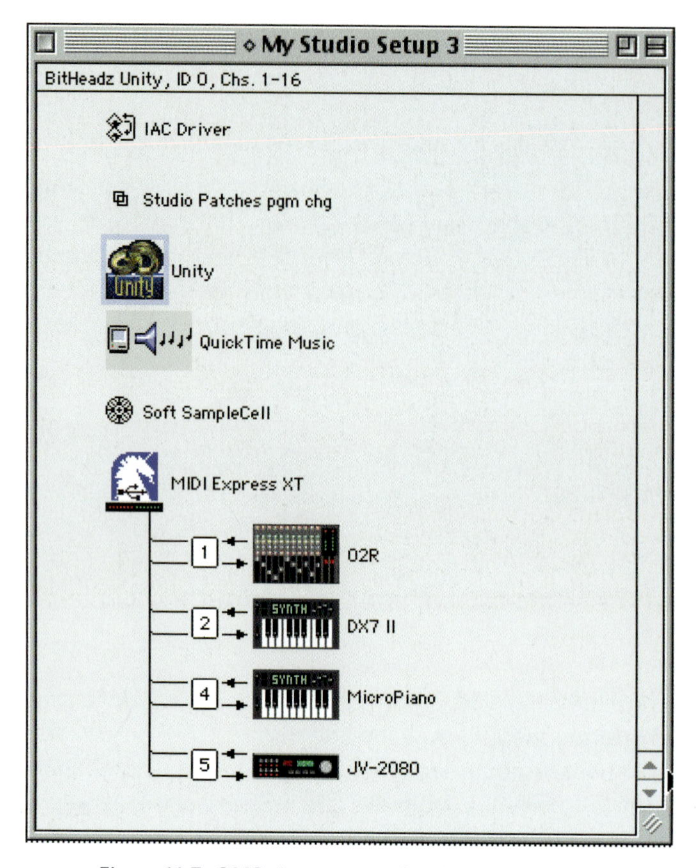

Figure 11.7 OMS document updated to include Unity.

When you save and 'make current' this OMS setup document, any application that uses OMS will list the Unity as a MIDI output destination.

You should also run the Unity OMS Publish utility once after installation to make sure that the Unity names will get published correctly. Now, in addition to sending MIDI to the synthesizer, the OMS-aware application can also get all of the bank and program names via the OMS Name Manager.

To select the default MIDI input, open up the Unity control panel. Select the MIDI tab on the top of the window to get to the MIDI page. The default MIDI input selection is in the lower right of the window.

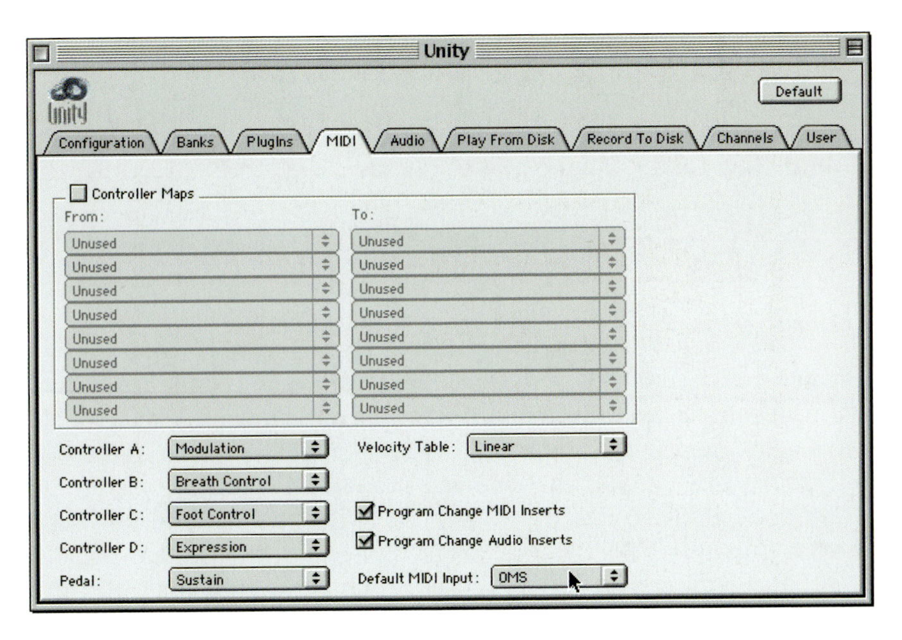

Figure 11.8 Unity Control Panel MIDI page.

While you've got the Unity control panel open, you can also select the audio output you want to use from the Audio page. There are lots of ways to get audio out of Unity Session. You can talk to hardware directly via interfaces like Sound Manager, ASIO or DirectIO, or you can talk to third-party sequencers via interfaces such as ReWire, MAS 2.1, DirectConnect or VST to route the digital output from the synthesizer back into the digital audio sequencer. The advantage of this is that once the audio output from the synthesizer is brought into the sequencer, it can be processed like any other audio track. You can apply plug-in effects, record the audio, and automate the mixdown.

Tip If you are using Unity Session without any other application requiring the sound card, then ASIO is a good choice. Bear in mind that not all ASIO drivers can support more than one client application talking to the hardware at the same time. So, for example, if you try to use ASIO output from Unity Session at the same time as Cubase is using this hardware, there would be a conflict. In this case, you should use ReWire or the Unity VST plug-in so that the audio from Unity is routed directly into the Cubase mixer.

Note Each ASIO hardware device has a software driver that comes with it. Unity Session requires that the ASIO driver be in a folder called 'ASIO Drivers' in the Extensions folder of the System folder. The installer creates this folder for you, but you have to place suitable ASIO drivers in it. These should be supplied with whichever card you are using.

Unity Session can output audio into Digital Performer using either ReWire or MAS 2.1. With Cubase you can use either ReWire or the Unity VST plug-in. ReWire also works with Logic Audio and Studio Vision 4.5. With Cubase, the Unity VST plug-in is installed by default, so if you want to use ReWire, you need to put the Unity ReWire 1 extension into the Extensions folder and remove the Unity VST plug-in from Cubase's VST Plug-ins folder. For some Cubase and Logic users, VST is a better choice than ReWire, allowing you to select Unity Session programs much more easily than with using OMS and ReWire. OMS is not involved, as the VST plug-in handles both MIDI and audio.

If you are using Digidesign hardware, DirectIO lets you output from Unity Session to the Digidesign hardware directly - without using the Pro Tools software.

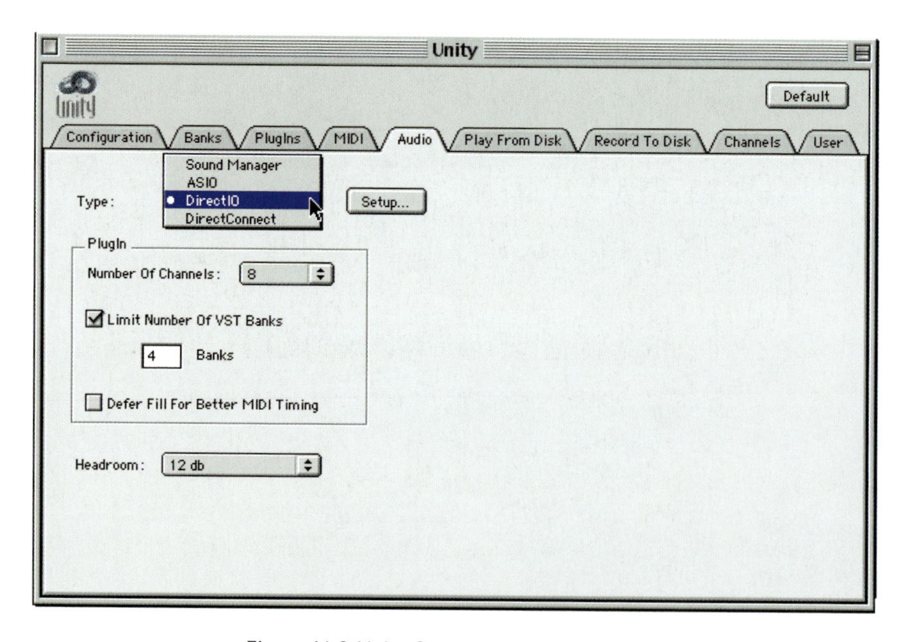

Figure 11.9 Unity Control Panel Audio page.

Pro Tools users can choose DirectIO when working with Unity in stand-alone mode or either RTAS or DirectConnect when running the Pro Tools software.

As with the ASIO drivers, DirectIO can only support a single client application, so, if you try to use DirectIO from Unity Session at the same time that the Pro Tools software is using the Pro Tools hardware, there would be a problem. The work-around here is to use either DirectConnect or RTAS in Pro Tools so that the audio output from Unity Session is routed to the input of an audio track in Pro Tools. OMS is used to route MIDI from Pro Tools to Unity Session.

The Unity RTAS plug-in is installed in the Digidesign DAE plug-ins folder by default. In Pro Tools you then need to have two tracks, a MIDI track that sends MIDI to the Unity OMS driver, and an Audio or Auxiliary track to receive the audio from Unity Session into Pro Tools. For example, insert the Unity RTAS plug-in into your Pro Tools track and open its window. Here, you can select which audio output from Unity Session you want to listen to on the specified Pro Tools audio track. You can select the main outputs or individual channels, and once a channel is selected individually, it is removed from Unity Session's main output mix.

Figure 11.10 The Unity RTAS plug-in window in Pro Tools.

At the left of the Pro Tools Edit window you can select which Unity Session bank and program you want to use on the MIDI track.

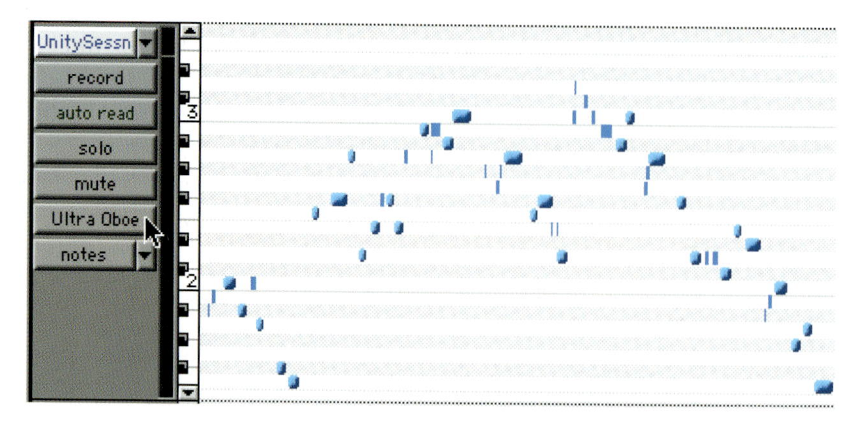

Figure 11.11 Pro Tools MIDI track with Unity Session as output showing the arrow tool about to click at the left of the Edit window to bring up the list of programs for Unity.

A large program select window appears with all the programs listed for you to choose from.

Figure 11.12 Unity programs listed in Pro Tools.

If you want to use DirectConnect, you will have to put the Unity DC plug-in from the Unity Session 'Utilities' folder into the DAE plug-ins folder, remove the installed Unity RTAS plug-in, and select 'DirectConnect' on the Audio page of the Unity control panel. You will also need to update your OMS setup and run Unity OMS Publish once after installation to get the Unity bank and program names into Pro Tools. As with RTAS, you also need to set up both a MIDI and audio/auxiliary track. On the MIDI track, select which MIDI channel of the Unity Session you want to communicate on. On the audio track, select Unity from the plug-in menu. You should now see the DirectConnect window showing the Unity Session logo. For each Pro Tools audio/auxiliary track, you can specify which output from Unity the track is listening to.

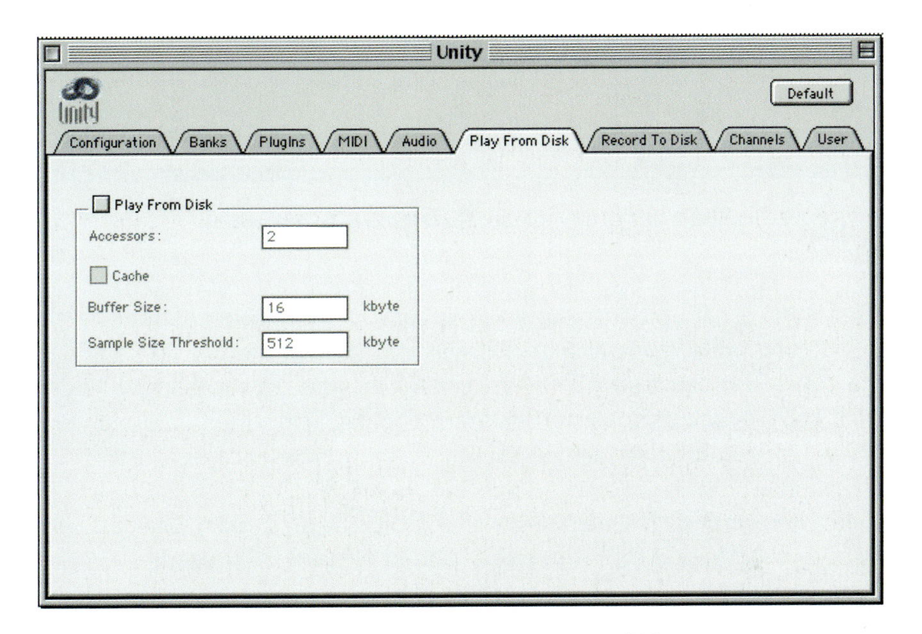

Figure 11.13 Unity Control Panel Play From Disk page.

Unity normally plays samples back from RAM, but there is also a 'Play From Disk' feature that lets you play back very large samples that won't fit into RAM – which is the case with many Gigasampler files. With playback from RAM you will get better polyphony and fewer conflicts with hard disk playback sequencers. But if you've tried increasing your RAM and still need access to more sounds, you can turn on the Play from Disk option from the Play From Disk page in the Unity control panel. Here you can also set the 'Accessors' parameter that controls the maximum number of simultaneous notes for a single sample and set a 'Sample Size Threshold', i.e. a sample size below which samples will be played back from RAM and above which samples will be played from disk.

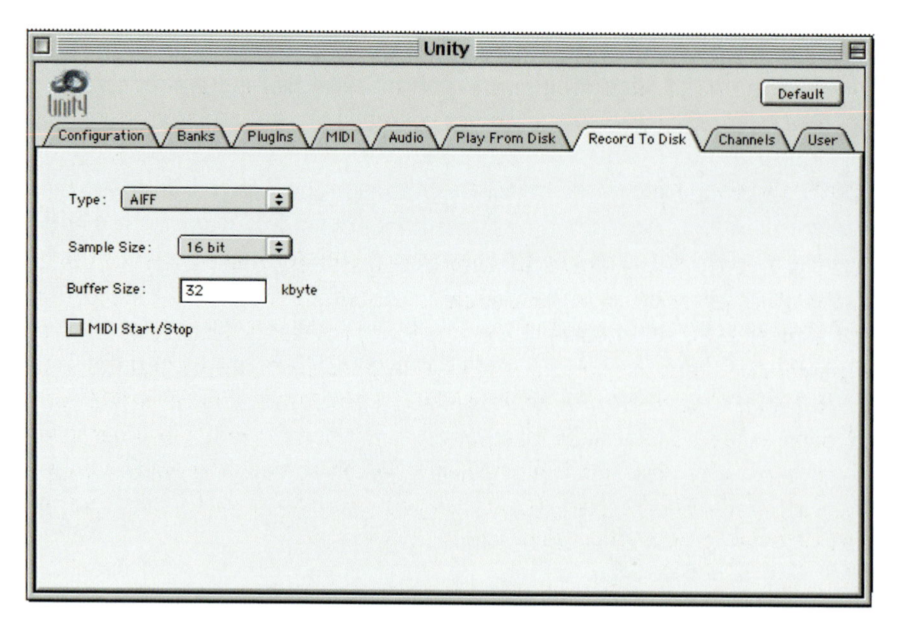

Figure 11.14 Unity Control Panel Record To Disk page.

Once everything is set up correctly, you can play Unity Session sounds 'live' from your MIDI keyboard or from your sequencer and simultaneously and immediately record to disk. You can also set the recording to start on receiving a MIDI start message if you like and choose the bit-depth and file type - AIFF, Sound Designer II, or WAV - of the file that will be saved.

If you need more sounds BitHeadz also offer Unity Topaz Studio Kits and Ultimate Acoustics for the Unity 3.0 product line, priced around $150 each.

Unity Topaz Studio Kits is based on the Sonic Emulations title that was released previously in Gigasampler format. The Unity title offers over 300 Mb of multi-sampled drums performed by New York session drummer, Zach Danziger. Unity Topaz Kits provides realistic sounding cymbals, hi hats, kicks, snares and toms that allow the user to build their own custom kits.

The second new title, Ultimate Acoustics, was created in conjunction with Gibson, Audio Technica and Bias Peak. The result is nearly 450 Mb of quality sounding acoustic guitar and bass samples in Unity 3.0 format. All of the samples in Ultimate Acoustic were performed by Bryan Blumer, and produced and programmed by David Das. The 24-bit, 44.1 kHz samples were created using several guitars and playing styles to choose from.

Both of these sound modules can be used stand-alone with the Unity sample player engine included or within most sequencers. Demos are available at the BitHeadz web site at http://www.bitheadz.com.

Chapter summary

BitHeadz have come a long way with their Unity system, and have developed it far beyond the original concept of a couple of separate plug-ins - one for sample playback and one for synthesis. Unity Session provides a wealth of high-quality sample playback, synthesis, and modelling plug-ins that integrate quite well with just about everything else you might want to plug into. There is still room for improvement as far as seamless integration is concerned, and the user interface, though good in many ways, still has a long way to go to really make life easy when using the software. Nevertheless, Unity Session is undoubtedly a new 'force to be reckoned with' on the virtual instruments 'scene'.

If you are a master of the arts of synthesis you will feel right at home with Unity as soon as you start programming the analogue synthesizer - in stereo - and especially the modelling synthesizers. If you simply want to play presets, the library sounds are first rate, and you can play just about any other sample format back using Unity Session. Do make sure that you have the fastest computer you can get, though. You'll need it!

12 Propellerhead

One of the first successful software synthesizers was Propellerhead's ReBirth, which simulates the classic TR808 and TR909 percussion synthesizers and the TB303 monophonic analogue synthesizer. Propellerhead has also developed a specialized tool favoured by many 'dance' re-mixers called ReCycle. ReCycle lets you set up drum loops to match the tempo and feel of your song. More recently, Propellerhead has developed a software simulation of a complete rack of synthesizers, samplers, drum machines, sequencers and effects – called Reason. In this chapter you will find an overview of each of these software applications with tips on using these and notes on various topics summarized from the manuals.

ReBirth RB-338

With ReBirth RB-338, the wonderfully named Propellerhead software company came up with a brilliant idea – the Roland TR808, TR909 and TB303 (arguably the most popular analogue drum and bass synthesizer units ever made) reborn as software!

Widely used in dance music and rap, the TR808 is revered for its bass drum and sought-after for its quirky snare, toms, congas and other sounds. Similarly, the TR909 sounds are used on every other 'house' record you hear. The TB303 can play basslines or melodic lines – and can easily be persuaded to make all the sorts of 'squeeks' and 'blats' you hear on 'acid house' or 'trance' records. You actually get two TB303-type synthesizers, so you can do a bass-line with one and accompanying parts with the other. These are immaculate simulations with all the original analogue synthesizer controls such as resonance and filter cutoff – and the original programming method is used to write patterns in real time or step time. Now the TR808 is just beautiful! All the controls you have on the original 808 are there for you, including the all-important decay control for the bass drum that lets

Figure 12.1 ReBirth screenshot.

you create those low booming sounds used to such great effect on records. Again, you program this exactly like the original 808 – just clicking on the simulated controls onscreen with the mouse. Ditto for the 909.

And it gets better! Next to each instrument section you get audio mixing controls including a Mute switch, level and pan controls, a distortion on/off switch and a delay control. Distortion makes the sound 'grungier' – although you can only use this on one section at a time. The Delay feature lets you feed the output of any or all of the three sections into the delay unit where you get controls for the number of delay repeats (the Feedback amount) and to pan the output in the stereo mix. Delay amounts are set using a switch for 16th-note or 8th-note triplets and a pair of up/down arrows to set the delay length in multiples of these (from 1 to 32).

The stereo audio mix plays back through the Mac's stereo audio outputs which are not of the highest quality, so professional users will need a way to get

CD-quality output. If you are using Pro Tools hardware and have the Digidesign Sound Drivers installed on your Mac, you can switch the Apple Sound Manager output to 'Digidesign' using the Sound Control Panel available from the Control Panels folder under the Apple menu. This directs the output from the Apple Sound Manager to play back via your Pro Tools hardware. Alternatively, you can save looped sections as audio files to disk using the Export Loop as AIFF or WAVE commands. So, if you wanted to be able to mix the different instruments individually, for instance, you could export each instrument pattern separately and then reassemble these in Pro Tools or other audio software. And you can always run your favourite Midi sequencer alongside ReBirth on your Mac and sync these together internally using OMS. Alternatively, you can sync ReBirth with external Midi devices using Midi clocks.

Using any combination of the synthesizer and drum-machine sections, you can write song patterns or record front-panel control movements in real time or in step time, to create filter sweeps or whatever. Programming the synthesizers and drum-machines in ReBirth is very easy if you already know the original TB303, TR808 and TR909 because the methods are exactly the same when it comes to writing patterns, choosing sounds, setting levels and so forth. However, if you aren't already familiar with these, the ReBirth manual provides excellent step-by-step guides.

The most basic way to program ReBirth is using the mouse to click the onscreen controls. You can also use the computer keyboard for menu shortcuts; for Rhythm Tap recording; for the Transport controls (Play, Stop, Rewind); for fast programming of the 303 synthesizers (with 'Program Synth from Keyboard' active on the Options menu); and for Pattern switching (with 'Select Patterns from Keyboard' active on the Options menu). Alternatively, if you have a suitable MIDI hardware controller with knobs, switches and faders, you can use this to control ReBirth remotely.

The 'focus'

One thing you need to get the hang of from the outset is called the 'focus'. For example, the computer needs to know which of the ReBirth sections should receive the keyboard commands. There are not enough keyboard commands available to have one set that covers all the ReBirth controls, so each section's controls become activated, and therefore controllable from the computer keyboard, when you click with the mouse on the relevant section. Alternatively you can select a Pattern in a particular section, or use the up and down arrow keys on the computer keyboard to set the focus to the section of interest.

To see which section 'has the focus', look for the vertical orange bar to the right of the Pattern selectors. In the example shown, the 808 rhythm section has the focus, as indicated by the orange bar.

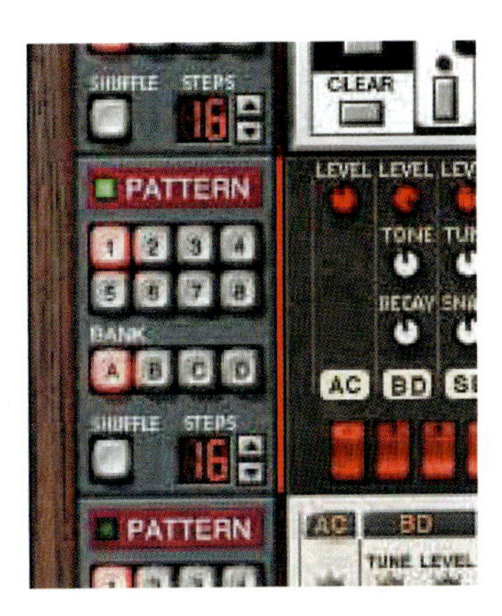

Figure 12.2 ReBirth detail showing the 808 rhythm section with the focus selected, as indicated by the orange bar.

Remote MIDI control

All the controls and switches in ReBirth can be controlled in turn by MIDI controllers or notes. Various MIDI Controller messages are used for the different controls in the Mixers, Synths, 808 and 909 Rhythm Sections, Master, Effects and Shuffle Sections. Various functions on the panel, such as those used for the transport controls or to turn distortion or compression on or off, can be controlled using MIDI note messages. These can be used in all modes.

Further control possibilities depend on the mode selected using the Options menu:

- When 'Select Patterns from Keyboard' is activated, MIDI notes can be used to select Patterns and Banks in the drum-machines and synths.
- When 'Program Synth from Keyboard' is activated, and the focus is set to one of the synths, the Synth Section Switches can be activated via MIDI note messages. Here you can select Pitch Mode On/Off, pitches of the notes, Octave, Step and so forth.
- When 'Program Synth from Keyboard' is activated, and the focus is set to the 808 Rhythm section, the 808 Rhythm Section Switches can be activated via MIDI note messages. Here you can select the Instrument and the Step On/Off.
- When 'Program Synth from Keyboard' is activated, and the focus is set to the 909 Rhythm section, the 909 Rhythm Section Switches can be activated via MIDI note messages. Here you can select the Instrument and the Step On/Off.

Once your Remote MIDI Control connection is up and working it can be used in various ways. For example, you could completely control ReBirth from a hardware controller with many buttons, knobs and faders mapped to control ReBirth functions. Or you could record parts of, or all of your Song automation in another software application. You would then remain in Pattern mode in ReBirth and send Pattern changes and control messages to ReBirth to program your Song from the other software application.

In this case you would need to set up your other application so that the Tracks you want to use for controlling ReBirth transmit to a MIDI port and MIDI channel that ReBirth can receive on. Then you activate Remote MIDI Control from the Preferences dialog in ReBirth and select the MIDI Input port and MIDI channel in ReBirth that the other application is transmitting on.

There are two major modes for Remote MIDI control, which you can select between from ReBirth's Preferences dialog: Standard Mapping is the preferred mode for those with advanced programmable MIDI controllers while Quick Mapping is the preferred mode for controlling the most common parameters using a MIDI keyboard's wheels and pedals.

In the Standard Mapping mode practically all ReBirth's controls can be remotely controlled. However, for best results you need a special hardware or software control panel that can be programmed to send all the different MIDI messages. If you only have a MIDI keyboard with a few switches and faders that can be used as remote controls for ReBirth, it is not practical to use this in Standard Mapping mode. In this case, you can use Quick Mapping.

Even if you only have access to a modulation wheel on your synth, this can be put to good use when recording knob movements in Song mode or when playing live. You can set this up using the Quick Mapping Edit button in the Preferences dialog. Here you select a function you want to use from the popup menus, click in the field to the right of the pop-up, and type in the number of the controller message you want to use. Or you can simply move the controller you want to use for this function: the field at the bottom of the dialog will show the number for this controller; then, when you click the Learn button associated with the function you want to use, this controller number will be automatically entered into the corresponding value field.

Sync with other sequencers

You can run another sequencer such as Cubase VST, Logic Audio or Digital Performer at the same time as ReBirth and synchronize the timing of these via OMS IAC on the Mac. And if the sequencer supports ReWire, this enables you to output ReBirth's audio into the audio mixer of your sequencer to provide even better integration.

You can also run ReBirth on one computer and send sync messages from a stand-alone sequencer or from another sequencer running on a different computer if you like. Yet another option is to use your favourite MIDI sequencer to

Note If you are using the Digidesign Sound Drivers to output audio from ReBirth via the Sound Manager, you will not be able to run Pro Tools at the same time because ReBirth will have 'grabbed' the Digidesign hardware - which won't allow two applications to control it.

directly control ReBirth with both running on the same computer. In this case you still have to program the patterns into ReBirth, but you can make these into songs by sending MIDI notes from your sequencer to select Patterns and Banks in ReBirth's drum-machines and synths. This is necessary because you cannot select Songs remotely.

Note Different music software applications handle tempo in different ways. Some derive their tempo from the digital audio clock while others derive tempo from the computer's clock - so even if you set the tempo in each application to the same value, they may run at slightly different tempos. This is not a problem for the MIDI if you synchronize one application to another, as only one will serve as the Master timing clock for playback in sync. However, if you have recorded audio in one software application which was set to, say 120 bpm, then you record audio in a different application, not synchronized to the first application, but still at a nominal tempo of 120 bpm, you can run into problems. If there is any discrepancy in the accuracy with which either software application clocked the tempo, then one, or both may not have been running at exactly 120 bpm - so the audio from each will drift out of sync. Obviously you need to be aware of this possibility and be prepared to compensate if necessary.

Using ReBirth with ReWire

Propellerhead and Steinberg developed their ReWire technology to allow real-time streaming of up to 18 separate ReBirth audio channels into any ReWire-compatible MIDI + Audio sequencer's audio mixer. ReWire also provides automatic, sample-accurate synchronization between the audio in the two programs and linked transport controls that allow you to control transport functions from either program. For example, with Cubase VST, when you run ReWire, the transports in the two programs are completely linked. It doesn't matter in which program you Play, Stop, Fast Forward or Rewind. Recording, however, is still completely separate in the two applications.

Note ReBirth RB-338 is limited to 44.1kHz audio playback. If you have Cubase VST set to any other sample rate, ReBirth will play back at the wrong pitch.

The Loop in ReBirth and the Cycle in Cubase VST are also linked. This means that you can move the start and end point for the Loop/Cycle or turn the Loop/Cycle on/off in either program, and this will be reflected in the other.

If you are using the Master Track in Cubase VST (i.e. with 'Master' activated on the Cubase transport), the tempo on the ReBirth transport won't have any effect on playback. However, if you are not using the Master Track in Cubase VST, you can adjust the tempo on the transport in either program, and this will immediately be reflected in the other.

Using a MIDI + Audio sequencer linked via ReWire with ReBirth RB-338 allows you to expand and integrate your ReBirth Songs with any other music - whether MIDI or acoustic recordings. Get your grooves going with ReBirth first, then record these onto audio tracks in the MIDI + Audio sequencer and add further MIDI and audio parts before your mixdown.

Tip If you want to record your ReWire channels onto hard disk tracks, use the Export Audio feature in your MIDI + Audio sequencer to bounce these to disk. Don't forget to mute all the other audio tracks first, or you will mix these together with the audio from your ReWire channels. The best way is to work on just one ReWire channel at a time - putting this into Solo to make sure no others will be audible. Typically, there will be options for automatically re-importing the created audio file onto a Track. Of course, once you have these ReWire channels as audio, you can go back and develop more material in ReBirth, and so on, virtually ad infinitum.

Note When running ReWire, multi-processor support should be disabled in Cubase VST.

To run a MIDI + Audio sequencer and ReBirth RB-338 at the same time you need a fast computer with plenty of RAM. Exactly how fast the computer needs to be depends on how many audio channels and plug-in effects you want to use in your MIDI + Audio sequencer. Exactly how much RAM you need depends on the number of audio channels that you plan to use in your MIDI + Audio sequencer and on how many Mods you want to be able to use in ReBirth.

Note When using Rewire, the launch and quit order is very important: first launch your MIDI + Audio sequencer, then launch ReBirth. When you are finished, you also need to Quit the applications in a special order: ReBirth first, then your MIDI + Audi sequencer.

ReCycle

Overview

ReCycle is an indispensable audio software 'tool' to have at your disposal if you regularly work with sampled drum loops. Every drum-loop sample has its own tempo, bass drum pattern, snare sound and so forth. But what if you want to change the bass drum pattern or change the tempo? You can always change the tempo by speeding up or slowing down the playback, but the pitch and timbre of the drums change as well, so this may not sound good. You can use time-stretching software that makes the tempo adjustment while endeavouring to leave the audio quality unaltered in other ways - but this doesn't always produce perfect results.

ReCycle lets you prepare your sampled drum loops for playback in any popular sampler, hardware or software based, or for playback within any compatible MIDI sequencer such as Cubase VST, Logic Audio, Digital Performer or Reason. Then, when you play back your MIDI sequences and decide that you want to make the music faster or slower, all you need to do is adjust the tempo of your MIDI sequencer and the tempo of the sampled loop will automatically follow.

When you load in a drum loop sample, ReCycle analyses the sample to find the individual drum 'hits' and also lets you 'slice' up the loop into individual sections. At the same time, ReCycle automatically produces MIDI files which perfectly recreate the original drum rhythms when you use these MIDI files to play back these samples. You can use these MIDI files in any MIDI sequencer to double the drum rhythms with other instruments, or to allow for further editing of the digital audio via MIDI in ways not previously possible. And if you want to change the sequencer's playback tempo, the samples will play back at the new tempo without the need for any time-stretching or compression.

Typically you load in a Sound Designer II file containing a drum loop, or transfer a sample directly from an Akai S3000 series sampler. Then you use the Sensitivity control at the right hand side of the waveform window to enter the 'slice' points to correspond with the drum 'hits' in the waveform display. You can edit these manually if the semi-automatic process does not get it exactly right. Left and Right locators are provided which you can set to specify the loop points. If you choose, say, one bar of music, you then specify the number of bars you have looped (i.e. 1 bar) in the 'Bars' popup, and the actual tempo of the loop will be automatically calculated for you. Next you set the 'Stretch' amount you want - which you set according to the amount by which you want to be able to slow down your drum-loop. Once you have a perfect loop, a tempo, a good set of slices and a suitable stretch setting, you send all this stuff to your Akai sampler, or save the files to your hard disk ready to load into SampleCell. Now you can launch your MIDI sequencer and connect your MIDI keyboard through to your sampler to play your new material. You will notice that each key plays a slice. What was previously one complete drum-loop recording is now a series of short snippets (the slices) spread out over the keyboard. If you now load up the MIDI file you just created with ReCycle, this will play the drum loop back exactly as it originally was. The difference is that now you can change the tempo on your MIDI sequencer and it will change the tempo of the drum loop within the limits you have set using the Stretch setting.

When you slice up a loop and play it back from your sequencer, each sample will play back in succession to recreate the original loop. At the original tempo, each slice will end exactly as the next starts. When you lower the tempo, there will be small gaps between the slices, which would normally disrupt the flow of the audio playback. Stretch is used to add an extra 'tail' of audio to each slice to fill out the gaps between the slices when the tempo is lowered. This tail is derived from the decay portion of the sound to achieve a natural-sounding result. This is not time-stretching in the same way as done by dedicated software like Time Bandit, but it achieves a result that can be superior with particular loops.

One of the things that makes ReCycle so powerful is the Effects processing which can be applied using control strips (panels) that you can optionally display underneath the menu bar, just above the waveform display. The Envelope, Equalizer and Transient Shaper commands in the View Menu open the control panels for each of these effects.

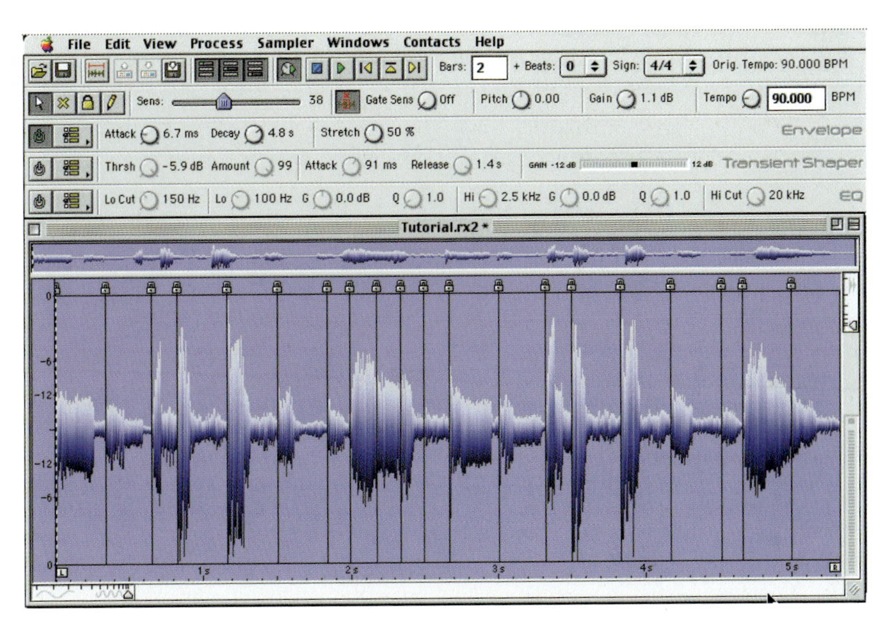

Figure 12.3 ReCycle Window showing sample 'slices' in the display, with EQ, Transient Shaper and Envelope control strips open above the display.

The Equalizer is the easiest to describe – this simply allows you to cut or boost selected frequencies to shape the overall sound quality.

Figure 12.4 ReCycle EQ controls.

The Transient Shaper takes a little more explaining. As the manual puts it, it is a type of attack/release envelope control that produces a result that can be likened to compression. Compressors level out the audio, making loud sounds softer and vice-versa. The result is that the levels become more even and individual sounds can get more power and 'punch'. Normal compressors are triggered by peaks and volume changes in the actual audio. The Transient Shaper is triggered instead by the individual slices in the loop. If you have ever used a conventional compressor, you will find that the Transient Shaper will affect the sound in a similar way. However, the Transient Shaper won't work as intended unless the file contains slices. To compensate for the volume loss that can be caused by this compression

Figure 12.5 ReCycle Transient Shaper controls.

effect, the Transient Shaper has automatic gain compensation to raise the overall level by a suitable amount.

Figure 12.6 ReCycle Envelope controls.

The Envelope controls also deserve a more detailed description. Again, the manual describes these well: the Envelope Attack and Decay parameters govern how the volume of a slice should change over time, from the time it is triggered (the slice note starts) until the slice note ends. This can be used to make a loop more distinct (by having a snappy attack and a short decay time) or more spaced out (by raising the attack time). The Stretch feature is used when you know you might want to lower the tempo of the loop in your sequencer. When you slice up a loop and play it back from your sequencer, each sample will play in succession. At the original tempo, one slice will end exactly where another starts. Of course, if you then lower the tempo, there will be small gaps between the slices, which disrupt the flow of the audio. Stretch is used to add an extra tail of audio to each slice, to lengthen it. This is derived from the natural decay of the sound. This extra tail of sound then fills out the gap between the slices when the tempo is lowered. The Stretch knob is used to set the amount of stretch as a percentage value. The percentage values tell you how much longer the entire sample will be after stretching. If you select the largest value (100%), for example, the slice will become twice its original length when transmitted to the sampler. To decide how much stretch you need, lower the Preview tempo by the same amount as you intend to use, and adjust the Stretch parameter as necessary. If there is a discernible gap of silence between slices you should raise the Stretch setting. And remember that Stretch is not Time Stretch - the Stretch feature doesn't lengthen the whole slice, it only adds a portion of sound to the end of it.

Using ReCycle

ReCycle was originally designed to be used with hardware samplers such as the Akai S3000. This is described fully in the manual and is fairly straightforward to set up.

The basic procedure is to open an audio file from hard disk or transfer a sample into ReCycle from your sampler, split it into slices, process as required, then transfer these slices back to your sampler.

You also need to create a MIDI file to use in your sequencer to playback the slices correctly from your sampler. Of course, if you have SampleCell or any of the new

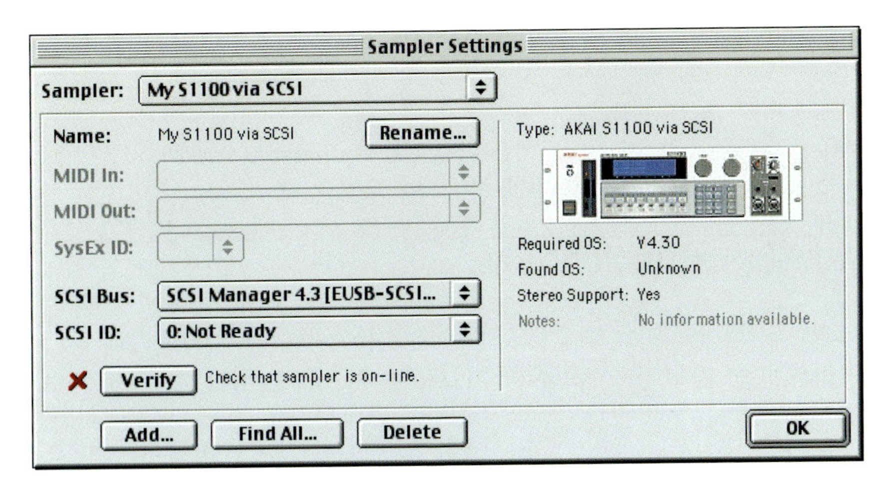

Figure 12.7 ReCycle Sampler Settings window.

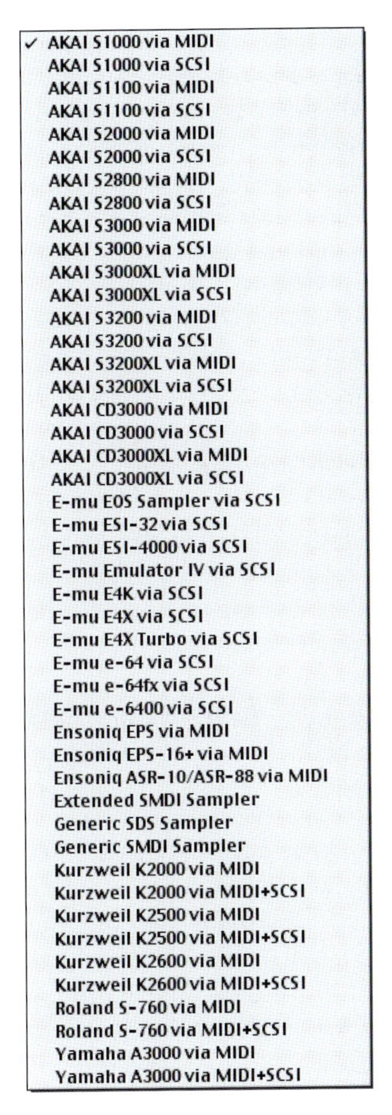

Figure 12.8 ReCycle Samplers menu.

software samplers, you can save the audio in an appropriate file format and control this from your MIDI sequencer in a similar way.

Using ReCycle REX files in Cubase

More recently, Propellerheads developed the REX file format that allows you to use ReCycle files directly within software such as Cubase VST and Reason. When you save your loop in ReCycle, this operation produces a REX2 file - the native file format used by ReCycle. REX2 files contain both the audio data of the loop, the slice information and the original (calculated) tempo.

When you import the REX2 file into an Audio Track in Cubase VST 5.x, the slices will now appear in their correct order and positions, and you will be able to change tempo and manipulate the slices much as you do when using a sampler. In Cubase, select the Audio Track into which you wish to import the file and set the Left Locator at the position where you want the file to appear. Depending on the type and complexity of the audio, you may want to set the Audio Track to channel 'Any'. Select 'Import ReCycle file', from the File menu. Locate the ReCycle export file you just saved, and open it. The file is added to the Pool. A number of Segments are created for the file, each one corresponding to a slice in Recycle. A Part which will play these Segments is automatically created on the active Track, starting at the Left Locator position. Now you can play back the ReCycled file at any tempo - just as if using a sampler. You can also edit it in detail, quantize, and so forth.

How to make drum loops fit the tempo of your Pro Tools session

Besides creating REX2 files for use in audio sequencers, or transmitting separate slices to samplers, ReCycle can also be used to simply change the properties of a file, such as the tempo and/or the pitch. You can then export the loop as one sample (not as separate slices) and use it in any audio application or sampler.

Now imagine that you have found a perfect loop to use in your song - but the tempo of the loop is wrong, and you don't want to change the tempo of the song to match the loop. You are using a hard disk recording system such as Pro Tools that does not support REX files - and you don't have a sampler. You might think that in this case ReCycle would be no use - as you don't have Reason or Cubase VST to transfer REX files to, and you don't have a sampler to accept MIDI messages to play back the ReCycle slices anyway. But you would be wrong! If your loop exists as an audio file on hard disk in a format supported by ReCycle, such as SDII, then you can open this in ReCycle, slice up the audio, make your changes, and then export the file in the appropriate format back to Pro Tools, or whatever.

You can use all the normal techniques to slice up the loop and set the number of Bars and Beats to calculate the original tempo. In Preview mode, simply type in the Preview Tempo to match the song tempo, changing the pitch to fit the song if necessary and adding any effects or processing if you wish. Activate 'Transmit as One Sample' on the Process menu then use the Export File dialog to save the file in its original format. Import this into your hard disk recording system (e.g. Pro Tools) and it will now match the song's tempo and pitch perfectly!

What else can you do with ReCycle?

ReCycle also allows you to effectively 'retune' a sample without time-stretching to make the tempo faster or slower. You can retune a sample in any sampler, but you have to do some nifty mental arithmetic to work out the correct pitch shift to achieve a desired tempo. ReCycle lets you specify the actual BPM that you are after and then communicates with your sampler to apply the appropriate tuning offset.

Another neat trick is to slice up your drum loop, create two versions, one containing everything but the snare hits, and one containing only the snare hits. Then send both samples to your sampler, set up your sequencer to play both samples at the same time, and route each sample to a different output of your sampler. This way, you can easily change the snare samples or EQ them differently, add effects or set the levels differently.

Yet another option is to just use ReCycle to extract the groove from the drum loop as a series of notes in a MIDI file, then use this in your MIDI sequencer with the sounds of your choice replacing each drum instrument, and maybe double the bass drum pattern with a bass guitar sample, or whatever.

So, to summarize, when working with drum loops in ReCycle you can change tempo without affecting pitch; change pitch without affecting tempo; quantize drum loops (either to straighten up the timing or to change the feel, for example by applying a 'groove map'); extract the timing (a groove map) from a drum loop and apply this to other sequenced parts or other loops; replace individual sounds in a drum loop; edit the actual playing in the drum loop without affecting the basic feel; extract sounds from loops; process the loops with various effects.

All-in-all, ReCycle is a first-rate tool which I would not want to be without, especially for working on dance or rap music, or anything which makes heavy use of sampled drum loops.

Reason

Reason is a complete music production system emulated in software. Visually, it resembles a hardware rack containing MIDI synthesizers, samplers, sequencers, drum machines, effects devices and a mixer. When it first became available it caused something of a stir – users were impressed at how good it sounded and how much it could do. Despite having its own reasonably powerful on-board sequencing capabilities, it is also possible to use Reason simply as a rack of MIDI devices that you can use with your favourite sequencer – Cubase, Logic, Performer, Pro Tools, Nuendo or whatever – using ReWire or OMS IAC.

All of Reason's rack devices operate in a very similar way to their hardware equivalents, so if you are familiar with the hardware, as many people are, then you already know most of how to use the software. For example, the mixer closely resembles the Mackie 3204 rackmount model with its fourteen stereo channels, basic two-band EQ section, and four effect sends. Reason provides a selection of eight different effects units, including Digital Reverb, Digital Delay, Foldback Distortion, Chorus/Flanger, Stereo Phaser, Compressor/limiter, 2-band Parametric EQ, and Envelope Controlled Filter. MIDI-controllable sound modules in version 1.x include the Subtractor polyphonic synthesizer, the NN19 playback-only sampler, the Dr Rex Loop Player which plays REX files created in ReCycle, and the Redrum drum-machine module that lets you program drum patterns. There is a built-in Roland-style pattern sequencer, the Matrix Pattern Sequencer, and a main MIDI Sequencer that you can use to record notes, controllers, device parameter automation and pattern changes. You can use this to play sounds from Subtractor, NN19, Redrum or Dr Rex via MIDI. MIDI notes can also be sent to the Mixer for muting, soloing and activating EQ. And you can control virtually all the rest of Reason's controls remotely, using MIDI controller messages.

In version 2.0, Reason adds full support for Mac OS X and will playback samples from ReDrum and Dr Rex at 24-bit resolution. And there are two new modules – the Malström 'Graintable' Synthesizer and the NN-XT Sample Player. Reason 2.0 also has many smaller, but useful enhancements compared with previous versions. For example, the position and size of the windows and other GUI settings are automatically saved with each Reason 2.0 song document, so that it will open in exactly the same state as when you last saved and closed it. And you can now load SoundFonts samples into ReDrum – and both presets and samples into NN-19 and NN-XT. Even more usefully, the LFOs in the various synthesizers, samplers and effects, such as the Chorus and Phaser, can be synchronized to the song tempo, with any of 16 possible time divisions.

The Sequencer window has also been improved a lot. You can open this in a separate window from the rack so that it is in view at all times and you can resize it and position wherever you like onscreen. If you have a mouse with a scroll wheel, this can be used with your computer's modifier keys for various scrolling and zooming operations in the sequencer. The Sequencer window can be switched between the Arrange view, which shows tracks and events, and the Edit view, which lets you edit notes, pattern changes, controller data, and so forth - using the various editing tools in the Sequencer toolbar. For example, an Eraser tool lets you delete events, and with Snap activated, clicking directly on events in the Arrange view with the Eraser tool will not only delete the events you touch - it will also delete all events within the chosen Snap value (e.g. 1 bar). You can also draw selection rectangles encompassing several Events with the Eraser tool - and delete them all at once. In the Edit view, all notes of the same pitch within the range of the Snap value will be deleted. The Magnifying Glass tool provides an alternative to the magnification sliders - allowing you to zoom in or out, horizontally and vertically. The Line tool can be used in the Edit view in the Velocity and Controller lanes for drawing Velocity ramps and editing Controllers, respectively. The Pencil tool can be used for creating more irregular curves, and you can use the Hand tool to scroll the display.

There are two CD-ROM sound libraries supplied with Reason 2.0 - the Factory Sound Bank which has a selection of patches and samples for the original Reason modules and the new Orkester Sound Bank which is full of patches and samples of orchestral instruments for the NN-XT. Several of the demo songs use these to great effect - letting you know just how good they are!

Reason is a very reasonably priced, yet powerful music production environment which will appeal to anyone who likes its mix of simulated analogue synths, drum machines, samplers, sequencers, mixers and effects. And you get so much 'bang for your buck' with this 'rack' of software that there must be a whole lot of worried hardware manufacturers 'out there' right now!

Note Reason takes advantage of the Altivec Velocity Engine instructions used in the Macintosh G4 processors to enhance performance, so for best results, use a G4 (or a newer model) with plenty of RAM and hard disk space. More RAM will allow you to use more devices at the same time.

Figure 12.9 Reason 'Song'. The MIDI input and Audio output hardware interface can be seen at the top of the rack with the mixer and a couple of effects units below this. Underneath, you can see the sequencer, with three tracks visible, and the transport bar is at the bottom of the Rack.

Reason Songs

The Song is the main file format in Reason. A Song file contains the device setup and all settings in the rack, as well as everything you have recorded in the sequencer - and references to any samples or REX files that you have used. You can have several Reason Songs open at the same time and each will appear in a separate Reason window, complete with rack, sequencer and transport bar areas.

Note If you want to open your song on another computer or send it to another Reason user, you would also have to bring all samples and REX files used by the devices in the song. To make this easier, Reason allows you to create 'self-contained' songs. A self-contained song contains not only the references to the used files, but also the files themselves. You can choose exactly which files should be included in the self-contained song, although files that are part of a ReFill cannot be included in a self-contained song, so you will need to save these separately.

Reason patches and factory sounds

A Reason 'patch' contains settings for a specific Reason device and these patches can either be separate files on your hard disk or files embedded in a ReFill file. To select a patch for a device, you can either click the folder button in the Patch section on the device panel (next to the patch name display) or select the Browse Patches item from the Edit menu. In both cases, the Browser dialog appears, allowing you to locate and select the patch on the hard disk or within a ReFill.

Several device types use patches: Subtractor, Malström, NN19, NN-XT and ReDrum. You can save all the panel settings you make on the Subtractor synthesizer panel into Subtractor patches so that you can store and recall these, just like synth patches on a hardware synthesizer. Malström works similarly. NN19 sampler patches not only contain information about the parameter settings on the device panel - they also include information about which samples are being used and their settings. Keep in mind that these sampler patches don't actually contain any samples - they only contain information about which sample files are used. So you will need to copy the sample files along with the Reason files when backing up or transferring to other systems. Again, NN-XT works similarly. Redrum patches contain a complete 'drum kit' with information about which drum samples are used, together with the parameter settings for each drum sound. Again, the actual samples are not included in the patch, there are only file references to these. Also note that Redrum patches are separated from Redrum patterns - selecting a new patch will not affect the patterns in the device.

Note Patches do not include information about any routings made using patch cords on Reason's back panel.

A ReFill file can contain patches, samples, REX files and demo songs. The Reason factory sound bank consists of one large ReFill file – the 'Reason Factory Sound Bank' which contains a large number of patches, samples and loops – serving as your main supply of sounds (much like the sound ROM in a synthesizer). This can either be installed on your hard drive or kept on the Factory Sound Bank CD, depending on the choice you make during installation.

Note You need a spare 500 Mb free on your hard drive to install the Factory Sound Bank. You can leave the banks on the CD-ROM, but that means you need to have the CD-ROM inserted whenever you want to load any of these sounds and the CD-ROM drive is relatively slow to read from compared with hard disk.

If you have opened a patch from within a ReFill, modified it, and want to save it, you have to save this as a separate file, using a different name to avoid any confusion. You can use the ReFill Packer application supplied with recent versions of Reason to build your own ReFill files. You can also download ReFills from other Reason users on the Internet, purchase them from sample manufacturers, or whatever.

The rack

The MIDI In Device and Audio Out hardware interface are always included at the top of the rack, along with the main Sequencer and Transport Bar at the bottom. It's up to you to choose what modules to load into the middle of the rack.

So what modules can you put into the rack? Well, that all depends on what you want to use when you are working on a particular song. You simply choose your own combinations of what is available and save these setups as individual Reason Song files. Now let's take a look at the various rack modules.

MIDI input

Figure 12.10 Reason MIDI In Device.

Using the MIDI In Device, you can select which Reason device is assigned to which of the 16 MIDI channels, using popup selectors for each channel. If you need more than 16 channels, there are actually four external MIDI buses, each of which can carry a different set of 16 MIDI channels for a total of 64 MIDI input channels. You can switch between these buses using the Bus Select buttons at the top of the MIDI Input module.

Using the standard MIDI section of the Preferences dialog, you can make various settings for the MIDI Inputs. Here, you should select Sequencer input as the MIDI port if you are using the Reason sequencer. Once you have selected your MIDI interface on the Sequencer Port pop-up (and which channel it should receive on), you can direct incoming MIDI to any device by just clicking the 'In' column to the left of a track name in the track list.

The MIDI input can come from an external hardware sequencer or from sequencer software, such as Cubase, that is installed on the same computer as Reason.

External control

If you want to directly control your Reason Devices from an external sequencer, select the appropriate External Control Bus input as the MIDI port using the Advanced MIDI Preferences.

Note Remember that it is always better to use a multi-port MIDI interface with your computer setup, so that you can select separate ports for Reason and your other MIDI devices to use.

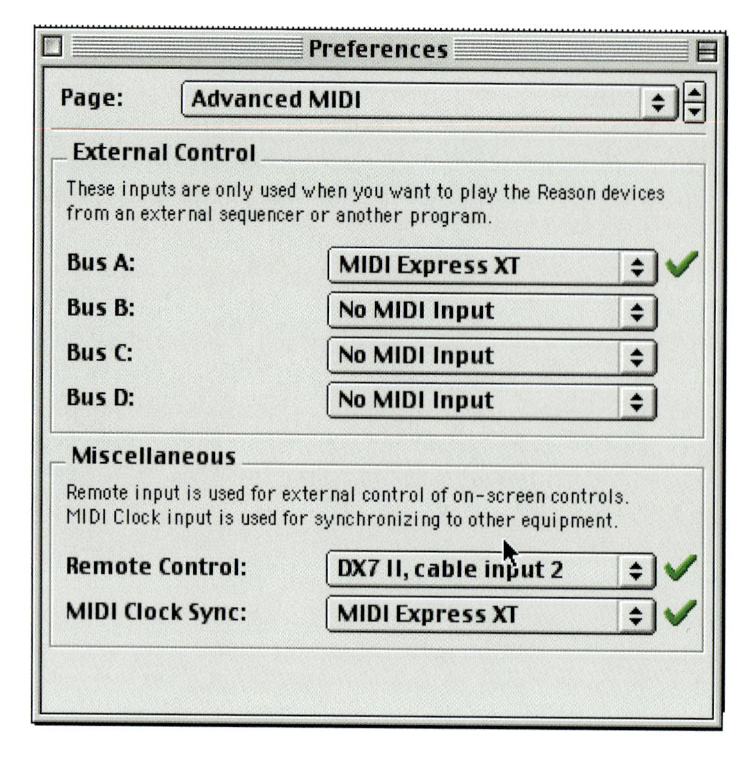

Figure 12.11 Reason Advanced MIDI Preferences.

Remote control

The Remote Control input is the one to choose to assign a MIDI port to receive MIDI Controller messages for 'live' remote control – using a dedicated hardware controller with lots of knobs, faders and switches, for example. There are several hardware controllers that you can buy that can make programming Reason much easier than by using your computer's keyboard and mouse. Of course, you will have to define which MIDI message to use for each Reason control – which could take a while to set up first time around.

Direct MIDI control

It is also possible to send controller data from an external sequencer to Reason parameters to achieve direct MIDI control. Just set up your external sequencer to transmit the correct MIDI controller messages on the right MIDI channel. To find out which MIDI Controller number corresponds to which control on each device, check out the MIDI Implementation Charts .pdf document supplied with Reason. Once you have located the controller numbers and set everything up, you can

Tip Try connecting a MIDI controller to a single MIDI input and routing this to different devices in the rack via Reason's sequencer. This way you will be able to play and control one device at a time. To set this up, open the Preferences dialog from the Edit menu and use the pop-up menu at the top of the window to select the MIDI page. Pull down the Port popup menu in Reason's Sequencer section and select the MIDI input to which your MIDI controller is connected. Check on which MIDI channel your MIDI controller sends, and set the Channel popup menu to this value. Reason's sequencer will only accept MIDI data on one channel at a time, so you can play notes and patterns from the sequencer and use MIDI Remote Control at the same time - even if you have a MIDI interface with a single input. For now, make sure the other popup menus are set to 'No MIDI Input'. Now, the Reason sequencer will receive MIDI data on the specified input port and MIDI channel.

record and edit the controller data in the external sequencer as you would normally, and the Reason parameters will react correspondingly.

Note Do not confuse direct MIDI control with Remote Control. Remote Control allows you to map an external MIDI Controller to any control on the front panel, and is primarily intended for 'live' tweaking of parameters during playback.

Audio output

Figure 12.12 Reason Audio Out hardware interface.

To play back the digital audio via your audio hardware, Reason uses the driver you have selected in the Preferences dialog. In the rack-on screen, this connection is displayed at the left-hand side of the Audio Out Hardware Interface. To the right of this can be seen 64 meters, one for each of the 64 possible outputs. Of course, your audio hardware may not support this many outputs - or you may choose to mix all the sounds from Reason internally and output these using just one pair of outputs. For example, if you are using a standard sound card with stereo outputs (or the built-in audio hardware on the Mac), only the first two outputs will be available. In the Hardware Interface device, the green indicators above each meter are lit for all currently available outputs. If a standard stereo audio card is used, only the first two outputs (marked 'Stereo' on the device panel) are available.

If you are using ReWire, Reason will feed the digital audio to the ReWire master application instead (typically an audio sequencer program), which in turn handles communication with the audio hardware. In this case, the meters indicate the audio routed via the ReWire channels.

To configure Reason to use the DirectIO driver, choose Preferences from the Edit menu and use the popup menu at the top of the Preferences window to choose the Audio Preferences page. Here you can use the Audio Card popup menu to choose the 'ASIO DigiDesign DirectIO' option to playback the audio from Reason through your Pro Tools hardware.

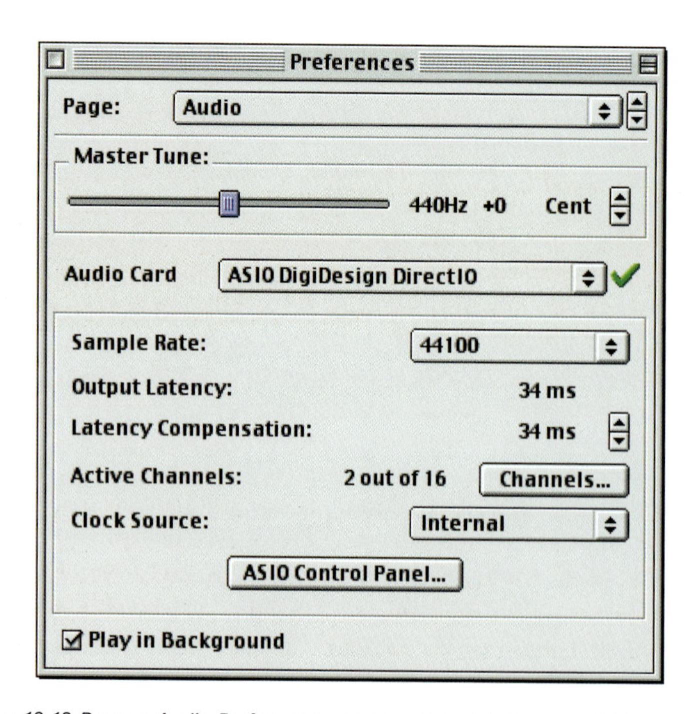

Figure 12.13 Reason Audio Preferences showing Output Latency and Compensation.

Note Reason can use the Pro Tools TDM or Digi 001 hardware for its audio input and output via DirectConnect. For this to work, you need to have an 'ASIO DigiDesign DirectIO' file inside the 'ASIO Drivers' folder inside the Reason folder.

I have a Pro Tools MIX card so I normally choose the ASIO Digidesign Direct I/O driver to use with Reason. There are currently 16 outputs available from my Digidesign ADAT Bridge Interface, but I often just use two of these to provide a stereo mix output from Reason. I use the Mixer in Reason to mix the sounds from the modules internally to the Mixer's stereo Master outputs, and hook these up to the first pair of outputs from my Digidesign hardware.

There is a readout of the current Output Latency on this Audio Preferences page and this shows a value of 34 milliseconds with my Pro Tools MIX card. Latency in this case is the delay you hear between when you play your Master Keyboard and when you actually hear Reason playing the sound. This 34-ms delay is only just usable as far as I am concerned. I do find this delay offputting while I am playing the keyboard and find myself wishing I had one of the newer low-latency cards, such as the RME Hammerfall models which Steinberg are marketing as part of their Nuendo range.

Note There is also a setting called Latency Compensation. This value is used internally in Reason to compensate for the latency when synchronizing Reason to another MIDI sequencer. Usually, Latency Compensation is set to the same value as the Output Latency.

Routings

You can always start out with an empty rack and add devices as you need these. The default song opens with a useful selection of devices already there for you to work with – all hooked up automatically. But what if you want to change the routings, perhaps to send the output of a particular device in the rack to a specific output on your audio hardware? Select Toggle Rack Front/Rear from the Options menu, or simply press the Tab control on your computer keyboard, and the rack will turn round to reveal the back panels of the equipment. Here you will see the connections between devices indicated by 'virtual patch cables'. Connections between instrument devices and mixers use red cables, connections to or from

Figure 12.14 Reason back panel - upper half - showing the various patch cords.

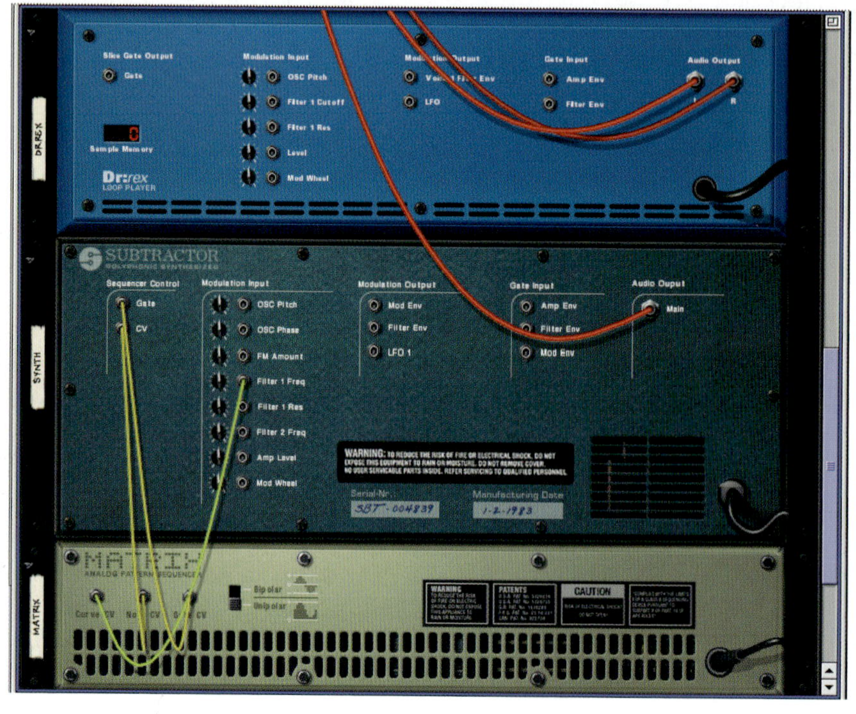

Figure 12.15 Reason back panel - lower half - showing the various patch cords.

effect devices use green cables and CV connections use yellow cables. Simply make your connections by clicking and dragging from one 'socket' to another on the back panels – and watch those cables swing.

Mixer and effects

To mix the audio outputs from the rack modules you can use the rack mixer, based on the popular Mackie 3204 rackmount model. This has fourteen stereo channels, a basic two-band EQ section, and four effect sends.

Reason provides a selection of eight different effects units, including the RV-7 Digital Reverb, the DDL Digital Delay Line, the D-11 Foldback Distortion, the CF-101 Chorus/Flanger, the PH-90 Stereo Phaser, the COMP-01

Figure 12.16 Reason mixer.

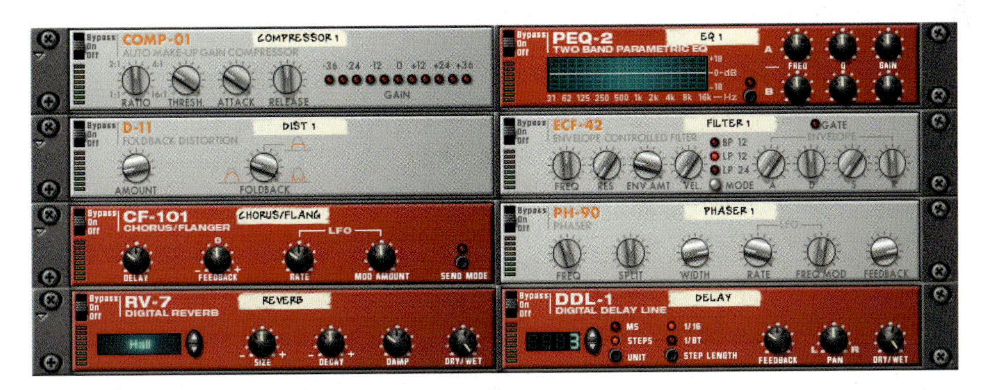

Figure 12.17 Reason effects.

Compressor/Limiter, the PEQ2 Two Band Parametric EQ, and the ECF-42 Envelope Controlled Filter. The latter is a synth-style resonant filter with three different filter modes and you can use a drum-machine or the Matrix sequencer to trigger its envelope to get some truly 'nasty' sounds!

Subtractor – the analogue synthesizer

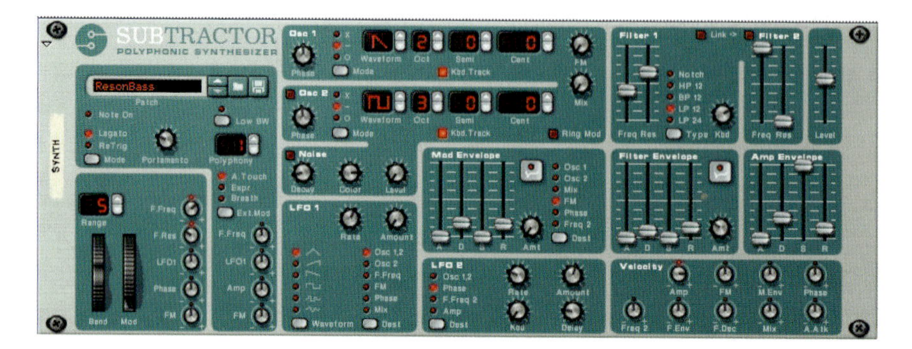

Figure 12.18 Reason Subtractor synthesizer.

The Subtractor is a polyphonic synthesizer laid out much like an advanced analogue synth. This features two oscillators, two filters and a host of modulation functions, allowing for everything from fat basses to swirling pads and screaming lead sounds.

Malström – the 'graintable' synthesizer

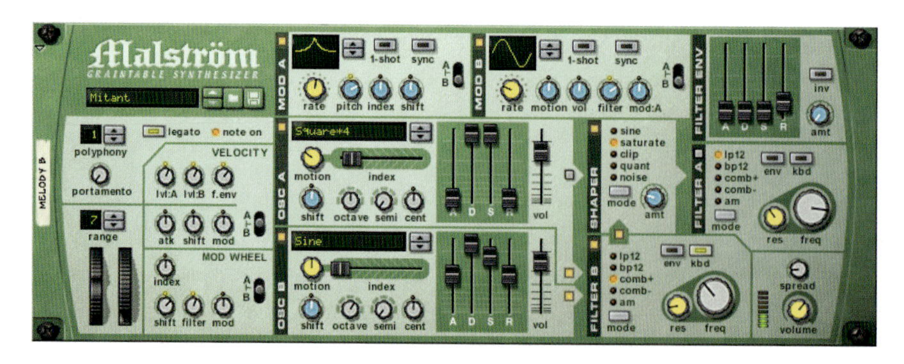

Figure 12.19 Reason Malström synthesizer.

The descriptively named Malström is a so-called 'graintable' synthesizer that can whip up a storm of interesting new sounds, quite unlike any you will have heard before. It is a polyphonic synthesizer featuring two oscillators, two modulators, two filters, a waveshaper and a large number of modulation and routing options.

NN19 – the sampler

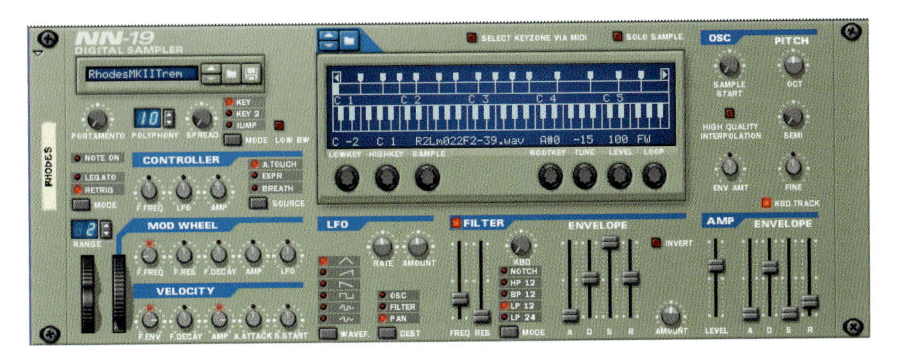

Figure 12.20 Reason NN19 Sampler.

The craftily named 'NN19' (wonder where they got that from?) is a playback-only sampler that lets you load in sample files in Wave or AIFF format and create multi-sample 'patches' by mapping these samples across the keyboard. Once you have loaded one or several samples, you can modify the sound using synthesizer-type parameters such as a filter, envelopes and an LFO.

NN-XT – the sample player

The NN-XT Sampler, like NN19, lets you load samples to create a multi-sample key map and lets you modify the sound using LFOs, envelopes and filters. The NN-XT also lets you create layered sounds and features velocity-switched key maps. Velocity switching sorts out which samples in a layered key map actually sound, according to how hard or soft you play your MIDI keyboard.

Dr Rex – the ReCycle loop player

There's also a loop player called Dr Rex that plays REX files created in ReCycle and a selection of REX files are included in the Reason Factory Sound Bank to get you started.

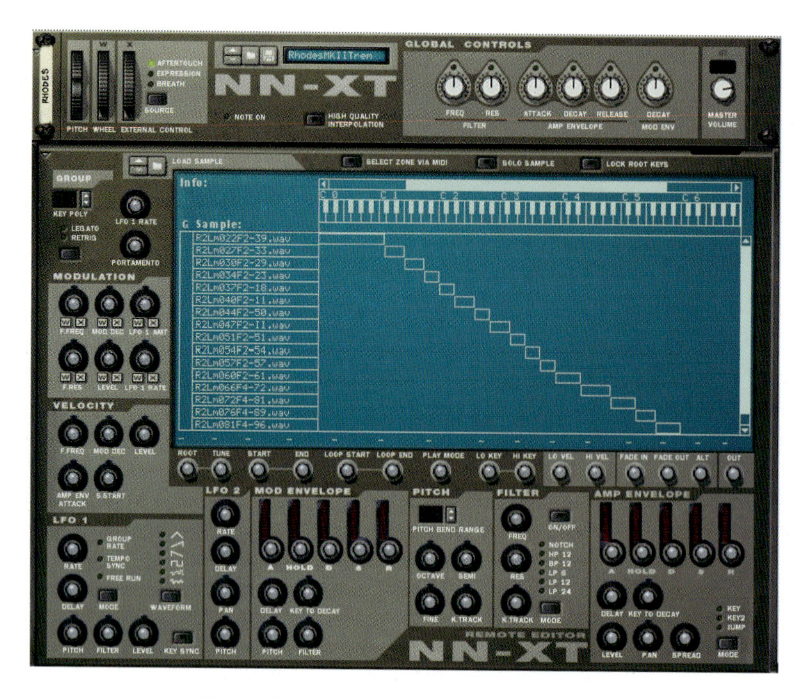

Figure 12.21 Reason NN-XT Sample Player.

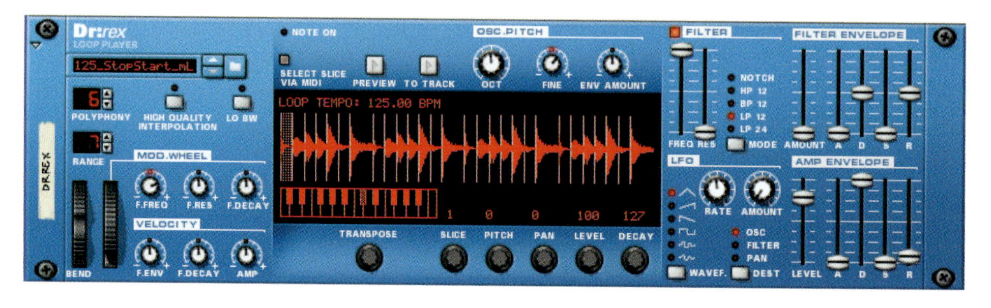

Figure 12.22 Reason Dr Rex ReCycle loop player.

Note ReCycle works with sampled loops. By 'slicing' a loop and making separate samples of each beat, ReCycle makes it possible to change the tempo of loops without affecting the pitch and to edit the loop as if it were built up of individual sounds. Just what you need to get your beats together!

After loading a REX file into the Dr Rex Loop Player, you can play it back at virtually any tempo, make settings for individual slices, extract MIDI playback data and process the loop with the built-in filter, LFO and envelopes. You can also play the individual slices via MIDI or from the sequencer; each slice has a specific note number (C1 for the first slice, C# 1 for the next and so on).

Redrum - the drum-machine

Figure 12.23 Reason ReDrum drum-machine.

And it doesn't stop there. Reason has its own drum-machine module called Redrum. This is a sample-based drum machine with ten drum sound channels into which you can load the excellent set of factory samples, or your own sounds in AIFF or WAVE format. As with ReBirth, there is a built-in Roland-style pattern sequencer, allowing you to create classic drum-machine patterns. You can also use Redrum as a sound module - playing it live from an external MIDI controller or from the main Reason sequencer.

At first glance, Redrum looks to be styled after pattern-based drum-machines, like the legendary Roland 808/909 units. Indeed, it does have a row of 16 step buttons that are used for step programming patterns, just like the aforementioned classics. There are significant differences, however. Redrum features ten drum 'channels' that can each be loaded with an audio file, allowing for completely open-ended sound possibilities. Don't like the snare - just change it. Complete drumkits can be saved as Redrum Patches, allowing you to mix and match drum sounds and make up custom kits with ease.

Matrix – the pattern sequencer

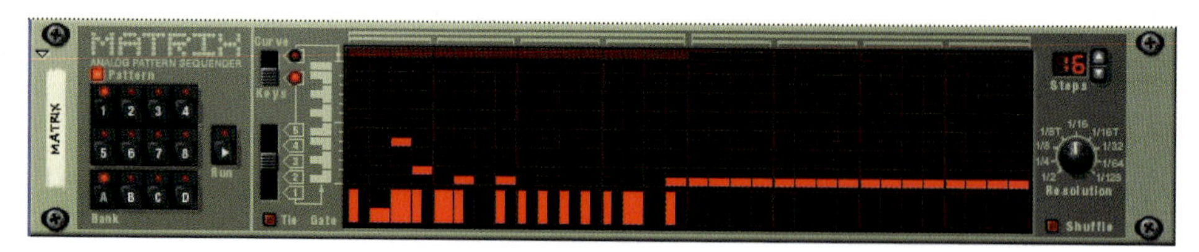

Figure 12.24 Reason Matrix pattern sequencer.

Talking about sequencers, there is also a stand-alone monophonic pattern sequencer, the Matrix, which is similar to a vintage analogue sequencer. Just connect this to any of the MIDI devices in Reason and it sends simulated CV (pitch) and Gate CV (note on/off plus velocity) or Curve CV (for general CV parameter control) signals to the device or device parameter. Before MIDI was invented, monophonic analogue synthesizers could be hooked up to a hardware sequencer using patchcords. Now Reason simulates this type of sequencer – even down to the patchcords!

How pattern devices integrate with the main sequencer

Although Reason's two pattern-based devices – the Redrum drum computer and the Matrix pattern sequencer – are very different in most ways, they handle patterns in the same ways.

A pattern device contains a built-in pattern sequencer. Unlike the main sequencer in Reason, a pattern sequencer repeatedly plays back a pattern of a specified length. A typical example of a pattern sequencer would be a drum-machine which plays drum patterns, usually of one or two bars in length. Having the same pattern repeat throughout a whole song may be fine in some cases, but normally you will want some variations. The solution is to create several different patterns and switch between these at the desired positions in the song.

In Reason, you can run each pattern device separately without starting the main sequencer, by clicking the Run button on the device panel. If you are running a pattern device separately and start playback of the main sequencer, the pattern device will automatically restart in sync with the sequencer. Patterns can be switched by pattern change events in the main sequencer, so you can record or create pattern changes in the main sequencer and have them occur at the correct position on playback – just as with a drum-machine sequencer. You can even combine the built-in pattern playback with playback from the main sequencer to add variations or fills to a basic pattern.

Note The tempo that you set on the transport panel is used for playback of both the pattern devices and the main internal sequencer.

ReBirth Input Machine

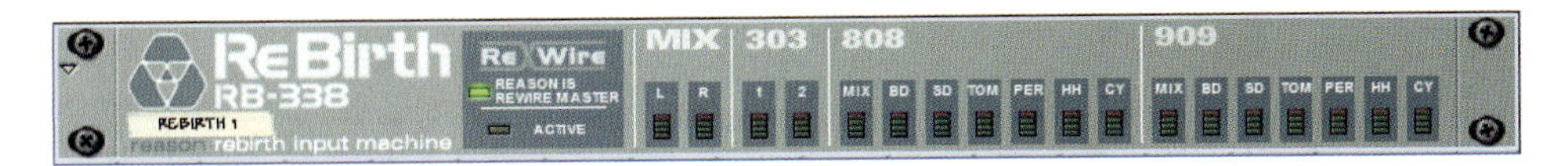

Figure 12.25 Reason ReBirth Input Machine.

If you already have Propellerheads ReBirth RB-338 software, you can run this at the same time as Reason, with their clocks synchronized with sample-accuracy. In this case, you can use Reason's ReBirth Input Machine module to feed up to 18 channels of ReBirth RB-338 audio outputs into Reason via ReWire. This way, both programs can share the same audio card and take advantage of the multiple outputs on that card. Another benefit of using Rewire is that either device will control the other device's transport.

Tip When you are setting this up, the launch and quit order for Reason and ReBirth is very important. You need to launch Reason first and make sure this is set up correctly to work with ReBirth. Make sure that there is a Mixer module loaded, otherwise ReBirth's L/R Mix channels will be routed directly to the Audio Hardware Interface. If you have a Mixer, the L/R Mix output from the ReBirth Input Machine will be automatically connected to the mixer's first available audio inputs. Next, create a ReBirth Input Machine module if necessary. Launch ReBirth, then select Reason as the application in focus. If both the 'Reason is Rewire Master' and the 'Active' indicator on the ReBirth Input Machine are lit, this indicates that the launch procedure was correct and that Reason and ReBirth are now locked and in sync. When you want to quit, do this in reverse - quit ReBirth first, then quit Reason.

Once the two programs are synced, you can route any of the 18 available outputs in ReBirth to separate channels in a Reason Mixer, or to the Hardware Interface for direct connection to a physical output on your audio card. If you flip the rack around, you will see a row of 18 audio outputs emerging from the back of the

ReBirth Input Machine module, with the Mix-L and Mix-R outputs either routed to the Reason Mixer or directly to Reason's Hardware Interface. These Mix-L and Mix-R outputs carry a stereo mix of all the sounds from ReBirth. If you hook up any of the individual outputs, its sound will be removed from the stereo Mix outputs.

The internal MIDI sequencer

The sequencer is the main composition tool you will use to record notes, controllers, device parameter automation and pattern changes in Reason. It's easy to get stuff happening right away with Reason's MIDI sequencer. Just hook up a MIDI keyboard and record into any of Reason's internal sequencer tracks. You can actually use up to seven different MIDI inputs if you have a multi-port MIDI interface. This makes it possible to use several different MIDI controllers and play and tweak each device in the rack independently – or you can play the devices in Reason from an external sequencer. You can play sounds from Subtractor, Malström, NN19, NN-XT, Redrum and Dr Rex, using Aftertouch and even Pitch Bend on all devices where it makes sense to be able to bend notes via

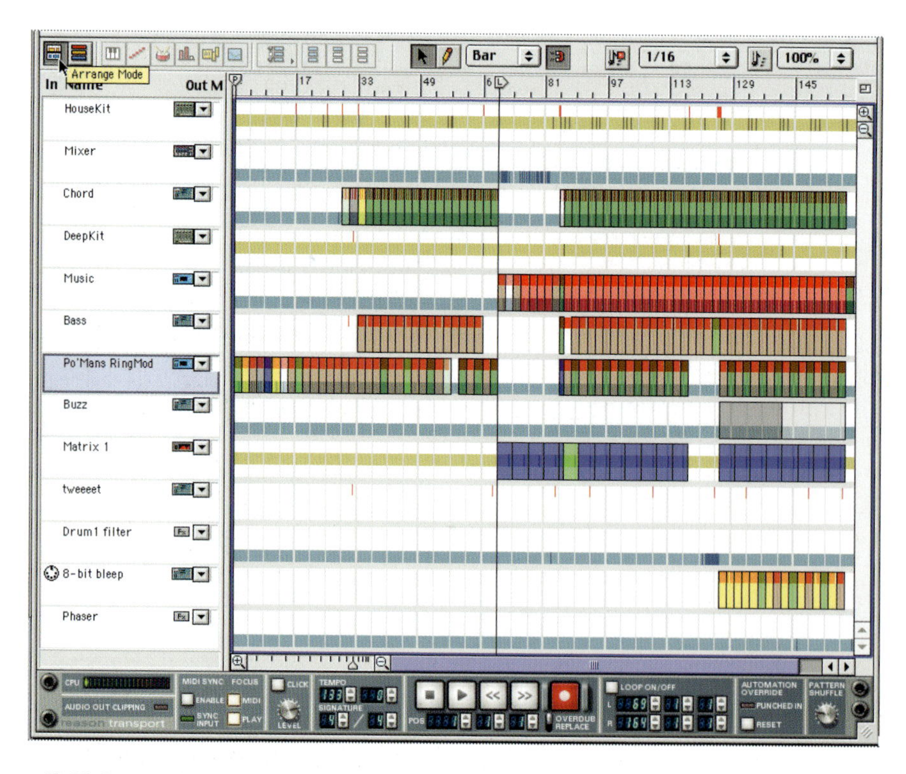

Figure 12.26 Reason Arrange view shows a number of tracks with coloured sections showing where the MIDI events are located.

MIDI. MIDI notes can also be sent to the Mixer for muting, soloing and activating EQ. And you can control virtually all the rest of Reason's controls remotely - using MIDI controller messages.

The sequencer can be found in the area below the rack, just above the Transport bar, which is located at the bottom of each song document window. The left part of the sequencer window shows the track list with the names of the sequencer tracks. The columns in this list allow you to connect tracks to devices, route MIDI and mute or solo tracks. The right part of the sequencer window has two main modes, the Arrange view and the Edit view.

With the Arrange view selected, you see the tracks lined up vertically with recorded events indicated as coloured bars (red for notes, yellow for pattern changes and blue for controllers). Here you can cut and paste patterns to arrange your Song.

The Edit view offers more detailed control for editing notes, pattern changes, controller data, and so forth.

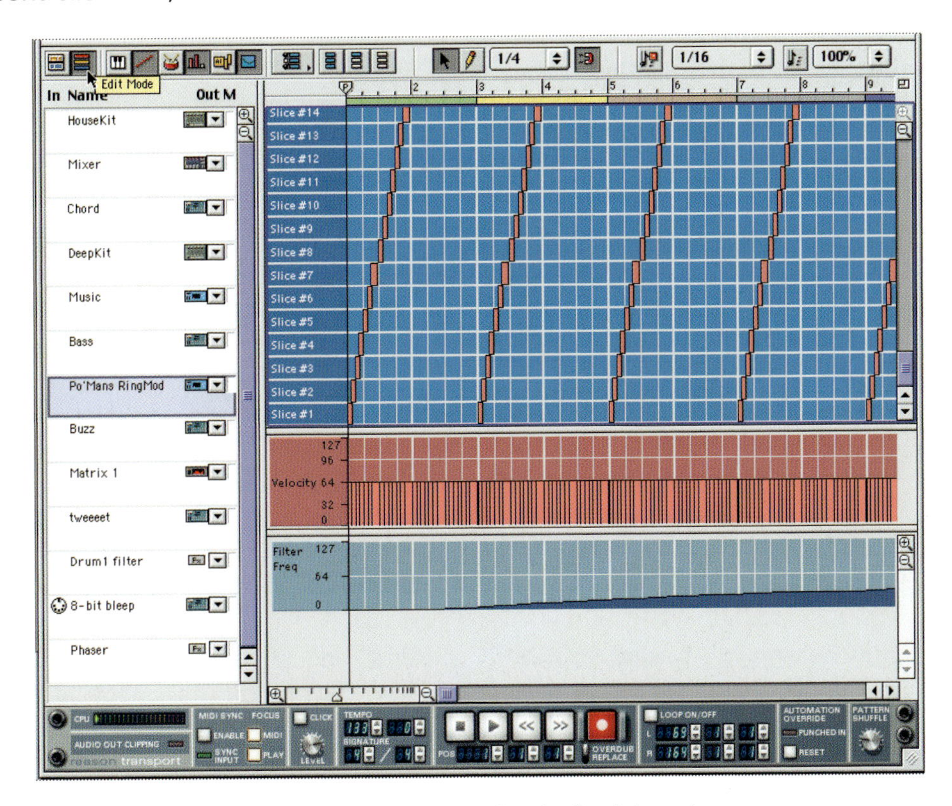

Figure 12.27 Reason Edit view showing details of the main sequencer.

The transport panel

Figure 12.28 Reason transport panel.

The panel at the bottom of each song document window is called the transport panel. This contains transport controls and settings that are global for the song, such as tempo and time signature, shuffle amount, CPU load indicator, and so forth.

Playing Reason from another sequencer

Tip First launch the host application, then launch Reason. When quitting, do this in reverse – quit Reason, then quit the host.

Understanding how to program the sequencer and pattern generators in Reason involves something of a learning curve, and, ultimately, you have more flexibility if you use a professional MIDI + Audio sequencer such as Cubase VST, Digital Performer or Logic Audio.

To route MIDI data from these sequencers into Reason, at present you must use Opcode's OMS Inter Application Communications (IAC) bus on the Mac or software such as Hubi's Loopback Device (HLD) on the PC.

Note When ReWire 2 becomes available for all these applications, this will allow bi-directional MIDI communication of up to 4080 MIDI channels (255 devices with 16 channels each) between compatible applications – so you won't need to use IAC or HLD.

Setting up Reason to use with your MIDI + Audio sequencer is straightforward enough. First you need to deselect the Sequencer port in the MIDI page of the Preference dialog.

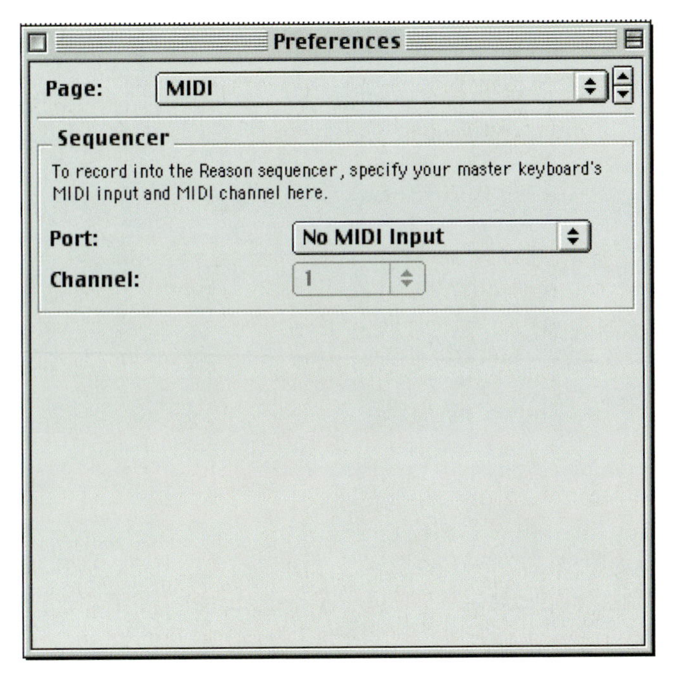

Figure 12.29 Reason MIDI Preferences.

Then set up the External Control buses to receive MIDI, assigning each of the four MIDI buses to the appropriate MIDI port. Using these, you can access more than 16 of the instrument MIDI channels in Reason, or distribute the incoming data through multiple ports to spread the load.

Note Depending on your MIDI interface, up to four separate ports, each with 16 channels, can be routed to Reason's four External Control inputs. Each port/device can be routed to a separate External Control bus input using the appropriate bus popup menu in the External Control section and any port/device can be routed to several bus inputs. Once you have routed several MIDI ports/devices to corresponding external buses you can use the Bus Select switch in Reason's MIDI In Device to select one of the four buses, identified as A-D, to change the channel-to-device routing in Reason.

In Reason, set up a rack containing whichever modules you wish to use - choosing synthesizers, effects processors and a mixer. Using the MIDI In Devices module at the top of the rack you can select each bus and make the appropriate

Figure 12.30 Reason Synthesizers' popup selector for Channel 1.

assignments. Using the small arrow to the right of each channel name you can open a popup menu containing the list of synths you are using in Reason. Use this to select the module for channel 1. You also need to set up the MIDI channels for each instrument here. Continue this procedure until you have set up all your channels, using additional buses as required.

Tip If you are using another sequencer with Reason – then you don't need Reason's sequencer cluttering up your screen. Once you have Reason set up to run from the other sequencer, just click on the 'maximize rack' button at the top of the rack's vertical scrollbar to hide Reason's sequencer from view.

Next you need to set up your audio outputs. Audio data from Reason (or any other ReWire compatible applications such as ReBirth) can be routed into the Audio mixers of these sequencers via ReWire to provide a tightly integrated working environment in which you can use your Reason devices purely as 'sound modules' – bypassing the Reason sequencer completely.

When you enable ReWire channels in Cubase VST's ReWire window, the corresponding number of input channels is automatically created in Cubase VST's channel mixer.

In Fig. 12.32, three stereo pairs of channels are showing, carrying the main stereo outputs from ReBirth, Reason and BitHeadz Phrazer. These channels are identical to regular VST audio Channel Mixer strips, so they have all the FX, EQ, busing and automation facilities, although they don't have any external inputs.

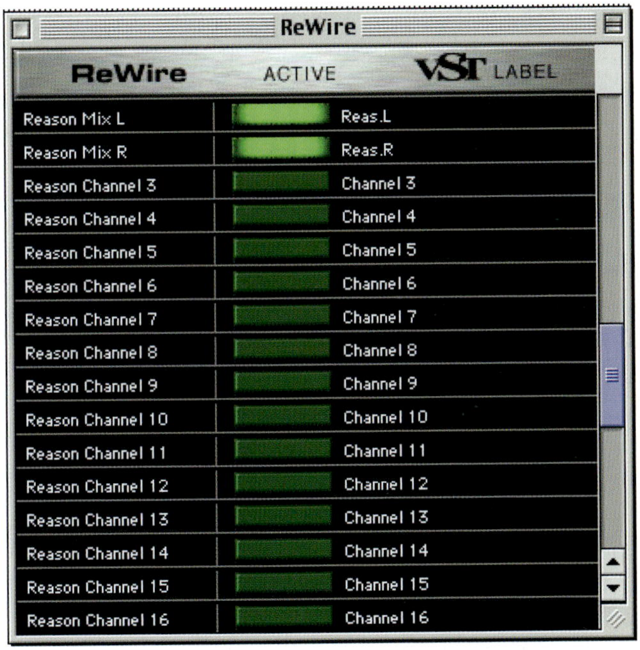

Figure 12.31 Cubase VST ReWire window showing inputs from Reason with the main stereo mix pair made active.

When you are using ReWire, the transports in the two programs are completely linked. It doesn't matter which program you use to engage Play, Stop, Fast Forward or Rewind. Recording, however, is still completely separate in the two applications. The Loop in Reason and the corresponding feature (Loop, Cycle or whatever) in the host application are also linked. This means that you can move the start and end point for the Loop/Cycle or turn the Loop/Cycle on/off in either program, and this will be reflected in the other.

Note With ReWire, as far as tempo goes, the host application is always the Master. This means that both programs will run at the tempo set in the host application. However, if you are not using automated tempo changes in the host application, you can adjust the tempo on the transport in either program, and this will immediately be reflected in the other. Of course, if you are using automated tempo changes in the host application, this will control both applications, so making adjustments to Reason's tempo will have no effect.

With Cubase VST/32 5.1 for the Mac which supports ReWire 2.0, the software will recognize the actual Reason devices and allow you to select these as MIDI Output destinations in Cubase VST's Arrange page - so you won't need to use Opcode's IAC.

> **Tip** If you want to manually play Reason devices from your MIDI keyboard when you are working inside another sequencer application, MIDI thru must be activated so that incoming MIDI is echoed out via the MIDI output. You will also need to make sure that the other application is 'thruing' its data to the correct MIDI port and on the right MIDI channel.

02R
DX7 II
IAC Bus 1
IAC Bus 2
IAC Bus 3
IAC Bus 4
JV-1080
MicroPiano
MIDI Express XT
QuickTime Music
Reason Badaid
Reason BASs
Reason Bass-Matrix
Reason DELAY
Reason Drums
Reason Hardware Interface 1
Reason MATRIX melodi
Reason Melody
Reason MELODY DIST
Reason Mixer 1
Reason REVERB
Reason Rhodes
Reason RnB-loop
Soft SampleCell
Studio Patches pgm chg
•Local•
◇Arpeggio
◇Cubase
◇Echo

Figure 12.37 MIDI Output Destinations popup from Cubase VST Arrange page showing Reason devices (as well as IAC buses).

Now you can play anything within Reason by routing MIDI data from your MIDI + Audio sequencer into Reason!

Triggering Reason patterns or sequences from an external sequencer

If you want to trigger Reason's pattern generators or internal sequencer from your external sequencer, you need to make sure that the MIDI Clock Sync is set up to synchronize these. You may have some useful patterns set up using Reason's pattern generators that would not be as easy to program in a typical external sequencer, for example. You can trigger patterns in Reason using MIDI Controller #3 messages that you insert into an appropriate track in your external sequencer. Bear in mind that pattern changes activated this way occur immediately - not necessarily at the end of the bar - which may or may not be what you want.

You may also want to use Reason's internal sequencer as well. Why? Well, the timing of tightly sequenced drum-machine and synthesizer parts can get 'sloppy' when sending data from any sequencer to any sound module - and Reason is no exception. If you need the tightest timing for particular parts, you may find that it is better to program these using the sequencer in Reason and simply send sync messages from your external sequencer to Reason to run these together. To set this up, open the Advanced MIDI preferences dialog. In the 'MIDI Clock Sync' field, make sure that the correct MIDI port is selected to receive the incoming clock signals. You also need to select MIDI Clock Sync instead of Internal Sync in the Options menu. Once this is done, you will be able to use the transport controls in your external sequencer to control Reason's Matrix and Redrum pattern sequencers and Reason's main internal sequencer.

Input and output

Tip If you want to 'convert' your ReWire channels into hard disk tracks, you can either record the audio in real time onto audio tracks in the host application or export your song as an audio file from the host application - using a feature named Export Audio or something similar.

When you have created a complete song, you can record your mix to a tape, CD or DAT recorder, or mix down to an audio file using the Export functions. This can be your final mix, especially if it is an instrumental or if you have used vocal samples, or it can form part of a more ambitious production that you are working on with your favourite MIDI + Audio sequencer. Of course, you can go the other way as well - starting out using your MIDI + Audio sequencer and mixing short sections to use as samples into the sampler or drum machine in Reason.

Reason supports a wide range of sample rates and resolutions, from 22.05 kHz/8-bit up to 96 kHz/16-bit, and you can even use files of different formats in the same device. So one drum sound can be an 8-bit sample, the next a 16-bit sample, and so forth; Reason will play back all of these samples at 16-bit resolution, regardless of their original resolution.

Summary

Reason has to be the best value for money around when it comes to software for synthesizing popular MIDI and audio devices. The sounds are great and you can incorporate your own samples. You can also control external devices or sync up with a conventional MIDI + audio sequencer. OK - so it isn't really going to replace hardware racks overnight, but it does mean that just about anyone can now afford to get a game started with their own hi-tech MIDI rack. Go for it!

humble 14'' computer monitor is much, much larger than the tiny parameter displays you get on typical MIDI devices.

Figure 13.1 Sound Diver Setup window.

The user interface resembles that of Logic Audio and there is a Key Commands window just like the one in Logic, so you can customize the computer keyboard commands to your liking. Remote control via MIDI keyboard is also available. Another possibility is to use a Fader Box for remote control. Excellent support for these is provided, with extra features provided for a couple of popular models such as the CM Automation Motormix to allow parameters and values to be displayed.

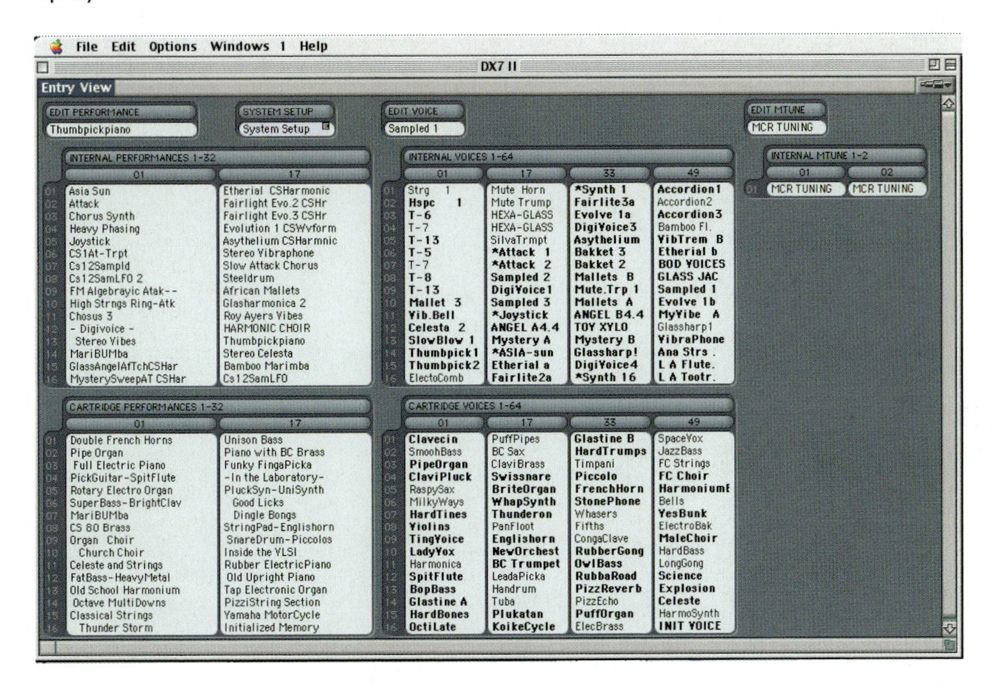

Figure 13.2 Sound Diver Device window.

The integration with Logic Audio is first-rate, as you would expect. For example, you can remotely control Logic's transport from within Sound Diver so you can playback sequences from Logic while altering your synth sounds in Sound Diver for instant feedback as to whether your edits are working. You can even synchronize switching of the 'screensets' - the selections of windows and sizes that you can store and recall within Logic or Sound Diver - for even greater convenience. That isn't to say that you cannot use Sound Diver with other sequencers, though. Courtesy of Opcode's OMS you can always pass timing and other MIDI data between different applications running simultaneously on your Mac.

I tested Sound Diver using a Yamaha DX7II synthesizer. The first thing I tried was to balance the volumes of a dual Performance that used two slightly different Voices (patches). As I moved the Balance control in Sound Diver's

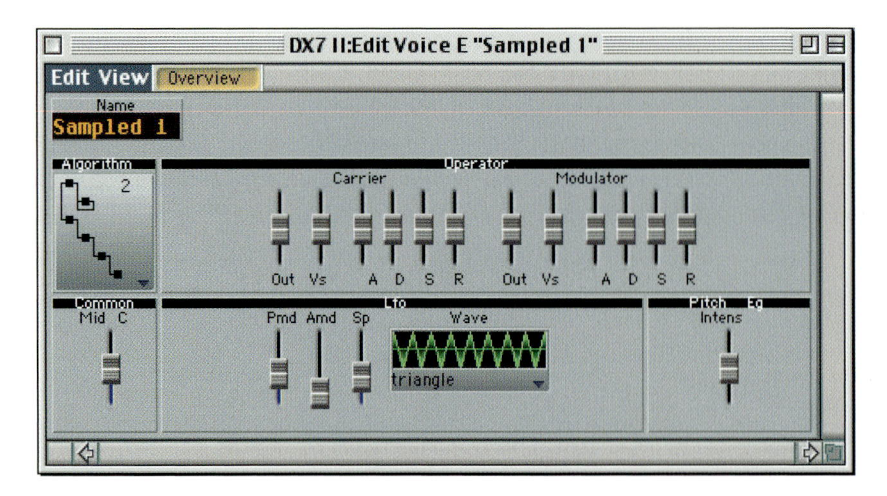

Figure 13.3 Sound Diver Edit Voice window.

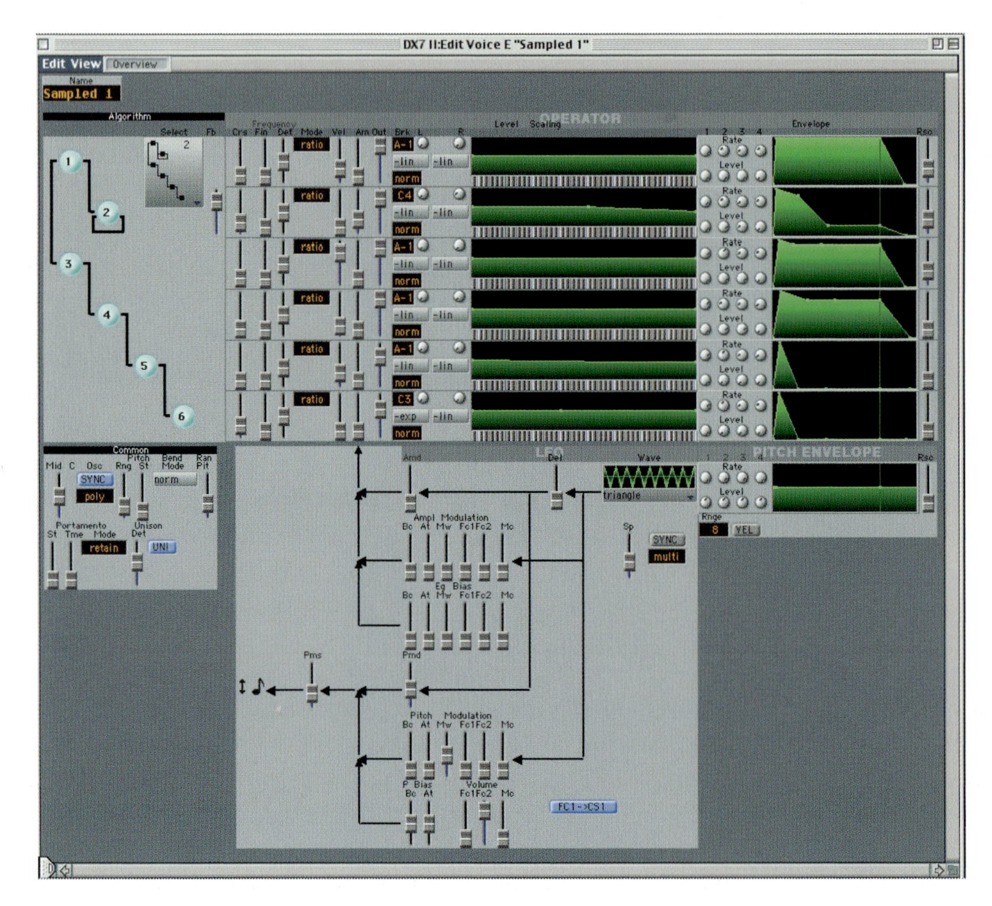

Figure 13.4 Sound Diver Edit Detail window.

Edit Performance window, a note played repeatedly on the synthesizer so that I could hear the effect of altering the balance without having to actually reach over and play the keyboard with my other hand. I found this AutoPlay feature very convenient most of the time, although I did need to turn it off sometimes so that I could play something more appropriate than the simple repeating note.

Sound Diver works on any PowerMac running any OS greater than 7.0 – with OS 9.0 or greater recommended.

The only real competition for Sound Diver is Unisyn from Mark of the Unicorn, especially now that Opcode and its once-popular Galaxy software are no longer 'alive and well'. I prefer Sound Diver – which now has the better user interface.

Celemony Melodyne

Melodyne is a revolutionary new editing tool that lets you edit pitch, formants and timing of monophonic vocal or instrumental recordings with far more detail than any other editing software. Available for both Mac and PC, Melodyne runs on all the popular Windows and Mac operating systems.

You can use Melodyne to pitch-shift monophonic audio by more than an octave, correcting formants if necessary, and the character of the sound will remain essentially the same. You can even change the rate of vibrato at the end of a sung note, or change the formant frequencies of an instrument to make a trumpet sound like a trombone or to make a male singer sound like a female singer. Although pitch-shifting does not alter the length of time that your audio lasts, you can always apply time-stretching, whether to a single note or to the whole 'arrangement', to make it shorter or longer. Melodyne also analyses the music to discover the rhythmic placement of the notes. These can then be synchronized to the rhythm of some other piece of music. You can easily create new melodies or build up arrangements of harmonies by copying and pasting, then pitch-shifting the copies.

Although you will not need to configure any audio hardware to get started with Melodyne if you are using the default built-in audio settings, you can use ASIO (on Windows or Mac OS9) or HAL (on Mac OSX) if you want to work with more than two audio channels, lower latencies and sampling rates other than 44.1 kHz.

Configuring the audio hardware is straightforward. Choose 'Preferences' from the Edit Menu and select 'Hardware' in the popup at the top left of the window. The next popup, at the top right, allows you to choose the kind of hardware driver to be used. By default the most common driver is selected: DirectX on Windows, Sound Manager on Mac OS 9 and HAL for the built-in device on Mac OS X.

On Windows, if any ASIO driver is installed on your system, you can choose between DirectX and ASIO. For recording or for multi-channel I/O, and for best audio performance, you should use ASIO drivers. When you choose ASIO, a second pop-up button allows you to choose the specific ASIO driver.

On Mac OS 9, you can choose between Sound Manager and ASIO. Of course, if you have Digidesign hardware with the Digidesign Sound Drivers installed, you can always redirect the output of the Apple Sound Manager to play back through your Digidesign hardware.

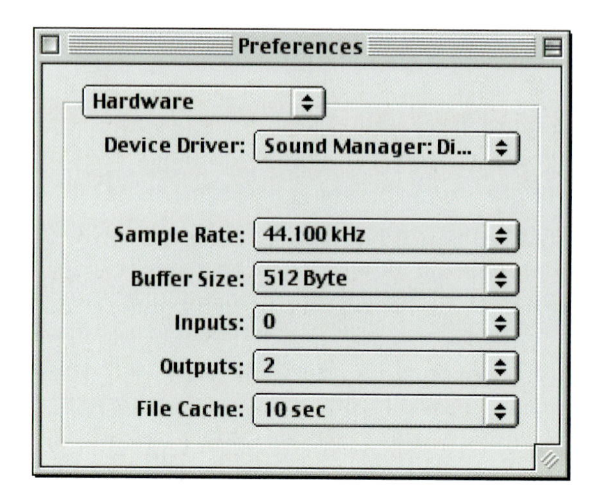

Figure 13.5 Melodyne Sound Manager preferences.

Nevertheless, for recording or for multichannel I/O, and for best audio performance, you should use ASIO drivers. When you choose ASIO, a second popup button allows you to choose the specific ASIO driver.

Unlike other ASIO applications, Melodyne does not have an 'ASIO Drivers' folder. Instead you use the '+' button that appears to the right of the ASIO Drivers pop-up to choose the ASIO driver that was shipped with your hardware directly from its normal location inside your main audio application's ASIO Drivers folder.

On Mac OS X, the first popup button allows you to choose any HAL (Core-Audio) driver that is installed on your system directly.

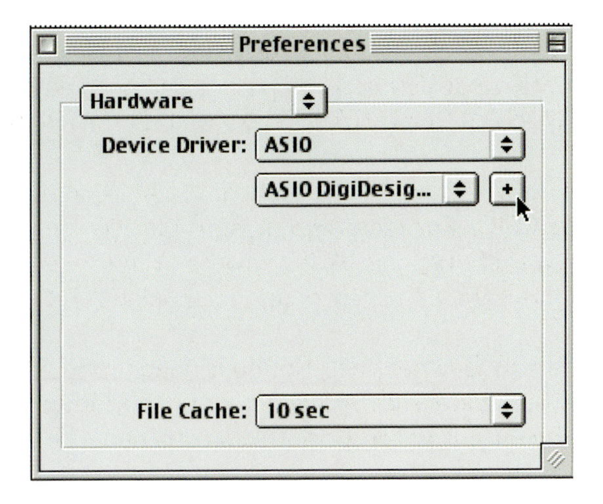

Figure 13.6 Melodyne ASIO preferences.

Note ASIO and Sound Manager are not supported on Mac OS X.

Most other audio editing software will allow you to change the pitch of a note or alter the tempo independently – so why use Melodyne? Well, Melodyne lets you do all this and much more – within a software environment designed specifically for this purpose. Say you want to edit a lead vocal track. Typically, the singer may be a little flat or sharp in pitch here and there, or the length of a phrase may be too short or long. Pro Tools plug-ins such as Serato's Pitch 'n' Time let you edit and correct situations like this reasonably well, but this type of plug-in's user interface is quite small and it can be difficult to work with long selections, such as a complete vocal track. Melodyne's user interface lets you work with up to 24 tracks at once, editing the timing of individual phrases (or even individual notes) till they are exactly as you want them. You can carry out detailed editing much more easily using the full-screen Edit window display, and Melodyne goes much further than any of its competitors, allowing you to adjust the vibrato that a singer typically uses at the end of sustained notes, for example. You can stretch or compress a note to make it longer or shorter and alter the rhythm of the phrasing at will. You can change the amplitude or timing of any note, lengthen or shorten a note, and shorten or lengthen the attack of any note to make it sound more staccato or legato. Formants are also handled intelligently, and you can either keep these static or shift them to create special effects. You can do plenty with tempos as well – and not only for vocals: you can also use Melodyne with percussion tracks.

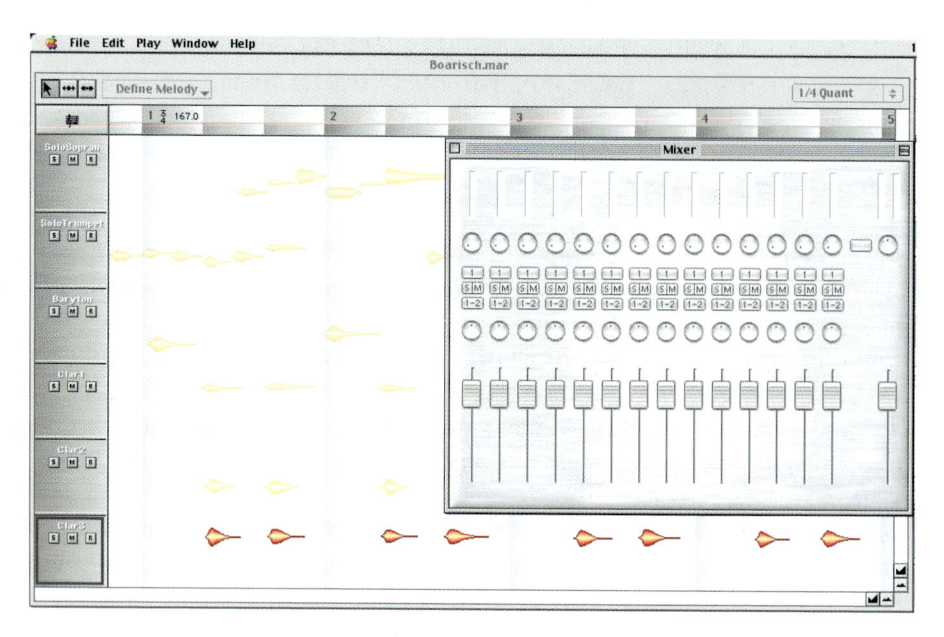

Figure 13.7 Melodyne Arrangement window showing several tracks with the Mixer window open in front of this.

You can edit your music while it plays and you will hear your changes as you move the notes. You can also make the moved notes snap to the next semitone or to a defined musical scale, Handling time changes is just as easy - you just grab the note and move it wherever you want. Depending on the tool you choose, the subsequent notes can be shortened, or they can also move in time. If you stretch a note, it will not be simply stretched linearly: the start of the note, e.g. the consonant at start of a syllable or the noisy part of plucked string, will be left unchanged. Or you can change the character of the note starts without changing their overall duration to make the melody sound more percussive or more legato.

Melodyne is actually quite a capable multi-track audio recorder, editor and mixer itself. Melodyne's Mixer window is very basic, though. It is just designed to let you select track inputs and outputs and set levels and pans. Nevertheless, this is sufficient to allow you to mix together processed files before transfer to a more capable system such as Pro Tools. Supported audio file formats include WAV, AIFF, SD2, SND, and AU, so you can transfer audio tracks to and from Pro Tools, Cubase VST or whatever you use as your main MIDI + audio sequencer.

When you open an audio file the first thing you do is to analyse the melody to work out the pitch, rhythm and tempo. When I tried this with a guitar melody, the results were good, but not perfect - so some manual editing of the pitches was required, using the Melody Definition window. On the other hand, with vocals and percussion tracks, Melodyne detected the tempos extremely well. Interestingly, Melodyne lets you quantize any varying tempos to fixed values, or match the tempo of one track to follow that of another. Re-mixers will immediately appreciate the potential here.

In the Arrangement window, you can open up to 24 tracks, load in (or record) several vocal or instrumental tracks, edit these and then mix them together. When you are happy with your results, you can export these as individual sound files or as mixes of combinations of any or all of these tracks - muting those you do not want to include in your mix. Arrangement window editing tools let you shift a track in time, alter the tempo of the arrangement, or cut and paste between tracks. So in the Arrangement window you can copy a vocal from one track to several others and edit the pitches of these to create as many harmonies as you like. You can also save the melodies you have analysed as MIDI files to view them as scores in a notation program, ideal for transcribing monophonic instrumental parts. Or you can export a MIDI file to a sequencer, complete with pitch and amplitude information - to make a synthesizer accurately 'double' the melody of a vocal, for instance.

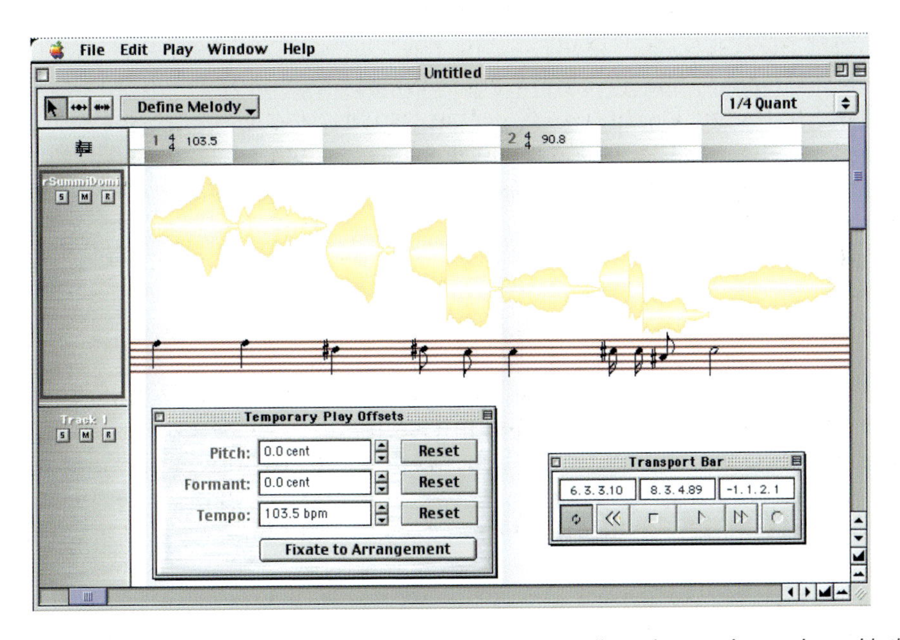

Figure 13.8 Melodyne Arrangement window showing notes as audio and as music notation, with the Temporary Play Offsets window and Transport Bar open in front of this.

The Arrangement window has a set of basic editing tools that let you shift the whole of a track in time, alter the tempo of the whole arrangement, or cut and paste sections or whole melodies within or between tracks. You can also display the notes in music notation once you have analysed the files. When you open, say, a vocal recording into track 1, the envelope of the audio is displayed without any assignment of notes to pitches. Choose 'Detect Melody' in the 'Define Melody' action menu in the inspector bar and the Definition window appears. Press the 'Detect' button to assign the notes of the detected melody to their correct pitches. If this produces an accurate result, which with less complex material it will, then you can start working on your arrangement right away.

For example, you could add a percussion track to the arrangement. The tempo of this may not match that of the melody - especially if it comes from a different recording. To fix this, select the percussion track, then select the second tool at the top left of the window. An 'Inspector' bar appears next to this, with a popup menu called 'Edit Timing'. Choose the first menu item here, 'Adapt Time', and the percussion track will automatically be processed to match the tempo of the vocal track. Other commands let you double, triple, halve or reduce the tempo to a third of the original. If you select the third tool you can change the tempo of the entire arrangement. Blue lines appear at the beginning of each bar of the arrangement to show where the tempo changes occur and a popup menu called 'Edit Tempo' appears at the top of the window. You can use its 'Equal Tempo' command to remove all the tempo changes and set one tempo for the whole arrangement. Now you can type any new tempo you like into the Tempo field. Other commands include Halve Tempo, Double Tempo and Tempo from Melody. This last command can be used to pick out the selected melody as a reference for the entire arrangement. The melodies in the other tracks then adapt to the original tempo of that melody. Or, alternatively, you might load a percussion track as your first 'melody', analyse this, and then constrain the tempo of subsequent vocal or instrumental tracks to follow this tempo.

Now if the melody definition is not absolutely accurate, you will need to make some corrections before doing any serious arranging. Choose 'Edit Definition' from the 'Define Melody' action menu to open the Melody Definition window.

The Melody Definition window is similar to the Editor window but has different tools. This window is used solely to define the pitch and timing to accurately represent the original melody - not to create new melodies. First, any wrongly detected notes should be corrected for pitch; then the note separation should be checked. Finally, the tempo should be defined. The first tool at the top left of the window lets you select notes and the second tool lets you correct pitch by dragging a note to its correct position. The third tool lets you edit the Note Separation.

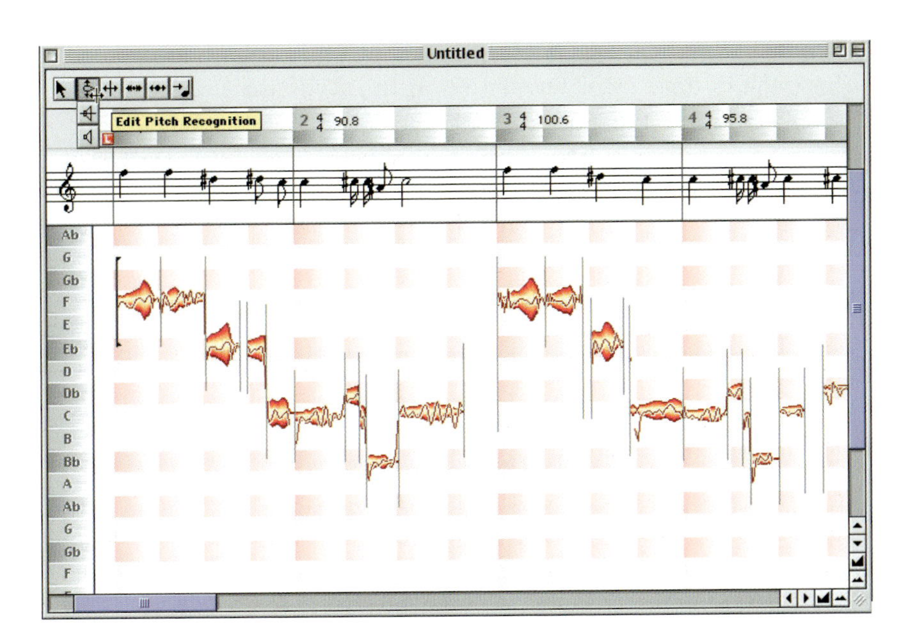

Figure 13.9 Melodyne Melody Definition window showing recognized pitches available for editing or correction.

Just click on a note to divide it into two parts. To adjust the position of a split, just grab and drag the separation marker. If you mess up, just drag the selection rectangle over the separated notes to stick them back together again. The fourth tool lets you define the tempo by moving a coloured 'anchor line' that appears in the display. Double-clicking at the start of a new bar sets or removes a new tempo definition. You can use the commands in the Action menu to 'Double Tempo' or 'Half Tempo', or choose 'Equal Tempo' to cancel all manually entered tempos. 'Optimize Tempo' attempts to adapt the tempo to that of the actual notes if the tempo was manually approximated, and 'Tempo from Arrangement' sets the tempo to that of the arrangement from which the melody definition was opened. You can also manually enter the nominator and denominator of the time signature and set the tempo in BPM using the fields that appear in the window when this tool is selected. The fifth tool lets you edit the Rhythm Assignment to move the notes to the left or right to correct their individual timing, while the sixth tool lets you edit the Pitch Assignment in semitone steps by dragging the notes up or down.

Double-clicking on a melody in the Arrangement window brings up the Edit window where you can carry out more detailed creative editing on individual notes. Here, the Edit Pitch tool lets you grab a note and move it up or down in pitch. It is very easy to see what you are doing in this window and you can zoom the display to help. You can either move the pitch freely or you can constrain it to

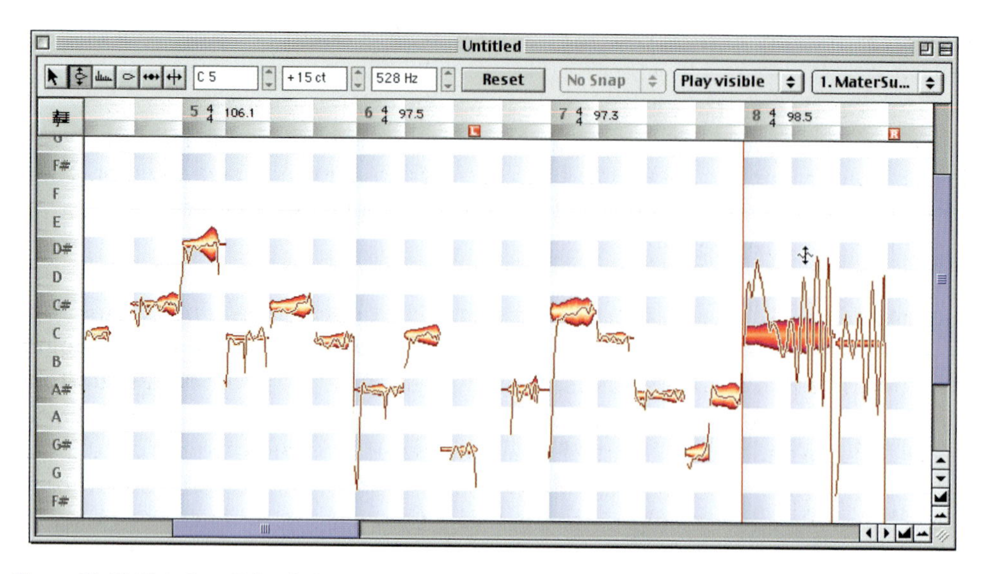

Figure 13.10 Melodyne Edit window showing the Pitch Align tool in action – increasing the vibrato depth of the last note.

move between the notes of a predefined musical scale that you select. Most of the tools in Melodyne have 'sub-tools' that you access by moving the mouse below the original tool on its popup selector.

The first pitch sub-tool is 'Pitch Align'. When you grab a note with this and move it up or down, the depth of the vibrato is increased or decreased. You can use this to make a pop singer sound like an opera singer with an exaggerated vibrato depth – or to remove the vibrato if you prefer.

The second sub-tool, Pitch Transition, lets you alter the speed of the pitch transitions between notes. A singer often slides in pitch between notes – rather than stopping one note and starting another at a different pitch. The speed of this transition might be too slow or too fast; now you can edit this. If you cause an immediate transition between two notes, this sounds like the effect on the Cher record that was the first commercial recording to feature this effect some years ago. As far as I know, Auto-Tune was used on the Cher recording, as Melodyne was not available back then. It was an accident that sounded 'cute' – so they used it. Melodyne makes it 'a breeze' to create this type of effect – or you can use it for the purpose for which it was intended, i.e. to make sensible corrections or edits to the phrasing.

Using the Edit Formant tool you can shift the frequency of the main formant in a singer's voice to make the voice sound deeper or higher – e.g. in order to make a soprano pitch shifted into 'tenor territory' sound like a tenor as well. As with the Pitch Tool, a sub-tool lets you change the transitions between formants applied to different notes. This tool, labelled Formant Transition, changes the speed of transition of the formant position between any two notes that are set to different values. Using the next tool popup, you can edit the amplitude and any amplitude transitions using the Edit Amplitude tool and its Amplitude Transition sub-tool to make a note softer or louder.

Quite often, you will want to change the timing of notes. The Move Notes tool lets you move one or more notes by grabbing them with the mouse and dragging them horizontally, i.e. in time, to the desired position. The Stretch Notes sub-tool lengthens or shortens a note in time. If you grab a note before its midpoint you change its length from the front. The reverse happens if you grab the note in its endpoint. The Edit Time Handle sub-tool lets you change the initial speed of a note, shortening or lengthening the attack of the selected note. This allows you to give the note a softer or harder attack, to make it sound more staccato or legato.

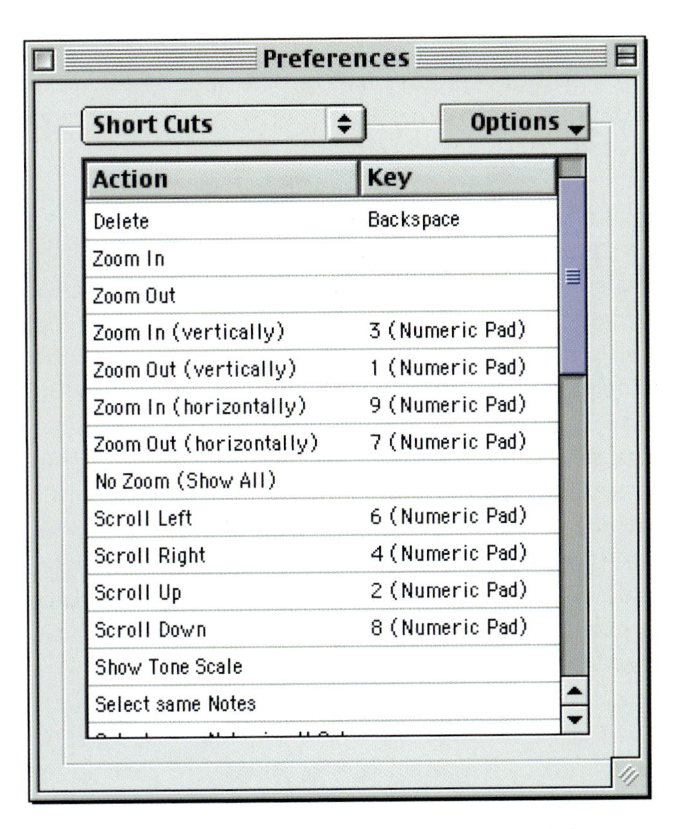

Figure 13.11 Melodyne Short Cuts Preferences window.

The final tool, Note Separation, lets you change the separation of notes or define which segments of the recorded audio material are to be considered as single notes. Melodyne automatically separates the notes, but you can use this tool to change the settings if necessary. The Segment Separation sub-tool cuts parts of the melody into free-standing segments. Two notes that follow each other directly will always be treated by Melodyne such that if the second one is moved the first one is stretched accordingly in order to keep the audio material flowing.

Melodyne version 1.1 adds a configurable multilevel Undo/Redo function that makes life a lot easier. Every editing software application should have this feature! Keyboard shortcuts can be defined using the appropriate Preferences window.

Even at version 1.1, it is still 'early days' for Melodyne. For example, the analysis function works well - but not flawlessly. However, features are being added all the time, so version 1.1 will now work with stereo soundfiles, and new functions and tools are being developed constantly. Nevertheless, Melodyne is capable of delivering professional results right now. Melodyne is fairly expensive - but it can 'work miracles' and the processed audio sounds very good. And nothing else available allows this detailed level of editing of melodies. So, for example, if you are producing vocalists or working on remixes you should definitely consider using Melodyne.

BitHeadz Phrazer

BitHeadz software's Phrazer is the Mac's answer to ACID for the PC. Like ACID, Phrazer is a loop-based composition tool. You can use your own loops, third-party loops, or the factory loops that come with Phrazer. Loops or phrases are automatically tempo-mapped and pitch-stretched so even though the loops may be of different tempos or in different keys, Phrazer sorts all this out for you. You can record and edit your own samples within Phrazer, and MIDI or computer keyboard triggers can be used to mute and unmute tracks so that you can play Phrazer in real time. You can even trigger one-shot samples in real time and MIDI can also be used for real time track volume and pan changes. You can use Phrazer together with your favourite MIDI + Audio sequencer, feeding the audio output from Phrazer directly into your software using Sound Manager, ASIO, DirectIO, DirectConnect, ReWire or MAS.

The control section at top left of the main window lets you vary the tempo, volumes, balance, sends or key of the song globally. Select any tempo and key and Phrazer will automatically map all the different samples to this tempo and key. The track data view at the right-hand side of the main window is where you

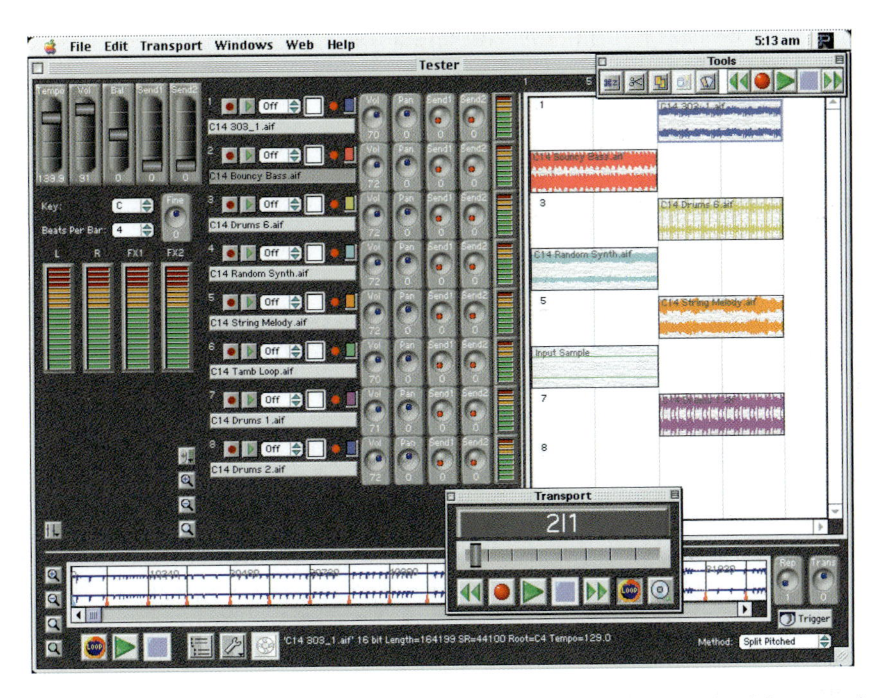

Figure 13.12 Phrazer main window showing tracks in the centre, track data to the right, control section to the left with the sample editor open at the bottom.

arrange your audio samples. Here you can drag samples around with the mouse, and edit using cut, copy, paste and clear. The editor at the bottom of the main window displays the five editors one at a time, depending on which is selected in the Edit menu. Here you can edit the samples in the tracks or edit the effects parameters. There are tools to edit the sample data graphically and you can also use the popup 'munge' dialogs to do all the usual stuff like fade, loop, normalize, reverse and so forth.

The effects are one of the neatest things to play with and you can change these while the sequence is running 'live', swapping chorus effects for reverb or whatever on-the-fly. Each track can incorporate several different effects that you can apply individually to different sections. You can also use global effects if you want an effect on all the time.

Once you have finished your masterpiece in Phrazer, you can record this to disk using the 'Render' command from the 'File' menu. This plays back your song – faster than real time – and creates a new AIFF file on your hard disk. You can also record a live performance to disk using the 'Record To Disk' command from the 'Transport' menu. The next time you play your song, a dialog will come up asking

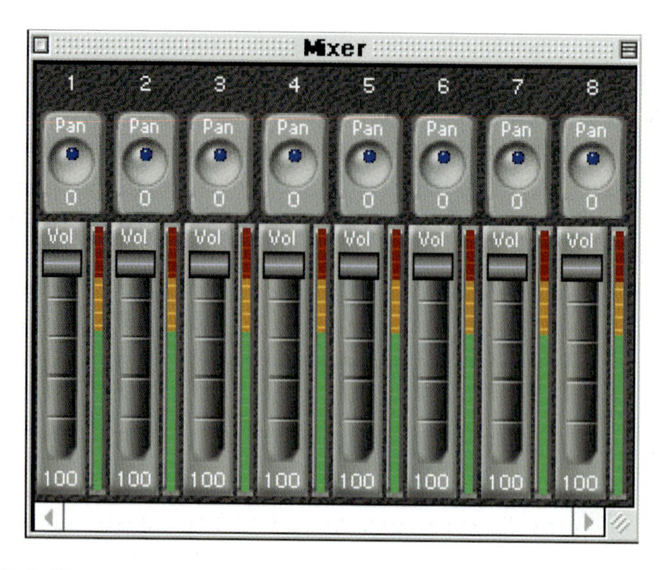

Figure 13.13 Phrazer Mixer palette showing pan and volume controls for eight tracks.

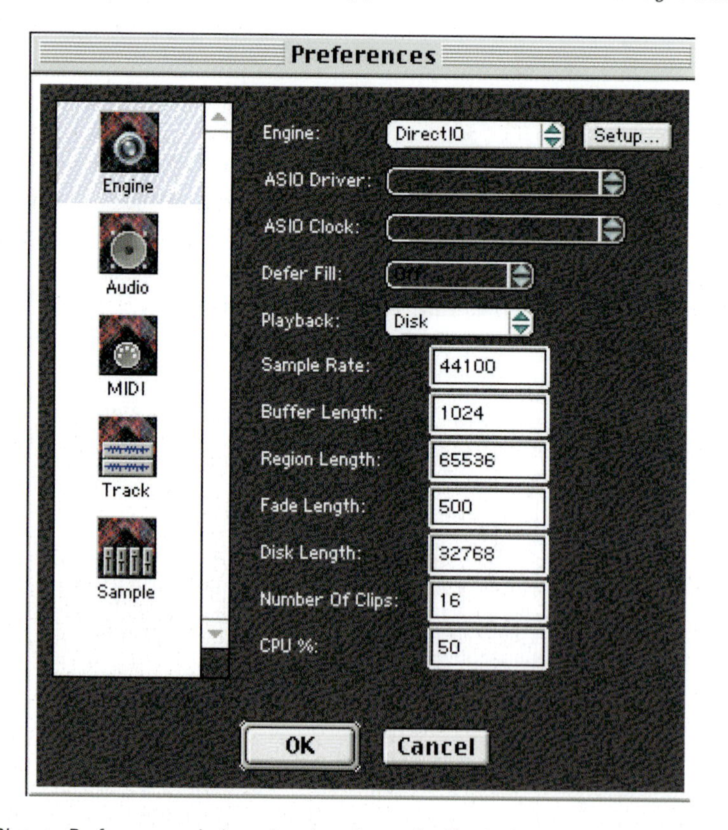

Figure 13.14 Phrazer Preferences window showing the audio 'Engine' set up for Digidesign DirectConnect output via a Pro Tools card.

you where to save the audio file. As the song is playing, you can turn tracks on and off, alter effects and so forth and your 'performance' will be recorded into a new AIFF audio file.

Phrazer felt a bit 'clunky' to me in its first version, but it is definitely a 'fun' thing to have around if you like messing about with loops. Loops or phrases are automatically tempo-mapped and pitch-stretched, making it extremely easy to put grooves together from different sources. Phrazer is not quite as advanced as ACID in terms of zooming in, for example. But it has MIDI control and allows plenty of sample manipulation within the application.

Audio Genetics Alkali

A friend asked me 'Why does my expensive Pro Tools system not have tools to easily change the BPM and pitching of drum loops – so you can make rhythm tracks directly in Pro Tools rather than having to use a sampler?' A damn good question! And then I found it! The Alkali Loop Laboratory running with Pro Tools.

Alkali works collaboratively with ReCycle, allowing you to instantly change the tempo of a loop in REX file format without pitch-shifting or time-stretching. Alkali also has a varispeed control which alters pitch without changing tempo. You can apply quantization to tighten up the beats and apply a 'swing' amount. Alkali works great with both drum and bass loops, letting you select the correct tempo and key for your song by simply moving a slider. Alkali also supports Digidesign's Direct Connect technology so the audio can be previewed within Pro Tools whilst

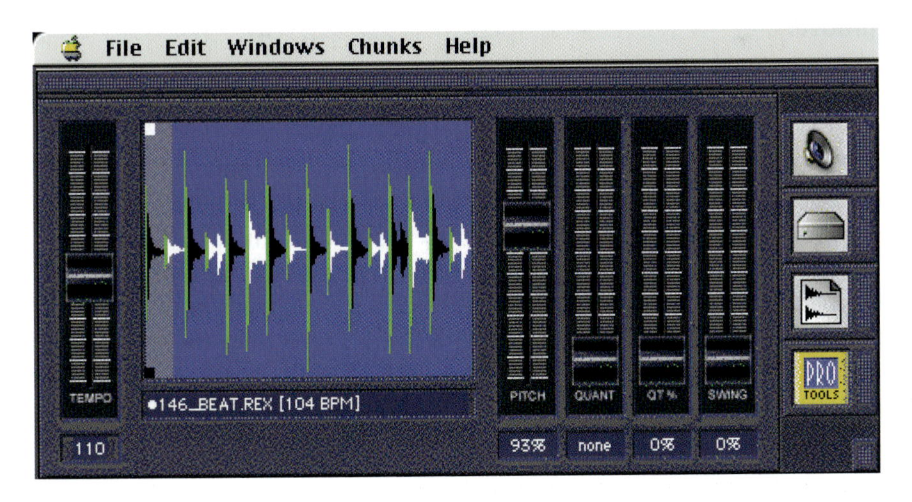

Figure 13.15 Alkali main window.

transcribe the notes by ear - and if it is still too fast you can always process the file a second time to halve the speed again.

If you need to make some last minute edits or add some effects to your audio files you can use the 'Open Selection In Editor' command, or double-click on any file in the batch list, and your selected files will be opened directly into whichever third-party editing software package you have designated. sonicWORX Essential, a cut-down version of Prosoniq's sonicWORX editing software, is bundled with TimeFactory and offers a range of useful effects along with good waveform editing features.

To test the software, I took a 100 bpm four-bar drum pattern and changed the bpm - first down to 50 and then up to 200 BPM. The results were not quite as good as those with Wave Mechanics Speed for Pro Tools - a sort of 'thwocking' sound appeared on the snare at 50%, while at 200% the congas sounded 'slappier' than on the original and the hihat sounded louder. When I pitch-shifted an entire mix, results were comparable with those from Speed. Then I pitch-shifted a solo voice and a solo guitar, each time choosing the 'Preserve Formants' option in TimeFactory. The results with the voice sounded faultless to my ears, while the results with the guitar shifted down from F to E♭ were much more convincing than when I used Speed - although my guitarist's ear still told me that there was something unnatural about the chords (which, of course, there was - you could not play these particular chord inversions in E♭ normally).

If you need to pitch-shift voices or instruments with prominent formants, TimeFactory is the package to use, although Speed, for example, makes a better choice for time-stretching drums and is quicker and more convenient to use if you are working in Pro Tools. Nevertheless, TimeFactory will appeal to more potential users than Speed - as it is available for both Mac and PC - and it delivers on its promise of high-quality results.

IK Multimedia T-RackS mastering software

When I think of audio signal processors, such as compressors and limiters, the first thing that springs into my mind is a picture of one of those classic tube models with the glowing arrays of tubes, chunky bakelite knobs, metal toggle switches and large analogue meters. That was the first kind I came in contact with when I started out in recording studios - too long ago to dare think about!

Of course, these days, like many of you 'out there', I am working mostly from my home studio with Pro Tools and an 02R. The 02R has a usable dynamics section - but it's not exactly 'to die for'. The situation with Pro Tools is a lot better in many ways, as there is a goodly selection of third-party TDM plug-in compressors available. You can choose from Focusrite, Drawmer and several others - my current favourite being the Waves Renaissance Compressor.

When it comes to mastering, you could use the processors in your console or in Pro Tools, but it is often better to deal with the final mastering as a completely separate stage in your production process, ideally using specialized tools for the job, such as the aforementioned classic tube devices. But we are talking serious bucks here. Now, if you are still hankering after some of those classic units, but your budget won't allow, you should check out T-RackS from IK Multimedia, a company based in Modena, Italy. T-RackS is a stand-alone software package which emulates these high-end analogue mastering tools - bundling together a state-of-the-art six-band parametric equalizer; a classic mastering stereo tube compressor/leveller; and a multi-band master stereo limiter, all at an extremely attractive price. You can get Pro Tools plug-ins which offer most of these features, but the big difference is that T-RackS offers a dedicated environment for final mastering - so you can boot it up quickly, load in a file in seconds, and concentrate solely on the important final stages of your mastering.

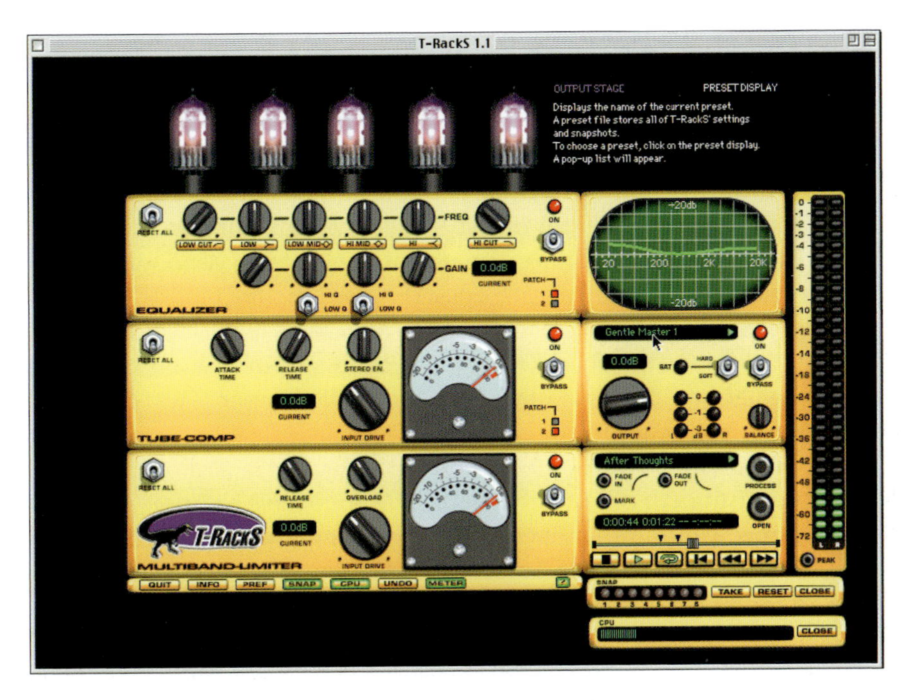

Figure 13.18 T-RackS main window.

Earlier I talked about the glowing tubes you see on classic units. Well, guess what - T-RackS has tubes which glow pretty much like the real thing as well! If you have ever watched tubes at work, you will have noticed that the colours change from purple to golden and the brightness varies according to the program material being processed. T-RackS's tubes do a fair job of emulating this behaviour, so, for example, if you switch the processors on in turn - EQ, compressor, limiter - the tubes glow more brightly as you add processors. A bit of fun really - but why not! The next thing that struck me was the sheer simplicity of the user interface. This resembles a typical rack of hardware with all the typical hardware-type controls you would expect to see laid out pretty much where you would expect to find them. So I took a quick peek at the menus and was very surprised to find that there was almost nothing there apart from Quit and Undo! It turned out that all the computer-type controls were spread around the T-RackS panels disguised as jack sockets and suchlike or discreetly available under neatly labelled buttons which bring up dialog boxes or floating windows from a 'console' at the bottom of the racks.

This console strip contains seven buttons and a question mark icon. At the left is the Quit button, followed by an Info button that brings up a Credits window. The next button brings up the Preferences window where you can enable Dithering, enable Realtime Processing, set the audio buffer length and choose the Interface Material - Gold, Copper or Chrome. This last setting actually lets you choose one or other of these three colour schemes for the interface - just to brighten up your day! The next button, labelled Snap, brings up a floating window which allows you to take up to eight snapshots of your T-RackS settings and quickly recall these - very useful for comparing different settings 'on the fly' during a song. The next button brings up another floating window containing an indicator for CPU activity and the button to the right of this lets you Undo your last action. Clicking on the last button, labelled Meter, opens up another floating window containing a large high-resolution sample-accurate peak program meter. Finally, the question mark icon at the far right of the console is actually a Help button. With this enabled, whenever you position the mouse over any of the controls a 'help' text is very neatly and conveniently displayed at the upper right corner of the screen describing the action of the controls.

At the bottom of the window to the right of the console you will find a set of transport controls with buttons for Stop, Start, Loop Playback, Return to Start, Jump To Next Marker and Jump To Previous Marker. You can set loop start and end points in your file by dragging the two small loop markers which appear on the timeline slider located above the transport controls when you enable loop playback - and the positions of these markers are also indicated in the numerical display above the timeline. An icon resembling a small jack socket, labelled Mark, is actually a button which lets you insert a Marker in your file wherever the timeline is positioned when you click on this button. Using these Markers you can jump quickly between various points, such as verse and chorus, using the Jump Marker buttons – or you can just drag the timeline slider to quickly position your playback point anywhere in your audio. Above the Mark button you will find two similar buttons labelled Fade-in and Fade-out. Clicking on either of these brings up a small floating window where you can set the fade time and choose a logarithmic or linear fade curve. At the right of this panel a couple of jack socket icons marked 'Open' and 'Process' let you load in a file and save it. I guess this is a bit like plugging your input in and patching your output to your master recorder, but in this case you are opening a file and then processing it and saving the result. You can preview before loading and you can even apply a preset to the audio file while previewing. This makes T-RackS very fast to work with. For instance, you can use this feature to make comparisons on all the songs of an album with the appropriate master before processing.

Typically, you might start out by equalizing your mix. Each of T-RackS's six filters has a different filter shaping, all specially developed for mastering. Looking from the left you will find frequency controls for the Low Cut, Low Shelf, Low-mid Peak, High-mid Peak, High Shelf, and High Cut filters. A row of on/off buttons lies underneath, with the Gain controls for the shelving and peaking filters below these. And at the bottom of this section a couple of toggle switches let you choose between low and high Q settings for the two peaking-type filters. A small numerical display shows the gain in dB or the frequency in Hz or kHz of the currently selected knob and you can easily check how these filters are altering the frequency response by looking at the oscilloscope-type display to the right of the EQ section. A Patch switch is provided to let you position the equalizer section either before or after the compressor in the audio chain.

Once you have your EQ set, you can apply compression to make your mix sound louder, more cohesive, and more 'punchy'. The usual Attack and Release Time knobs let you set the time constants of the compressor and a Stereo Enhancement control is provided which affects the Stereo width of the Mix – enlarging or reducing it. This is not a conventional 'threshold and ratio' compressor though – it is a 'soft-knee' unit with a 'no threshold' characteristic – so you raise the input drive to increase the compression and lower the input drive to reduce it. There isn't a particular level at which the compression begins. Instead, you will see compression on the Gain Reduction VU even at very low levels. The idea here is that this works like an old tube compressor – providing a gentle compression that is ideal for mastering.

If you want to bring the level of your mix up even higher, you can use T-RackS's multi-band limiter to tame the peaks which could be happening at low, mid or high frequencies anywhere in your mix. The simplicity of the controls provided hides the complexity - which I reckon is a good thing! You get an Input Drive control that makes your mix louder as you limit more peaks, with an analogue VU to show the amount of gain reduction in dB. Although there are actually three frequency bands within which limiting takes place independently, no controls are provided for these, so the VU just shows the average of the limiting by the three bands. As you would expect, the Release Time knob lets you set how long the limiter will take to return to unity gain after peak limiting, while the Overload control needs a little explanation. As you increase this control, the limiter will let progressively more of the peaks 'through', and these will then be clipped in an analogue-like fashion - as would happen with a classic tube device. A neat touch!

When the three processors are set pretty much the way you want, you can use the Output section to make more 'tweaks'. To the right of the Output control knob, three pairs of LEDs indicate when the level reaches −3 dB, −1 dB and 0 dB. Another LED above, labelled 'Sat' for saturation, indicates when clipping occurs, and an associated toggle switch lets you choose between hard (digital) or soft (analogue) clipping. Obviously, if you don't want any clipping you need to set the output control lower until the clipping LED never lights up. You can use the Bypass switch on the Output section for A/B comparisons with your original audio and if the original mix has L/R level calibration problems you can use the BALANCE control to correct these. Above the output controls you will find the Preset display showing the name of the currently selected preset. Clicking and holding on this brings up a popup menu containing the factory presets such as Brickwall Master, Gentle Master, Flat EQ + Hard Compression, Low & High Boost + Compression and so forth. Of course, you can save and load your own settings as presets - using another popup menu that appears when you click on the arrow to the right of the Preset display.

Once you have everything set up, you just click on the Process button to create your output file. You'll be asked to name your new file in the dialog box that appears and processing will start immediately after clicking OK. If you have enabled Realtime Processing in the preferences menu, you will hear your new file playing as it is being written to the hard disk and you can 'tweak' any of the controls in real time, or recall any of the snapshots, to make adjustments while this is happening. If you don't need to hear it or adjust it, just disable Realtime Processing and your file will be processed faster than real time. And that's it! You're all done!

I tried out a solo jazz guitar piece - a fairly dry recording of an acoustic Gibson L5 guitar. I had already EQ'd this to take out some of the boominess and some of the presence boost it had as a result of being recorded with flat EQ but using an AKG C12 VR microphone which emphasizes some of the higher frequencies. I had also applied some gentle compression using the Waves Renaissance Compressor. T-RackS comes with an excellent selection of presets and I found that 'Gentle Master 1' added back some of the highs and lows which were getting a little 'lost' as a result of my previous EQ'ing and made the overall sound a lot 'punchier' without emphasizing any boomy or scratchy frequencies - just what I was looking for!

To see how versatile T-RackS would be, I checked out an 'Alanis Morrisette' sound-alike mix next, packed full of acoustic and electric guitar parts, bass, kit drums, drum loops, synth sounds and suchlike. This time I decided not to bother with the presets - just to go for it by tweaking the knobs myself. The bottom end on the mix was a little heavy so I started out using the EQ section to reduce some of the lower frequencies. The 'scope' came in really useful here - adding visual feedback to what my ears were hearing. I ended up just dipping a fairly narrow range of frequencies around 150 Hz by a little over a dB using the low-mid filter to get the effect I was looking for - rather than rolling off the low frequencies with the low filter as I had first intended. Then I decided that the bass and drums weren't sounding raunchy enough. A job for the compressor section! So I used fairly fast attack and release settings (32 ms and 320 ms if you really want to know) and pushed up the input drive knob till the track really started to rock!

I did find it more difficult to control the knobs using the mouse instead of grabbing the real thing - but you would expect this. On the positive side, sweeping through the frequencies with the mix playing through the EQ section, sounded just as smooth as on the best hardware I have used.

T-RackS is also available for RTAS and HTDM for Pro Tools systems and in VST format for Cubase and Nuendo systems. The 'racks' have been re-designed and improved for these versions. With the RTAS version, for example, the various modules are all available as separate plug-ins - the Equalizer, Multi-band Limiter, Soft Clipper, and Tube Compressor - or you can load them all at the same time, as the Mastering Suite plug-in.

Several new features have been added and improvements have been made compared with the original stand-alone version. The Equalizer has sweepable mid bands and both the peaking filters can now cover the entire 20 - 20 kHz audio spectrum. The high-end response has been refined to be even more musical and transparent giving the characteristic smooth high-end associated with classic analogue equalizers. In the Compressor, a variable high pass filter has been added to the detector stage – where the compressor analyses the incoming signal and determines how gain reduction must be applied. This helps to prevent the 'pumping' typically caused by low-frequency signal content. Labelled 'Sidechain HPF', higher settings will make the compression effect softer and less audible while lower settings will make the effect more audible.

> **Tip** If you want to hear the compressor working, go ahead and use the lower settings, but if you are working on acoustic material you will appreciate the gentler compression effect that you will get using the higher settings.

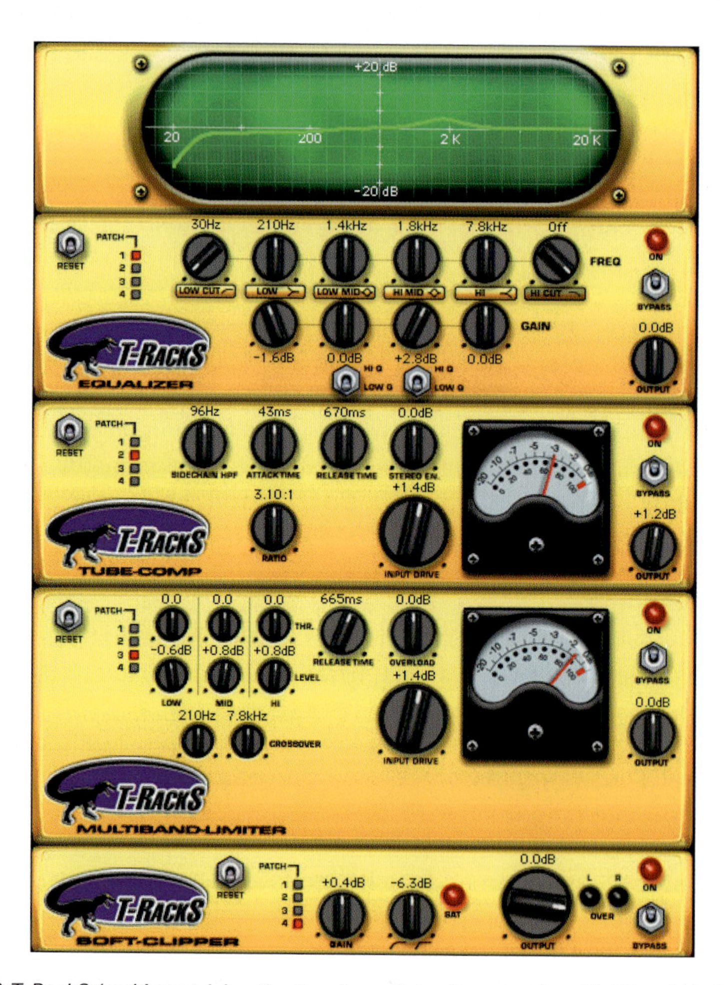

Figure 13.19 T-RackS 'rack' containing the Equalizer, Tube Compression, Multiband Limiter, and Soft-Clipper.

In the Multiband Limiter, all three limiter bands, low, mid and high, are individually adjustable for threshold and level. This allows you to limit the low end without applying unwanted compression to the mid-range, for example. The frequency crossover points between the bands can also be adjusted, providing even more flexibility. The output stages of both the Limiter and the Soft Clipper have also been modified. So, for example, if the output level control of the Limiter is set to 0 dB (the default) the output of the limiter will never go beyond −0.05 dBfs. You can just increase the drive control until you hear too much distortion or limiting, secure in the knowledge that the output is being held to a maximum level of −0.05 dBfs, without any overs.

T-RackS is a very straightforward, yet powerful, package of software mastering tools that does what it says it will, simply and effectively. I particularly liked the user-interface design and was quite impressed with the sonic results - which compare very favourably with plug-ins costing many times more!

IK Multimedia AmpliTube

AmpliTube, available for both Pro Tools and VST systems, not only models vintage and modern guitar amps and cabinets, but also models the 'stomp-boxes' that guitarists often use, the microphones used to record the sound, and various effects that are typically applied during post-production.

You can switch the interface between the Amp Module, the Stomp Module and the Post FX Module by clicking on the buttons in the bottom right-hand corner of the plug-in's window. The input signal is fed to the stomp boxes section first, then to the Amplifier and then to the mono-to-stereo Post FX section.

The Stomp effects include Wah-Wah, Delay, Chorus, Flanger and Overdrive - effective simulations of typical popular pedals. All the usual controls are provided for these, so guitarists will have no trouble operating this module.

The vintage wah pedal offers both manual and auto wah options and the analogue delay delivers the warm tone typical of early analogue delay units - but without the hiss! The analogue chorus is modelled after one of the most loved 80s chorus boxes and the analogue flanger is capable of a wide range of flanging effects. The solid-state overdrive is perfect for driving the first stage of the AmpliTube pre-amps and can saturate itself as well as overdrive the amp.

The Amp module has four large knobs that let you choose between the seven pre-amplifier types, five EQ types, four power amp types and nine speaker cabinet

Figure 13.20 AmpliTube Stomp Effects.

types. Another pair of knobs lets you set tremolo rate and depth and a single knob is provided to control the spring reverb. There are yet more knobs for low, mid and high EQ, presence, gain and volume, and there is a noise gate control next to the main output level knob. Switches are provided to let you select dynamic or condenser microphone types and choose on-axis, off-axis, near or far microphone positions.

Pre-amp types include Solid State Clean Modeled after Fender Solid State combos, Vintage Clean (tube) Modeled after the Vox AC-30 top boost, Tube Clean Modeled after the Fender Super Reverb, British Crunch Modeled after the Marshall JCM-800, Modern High Gain Modeled after the Mesa Boogie Dual Rectifier, Solid State Lead Modeled after Fender Solid State combos, and Fuzz Modeled after a 60s fuzz pedal.

Choices of cabinet include Small combo Fender small 15W solid state combo (80s), Open Back 1×12 Ampeg AX-30 100W Tube Combo (90s), Open Back 1×12 II VHT Pitbull 50W Tube Combo (90s), Vintage Open 4×10 Fender Super Reverb vintage Amp (60s), Modern Closed 4×10 Marshall modern 4×10 closed cab (80s), Vintage 4×12 Closed Marshall vintage 4×12 cab (70s), Modern 4×12 Closed Marshall modern 4×12 cab (80s), British 2×12 Vox top boost AC-30 (60s) - or No Speaker D.I. line out (after the power amp).

The AmpliTube tremolo is an exact recreation of the Opto-Tremolo fitted to 1960s Fender amps and AmpliTube's spring reverb is the closest thing I have heard to the peculiar (but attractive) sound of the spring reverbs used in these.

The main set of amplifier presets provides plenty of settings to get you started, with a bias toward 'rock' type sounds. Interestingly, the presets also include a set

Figure 13.21 AmpliTube pre-amps and cabinets.

for bass players, a set optimized for use as insert or send effects, and various 'VIP' player settings.

When you have all this set up, you should have the guitar signal coming out from the microphone, which has been placed in front of the cabinet, with stomp effects feeding into the preamp. By this time your sound should be building nicely, and to round everything off you can apply some further effects using the Post FX

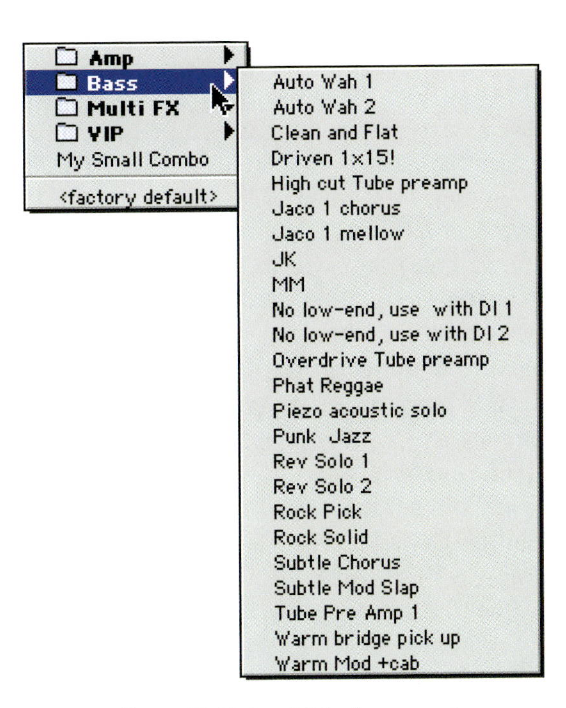

Figure 13.22 AmpliTube bass presets.

module. This has a three-band parametric equalizer, a digital delay and a high quality 'hall' type digital reverb – all of which produce a stereo output signal. So after this last stage the signal becomes stereo because of the Reverb and Delay processing.

Figure 13.23 AmpliTube Post FX.

Note The input signal to AmpliTube is always mono. If AmpliTube is inserted on stereo channels or buses it will sum the left and right input to get the mono input. The output signal is always stereo, so while it is not possible to open AmpliTube as a 'mono to mono' plug-in, it can be opened as a 'mono to stereo' or 'stereo to stereo'.

With TDM systems you can insert the HTDM plug-in on the track you intend to record onto, record enable the track, play your guitar and you will hear the plug-in. There is a short latency delay that you will have to put up with. Using a TDM MIX card this was just about acceptable for me – but not ideal. The best way is to record the audio using a straight sound, then change this afterwards. But then you don't get the 'vibe' of the actual sound – and you have to play the notes with a straight, clean sound for best results, which can affect the way you play. So putting up with the latency is probably the better compromise. You cannot use a 'live' input with the RTAS version on a TDM system because Pro Tools TDM bypasses RTAS plug-ins on a record-enabled track – although you can use AmpliTube RTAS with Pro Tools LE.

Cubase VST lets you record through AmpliTube – but you do have to get your system set up for lowest latency to avoid the annoying processing delay. If you use a low-latency card such as the RME Hammerfall or the Steinberg-badged Nuendo cards based on this, you should have no problems with latency. Logic

Audio versions 4.× work the same way as Pro Tools TDM RTAS systems – they won't let you hear the plug-in while recording – although the manual explains a somewhat cumbersome workaround for this.

AmpliTube's competitors are the long-established Amp Farm which is only available for Pro Tools TDM systems, Steinberg's Warp which is only available for VST systems, and Universal Audio's Nigel which is only available with the UAD-1 Powered Plug-in DSP card. These all have their own particular characteristics – just like when you buy different guitar amps. AmpliTube has the advantage of being available for both Pro Tools and VST systems and compares favourably with the competition.

Synchro Arts VocALign

VocALign does what its name suggests – it lets you align vocals. Developed originally for automatic dialogue replacement (ADR), it can also be used to match the timing of any audio signal to that of any other. It provides automatic analysis and time alignment of new or replacement audio to match the timing of original audio – leaving the pitch of the aligned audio unchanged. VocALign is the only software available that can automatically align the timing of double-tracks or other over-dubs on music recording sessions. VocALign works by applying varying amounts of time-stretching or compression to one signal, called the 'Dub', to make its energy peaks and troughs align in time with a 'Guide' signal.

ADR is particularly useful for foreign dialogue replacement or for any post-production work where original dialogue has to be replaced while maintaining lip-sync. VocALign really 'takes the sweat' out of this process, allowing dialogue to be replaced much more accurately and quickly than by other techniques. 'Time is money', as they say, and when highly paid actors have to be called in to replace dialogue, this can severely impact on any production budget. Anything that can be done to speed this process up is greatly valued. At least one UK film studio has a mobile studio van permanently parked on-site with a Pro Tools system and VocALign and a small voiceover booth installed. Whenever unusable location dialogue has to be replaced, this can be recorded in the van before the actors leave the studio lot. This avoids the logistics, and the accompanying expense, of bringing the 'talent' back into a studio at a later date to replace the dialogue.

VocALign can also be used in music production to accurately double-track lead vocals or to match up the backing vocals with the lead vocal. Similarly, you could use it to match instrumental tracks, or for a number of other alignment tasks such as sync'ing a drum machine's audio output to a real drummer's playing (or vice versa).

The VocALign Pro package includes a stand-alone version of VocALign; the AudioSuite plug-in, VocALign AS; and TimeMod, a high quality, linear time stretching and compression algorithm. In its present form, VocALign Pro has serious compatibility problems with Pro Tools and Mac OS X. There will be a new Pro version but its release date and feature set are not decided. Currently Synchro Arts encourages users to purchase either of the VocALign Project packages - neither of which includes TimeMod.

'VocALign Project for Pro Tools' is the Audio Suite version which is compatible with Pro Tools 5.0 and higher. It supports both PC and Mac versions of Pro Tools. The installer disk contains both versions and the same Product Key is used for both versions.

'VocALign Project' is for Pro Tools 5.0 and earlier, and Digital Performer 3.0 and higher. It also works on the PC, and supports Wave, Broadcast Wave, and operation via the Windows clipboard. The installer disk contains both PC and Mac versions and the same Product Key is used for both versions.

VocALign project

The stand-alone version of VocALign is available for both Mac and Windows computers. On the Mac it uses Sound Designer II files at 44.1 or 48 kHz and works with OS8.6 or higher. You can also use it with Digital Performer 3.0 or higher and Pro Tools versions 3.2-5.0.

Note VocALign Project is an application, not a TDM plug-in, and therefore does not require additional Digidesign hardware. It is not compatible with Pro Tools software later than version 5.0.

It also works stand-alone on any Windows systems, and in this case it uses Wave or Broadcast Wave files at 44.1 kHz and 48 kHz. It can also be used via the Windows Clipboard with Syntrillium Cool Edit Pro or Sonic Foundry Sound Forge, and can be integrated with Digigram systems using X-Track Audio Suite version 3.21 or higher.

Operation is similar in all cases. Load in both your Guide and your Dub (the one you want to match to the Guide) audio files. Then VocALign lets you create a new aligned audio file that you can use to replace the guide in your project. This new file can be automatically positioned in supported third-party editing software.

Using the Mac this is very straightforward. First you launch the VocALign application, making sure that the audio from your Apple Sound Manager is routed to your monitors.

Note If you have a Digidesign audio card installed, you can route the audio from the Sound Manager via this.

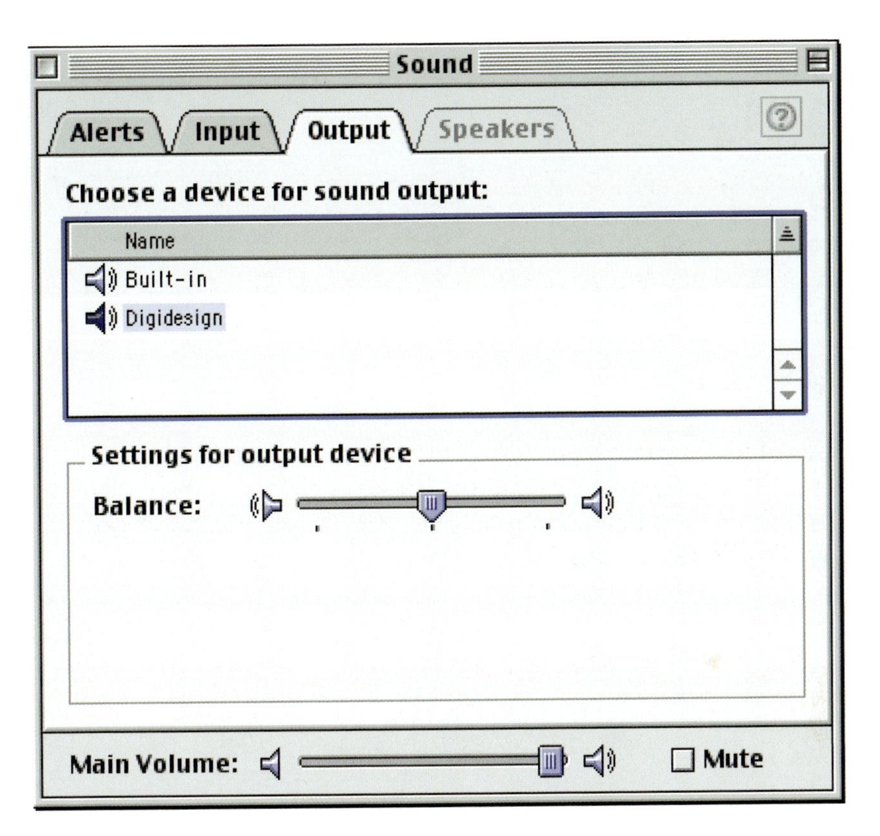

Figure 13.24 Apple Sound control panel.

From the Finder you can drag-and-drop the file you want to use as the guide onto the upper black panel in VocALign Project. Then drag-and-drop the file you want to match to this onto the lower black panel. You can audition the files that you have selected using the Play menu or by hitting the 'f' key on most Macintosh keyboards. To line these up, just hit the Align button and you will see a trace of the aligned audio above the Guide waveform. The peaks and troughs should match those of the Guide quite closely.

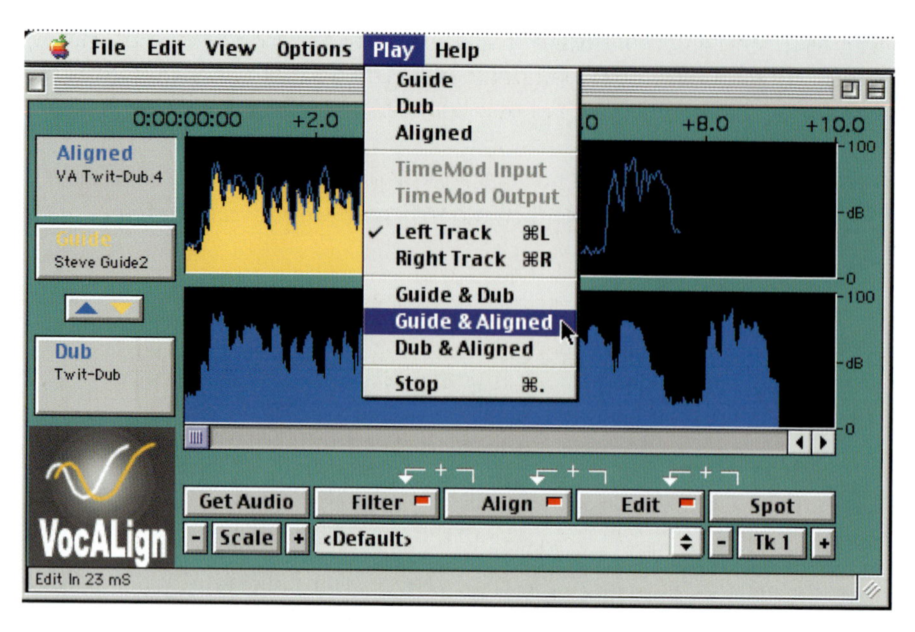

Figure 13.25 VocALign Project stand-alone software.

You can audition the Aligned and Guide files together, using the PLAY menu or by hitting the 'e' key on most Macintosh keyboards. To capture the aligned audio into a sound file, just press the Edit button. This creates a new sound file on your hard disk (normally in the same folder as the Dub audio is located). You can now replace the guide audio in your audio editing software with this new file. And if you are using Digital Performer 3.×, for example, this can be done automatically for you.

TimeMod

TimeMod is available as part of the Macintosh version of VocALign Pro and is designed to provide an easy to use, high-quality audio stretching and compression utility. TimeMod can work with Pro Tools or in a stand-alone mode on SDII files.

TimeMod is a non-real time, time compression/expansion process with the following features: it will process user-selected mono or stereo audio in Pro Tools and create new files with the required duration. It can automatically name the new files and spot them as regions back to user-selected Pro Tools tracks at the original signal's start or stop time. The user can quickly specify the required changes by setting a new start or end timecode, duration, ratio or change in the TimeMod window. TimeMod generates user-specified invisible 'handles' which will be added to both ends of the selected audio and processed with it, so cross-fades and extensions can be added later. Stereo is processed without phase shifts between the two output channels and no pitch change is introduced.

It is a linear process (as opposed to VocALign which is non-linear) - so the time modification is uniformly spread throughout the signal and maintains rhythms accurately.

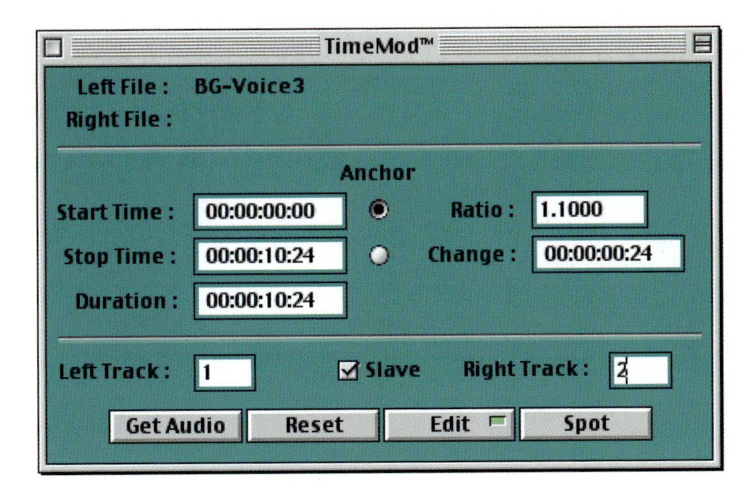

Figure 13.26 Synchro Arts Timemod.

TimeMod is operated similarly to VocALign. The user selects mono or stereo audio from within a region in Pro Tools and presses a macro key combination to capture the region information from Pro Tools and access the TimeMod window. There, the user enters the desired time modification information and presses a button to create a new audio file with the desired length audio. Hit the 'i' key to listen to the Input audio and hit the 'o' key to listen to the output audio. This file can then be named and spotted back into a selected Pro Tools track automatically.

In stand-alone mode, you can open any mono or stereo SDII format audio file into TimeMod using the Get Audio button. This open a standard 'Open File' dialog box that lets you navigate through your files and folders.

VocALign AS

VocALign AS is an AudioSuite plug-in that provides perfect integration with Pro Tools. With the audio source material already on hard disk, the user selects the original 'Guide' and new 'Dub' audio regions in Pro Tools and passes the regions to VocALign. VocALign aligns the timing of the Dub to the timing of the Guide then returns the Aligned audio directly to any chosen Pro Tools track with sample accuracy.

When you select the VocALign plug-in from the AudioSuite menu in Pro Tools, a window appears with an upper display area for the Guide track and a lower display area for the Dub track. To load a region into either of these, first select it in the Pro Tools Edit window then press the Capture button in VocALign – located below the display area. Buttons to the left of the VocALign display areas let you choose whether the Guide or Dub display will receive the audio region when you press the Capture button. With both Guide and Dub loaded into VocALign, the next step is to select a destination track for the Aligned audio using the popup selector just above the display.

Now you can press Align to generate an Aligned Dub trace that lets you visually check alignment. Assuming that this all looks ok, you just need to press Process and the Aligned audio will be sent to the selected destination track in Pro Tools. To check that this worked, just mute the Dub track in Pro Tools and play back the Guide and the Dub – which should now match closely.

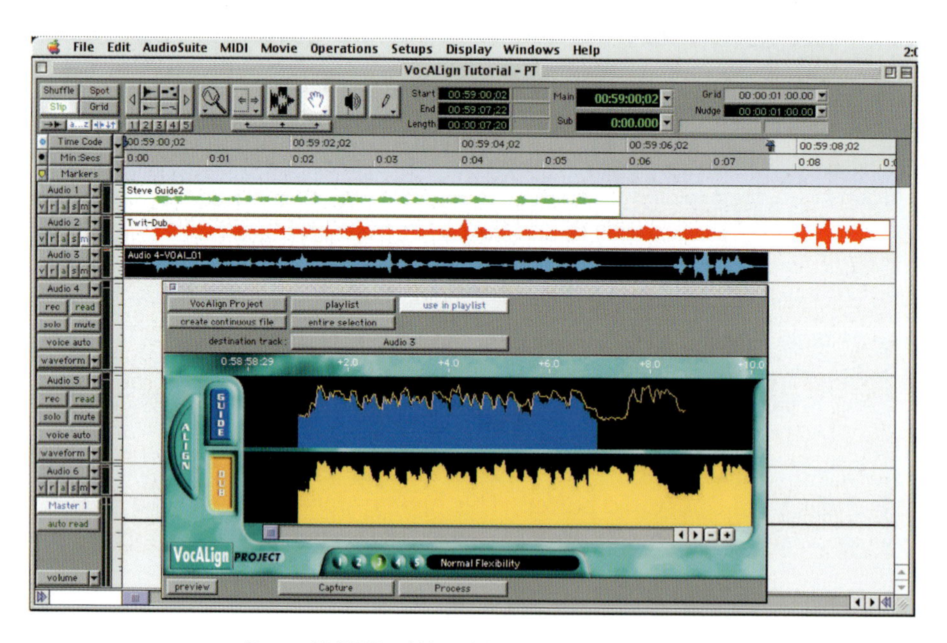

Figure 13.27 VocALign AS session in Pro Tools.

VocALign AS works with any Digidesign-approved Mac OS Pro Tools or Pro Tools LE system running 5.0 or higher software and it supports both 16- and 24-bit audio with sample rates up to 192 kHz.

Antares Kantos

Kantos - now there's a name to conjure with! But what is it? Well, it's a software synthesizer that uses the subtractive synthesis method that many people are familiar with. So it has oscillators, filters, LFOs and the like. So why the mysterious name? Well, this is a synth with a difference! With most synthesizers you have to hook up a MIDI keyboard to play them. This is fine if you play keyboards - but not a lot of good if you are a singer or a guitarist! Andy Hildebrand, the person who designed Auto-Tune and inadvertently unleashed zillions of Cher-style vocal clones onto an unsuspecting world, has now come up with something to let non-keyboard-playing musicians join in the synthesizer plug-in fun. Kantos works by extracting the pitch, dynamics, harmonic content and formant characteristics from audio input such as a vocal or a guitar melody. This information is then used to control the Kantos synthesizer - transforming the input into synthesizer melody lines.

Figure 13.28 Kantos user interface.

The Kantos synthesizer has two wavetable oscillators (each with its own resonant filter and chorus generator), a noise generator, two envelope generators, two LFOs, and a delay effect. The unique Articulator takes the harmonic content and formant information from the input and applies this to the output of the oscillators and noise generator, producing effects similar to a vocoder. The

Noise Generator is a broadband noise source with its own dedicated multimode resonant filter. It's useful for adding sibilance to a patch, and for passing through the Articulator to create a unique 'whispering' effect. Using the Modulation Matrix, any source can control multiple destinations, and any destination can be controlled by multiple sources. You can route the two multi-waveform LFOs using this matrix and the frequency of these LFOs can be set manually or using the Tap Tempo feature. Two ADSR Envelope Generators are also included as modulation sources that can feed destinations in the Modulation Matrix. And the flexibility doesn't stop there! Normally, a preset's dynamics are derived from the dynamics of the input audio, but the Amplitude Envelope also has an 'On' button that allows the envelope to control dynamics instead. To help to create rhythmic effects, Kantos includes a delay line with variable feedback. The delay time can be set in absolute time or, via the Tap Tempo function, in bpm. The Tempo Control section lets you set the delay time and the LFO 1 and LFO 2 frequencies. The submixer lets you balance the oscillators and noise generator with two additional sine wave oscillators, while the main mixer balances the main synth output, the delay line return, and the original unprocessed input audio.

Tips The manual has three excellent suggestions for how to use Kantos:

1. Use a drum loop as input. Tune one oscillator low and constrain to one note for a bass pulse. Constrain the other oscillator to three or four notes for rhythmic pseudo-arpeggiation.
2. Edit a bit of speech so that it loops in tempo with your song and use it to drive Kantos. Program some slow modulations in the Mod Matrix so that the loop evolves in interesting ways over time.
3. Ever record a synth lead or bass part that sounded fine at the time, but in the mix it didn't quite work with the rest of the parts? Set your synth bass to a clean, single-oscillator sine wave patch with the desired dynamic characteristics (e.g., velocity and envelope control). This type of audio is ideal for driving Kantos. Record this part into your hard disk recording program. When it's time to mixdown, set Kantos to create the desired sound, and drive it from the simple sine wave track.

Kantos is available as an RTAS, MAS or VST plug-in that you insert into the channel that is playing the audio you want to use to control Kantos. It works best with monophonic source material (one note playing at a time) and can follow pitch-bends. You can also apply the pitch constrain feature to limit the output to the notes of a particular key or scale to produce melodies from polyphonic or unpitched input such as chords or drum loops. Shades of Auto-Tune here!

The user interface is quirky but easy enough to use. Pick your preset, tweak a few controls to adjust the triggering, making sure there is enough input level, for example, and tweak the synthesizer controls to your taste. A noise gate is conveniently provided to let you clean up any background noise in the control audio.

I used a short melodic improvisation on jazz guitar as my test. This was expressively played with plenty of slides between notes, several pulloffs, and lots of dynamics in the playing. I chose the Mellow Tone preset and Kantos made a fair job of tracking the pitches and the slides - often referred to as 'portamento' in synthesizer jargon. But the faster notes either disappeared or mis-triggered the synthesizer. Then I tried a less well-defined preset - the WhisperTron (sounds like something Woody Allen might have invented). This worked better - because this sound didn't need to track the input so precisely. The moral of this story is that you will need to use very clean and well-defined input material for best results.

If you like to fiddle with software to find unique effects, then Kantos is for you. It won't magically transform everything you sing or play into an accurate synthesized version, but if you take care with the input material and settings it will. And you can definitely come up with unusual sounds.

Cycling '74 Pluggo

Pluggo is a VST plug-in that lets you open other plug-ins within it. A set of plug-ins created using the Max/MSP programming environment is included to get you started, along with the stand-alone Plug-in Manager software that lets you organize your plug-ins more effectively.

BASICBEAT1•
BASICBEAT2•
BASICBEAT3•
BASICBEAT4•
BASICVOCAL1
BASICVOCAL2
BASICVOCAL3
BASICVOCAL4
BASICVOCAL5
BASICVOCAL6•
BassMaker•
BeatKrickets
BigBottomBeat•
BreathyCluster•
BubbleDub•
Cataclysm•
CatScratch•
Cymtonosity•
DeepAmbience
DeepAndDoomy
DescendingSirens
DroneSolo
DroneyFifths
DrumBassMix•
EdgySynth
IndustrialRezPad•
JuicyVowelDrone•
LaughingSwarm•
MellowTone
MonsterOverdrive
NasalTracker
PlusStrings
PunchySawtooth
QuiveringPulse
RampWalkerBass•
RepeatHarmonics•
ResonantGrizz•
ScreemingMeemies
SpaceBassEcho
SpectraFlangey
SteamTransmission
Subwoofer
SuspendedSweeps•
SweepingPad
SweepingRez4ths•
TonalTexture•
UltraSubKick
UnderwaterReznanz•
VoiceOctaves
WallOfSound•
WhiplashEcho•
√ WhisperTron•

Figure 13.29 Kantos factory settings menu.

You can open any VST plug-ins that you already have into Pluggo to take advantage of its special features. For example, Pluggo enables plug-ins to send audio and/or control information to each other and various modulator plug-ins are included that you can use to control the parameters of your other VST other plug-ins. Also, Pluggo has an enhanced user interface for working with plug-ins that provides undo, parameter randomization, and display of all plug-in parameters as egg sliders with coarse/fine control. Pluggo will also open any plug-in made with the Max/MSP plug-in development tools - essential for plug-in developers using these tools.

There are actually two plug-ins - Pluggo and PluggoSynth - that appear in the VST Instruments menu in Cubase or Audio Instruments in Logic. Pluggo is for signal processing effects and PluggoSynth, as its name suggests, is for synthesizer plug-ins. Pluggo works from within any VST-compatible host software, not just Cubase and Logic, and a version is supplied that works with MAS-compatible hosts as well. So with Pluggo you can use VST plug-ins within Digital Performer! Pluggo installs its plug-in files into the VstPlugins folder inside the Digital Performer folder; these Pluggo plug-ins and any VST plug-ins that you put into Digital Performer's VstPlugins folder can then be accessed from the pop-up plug-ins menu in Digital Performer's Mixing Board.

Note Pluggo cannot host MAS plug-ins, nor can it modulate the parameters of MAS plug-ins.

There are more than 70 effects provided with Pluggo - everything from 'Rough Reverb' to 'Warble'. These are of the cheap and cheerful variety (which you ought to expect at less than a dollar per plug-in), but they certainly represent value for money - and you might just find that extra special effect you have been searching for among them!

You can use any VST plug-ins (or MAX files) with Pluggo, as long as they are in the appropriate VST plug-ins folder. When you insert Pluggo, a dialog box comes up from which you can open VST plug-ins, Max patcher files, or Max collective files. If the plug-in was made with Max/MSP, opening it in the Pluggo plug-in is no different from opening it directly as a plug-in from the sequencer's pop-up menu of effects. If the plug-in has its own edit window, this interface is displayed first when you open its Pluggo-hosted edit window. The parameters window is also available and you can switch between these using the View menu. If you switch to the parameters window, the plug-in's parameters are displayed using Pluggo's 'egg slider' interface.

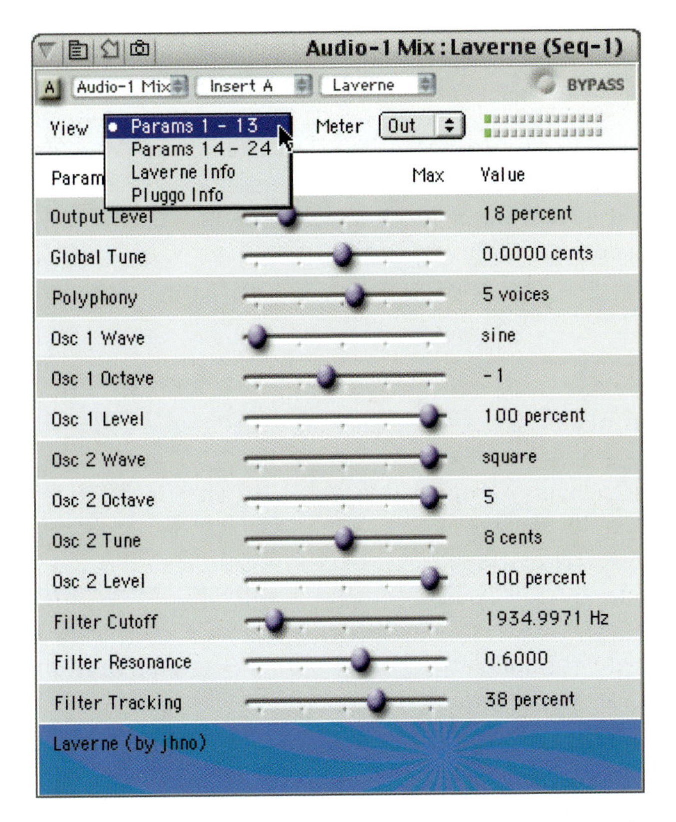

Figure 13.30 Pluggo Laverne plug-in showing the 'egg-slider' interface.

So Pluggo, priced around $75, not only provides you with a bargain bagful of potentially very useful plug-ins, but also offers useful additional features for anyone working with plug-ins, whether in VST or MAS host environments.

Cycling '74 Plug-in Manager

The Plug-in Manager software, supplied as part of the Pluggo package, is an extremely useful utility that lets you manage any plug-ins – not just Pluggo or VST plug-ins.

When you launch the software, select Manage New Folder from the File Menu, to open the main window. This is more than a little similar to the standard Extensions Manager software that you will probably be familiar with from Apple's OS 9.×.

In the main window, you can look at a particular folder full of plug-ins and turn on or off the ones you want to be active. The ones you de-select are moved to a

Figure 13.31 Plug-in Manager splash screen.

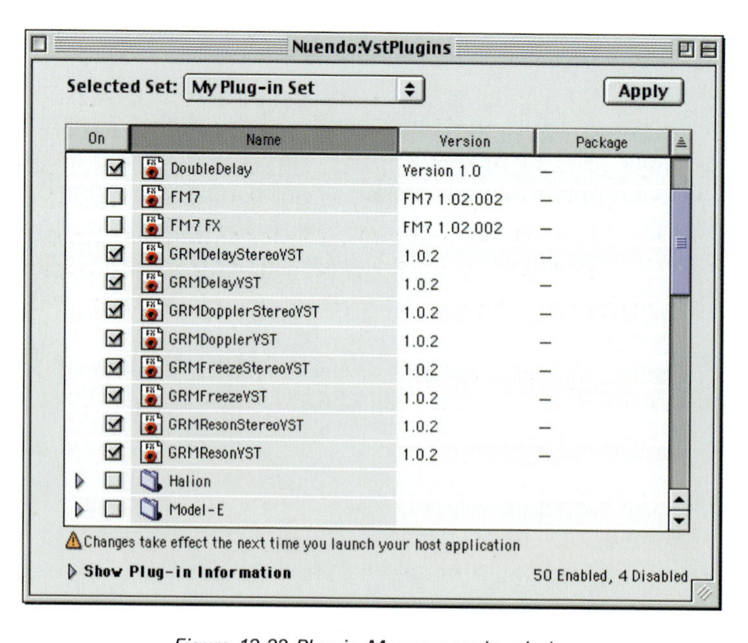

Figure 13.32 Plug-in Manager main window.

Plug-in Disabled folder. As with Extensions Manager, you can have various sets of these.

Plug-in Manager is a great utility to have - even though you can always create a Plug-ins Disabled folder and move your plug-ins manually. I used to move my system extensions manually until I got used to using Extensions Manager in OS9 - and I almost never do this manually nowadays. Having a utility to organize this really does save you time - which is always a precious commodity!

Chapter summary

So, if you use lots of synthesizers and you need synthesizer editor/librarian software, Sound Diver makes an excellent partner for your sequencing software. Then there are several 'tools' that remixers will find extremely useful: Melodyne lets you work wonders with vocals, while Phrazer and Alkali are ideal for working with loops. Time Factory lets you work with vocals or loops, changing the pitch or the tempo while taking formants into account when necessary.

When the time comes for mastering your mixes, T-RackS gives you a complete set of mastering tools in either stand-alone form or as a plug-in suite. And guitarists will love AmpliTube! Post-production engineers will probably be aware of VocALign already, but recording engineers and producers will also find this useful for doubling vocals or matching the timing of harmony vocals with lead vocals.

Kantos is going to be interesting to the more experimental musicians and composers, producers and remixers. Its synthesizer sounds can be intriguingly different, thanks to its unique Articulator and comprehensive modulation features, and it can track the pitch and timing of acoustic input extremely well. Cycling '74's Pluggo comes with a bunch of useful signal processing plug-ins, but is even more useful as a 'host' environment for VST plug-ins that can itself plug in to your favourite audio software - allowing VST plug-ins to be used within Digital Performer, for example.

14 DSP Cards

The more complex plug-ins can seriously occupy the attention of your computer's CPU, which is why some of these are designed to take advantage of the second processor in dual-processor computers. If you have a Pro Tools TDM system, the TDM plug-ins use the DSP on your TDM hardware instead. And both Logic Audio and Digital Performer can use TDM plug-ins with the TDM hardware. But Cubase VST and Nuendo cannot use the TDM cards, so two companies - TC|Works and Universal Audio - have developed PCI cards containing processors dedicated to running VST plug-ins.

Universal Audio are perhaps best known for their re-issues of classic hardware such as the Urei 1176 and the Teletronix LA-2A. What is not so widely known is that Universal Audio have entered the plug-ins market - and have taken over a software company, Kind of Loud, with a good reputation for developing high-quality plug-ins. The Universal Audio Powered Plug-ins system includes a PCI card to take care of the signal processing, along with a suite of VST plug-ins that run on this card to provide reverb, guitar effects and comprehensive channel strip effects, plus high-quality 1176, LA-2A and EQP-1A emulations.

TC|Works is the software division of TC Electronic, based in Hamburg, probably best known for TC|Spark editing software and TC|Tools plug-ins. The TC|Works PowerCore is a PCI card containing signal processors that comes with TC|Tools plug-ins (comprising TC's Megaverb, Chorus/Delay and EQ Sat) along with TC Voice Tools, TC Vintage CL, the TC Compensator and the TC 01 Virtual Synthesizer.

Universal Audio UAD-1 DSP card

Universal Audio's suite of Powered Plug-Ins run on the UAD-1 PCI card, available for both Mac and PC. As well as vintage compressors (emulations of UA's own classic 1176LN and Teletronix LA-2A compressors), the Powered Plug-Ins package also includes an emulation of the classic Pultec EQP-1A, along with Kind of Loud's RealVerb Pro, a Channel Strip specially developed for the card called the CS-1, a rack of guitar effects named 'Nigel', and the UAD-1 PCI DSP card itself. The plug-ins are supplied in VST format so they can be used with Cubase VST, Nuendo or Logic Audio, for example, and Powered Plug-ins for MAS is currently in development.

The main advantage of using the UAD-1 card is that it reduces the processing load on your CPU by providing a set of useful plug-ins that run on the card's DSP. With the card taking care of a significant part of your signal-processing requirements, your host application will be able to cope with more tracks, more automation, and more native effects. And there is plenty of DSP power on this card to share out between the plug-ins. For example, you could use 8 RealVerb Pros or 32 EQs, 16 compressors and 3 RealVerb Pros - lots more than you are likely to be able to use if you are just relying on your computer's CPU. The DSP on the card is optimized to provide even greater quality of signal processing than you are likely to get when using the computer's CPU, and there is artefact-free smoothing on all parameters - so you get no zipper noise when adjusting the controls. Technically speaking, the 7'' form factor PCI card uses bus mastering direct memory access (DMA) for zero host load and maximum sustained host-card transfer rate. It is fully PCI 2.1 compliant and supports 66 MHz operation with fast bus timing.

Note The plug-ins all support 24-bit 96 kHz operation, although you do have to bear in mind that plug-ins running at 96kHz use twice as much of the UAD-1's processing power as those running at 48kHz.

All the parameters can be automated and the card uses floating-point arithmetic for high quality results. And the quality 'shines' through when you use the 1176 or the LA2A. These respond with all the subtlety of the original hardware units - but with pristine digital quality. The Channel Strip contains a virtual 'rack' of processors that you can use for both technical and creative processing. These can be loaded as one large rack or as modules containing just the processes you want. Similarly, Nigel includes the most common types of signal processing that guitarists use and can be loaded as a rack of processors or as individual modules.

Installation of the card is very straightforward, with clear instructions provided. Once installed, you can check that everything is working OK using the supplied UAD-1 Performance Meter software application. This displays the current CPU and memory status of the UAD-1 DSP hardware card in real time. It also contains system information and configuration windows that enable you to confirm that the UAD-1 is functioning properly, to check the version of the software drivers, and to adjust the UAD-1 buffers. The Performance Meter user interface consists of a single, small floating window. You can keep this running at the same time as you are running the host application in which you will use the plug-ins - so you can monitor the resource load of the UAD-1.

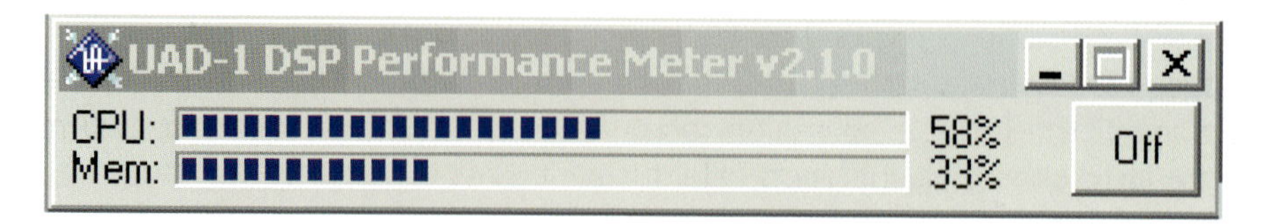

Figure 14.1 UAD-1 DSP Performance Meter.

Pultec EQP-1A Equalizer

The classic Pultec EQP-1A has been extremely faithfully modelled by Universal Audio. The front panel closely matches that of the original, so all the controls are where you would expect them to be. They could not be much simpler to understand and operate - just take a look at the front panel to see what I mean. You get high and low switchable-frequency filters with a single Bandwidth control, and two sets of Boost and Attenuation controls, an in/out switch, an on/off button with an associated LED - and that's about it!

The unique feature of this equalizer is the way that you can both boost and attenuate the filters - adding to its characteristic audio 'signature'. The EQP-1A can often work 'miracles', shaping the sound of the voice or instrument so that it blends well in the mix, yet sounds clear and well-defined.

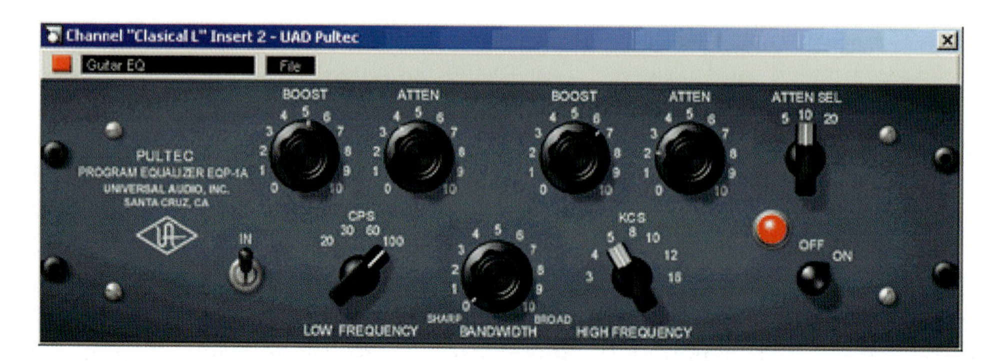

Figure 14.2 UAD Pultec EQP-1A.

UAD 1176LN Limiting Amplifier

The 1176LN Limiting Amplifier is one of the most sought-after analogue processors used in recording studios today. A 're-issue' hardware version is currently being manufactured by Universal Audio. The 1176LN is known for bringing out the presence and colour of audio signals, adding brightness and clarity to vocals, and adding 'bite' to drums and guitar. This plug-in is modelled after the 'blackface' D and E revisions. Rotary knobs are provided for Input and Output level and for Attack and Release times for the limiter. Pushbutton switches are provided to let you select the four compression ratios - 4:1, 8:1, 12:1 or 20:1. The VU meter also has a column of pushbutton selector switches.

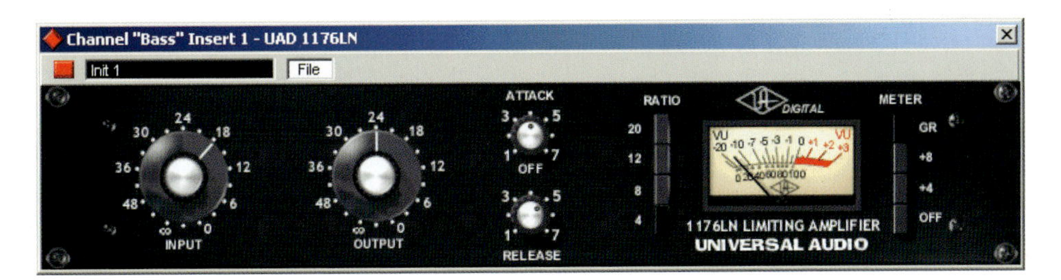

Figure 14.3 UAD 1176LN Limiting Amplifier.

The user interface is simple and uncluttered, although you do need to understand how the controls work before you will get the results you want. For example, the Input control actually adjusts the threshold and the amount of gain reduction. The Attack control lets you define how quickly the compressor responds once the threshold has been exceeded. Rotate this control clockwise to set faster attack times in its range from 20-800 microseconds. Release is the opposite - i.e. how fast the compression action ceases to be applied once the input signal falls below the threshold. Again, you rotate the control clockwise to set faster release times in the range from 50 to 1100 milliseconds. Slower release times can be useful for material with frequent peaks, but it is always a 'juggling act' to set the optimum value for release.

The four pushbutton switches to the left of the VU Meter determine the compression ratio. Ratios of 20:1, 12:1, 8:1, and 4:1 are provided. The 20:1 and 12:1 settings are typically used for peak limiting, while the 4:1 and 8:1 settings are used for general compression. Just like the hardware version of the 1176LN, you can have all the Ratio buttons switched in at the same time. In this mode, the ratio is around 12:1, the release happens faster, and the shape of the release curve changes. With lower amounts of compression the attack is delayed slightly, as there is a slight lag before the attack attenuates the signal, and the attack value

Figure 14.5 The UAD CS-1 Channel Strip.

Note The EX-1M is a monophonic version of the EX-1 that enables independent left and right EQ settings in master effects chains. For example, the EX-1M requires half the processing power compared to that of EX-1 when used on a mono audio track within Logic Audio. So the EX-1M should be used on monophonic audio tracks within Logic whenever possible to conserve the card's DSP resources for use with other plug-ins.

Figure 14.6 The UAD EX-1 Equalizer Compressor.

Figure 14.7 The UAD DM-1 Delay Modulator.

When the Mode is set to one of the delay settings, the maximum delay is 300 ms and when set to one of the chorus or flange settings, the maximum delay is 125 ms. In the Flanger modes, the L and R delay controls have slightly different functions than in the chorus modes. The high peak of the flanger is controlled by the settings of the L and R delay controls, while the low Peak is determined by the setting of the Depth control.

The Recirculation (RECIR) knob sets the amount of processed signal fed back into its input. Higher values increase the number of delays and intensity of the processed signal. The Damping knob controls a low pass filter that reduces the amount of high frequencies in the signal. Lower settings reduce the brightness; higher settings yield a brighter signal. Damping also mimics air absorption or the high frequency roll-off inherent in tape-based delay systems.

Note If you need delay times longer than 300 ms you can use the DM-1L plug-in instead. The DM-1L has a maximum time of 2400 ms per channel and includes a right-delay and left-delay line Ratio Link switch so it uses significantly more of the UAD-1's memory resources than the DM-1.

UAD RS-1 Reflection Engine

The UAD RS-1 Reflection Engine works in mono or stereo and provides up to 300 ms of delay per channel. Room size is adjustable from 1 to 99 metres. The wide range of delay presets includes single echo, pattern echo and spatial room simulations. These room shapes and simulations were developed in conjunction with NASA - so they should be good (and they are). Special effects include forward and reverse gated reverb.

The RS-1 Reflection Engine simulates a wide range of room shapes, and sizes, to drastically alter the pattern of reflections. While similar to that of the RealVerb Pro plug-in, the RS-1 does not offer the same breadth of features (such as room hybrids, room materials, morphing, and equalization). However, if you do not need the advanced capabilities that RealVerb Pro offers, you can use the RS-1 to achieve excellent room simulations, while also preserving DSP resources on the UAD-1 card.

Figure 14.8 The UAD RS-1 Reflection Engine.

The Delay control sets the time between the direct signal and the first reflection. The Size parameter controls the spacing between the reflections. The Recir control affects the amount of reflections that are fed back to the input and controls how many repeats you hear.

RealVerb Pro

RealVerb Pro, originally developed for Pro Tools TDM systems, is now available for MAS and as a Powered Plug-in for VST-compatible host software. You can design your own custom rooms, controlling the shape, size and materials, and adjust the room size from 1 to 99 metres. Real-time blending and morphing capabilities make this one of the most powerful reverb units for creating those extra special effects. Full control of early reflections and late-field reverberation is provided, along with diffusion control.

The user interface is one of the best I have seen for controlling reverberation parameters. It is divided into distinct panels, each with a different function. For example, the panel running along the top of the window has a popup menu to let you select from the 30 factory presets.

You can set the pattern of early reflections in the reverb using the room shapes and sizes parameters in the Shape panel. RealVerb Pro actually lets you specify two room shapes and sizes that can be blended to create a hybrid of early reflection patterns. There are 15 room shapes available, including several plates, springs and classic rooms, and all the parameters can be adjusted in real time without causing any disturbance in the audio.

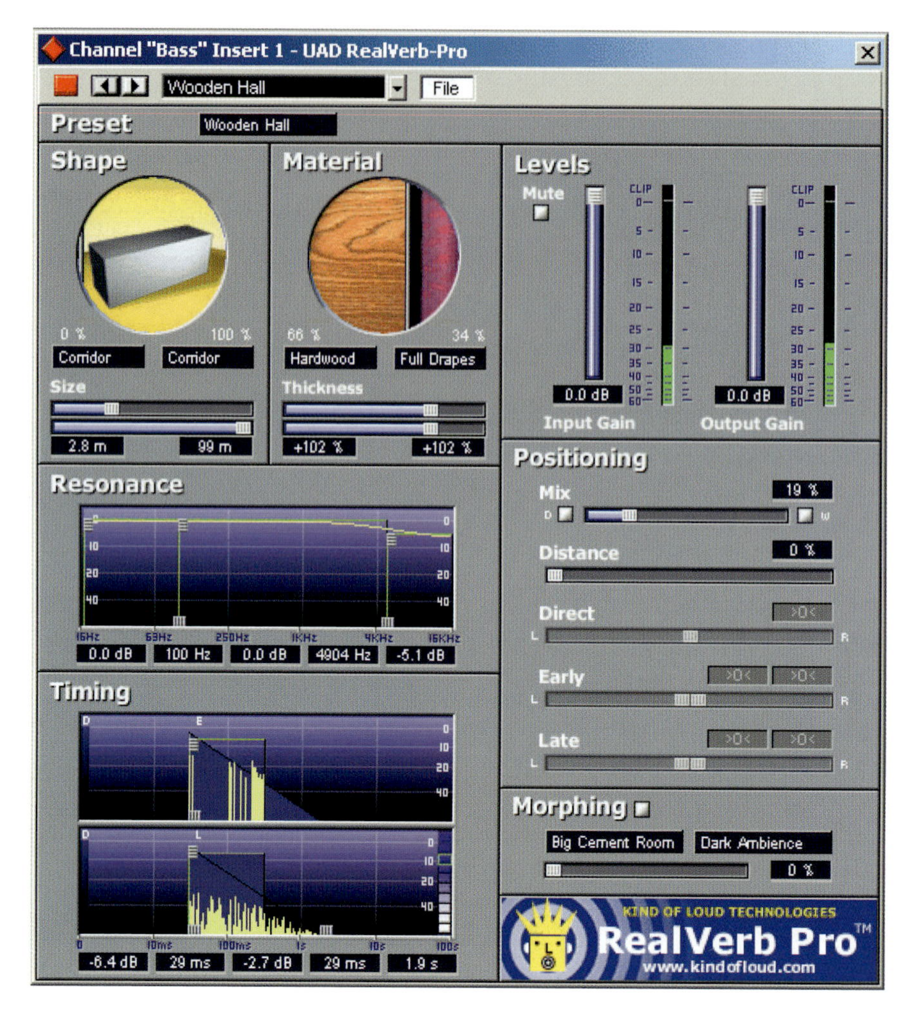

Figure 14.9 The RealVerb Pro.

The Material panel lets you specify two room materials with independent thicknesses that can be blended to create a hybrid of absorption and reflection properties. For example, to simulate a large glass house, a blend of glass and air could be used. There are 24 real-world materials provided, including brick, marble, hardwood, water surface, air and 'audience', along with 12 artificial materials – all with predefined decay rates.

The Resonance panel has a three-band parametric equalizer that can control the overall frequency response of the reverb, affecting its perceived brilliance and warmth. By adjusting its Amplitude and Band-edge controls, the equalizer can be configured as shelf or parametric EQs, as well as hybrids between the two.

The Timing panel offers control over the timing and relative energies of the early reflections and late-field reverberations. These elements affect the reverb's perceived clarity and intimacy. The early reflections are displayed at the top of the Timing panel, with controls for Amplitude and Pre-delay. The late-field reverberations are displayed at the bottom, with controls for Amplitude, Pre-delay and Decay Time. To illustrate the relation between both reverb components, the shape of the other is represented as an outline in both sections of the Timing panel.

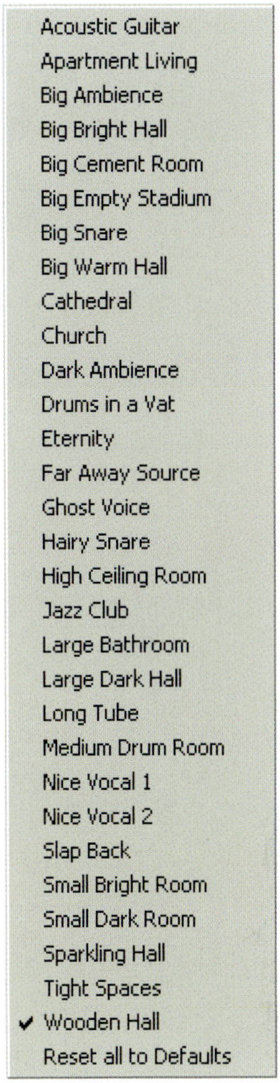

One of the unique features of RealVerb Pro is the ability to separately position the direct path, early reflections, and late-field reverberation. The Position panel provides panning controls for each of these reverb components and the Distance control adjusts perceived source distance - allowing much more realistic synthesis of acoustic spaces.

All RealVerb Pro controls can be varied continuously to create smooth transitions between selected values. This capability enables RealVerb Pro to morph among presets by transitioning between their parameter sets. This approach is in contrast to the traditional method of morphing by crossfading between the output of two static reverberators. The method employed by RealVerb Pro produces more meaningful intermediate states.

There are two popup menus in the Morphing panel - one to select each preset. Underneath these, a slider is provided to set the morphing balance between the two presets. Note that when RealVerb Pro is in morphing mode, the other RealVerb Pro spectral controls are grayed out and cannot be edited.

Figure 14.10 RealVerb Pro presets menu.

Finally, using the Levels panel you can check the input and output levels on the meters and adjust these using the sliders provided.

Powered Plug-ins 'Nigel'

'Nigel' is a guitar effects processor with modules for Gate/Compressor, Phasor, Mod Filter, Preflex (pre-amplifier) and Cabinet, Trem/Fade, Mod Delay, and

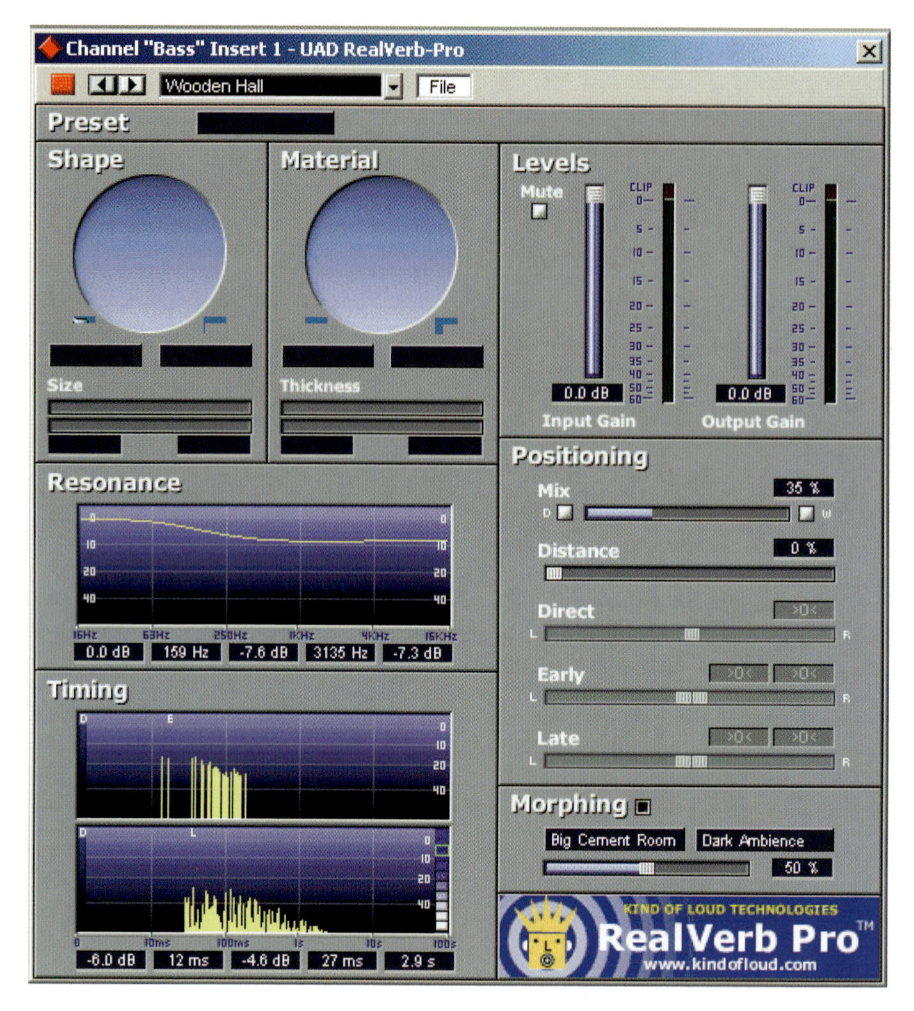

Figure 14.11 RealVerb Pro Morphing.

Echo. Rumour has it that this plug-in is named in honour of Spinal Tap's Nigel Tufnel - which sounds about right!

All of the modules are included in one plug-in - called Nigel. Various of these are also available as separate plug-ins, or in smaller combinations, as described below.

Preflex

The Preflex module can be thought of as the 'heart' of Nigel. Preflex features user-updatable amp models and offers continuously variable morphing between any two amp types. The Preflex module is subdivided into three sections - the

Figure 14.12 Powered Plug-ins Nigel.

Figure 14.13 Nigel Preflex and Cab.

Tip With the PreEQ Switch, you can change the order of the EQ and Compressor block. By default, the Compressor block is behind the EQ, but if you want EQing after you have compressed the signal, just switch on PreEQ.

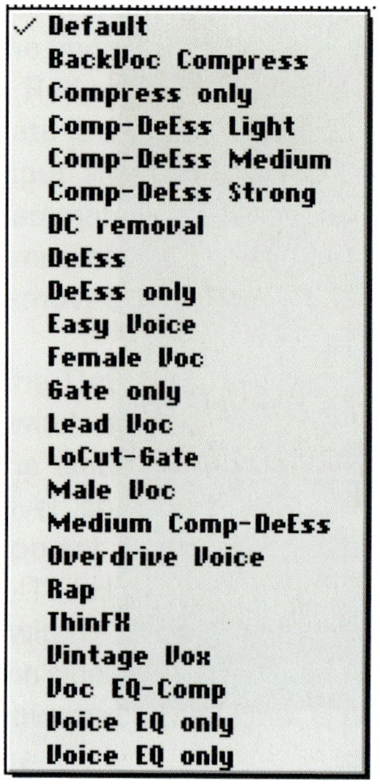

✓ Default
BackVoc Compress
Compress only
Comp-DeEss Light
Comp-DeEss Medium
Comp-DeEss Strong
DC removal
DeEss
DeEss only
Easy Voice
Female Voc
Gate only
Lead Voc
LoCut-Gate
Male Voc
Medium Comp-DeEss
Overdrive Voice
Rap
ThinFX
Vintage Vox
Voc EQ-Comp
Voice EQ only
Voice EQ only

Figure 14.31 TC|Works PowerCore VoiceStrip presets.

Once you have made all your settings, you can always save these and use them again and again if they suit. The VoiceStrip also comes with a bank of 22 presets that can serve as useful starting points. However, all the audio I have edited over the last 20 years has needed individual attention every time when setting the EQ or compression levels – and it's likely that you will find the same.

TC Vintage CL

The Vintage CL is a general-purpose compressor designed to have the characteristics of a typical 'classic' analogue model.

Note The Vintage CL uses a different compression algorithm from the one used in the TC VoiceStrip.

The user interface is stylistically similar to that of the VoiceStrip, with a pair of large meters at the top left of the window and rotary controls for input and output levels to the right of these. The Compressor section has a switch for Soft Knee mode as well as the usual controls for Threshold, Ratio, Attack and Release. The Limiter section has a switch for the SoftSat feature with an associated SoftSat amount control, along with controls for Threshold, Release and Ceiling.

Note You can run up to six mono or four stereo versions of the Vintage CL plug-in at 44.1 kHz on one PowerCore DSP chip.

Clip
In
Out
-16
dB
-14
dB
INPUT
OUTPUT
Left
Right
Left
Right
0.0dB
0.0dB
0.0dB
0.0dB
SOFT COMPRESSOR
FULL RANGE LIMITER
Red.
Threshold
-40.0dB
Ratio
2.5:1
Soft Knee
Mode
Soft Clip
Limit
SoftSat
0.0dB
SoftSat
Active
Attack
14ms
Release
140ms
Output
+8.0dB
Threshold
-0.1dB
Release
0.20s
Ceiling
0.00dB
TC | POWERCORE CL 96K
Ver.3.2
TC | WORKS
POWERCORE
DISABLED

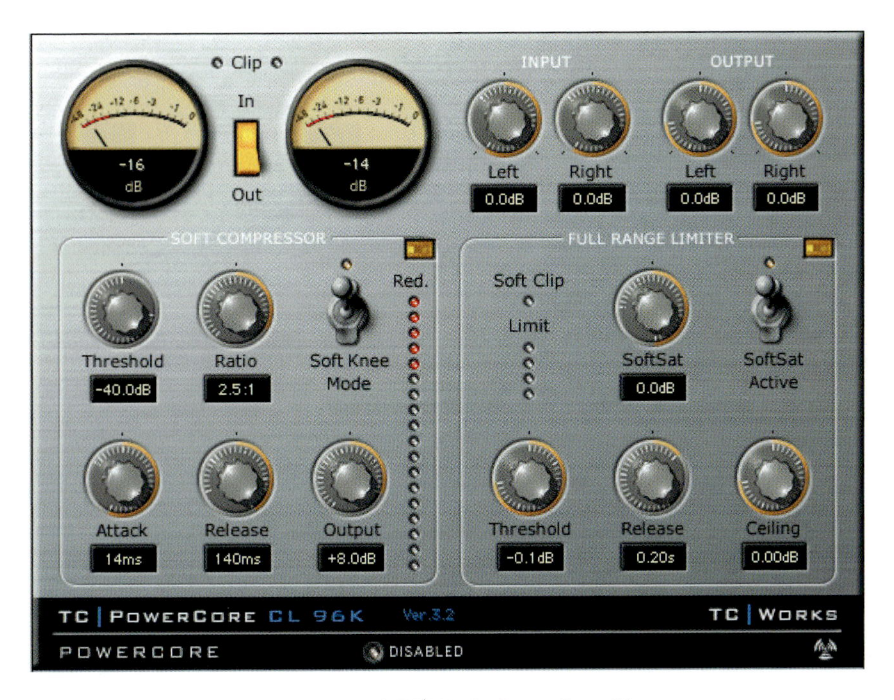

Figure 14.32 TC|Works PowerCore CL.

A selection of presets is provided that will put you 'somewhere in the right ball park' for different styles of compression. In action, this plug-in does exactly what it says it will – it compresses your audio, effectively and efficiently.

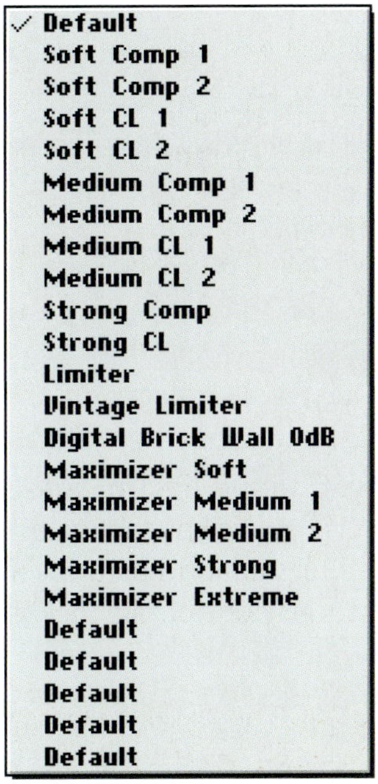
✓ Default
Soft Comp 1
Soft Comp 2
Soft CL 1
Soft CL 2
Medium Comp 1
Medium Comp 2
Medium CL 1
Medium CL 2
Strong Comp
Strong CL
Limiter
Vintage Limiter
Digital Brick Wall 0dB
Maximizer Soft
Maximizer Medium 1
Maximizer Medium 2
Maximizer Strong
Maximizer Extreme
Default
Default
Default
Default
Default

Figure 14.33 TC|Works PowerCore CL presets.

TC|Master X3

You can think of TC|Master X3 as a 'virtual finalizer', as it uses the same technology as the Finalizer to provide multi-band signal processing for mixing and mastering. Or perhaps eight finalizers would be a closer analogy – as you can run up to eight instances of this plug-in at the same time on one PowerCore board.

Master X3 integrates several essential tools for mastering into one interface – multi-band expansion, compression and limiting – and you get several

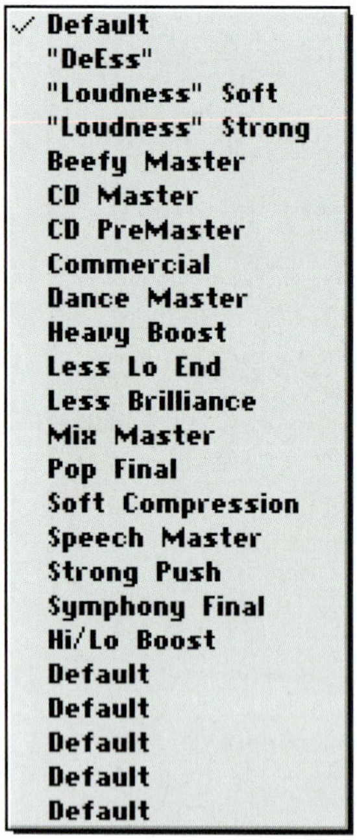

✓ Default
"DeEss"
"Loudness" Soft
"Loudness" Strong
Beefy Master
CD Master
CD PreMaster
Commercial
Dance Master
Heavy Boost
Less Lo End
Less Brilliance
Mix Master
Pop Final
Soft Compression
Speech Master
Strong Push
Symphony Final
Hi/Lo Boost
Default
Default
Default
Default
Default

Figure 14.34 TC|Works Master X presets.

useful presets such as CD Master, CD PreMaster, Dance Master and Beefy Master to use as starting points for your settings.

When you are mastering your mixes, you need accurate level metering. Master X3 provides highly accurate peak programme meters (PPMs) with a peak hold function to let you set your levels as close to 0 dB as possible without distorting the signal. Above these, the Consecutive Clippings display gives a numerical readout of single clips with visual feedback after the third consecutive clip.

Three bands of EQ are provided with user-selectable frequency ranges. These are extremely easy to use and give you instant graphical feedback while you drag the bars up or down to boost or cut the different bands. Graphical displays are also provided for the dynamics processing which you can apply to each band. A gain reduction meter below each display shows how much processing is being applied to the signal.

The dynamics modules include an expander, a compressor and a limiter with parameters available to set threshold levels and release times. You can also set a ratio for the first two, a range for the expander, attack times for the compressor and limiter, and a SoftClip amount for the limiter. This latter is a proprietary process that makes the limiting process less audible - and similar in character to a classic analogue unit.

The overall processing characteristics are controlled via 'Target Curves' that simplify the set-up process by allowing you to pre-select a global style for the bands. Interaction between the bands can be fine-tuned by applying 'Target Factors' which determine the frequency focus of each processing module, significantly reducing the number of user parameters required to set this complex process up. So - how does this work? Well, only one set of parameters is provided for each dynamics module and these are applied to the three frequency bands via the Target Curve and Target Factor settings. The displayed parameters always refer to the centre band and with Target Curves you can define the way the parameters are applied to the other bands. For example, the Linear curve applies identical values to all three bands, the Pink curve causes the Hi band to be processed less and the Hyped curve causes the Hi band to be processed more, while the Smiley

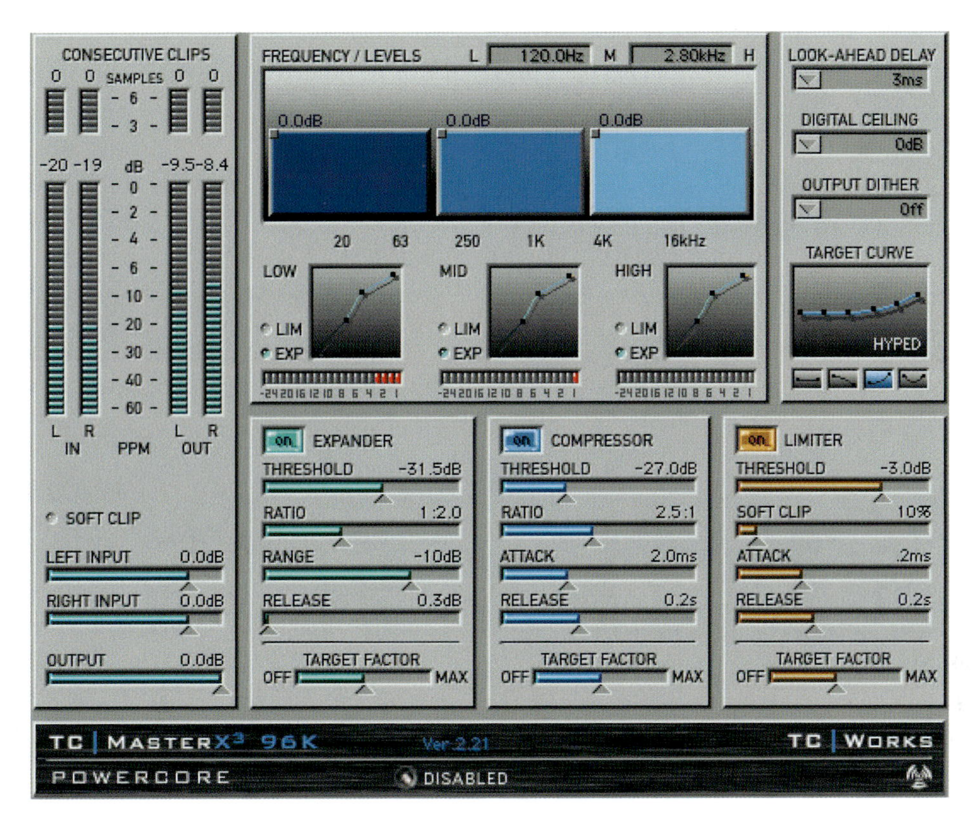

Figure 14.35 TC|Works Master X.

band causes the Lo and Hi bands to both be processed more so the diagram of the frequency response lifts at either end, forming a 'smile' shape. The Target Factor lets you control how much each band will be affected by the processing.

Finally, at the top right of the plug-in's window, there are three popup menus to let you set additional parameters. The first of these lets you set the Look-Ahead Delay that helps to cope with sudden peaks, so sudden anomalies in the program material won't create audible artefacts. You can also set a Digital Ceiling which you would typically set at some point less than 0 dBfs as the maximum value. This 'brick-wall limiting' feature is important to use when mastering to make sure that you can never exceed 0 dBfs. Master X3 features high-quality uncorrelated dithering and the third popup lets you choose the dithering bit-depth - 22-, 20-, 18-, 16- or 8-bit.

So what can you do with it? Well, the Compressor can 'pump up the volume' of your mixes to the max while the Limiter prevents any strong peaks from causing distortion, and the Expander can reduce low-level signals such as noise on vocal tracks. So, yes, it is like having a TC Finalizer in software that you can use to put the final touches to your mixes and get them to sound smoother and louder on CD. Recommended.

TC PowerCore 01 Synthesizer

Figure 14.36 TC PowerCore 01 Synthesizer.

The TC PowerCore 01 is based on the classic Roland SH101 and offers similar facilities. Conceptually simple, it has just one main oscillator plus a sub-oscillator that can be mixed with this – or you could just use the sub-oscillator to get those deep house tones that have been popular for some time now. There is a noise generator that you can blend into the mix to put a little edge on some sounds, or use as the main basis for other types of sounds. The filter section has that all-important resonance control along with other controls for the filter envelope and so forth, and this is followed by an amplifier section with ADSR envelope controls and a drive control. Pitch and Mod wheels are provided, along with an LFO section to let you create modulation effects, and there is a Glide (portamento) control sited next to the Volume fader. This is about the simplest monophonic synthesizer you are likely to come across and can be a very handy tool to have available to let you quickly whip up a bassline or leadline – or some basic synthesizer effects.

TC Compensator plug-in

Whenever you process audio using a plug-in, this will be delayed compared to unprocessed tracks. This delay is referred to as the 'latency' and is largely dependent on the settings of the buffer on the PowerCore DSP card. This can be adjusted manually to minimize this delay, for example when playing virtual instruments 'live'. Also, most host applications now feature Automatic Latency Compensation to compensate for this delay when playing back.

Note Every TC PowerCore plug-in also provides you with the No Latency Mode, which is a great way to record-monitor reverbs or to play a PowerCore Instrument without any effects of latency at all.

Nevertheless, there are situations where latency is not automatically compensated for, or when the No Latency mode is not the best choice to use. In these situations you can use the TC Compensator plug-in - a native VST plug-in that runs on the host CPU.

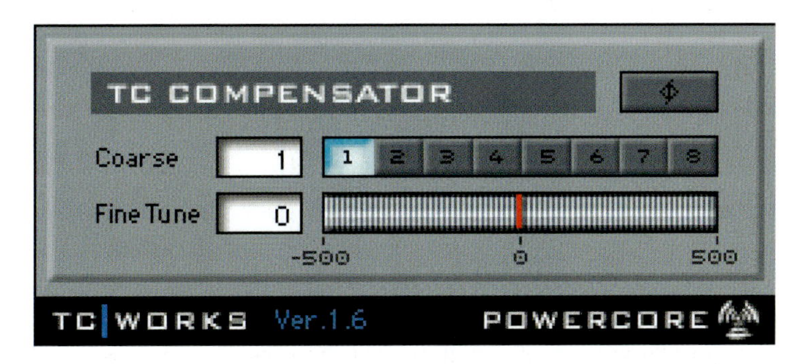

Figure 14.37 TC|Works PowerCore Compensator.

There are just three parameters to set: Coarse Tune, Fine Tune and Phase. The Coarse parameter lets you select the number of plug-ins that you wish to compensate for. The Fine Tune adjustment of +/−500 samples lets you compensate for plug-ins that let you introduce some additional delay, such as the Look Ahead Delay in Master X3. The Phase icon lets you invert the phase so you can audition any 'Fine Tune' adjustments you make and to make sure everything is perfectly in sync.

You need to insert an instance of the Compensator on all the tracks that are not being processed with the PowerCore to delay these by the correct amount to keep them aligned with the processed tracks.

Note If you need to apply the Compensator to several tracks, the easiest way is to bus the outputs of all these tracks to an auxiliary Group channel and insert just one Compensator on this. (See the Manual for a fuller explanation.)

Chapter summary

The idea of using a dedicated card to run your plug-ins is definitely a winner. The question is: 'does this particular card support the plug-ins that you want to use?'

The TC PowerCore has a very solid set of software bundled with it, with tools for tracking, editing and mastering. Also, the third-party support looks very good - the Sony Oxford EQ in particular will surely attract professional users. Bundling Spark LE with the package is something of a 'master stroke' as well - providing a host application for the VST plug-ins that you can use 'right out of the box'. Overall, PowerCore is stiff competition for the UA Powered Plug-Ins - even though it costs 25% more.

The 'Crown Jewels' of the Powered Plug-Ins are undoubtedly the vintage processor emulations. The 1176, LA-2A and EQP-1A, far from being meaningless product identifiers, are virtually 'household names' in the world of professional audio. The RealVerb is more remarkable for its interface than for its sounds, but certainly provides for many interesting creative possibilities - especially using the morphing feature. The channel strip is comprehensive and functional, and Nigel features just about every guitar effect you could ever want. But, in my opinion, the vintage emulations are so good that these alone will justify you shelling out for the whole package!

However, at the time of writing, multiple UAD-1 cards were not supported and multiprocessor support was not implemented, forcing users to disable MP support within the VST host application. (Obviously, multiprocessor support is not needed to run the UAD-1 plug-ins as these run on the UAD-1 card, not on the host computer - but it would be useful to run other plug-ins on the second processor.) Also, no third-party developers are offering software for this card.

Maybe you should get both! This, of course, depends on how many PCI slots you have in your computer and how many you have free. If you have one audio card and one SCSI card, you can probably fit a couple more cards in your machine, so, no problem.

Ultimately, it comes down to whether you want the plug-ins available for the TC card or those available for the Universal Audio card.

15 'Roll Your Own' Plug-ins

Have you got ideas about designing the next Auto-Tune or Amp Farm? Or maybe synthesis is your thing and you have your own unique 'take' on synthesizer design! If you want to develop plug-ins commercially, you will probably want to be talking to Digidesign about becoming a Third-Party Developer, or to Mark of the Unicorn about becoming a Motu Development Partner. You are likely to need plenty of high-level programming expertise, and there will be various criteria to meet to qualify for these development schemes. Steinberg is the other 'big name' for plug-ins, and you can download the Steinberg Development Kit (SDK) directly from Steinberg's website at www.steinberg.de if you want to 'jump in at the deep end'. I have heard very good reports about the SDK – although I was unable to download it in time to include a review of it in this book. But if you want to 'roll your own' plug-ins or virtual instrument without getting your hands too dirty with computer code, there are several solutions available, including DUY Spider, DUY Synthspider, Cycling '74 Max/MSP, and Native Instruments Reaktor.

Cycling '74 Max/MSP Pluggo

The mysteriously-named Cycling '74 distributes the Max/MSP package including the Max object-oriented graphical programming environment, with its MSP audio extensions and the Max/MSP Runtime application. Using Max and MSP you can construct MIDI and audio applications that you save as Max patches. You can open these for use either within the Max/MSP programming environment or using the Max/MSP Runtime application. This is useful enough, especially when you want to create any sort of 'stand-alone' application – just use a Max patch within the runtime application. But how does this help you write your own plug-ins? Well, the package also includes the Pluggo VST plug-in that lets you load Max 'patches' and other VST plug-ins into any supported host environment – such as Cubase VST. Pluggo is another 'runtime shell' that uses Max/MSP to provide additional functionality for VST plug-ins, and Pluggo also enables VST plug-ins to be run from within an MAS host-environment. Using Max and MSP, you can write your own MIDI and/or audio plug-ins and open these Max/MSP patches in Pluggo for use within your host environment. Even better, you can use the supplied Plugmaker application to turn Max patches into VST plug-ins! As it happens, Pluggo's plug-ins and objects themselves were written using Max/MSP - what goes around comes around!

Now just to make this all very clear - you can write a MIDI application using Max and you can write an audio application using MSP. These applications could be used to create special MIDI or audio effects, for example. You can run these applications inside the Max/MSP programming software that you use to create the applications, or using the special 'runtime' application that just lets you run the finished application (no tools to create software here). You can also run these applications from within Pluggo - which itself works as a VST or MAS plug-in - so you can use them with your favourite MIDI sequencer or audio software. And, even more usefully, Plugmaker lets you turn the applications into bona-fide VST plug-ins that you can use with any VST-compatible host software.

Max is a high-level, graphical programming language that uses the Macintosh's graphic capabilities and icon-based user interface to enable you to construct MIDI and audio software applications by simply connecting objects to each other on-screen. Some data still has to be typed in, of course, but this is nothing compared

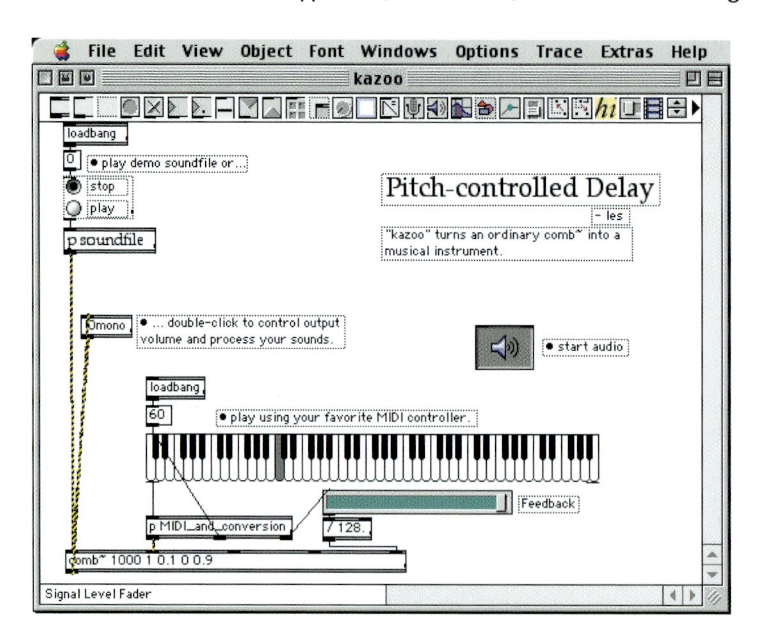

Figure 15.1 MAX programming environment patch window showing objects connected on-screen and data-entry fields open and ready to accept input.

to the hundreds or thousands of lines of code needed for a typical software application.

MSP is a set of specialized extensions for Max that includes over 170 Max code 'objects' that you can use to quickly build your own synthesizers, samplers and effects processors - without starting from scratch. You can specify exactly how you want your instruments to respond to MIDI control, and you can implement the entire system in a Max patch. MSP objects are connected together by patch cords in the same way as Max objects.

For audio playback, MSP supports the Apple Sound Manager, ASIO and ReWire, DirectConnect and VST interfaces. Unlike with previous versions, there is no direct support for specific hardware, but you can use ASIO instead. ASIO drivers exist for a large collection of audio cards, including the ASIO DigiDesign DirectIO driver for Digidesign hardware that can be obtained from the Steinberg website at www.steinberg.de. Note that you cannot use Direct I/O while Pro Tools is running. Digidesign hardware is dedicated to a single application at a time, so if you want to use Pro Tools with MSP, you should use the DirectConnect audio driver.

Figure 15.2 MSP audio drivers.

When you have finished your programming in Max, you can hide any connections that you don't need to see as part of the user interface and close any data-entry fields that are not required by the user.

You can save your work as a Max 'patch' file on disk and re-open this for editing later. If you simply want to use the patch, you can open it using the Max/MSP Runtime application. As its name is intended to suggest, this does not include the programming interface, it simply lets you run the patch.

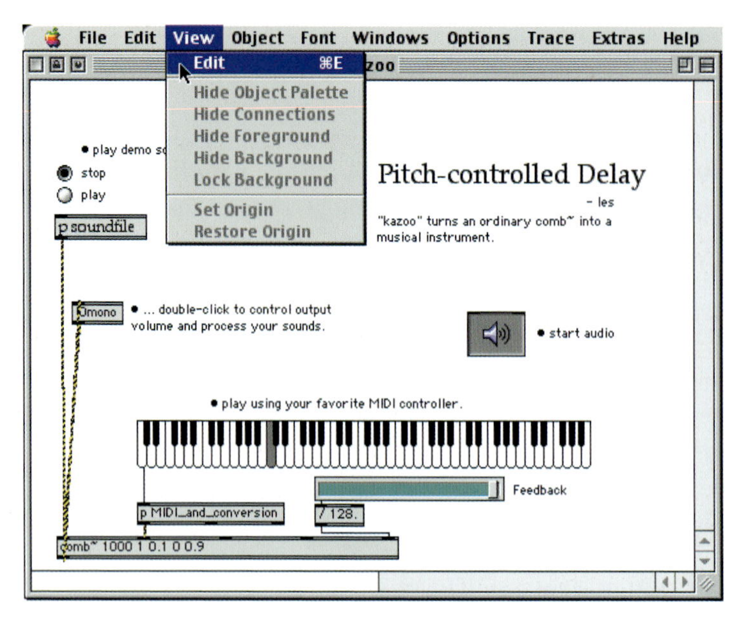

Figure 15.3 MAX programming environment patch window with connections hidden and data-entry fields closed.

Tip Check out the set of example Max patch files provided with the software to help you get started - you might find what you are looking for. I was surprised to discover the 'FM Surfer' patch which lets you produce bursts of randomly generated FM synthesized sound by simply tapping keys on your computer keyboard. It just so happened that I needed some sounds exactly like this for a movie soundtrack I was working on!

If you are using Max/MSP to build your own plug-ins, you will need to use an application called Plugmaker - along with Pluggo. Pluggo can open Max patcher and collective files so you can test your patcher directly within the host sequencer. Once you are happy with the way your Max patcher is working, you can use the Plugmaker application to turn it into a VST plug-in that you can then just drop into any VST plug-ins folder to open directly within the host software.

Note You don't have to use Pluggo to load these into your VST-host application - they are bona-fide VST plug-ins - but you may wish to take advantage of Pluggo's interesting features.

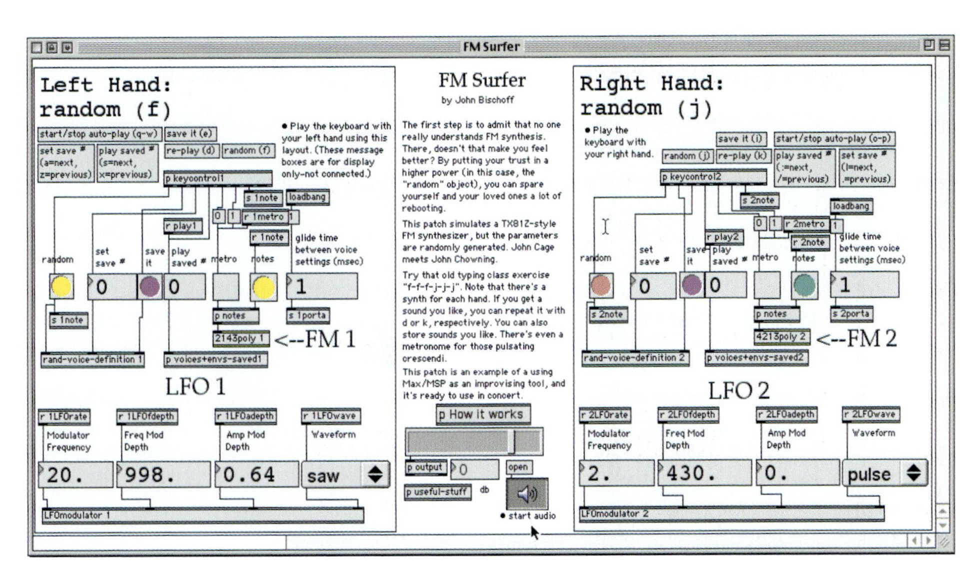

Figure 15.4 A Max patch that lets you produce random FM sounds.

The Max/MSP programming environment does have a fairly steep learning curve. OK, it is easier than C + +, but it is nowhere near as easy as the DUY 'spiders' or even Reaktor. The advantage is the power to do things that you cannot do using these other packages.

> **Note** In the USA there are at least 38 colleges and universities that use MAX/MSP to teach their music synthesis courses, and there are evening classes at Berkley where you can learn how to program using MAX/MSP.

The latest development from Cycling '74, announced just as this book was being completed, is called Jitter. This is a set of 133 new video, matrix and 3D graphics objects for the Max graphical programming environment. The Jitter objects extend the functionality of Max4/MSP2 with flexible means to generate and manipulate matrix data. This means any data that can be expressed in rows and columns – such as video, still images or 3D geometry, as well as text, spreadsheet data, particle systems, voxels or audio. Jitter is useful to anyone interested in real-time video processing, custom effects, 2D/3D graphics, audio/visual interaction, data visualization and analysis.

Native Instruments Reaktor

Reaktor is a complete sound design studio that lets you create synthesizers, samplers and effects processors. You can use Reaktor's Panels, Structures, Ensembles, Instruments, Macros and Modules to assemble your own unique designs.

A window, where modules are placed and interconnected, is called a Structure. Complex signal processing designs can be implemented by connecting Modules that carry out relatively simple tasks - as with the classic modular synthesizer systems. Macros are combinations of Modules and other Macros to form basic building blocks. Macros have an internal structure, just like instruments, but they do not manage MIDI data themselves, they have no snapshot automation, and they do not have a separate Panel. They are used to create functional blocks of Modules and other Macros to help to make things clearer when you are working on complex designs.

Instruments are high-level Macros that have their own MIDI settings and their own Panel. An Instrument can contain other Instruments and Macros arranged in a hierarchical fashion. A Panel, in case you were wondering, is the user interface of an instrument - like the front panel of a hardware synthesizer or effects unit. Ensembles and Instruments can have their own Panels containing various switches, knobs and faders.

The highest level in this hierarchy, i.e. the Instrument that contains everything else, is called the Ensemble. Instruments are added into the Ensemble Structure by loading them from the library that comes with Reaktor - or by creating a New Instrument if you prefer to start from scratch. The Ensemble's Structure window shows the Instruments, how they are interconnected, and the connections to the audio inputs and outputs for the sound card you are using.

When you open a new Ensemble Structure window it will just contain two modules - the Audio In and the Audio Out - by default. You can Control-click (right mouse-click in Windows) in the Structure window to bring up a 'context' menu that lets you insert modules into the window. Here you can select oscillators, filters, envelopes and all the other building blocks of analogue synthesizer, samplers and effects.

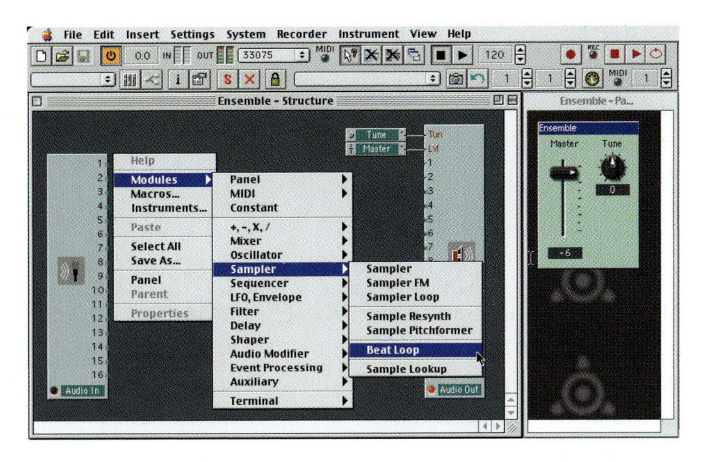

Figure 15.5 Native Instruments Reaktor showing various modules that may be chosen from the context menu.

For example, the Beat Loop module lets you play back a stereo multi-sampled drum loop. You can click on this module to open its Properties window where you can load in your loops and set various parameters.

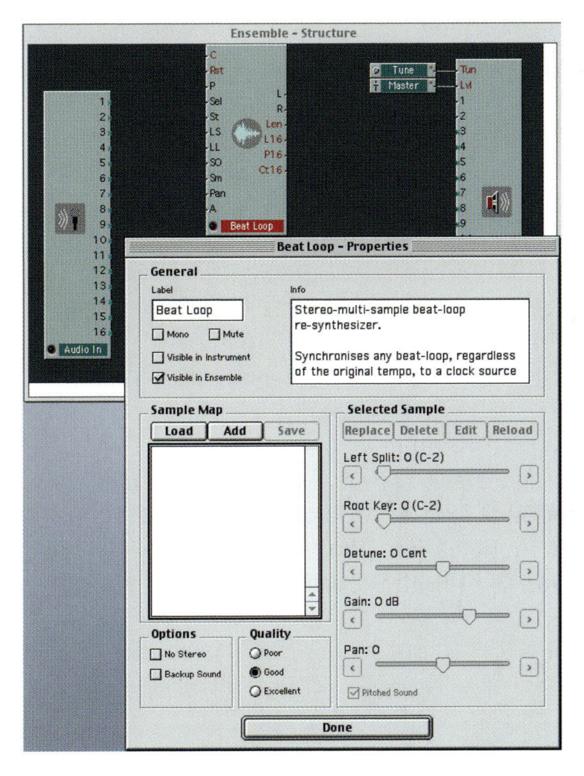

Figure 15.6 Native Instruments Reaktor showing the Properties dialog for the Beat Loop module.

To connect the modules together, you simply point, click and drag the mouse from outputs to inputs.

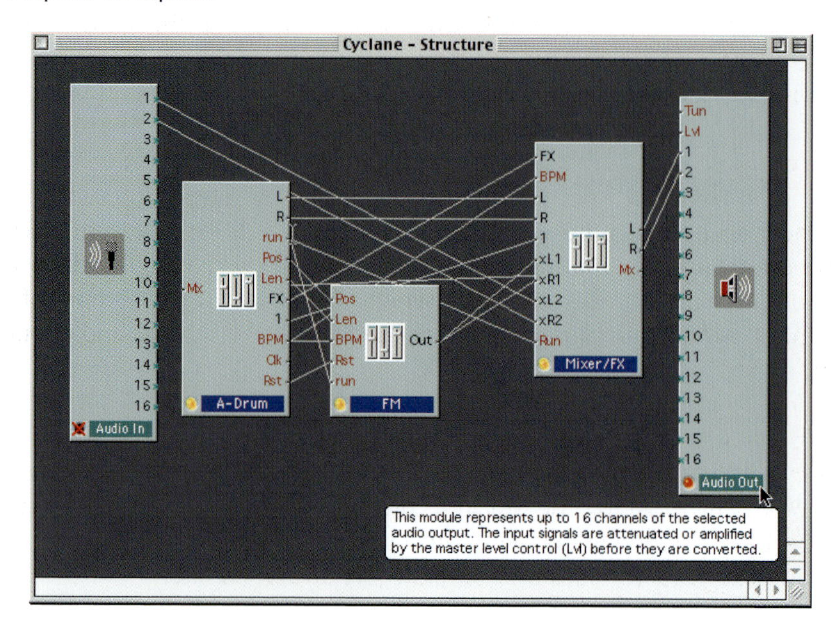

Figure 15.7 Native Instruments Reaktor Ensemble Structure for the 'Cyclane' preset.

With a simple effect or synthesizer, there is no need to use Macros and Instruments. However, as your effect or synthesizer grows in complexity, you will value Reaktor's ability to organize the various modules using these features.

So, there is still something of a learning curve with Reaktor - but it beats the heck out of programming all this stuff in C++. Reaktor's strength lies in the amount of work the programmers have put into developing the modules and making them available for you to simply hook up in this way. Of course, you still have to work out just how you are going to arrange things yourself, and this could take quite some time until you know your way around Reaktor thoroughly.

To help you get results more quickly, a range of Instruments and Ensembles are provided that you can use right away. In fact, one of the best ways to learn Reaktor is to take a look at how these presets have been put together.

Some of my favourites include Loopo, Plasma and Kompressor. Loopo uses a fiendishly clever method to make the best loops possible from your samples. You just load in your sample, use the Loop control to set the position where you roughly want to have the loop. The Resynth module within Loopo then cuts the sample up into small 'particles' and reassembles them ready for playback.

Before playing back the loop, you set the loop Length parameter to zero, as this will not now be used. On playback, once the loop point has been reached, the Resynth module starts repeating the same sample slice, i.e. the sound 'particle', over and over again. This method works well with material that is periodic - i.e. it has a unique pitch.

If the sample does not have a unique pitch, it will not have a unique period that can be used to make a loop. In this case you can use Plasma. Plasma takes a sample and freezes it in time wherever you like. You can then resynthesize this using the granular synthesis method of the ReSynth module to produce a sustained sound. This method works well when you have a short sample of some material that has no particular unique pitch, such as a choir or a sound effect, and you want this to last much longer.

The Kompressor was designed specifically for percussive sounds. This lets you make samples shorter or longer using the Resynth-Sampler module's Time Warp controls. As with most of the Reaktor presets, this can do much more than perform its basic functions and you can introduce some unusual surprises into your sound using the Signal Informed Re-Synthesis parameter.

There is not space here to do much more than 'scratch the surface' of Reaktor's powerful capabilities. Suffice it to say that if you are serious about exploring sound synthesis, then Reaktor is something you will not want to be without.

DUY DSPider TDM

Similarly to Reaktor, DUY's DSPider lets you build your own sound processors simply by interconnecting modules onscreen.

There are two operating modes - Reader and Advanced. Reader Mode is used for loading and running 'patches' which can be DUY presets, third-party presets or your own creations, while Advanced Mode is where you create, edit and run your own patches - or edit DUY or third-party presets.

DSPider comes with plenty of these presets so you can use it straightaway as a compressor, limiter, reverb, stereoizer, enhancer, maximizer, panner - or as a combination of all these - in Reader Mode. Advanced Mode will let you create your own auto panners, filters, chorus, flange and other delay effects. You can also design your own reverb or dynamics processors, implement custom filters and even create basic synthesizer sounds.

Of course, to design your own reverb is going to require a reasonable knowledge of digital signal processing techniques. If this is not you, then stick to using the presets. Nevertheless, for those who do possess the requisite knowledge, DSPider is an ideal tool to use to create, publish and commercially exploit plug-ins for TDM systems. And to encourage third parties to develop their own presets for sale commercially, DUY provide a 'Locked-Preset' option that allows users and developers to keep the algorithms they have developed hidden.

There are two different versions of the plug-in provided - the Reader and the Advanced versions. The basic Reader Mode plug-in has a relatively small screen and is primarily used to run presets. The Advanced Mode plug-in provides a large working surface that allows the user to edit or create plug-ins - as well as to run them.

DSPider Reader Mode

The DSPider screen in Reader Mode is split into two sections with the Save and Load and other controls at the left and a large 'blackboard' area to the right where the plug-in modules appear. Here you will typically see various graphical modules including sliders, plasma displays, scopes, text and numeric readouts - in fact, whatever the designer of the plug-in decided to let you see.

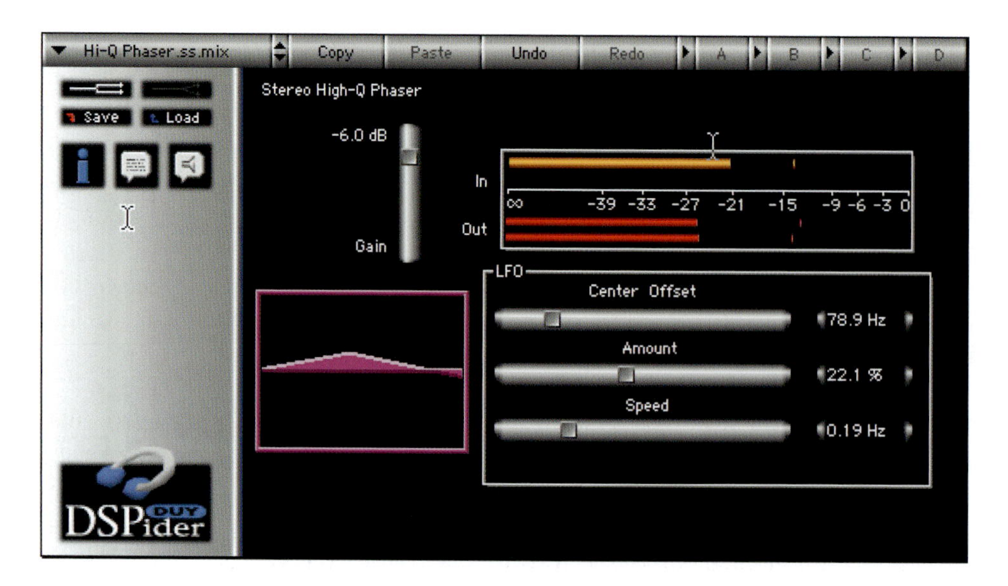

Figure 15.8 DUY DSPider Reader mode.

In the blurb supplied with the software, DUY suggest that DSPider might be the only plug-in you would ever need. That's probably a bit of wishful thinking on the part of their marketing department. Nevertheless, if you have to choose just a small selection of plug-ins because of budgetary constraints, then DSPider is definitely worthy of strong consideration.

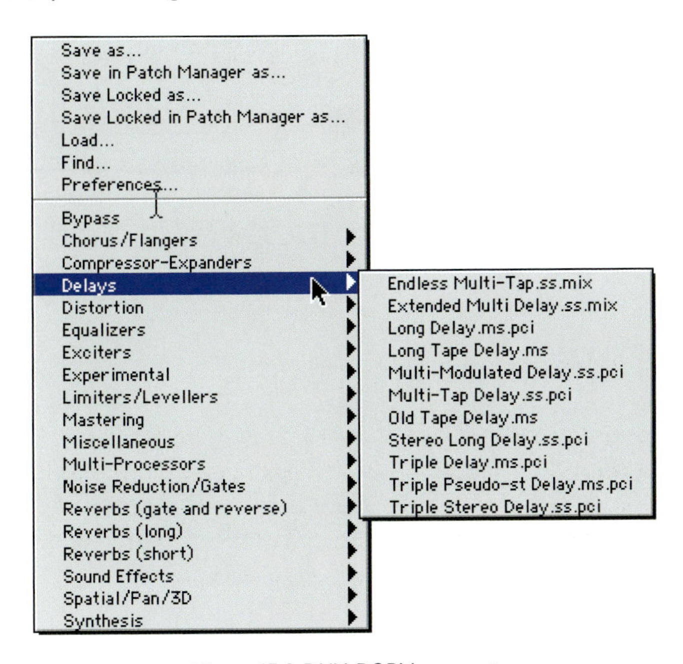

Figure 15.9 DUY DSPider presets.

An impressive library of over 200 presets is provided - including compressors, reverbs, equalizers, limiters, synths, noise reduction systems, 3D effects, de-essers, sound effects generators, and so forth. In fact, something for all occasions! And to help you manage this plethora of presets, a useful Patch Manager feature is provided that allows Pro Tools users to switch between presets automatically.

DSPider Advanced Mode

When you select DSPider Advanced from your list of plug-ins, you are presented with a palette of icons at the left of the window with a work area occupying most of the rest of the window. You use these icons to select from the 40 processing modules available. All the modules can be easily programmed and patched together using a simple 'drag and drop' procedure. If you have the time and the motivation, not to mention at least a basic knowledge of audio signal

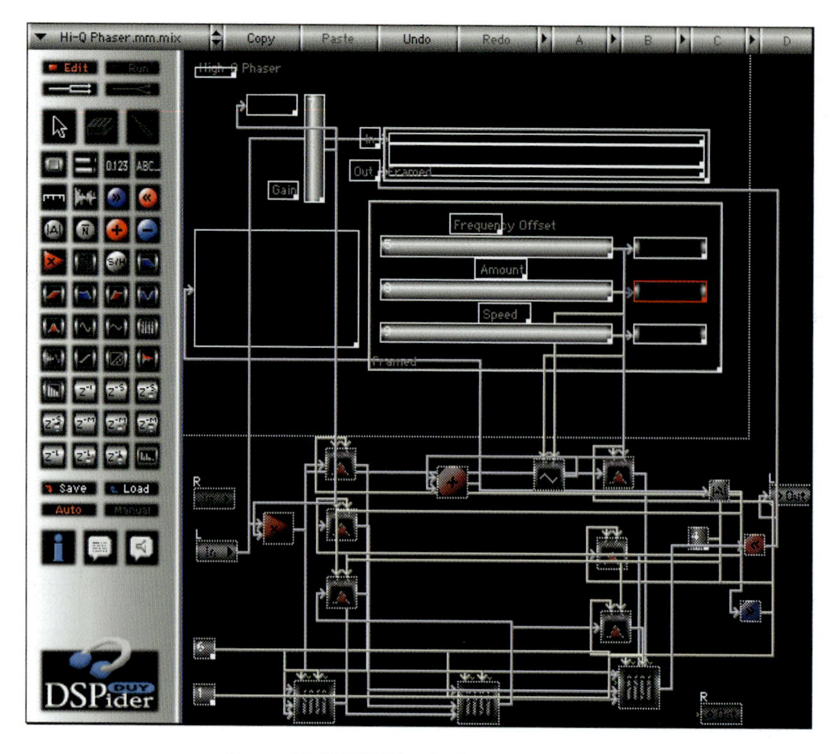

Figure 15.10 DSPider in Advanced mode.

processing, you can use DSPider in Advanced mode to create interesting new plug-in processors tailored to your specific application.

The list of modules ranges from simple mathematical operations such as addition, subtraction and multiplication, to multi-functional modules like filters, oscillators, waveshapers, filters, multiple delays, envelope followers, pitch trackers, ramp generators, sample-and-hold and spectral shapers. A flexible multi-reflexion chamber and multiple delay modules allow the creation of complex user-defined reverbs. Graphic modules including sliders, numeric readouts, plasma meters, scopes and so forth, allow the creation of complex graphical user interfaces.

In Advanced Mode there is a Run Mode view which looks similar to the Reader Mode screen, but this time there is a much more comprehensive palette of controls at the left and the 'blackboard' space is much larger. There is also an Edit Mode where you can change the position of modules, modify their parameters, make connections and so forth. Basically, you drag and drop modules from the palette at the left and then hook these up in various ways to implement your processor. You can choose to hide any of these modules from the user to create a much simpler, clearer display. You will probably want to do this with even a moderately complex plug-in - as the screen can get to look very cluttered very quickly. Nevertheless, this graphical approach to programming is much easier than using a conventional programming language such as C++ or Java to create a plug-in.

There is not space here to really do justice to the range of features you will find in DSPider. Suffice it to say that you could spend a very long time exploring all the possibilities - and DSPider would unquestionably make an excellent educational tool for anyone interested in processor design.

DUY SynthSpider

SynthSpider takes the concept of the DSPider and applies this to simulating synthesizers in software. As with DSPider, there are two versions of the plug-in - the basic preset Reader and the Advanced version that provides full editing and synthesizer creation facilities.

SynthSpider Reader

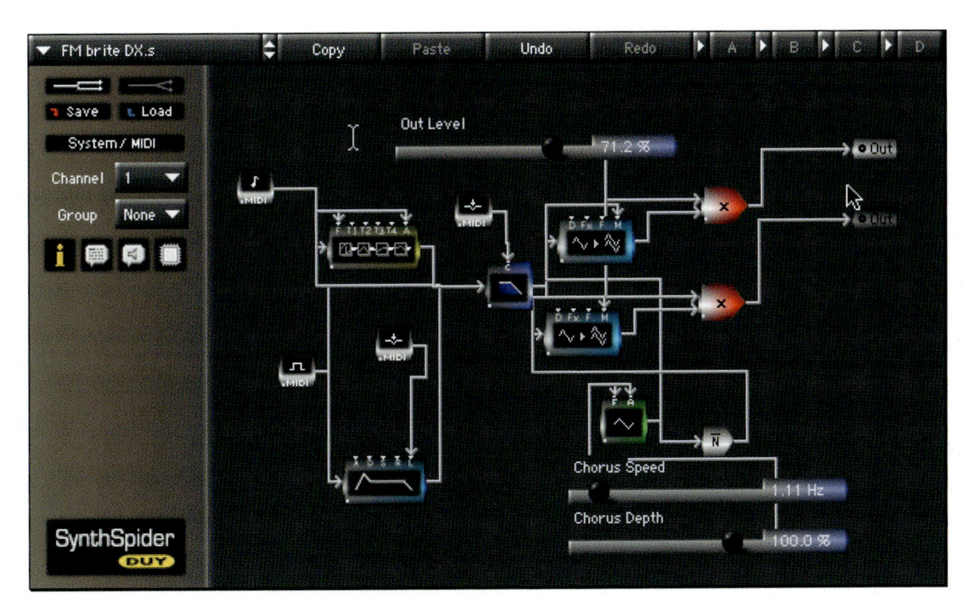

Figure 15.11 DUY SynthSpider in Reader mode.

You get more presets with SynthSpider than you can shake a stick at! Analogue mono and poly synths, bass, drums, keyboards, effects and lots more. Several demo files are included with the software and these use SynthSpider as the only sound source, to show off what it can do. The demo files include versions that use D-Verb or DSPider or ReDSPider reverb effects, so you will need to have one or more of these plug-ins installed before you will hear the demos as intended. (The ReDSPider Demo installer is included on the SynthSpider installation CD, and trial versions of the DUY software can be downloaded from the DUY website.) You will also need a Pro Tools Mix+ system with two cards to play these demos in their entirety.

Demo titles include E-Beats, Bach Invention in F, Cmin Bach Prelude, OnlySequences1, OnlySequences2. The latter is my personal favourite - with its layers of (mostly) monophonic sequences which are faded in and out using Pro Tools track automation to create a stimulating mix of electronic sound. These demos do not use audio files - they use MIDI tracks to trigger specially prepared SynthSpider presets. A couple of these - 'OnlySequences1' and 'OnlySequences2' - have no MIDI tracks. With these, one of the modules, the 'Sequencer', is doing all the MIDI-generation internally, inside SynthSpider. It is

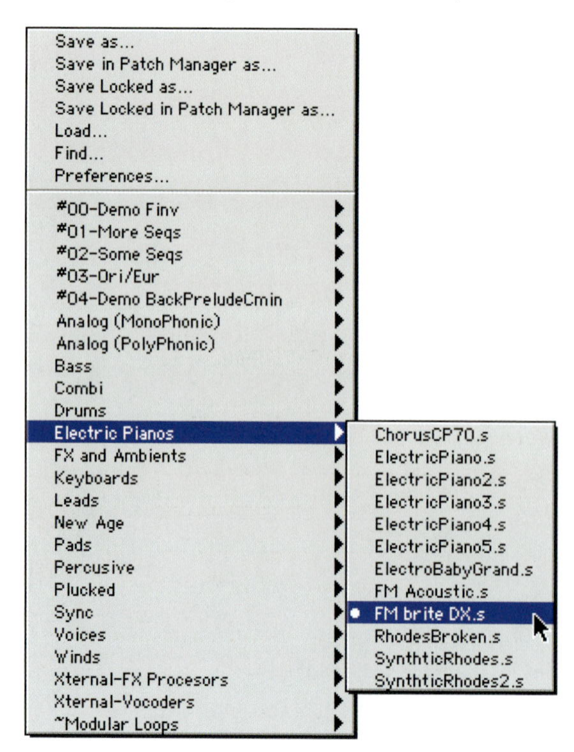

Figure 15.12 DUY SynthSpider presets.

worth spending some time analysing how these demos work to give you a good idea of SynthSpider's capabilities.

Several of the SynthSpider presets have sequenced patterns built in. For example, in the Analog (Polyphonic) folder there are six presets named Performance1.m to Performance6.m. Each of these has a different rhythmic pattern programmed into it so that when you hold any note, a series of notes at the same pitch are played back in rhythm.

To play a SynthSpider synth, for example from within Pro Tools, you need to make sure that SynthSpider is configured as a node in Opcode's IAC Driver. Open your OMS Setup document and double-click on the IAC icon. (If there is no IAC icon you need to install this option from the OMS installer first.) A window opens containing four text fields that allow you to define nodes for up to four IAC 'buses' by naming these appropriately. Name one of these 'SynthSpider'.

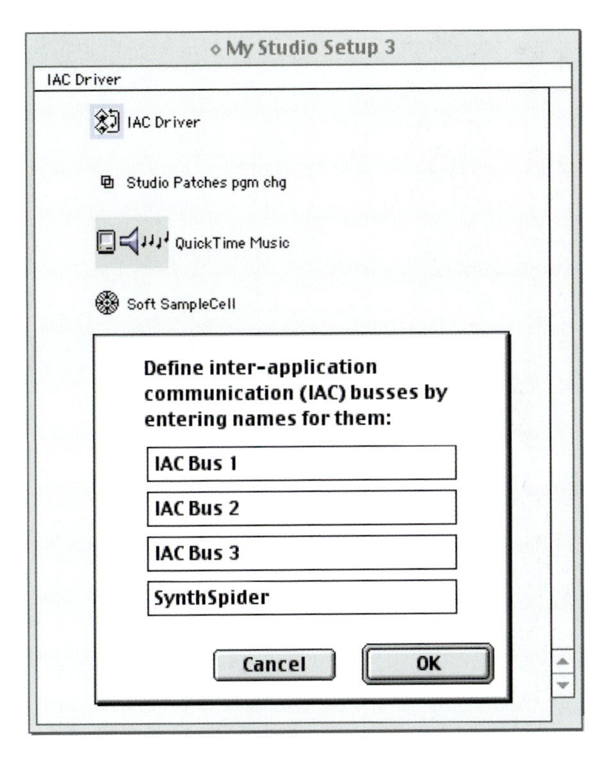

Figure 15.13 OMS IAC setting for SynthSpider.

Figure 15.14 DUY SynthSpider user interface showing System/MIDI button.

Once this is set up, launch your host application – Pro Tools, Logic Audio or whichever – and set up the MIDI for SynthSpider. Click on the System/MIDI button to bring up the settings window.

When the window opens, make sure that the OMS button is selected and that the 'Use Virtual Node' button is not selected. Now you can click on the 'Listen port' popup menu to select the OMS device that will send information to SynthSpider; If you want to play SynthSpider from your MIDI keyboard, you can select the MIDI keyboard here. More often, you will want one of your sequencer tracks to play SynthSpider, so in this case you would select the SynthSpider IAC bus node that you created using OMS.

With this setup, you can route any MIDI track from your sequencer to SynthSpider via the OMS IAC bus.

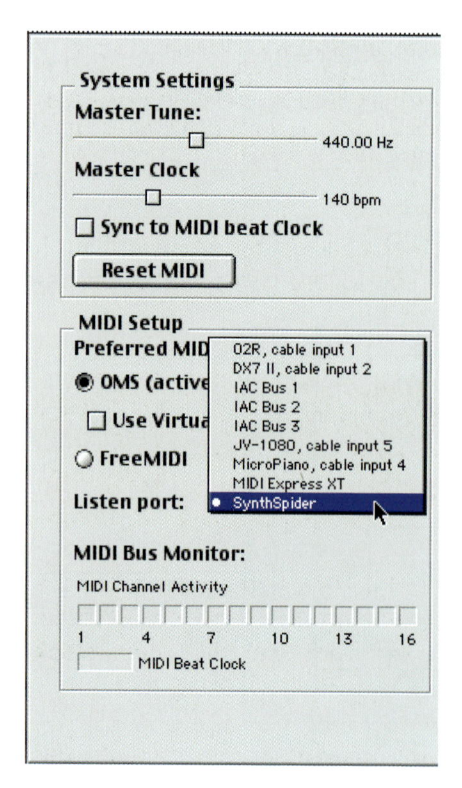

Figure 15.15 DUY SynthSpider System/ MIDI settings.

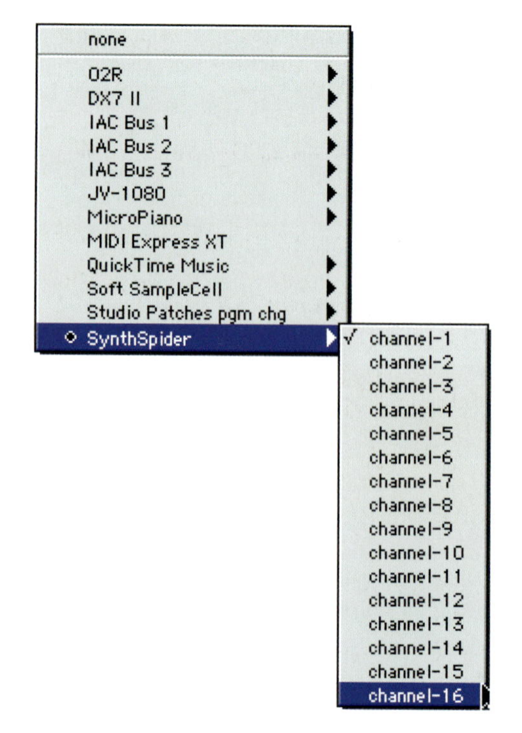

Figure 15.16 Pro Tools MIDI channel output popup showing SynthSpider channel 1 being selected from the list of destinations obtained from the current OMS Setup document.

Another way to set up SynthSpider for use with Pro Tools is to use a virtual OMS node. In this case, you need to select the 'Use Virtual Node' option and type the name of the node, in this case 'SynthSpider'.

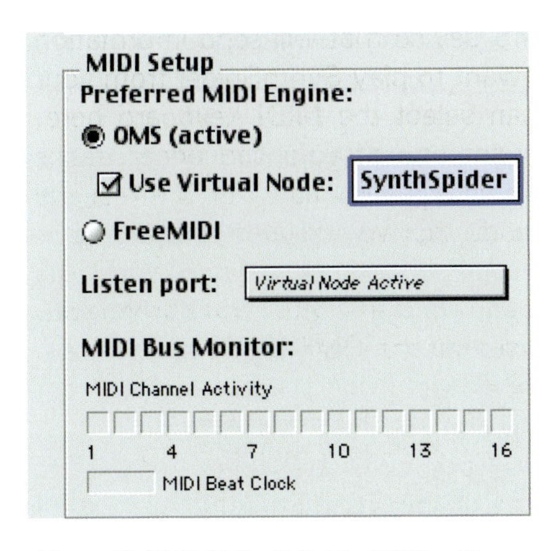

Figure 15.17 DUY SynthSpider MIDI settings.

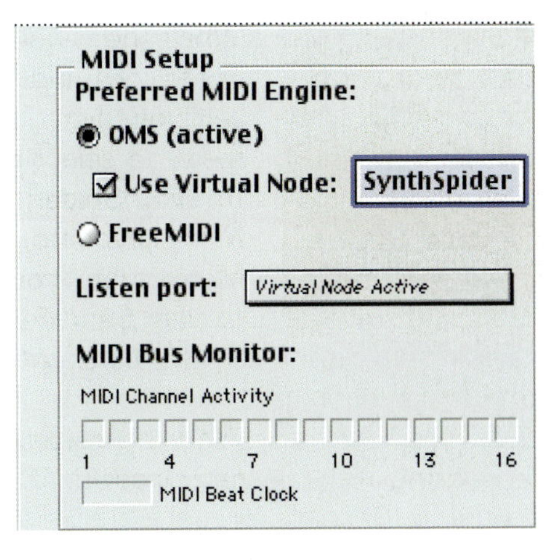

Figure 15.18 Pro Tools MIDI channel output popup showing SynthSpider virtual node destinations created by selecting SynthSpider's 'Use Virtual Node' option – with the IAC Driver node removed.

The Listen port becomes deactivated and the virtual node becomes active. A disadvantage here is that you cannot access any of the other OMS devices such as your MIDI keyboard.

Note If you decide to use the virtual node option, then you won't really need the SynthSpider IAC bus node created in your OMS Setup document – so you can remove this.

When you use the MIDI channel output popup in Pro Tools you will see listed the 16 available MIDI channels to the SynthSpider, along with an extra channel specifically provided to handle sync. This is used with the MIDI beat clock when synchronizing the sequencer modules.

The System/MIDI window has a large Voice Management section that lets you set up the various polyphony options for SynthSpider. Voice Management gets a little complex with SynthSpider, but it's easy enough once you get the hang of it.

Basically, SynthSpider can receive on any of up to 16 MIDI channels, which is the easy bit - but then each MIDI channel has four different voice modes and various trigger options.

The number of 'voices' available in each plug-in depends on how many instances of SynthSpider you are using and how these are configured. You can play each instance of the SynthSpider plug-in using either one or a combination of the 16 possible MIDI channels. If you play using just one channel on one instance, you will hear just one voice - although if you play back using several channels on one instance you will hear several voices.

The way SynthSpider is typically used is with multiple insertions of the plug-in - either with all being played back on different MIDI channels, or with several playing back on the same MIDI channel. The way this works is that if you want to play polyphonically, you have to insert as many instances of the plug-in as voices you wish to use and set these to the same MIDI channel. You also have to choose one of the polyphonic modes. Of course, you can have other instances set to other MIDI channels that will play monophonically, or polyphonically if you set several of these others to an identical channel and set their modes to Poly.

So what happens if you have several instances of the SynthSpider plug-in using mono mode, but set to the same MIDI channels? In this case, all the instances of the plug-in set to the same channel will behave as just one monophonic voice, and the sounds you have chosen for each instance will all play at the same time when you play a note.

So, if you have them all play on the same channel and they are in monophonic mode, the multiple instances behave as one voice - and you get a layered monophonic sound. This is good for building 'fat' sounds - whether all the layers use the same or a different patch.

Note The mono mode works in a special way with Multi-Trigger set to Off - it memorizes all the keys you press on a keyboard, plays one of these in the normal monophonic way, then plays the others when the first note is released.

To play polyphonically, you need to choose the Poly mode for each instance of the plug-in. The three polyphonic voice modes available are Poly 1 Mode (First Note) Poly 2 Mode (Last Note) and Poly 3 Mode (Rotative). Poly 1 mode assigns voices according to the order in which you play the notes. Its priority is to keep the first note playing until it is released rather than cutting it off when a subsequent new note is played. Poly 2 mode makes sure that the last note played will sound, even if an earlier note has to be cut off before it has finished. Poly 3 mode assigns voices in rotation - each new note played uses a new voice - even if you play the same note repeatedly. This mode is particularly useful with panned voices where every new voice can appear in the mix at a different panned position.

Options are also provided to let you decide how to handle 'stolen' notes. A note is said to have been 'stolen' when you play more notes than there is available polyphony - so one of the notes stops playing when you play the new note. With the 'Stole On' and 'Stole RTN' options both selected, when a note is stolen and subsequently becomes available again, it will start playing again. You can even force this note to use the velocity value of the last note you played by selecting the 'Stole Vel' option - although this is not always a good thing to do.

The final option - Prog. Change - lets you enable or disable MIDI Program Changes from the information in the Patch Manager.

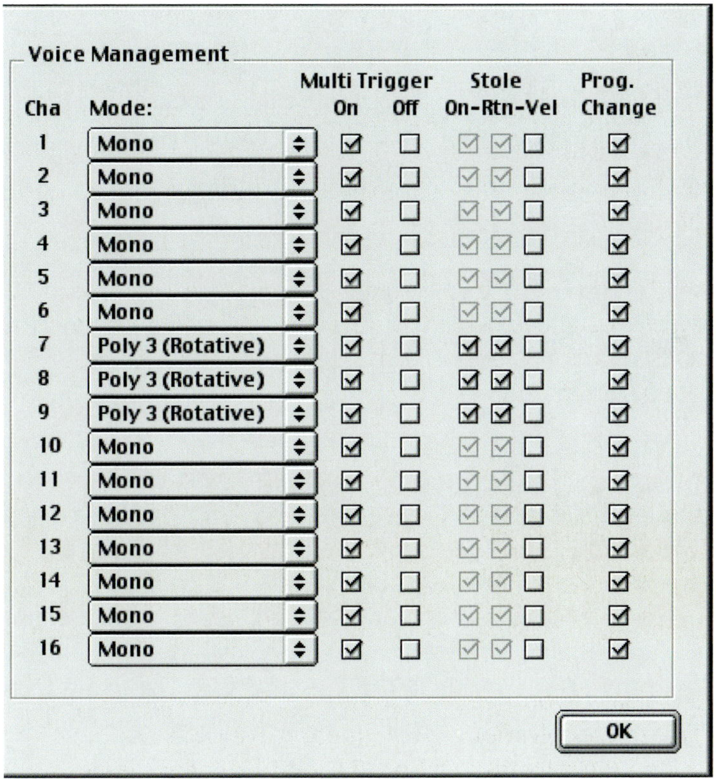

Voice Management

Multi Trigger On Off — Stole On-Rtn-Vel — Prog. Change

Cha	Mode:
1	Mono
2	Mono
3	Mono
4	Mono
5	Mono
6	Mono
7	Poly 3 (Rotative)
8	Poly 3 (Rotative)
9	Poly 3 (Rotative)
10	Mono
11	Mono
12	Mono
13	Mono
14	Mono
15	Mono
16	Mono

OK

Figure 15.19 DUY SynthSpider Voice Management.

Underneath the System/MIDI button, each SynthSpider instance has two popup selectors – one for the MIDI channel and one for the 'group'.

Figure 15.20 DUY SynthSpider controls.

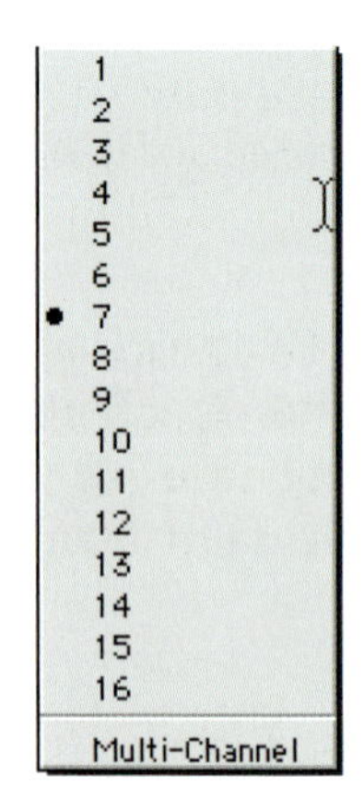

Figure 15.21 DUY SynthSpider MIDI channel popup.

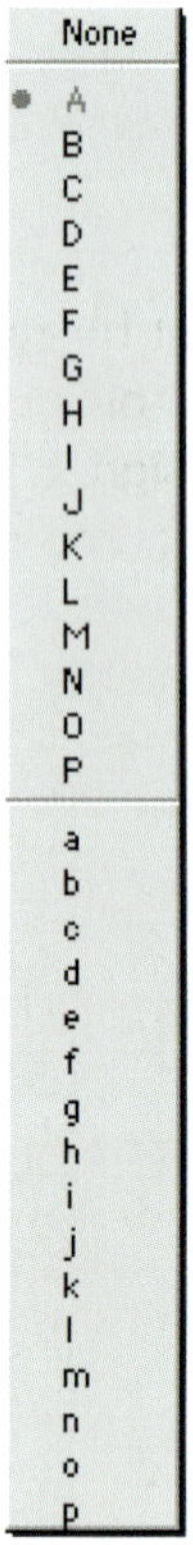

Figure 15.22 DUY SynthSpider grouping system popup.

You can select any of the 16 MIDI channels here, or you can use the Multi-channel option which lets you use more than one MIDI channel with one instance of SynthSpider.

The grouping system lets you organize multiple instances of SynthSpider into groups that can then be dealt with similarly. There are 16 main groups labelled 'A' to 'P' with capital letters that act as 'Master' groups and 16 sub groups labelled with letters 'a' to 'p' in lower case that act as 'Slaves' to these. Any Master can have as many Slaves as it likes and you can edit and automate any parameter on a Master group and the parameters on all its slave groups will follow. So, for example, if you have inserted several instances of the plug-in and you want to have all these use the same patch and play polyphonically, you would set the first plug-in to be the Master, say Group A. The others would then be set as slaves to this, set to a, b, c, d and so forth. In this case it is convenient to be able to use one plug-in's window to control all the plug-ins. This is what the Master/Slave system lets you do. Most of the controls in the palette at the left of the plug-in window are removed from the slaves because, of course, they are controlled from the Master. The only control remaining is for the MIDI channel.

Below the Channel and Group popup selectors there is a group of four small icons. The first of these, marked 'i', opens a window with info about the patch you have loaded - such as notes by the person who programmed the patch. The next icon turns on the 'balloon help' system that brings up a small window to explain whichever user-interface component you point at with the mouse. The third button speaks the help messages when the computer's Speech Manager is active.

The last button brings up the DSP Allocation Preferences window. Here you can enable or disable particular kinds of chips in the different varieties of the Digidesign DSP cards. If you have more than one Mix card, you can force these to accept only one instance of SynthSpider per card.

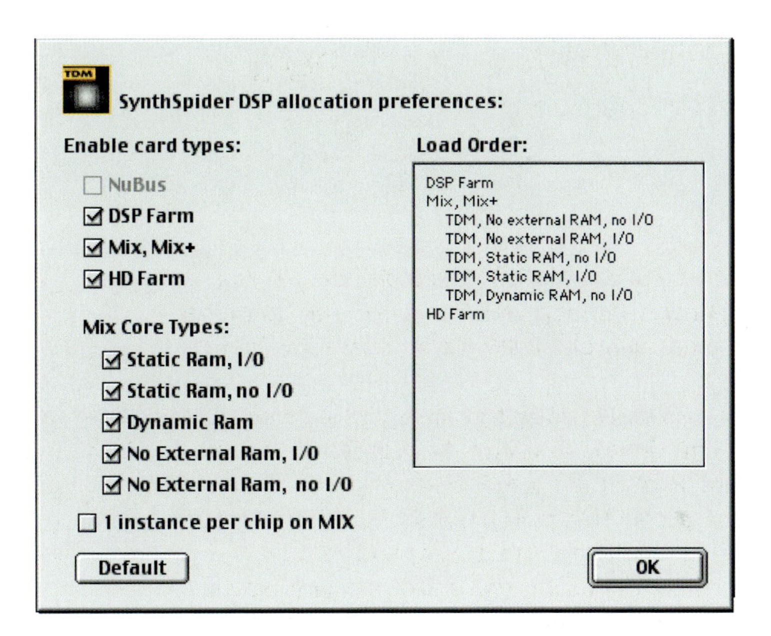

Figure 15.23 DUY SynthSpider DSP Allocation Preferences.

SynthSpider Advanced

The SynthSpider Advanced plug-in has 40 modules which you can drag and drop onto the working surface to the right of the window from the palette at the left of the window.

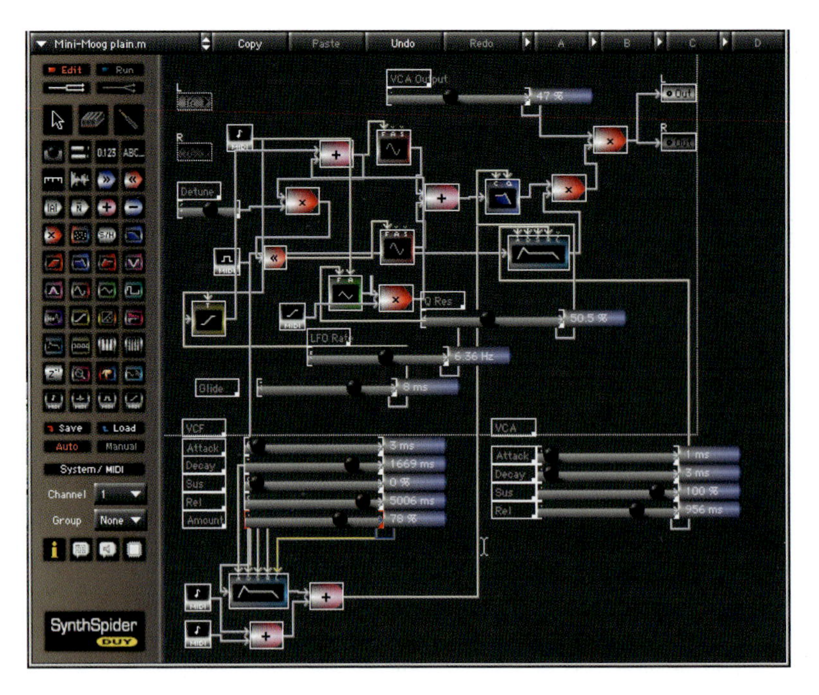

Figure 15.24 DUY SynthSpider Advanced mode.

> **Tip** Anything you place within the working surface area bounded by dotted lines will be visible when opened in the Reader plug-in. Anything placed outside the dotted lines will only be visible when opened using the Advanced plug-in.

The way you build the modules up is fairly obvious, especially if you are familiar with the basics of a synthesizer such as the Mini Moog. A first step would be to set up an oscillator to produce some sound. The SynthSpider Oscillator module has options for all the typical analogue waveforms and can also be used to generate FM sounds. The generated waveform can also be modified in a variety of ways using the options in the window that appears when you click on the module you have placed onto the working surface.

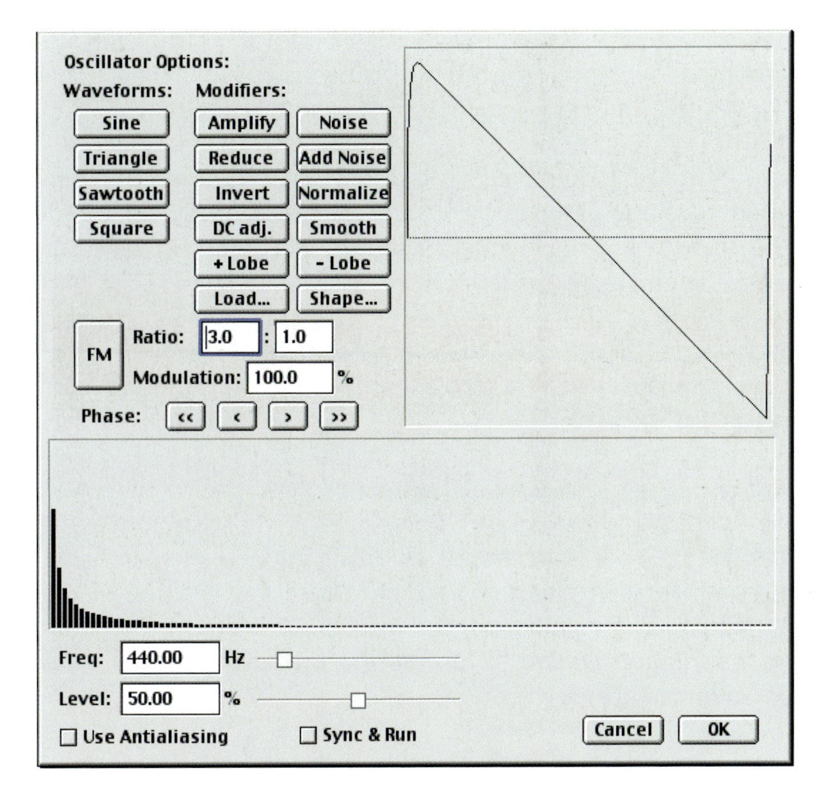

Figure 15.25 DUY SynthSpider Oscillator module options.

A next step would be to filter this using one of the filter modules, such as the two-pole low pass filter. SynthSpider's two-pole filter options let you set a suitable cut-off frequency and resonance value. This corresponds to the voltage controlled filter (VCF) section of a typical analogue synthesizer.

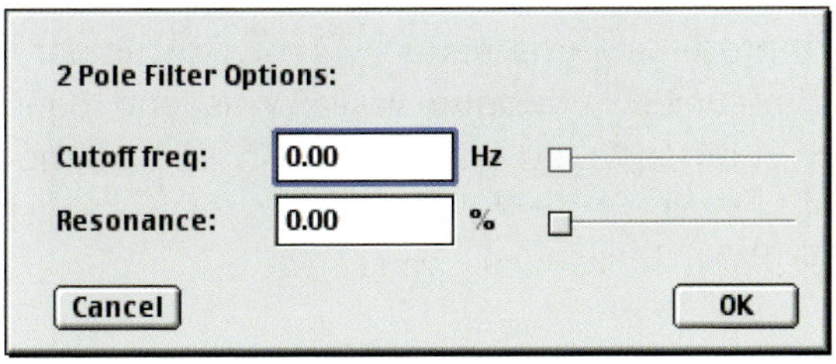

2 Pole Filter Options:

Cutoff freq: 0.00 Hz

Resonance: 0.00 %

Cancel OK

Figure 15.26 DUY SynthSpider two-pole Low Pass Filter module options.

You will usually want to apply an envelope generator to control the filter, so go ahead and insert one of these. SynthSpider's Envelope Generator module lets you set values for Attack, Decay, Sustain, Release and envelope level.

An analogue synthesizer would typically have a voltage controlled amplifier (VCA) and associated envelope generator. You can use a SynthSpider envelope generator combined with a simple signal multiplier, the SynthSpider Multiplication module, to achieve the same result.

To help you visualize the action of these envelopes, you can hook up an oscilloscope module to each one. The Scope modules display the waveform as an amplitude-against-time graph.

The next step would be to insert a MIDI Gate module to control the envelopes and a MIDI Note module to let you control the oscillator via MIDI. Once you have all your modules arranged on SynthSpider's working surface you can hook these up using the Patcher Tool. To use this you make sure you are in Edit mode, then point, click and drag at the centre of, say the MIDI Note module and drag to the Frequency control input marked 'F' on the top of the Oscillator module. When you let go, a connecting line will be in place.

To set the parameters of the envelope generators, double-click on the envelope generator module while you are in Edit mode and the options window will appear. If you want your synthesizer design to have particular fixed values for any of the available parameters, then set these here.

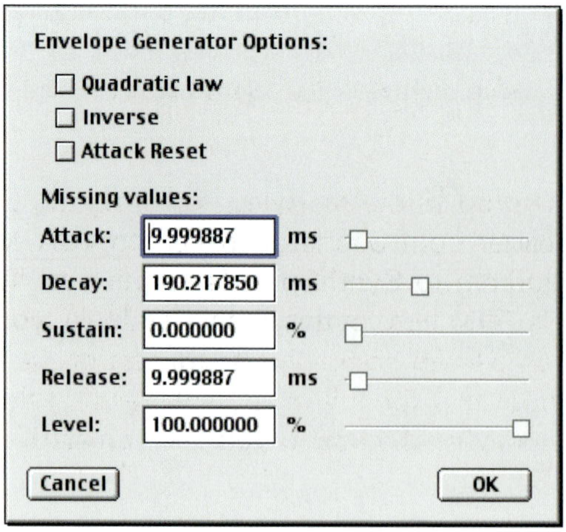

Envelope Generator Options:
Quadratic law
Inverse
Attack Reset
Missing values:
Attack: 9.999887 ms
Decay: 190.217850 ms
Sustain: 0.000000 %
Release: 9.999887 ms
Level: 100.000000 %
Cancel
OK

Figure 15.27 DUY SynthSpider Envelope Generator module options.

Alternatively, you may want to allow the user to vary these parameters using a control, for example to set the overall level of the envelope that will be applied to the filter. In this case, you can insert as Slider object onto your working surface and hook up the output of this to control any of the envelope parameters - in this case the level, marked 'L'. A quick way to do this is with the Patcher Tool in Edit mode is to hold the Command key, point click on and drag from the 'L' on the envelope generator to a free area on your working surface - to drop a new Slider control there, automatically hooked up to the 'L' input on the envelope generator.

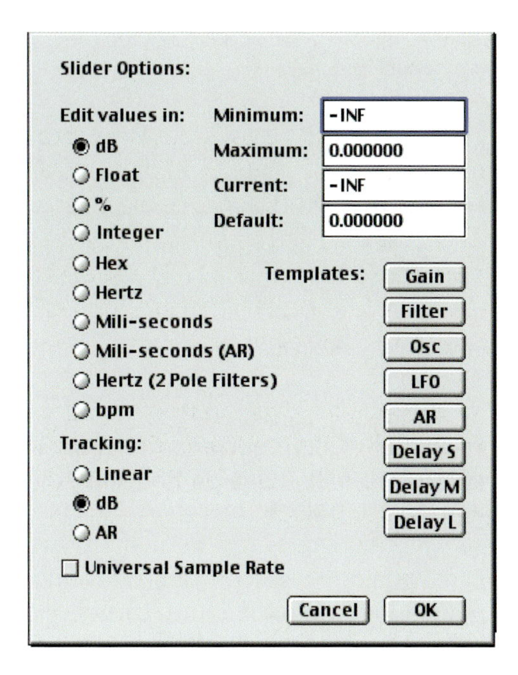

Figure 15.28 DUY SynthSpider Slider options.

When you have hooked up all the modules, following the typical signal flow for your analogue synthesizer from Oscillator to VCF to VCA, you need to connect the result of your algorithm to SynthSpider's output bus. This feeds the output signal to the Pro Tools TDM bus. In this simple example, you make a connection between the Multiplier used to combine the outputs of the envelope generators and, say, the left output bus for this mono synthesizer. Now you can hook up your MIDI keyboard and play the sound you have created.

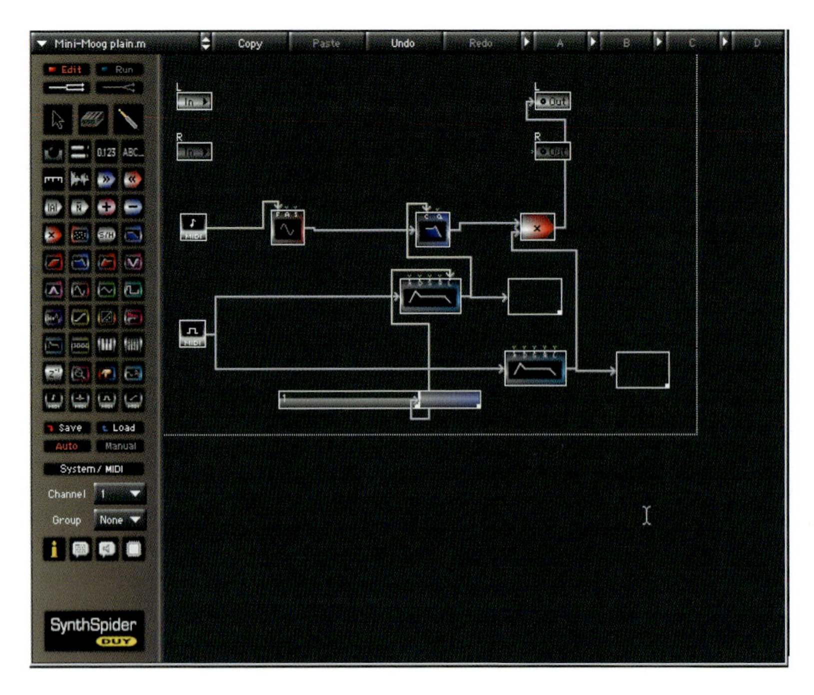

Figure 15.29 DUY SynthSpider - hooking up modules to create an analogue synthesizer.

So you can create a basic synthesizer using just 10 modules, as just explained. Of course, having 40 modules available provides an incredible range of flexibility here. There is insufficient space here to describe all these modules, but one more is particularly worth mentioning - the Sequencer module. This emulates the functions of the original analogue sequencers, providing 32 steps into which you can enter notes, silences and legatos. You can choose the note and the octave for each step and you can also define the velocity and gate values for each note. To make a note sound longer than one step, the gate sliders for the steps you want the sound to last for must be set at 100% and all the steps must be playing the same note. If you don't need 32 steps in your sequence, you can reduce the number of steps using the slider running along the bottom of the window.

You can run any sequencer module from its own internal timebase, or you can sync to a MIDI Beat Clock coming from some other software - such as Pro Tools. The internal base of each sequencer can also be divided independently of the clock source. Once you have set up your sequence, you can control SynthSpider modules internally by setting the MIDI output channel of the sequencer modules appropriately - using the Channel popup. It makes sense to avoid using any MIDI channels that you are using to control SynthSpider from external sources.

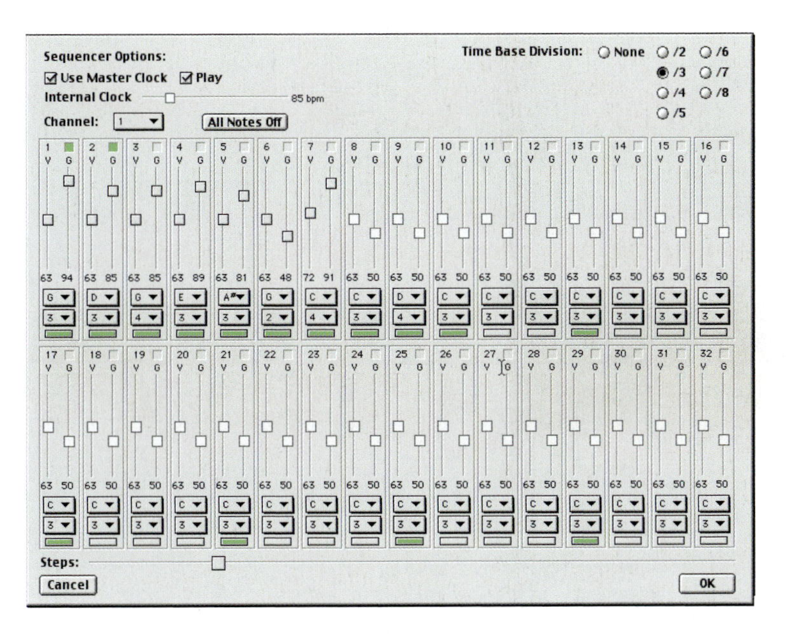

Figure 15.30 DUY SynthSpider Sequencer module.

The verdict? SynthSpider Advanced lets you put together customized synthesizer simulations with relative ease. It beats the heck out of plugging in endless patch-cords in a large modular synthesizer, and has the benefits of total recall and parameter automation that come with digital audio software. What you miss out on is the tactile control you get with real hardware, although you can use MIDI controller hardware to provide at least some of this. What you gain is the flexibility of configuration and use.

Chapter summary

Hopefully, this chapter will be of some assistance to you if you are trying to make a decision about which of these to buy. The DUY Spiders are probably the most user-friendly, while Reaktor had a steeper learning curve for me. Max/MSP is a bit more demanding technically, but has the advantage that you can develop plug-ins that will work immediately with VST- or MAS-compatible host software.

16 Key Issues

So – plug-ins and virtual instruments are happening! And it's not just 'techie nerds' who are using these things – even drummers and guitarists are having fun with these software goodies. Producer John Leckie has a theory to explain why: 'It's the way that lots of these plug-ins are animated – rather like computer games in some ways – that makes them even more appealing. This is what is helping anti-computer band members to accept these tools. It's the look of the things – especially the realistic simulations of the Hammond B3 or the Roland TR808 or the Fender amps – they just don't look like computer stuff any more. That's why the drummer in the band or the singer who hates computers can relate to these.' One of the biggest-selling plug-ins so far has been Antares Auto-Tune. Again, Leckie had interesting observations to make about this: 'You hear a lot of talk about Auto-Tune – with some people all for it while others are dead set against using it. The pitch correction thing has become much more accepted – with believers and non-believers. Some producers are now pitch correcting vocals as a matter of course. And the Cher thing has become a standard effect.' From a producer's point of view, of course, the voice is the most important element in the mix. 'What you do to the voice, the slightest tweak, affects the overall mix much more significantly than anything else you might do to the mix. Because we are human, that's what we hear first in a pop song. You recognize all the great singers because they are human beings. Every singer in the Top 20 sounds totally different,' says Leckie. Speculating about how plug-ins such as Audio Ease River Run and Celemony Melodyne might be used, Leckie wondered 'Could you turn Bowie into Rod Stewart using a bit of granular synthesis, or apply a Marc Bolan or a Paul McCartney vibrato preset? The vibrato and the way the notes are delivered and ended – the ability to change that would be like creating your own pop star. Often pop singers have very distinctive voices. There are surely some very interesting questions that would be asked if the producer were to mould the sound so directly! Now that there is so much power available to change all the vocal characteristics, will the singers appreciate and accept this?' So - interesting thoughts from a top producer to keep in mind!

Coming back down to earth, let's talk about costs and features, pros and cons and the like. Prices of software plug-ins are typically lower than those of their hardware equivalents. But are they as good and do they sound the same? Well, as far as software simulations of classic analogue hardware goes, I have only had a few brief opportunities to compare the software with the hardware side-by-side - but each time the software came up sounding very good compared with the hardware for me. Generally speaking, the software sounded cleaner, but sometimes lacked character compared with the hardware. In some cases the software sounds better - while in other cases, such as with the TC reverbs, the hardware sounds noticeably better. But this is because TC deliberately choose not to use their best algorithms in software that they know is likely to be 'cracked'! This is one reason why TC introduced their PowerCore cards, which produce superior results with TC's plug-ins compared with the Native or TDM versions - you can't 'crack' hardware like you can software!

Are plug-ins easier to use than their hardware equivalents? Often, the user interface is virtually identical to that of the hardware, although you have to control this using the computer's mouse and keyboard unless you have other hardware control surfaces connected to your computer. I find it easy enough to use the mouse and keyboard to control the plug-ins, but hardware controllers give you knobs and switches similar to those on hardware effects units. These are usually assignable controls whose functions change according to which plug-in they are controlling. So the hardware controls can be less intuitive to use than the controls on dedicated hardware effects units - which is why I actually prefer using the keyboard and mouse. A big advantage of the software interfaces is, of course, the way that you can view everything in a large window within the software environment you are using to record and edit your audio. And you always have the advantage of total automation and recall. With digital (as opposed to simulations of analogue) effects, the larger windows available on computer screens make it possible to display the parameters much more clearly than on the relatively tiny LCD displays found on most hardware units - a distinct advantage.

When it comes to the 'sound' of plug-ins, you do have to remember that any analogue circuitry, whether in analogue effects units or in the A/D or D/A converters and associated analogue amplification used in digital effects units, will always have its own (probably recognizable) audio 'signature'. Nevertheless, it is possible for software simulations to mimic the sound of these devices quite closely, and the various companies involved in developing these are very confident that this can be done successfully - if not with the first versions, then almost certainly in subsequent versions which they have had some time to refine. Interestingly, just about all of these companies, even those still making analogue equipment, agree that the future of signal-processing lies in software!

With virtual instruments, there are some disadvantages compared with the real hardware. Latency for one thing (although the latest systems are eradicating this); and virtual instruments do require a lot of processing power. But, again, the way that everything integrates into the software environment that you are using to record, edit and mix, with full automation and total recall, is powerfully seductive.

One thing is for sure - you will need the fastest, ideally dual, processors and plenty of RAM for best results, and if you try to do too many things at once on any computer you will have problems. This might be clicks and distortions in the audio from the virtual instruments, or a general slowing down of the various operations. So a dual 1.25 MHz G4 PowerMac with 1 Gb or more of RAM would definitely not be overkill.

If you have a Pro Tools TDM system you should strongly consider buying two or more additional processing cards if you want to work with lots of plug-ins and instruments. These are expensive cards, and you may need to spend even more money buying an expansion chassis to accommodate these - depending on what other PCI cards you need to use. For example, you may need a SCSI card, an additional card to run a second monitor, or a video capture card if you are working to picture. Two monitors are almost essential these days. Pro Tools works great with one monitor for the Edit window and a second for the Mix window. And with Logic or Cubase you could really do with three monitors to accommodate the vast number of windows that these programs require you to work with. The TC Electronic PowerCore and Universal Audio UAD-1 PCI cards both provide a useful selection of plug-ins that run on these cards, instead of using up your valuable CPU resources. However, until more plug-ins are available for these, they can only supply a partial solution. Also, when I asked why Waves have not introduced any PCI processing cards for their plug-ins I was told that, 'It's better to go with native processors. Or you can use all DSP as with Digidesign TDM systems - but hybrid systems are not a good idea.' So not everyone agrees that PCI processing cards are the best solution for plug-ins. Processor speeds are increasing with each new model of computer and the software developers can and do take advantage of these speed increases. For instance, Waves are continually re-coding and optimizing their software for the latest computer hardware: the latest Macs can run up to 17 instances of Renaissance Reverb and the new Pentium 4 can handle more than 20!

When thinking about the costs of plug-ins, it is also worth remembering that there are software upgrades to consider. And these are coming furiously fast these days. You can keep using your old plug-ins for, say, 12 months or so, but beyond that length of time you are likely to need to upgrade these - and the upgrades are likely to be chargeable.

At the time of writing, Pro Tools HD systems have recently been introduced and the Mac OSX 'Jaguar' version has just been released. But most people are still using the older Digidesign systems, and none of the MIDI + audio sequencers (including the Pro Tools software) works with Mac OSX as yet. One of the problems here is that the software developers have to write new code for OSX - not only for the sequencers but also for the vast range of plug-ins. Inevitably, some of the existing plug-ins are simply never going to make the transition. Antares MDT and JVP, for example, are never going to be HD-compatible. Auto-Tune and Mic Modeler will be - and Kantos is already HD-compatible as an RTAS plug-in. This more or less spells out the way it will be with other manufacturers as well. Some of the older plug-ins will not be ported to HD systems or to Mac OSX or to any other formats on Mac or PC for that matter - they have simply reached the end of their software 'lives', having been superseded by more recent designs or whatever. The more successful and more interesting ones will almost certainly make the transition, and in any case, any new plug-ins are likely to be developed in a wider range of formats, including Pro Tools HD and Mac OSX.

Do current computers have enough processing speed and RAM to do justice to the cutting-edge plug-ins? Only if you get the fastest machines and load them with extra RAM. The designs of the leading edge plug-ins are becoming much more demanding. For example, with the recently released Waves Native Masters bundle, the minimum specs are as follows: 'G4 Strongly recommended. G3 400 MHz minimum. 96 Mb RAM for basic operation, 256 Mb for multi-track work. 800/600 minimal display resolution. Masters TDM will work with MIX or HD core and farms only. No support for D24, Merle-DSP farm. Working at 96 kHz requires a Mac G4 867 MHz for a single instance of LinEQ or 500 MHz for LinMB.' The trends here are clear. The older Digidesign cards are no longer supported, you need lots of RAM and a recent computer running at the higher clock speeds, and if you want to work at 96 kHz sampling rate you will have to accept limitations at present.

What about OSX for the Mac? As I finished writing this chapter, an email arrived to say that Emagic had just released Logic Audio for OSX. Steinberg and Mark of the Unicorn were promising to release versions of Cubase and Digital Performer for OSX by the end of 2002, and Digidesign were simply saying that they were working on it. Meanwhile, Apple were announcing that new models of their computers would come with OSX pre-installed and available on a bootable System Installer CD-ROM. OS9.2 will still available for backwards compatibility, preinstalled on the hard drives and also available on the supplied Restore CD-ROM, but no longer available on a bootable System Installer CD-ROM. The message from Apple is clear - they want you to migrate to OSX and are pushing all their new customers in this direction. Yet no one, apart from Logic Audio users, can use OSX for their MIDI + Audio recording on the Mac at the time of writing (September 2002). It is not too surprising that Apple recently bought Emagic - they probably wanted to make sure that a leading MIDI + Audio sequencing package would be available for OSX to drive music and audio users in this direction. By the time you read this, it will be at least April 2003, and it will be interesting to see how many software companies have actually made their software available for OSX.

Other key issues are whether to go for a Mac or a PC, and whether to go for a desktop or a laptop model. I compared a Sony VAIO laptop with a PowerMac Titanium model. The clock speed of the Sony was 2 GHz and the clock speed of the Mac was 867 MHz. The Sony had a 16.1'' screen while the PowerMac had a 15.1'' screen. Apple have been publishing information saying that Photoshop filters running on a 1 GHz Mac run comparably fast to Photoshop filters on a 2 GHz PC. That is because Photoshop takes advantage of the G4 Altivec Engine to speed up this type of processing. I compared Waves and TC|Works plug-ins running on both machines. I soon discovered that I could work with more than twice as many instances of these on the Sony VAIO than on the Mac. I also found that the 16.1'' screen was large enough to work with very comfortably. I have to say that I far prefer the Mac's operating system to any variety of Windows, and I very much like the look of the latest PowerMac laptops with their 17.1'' screens but if I really needed to run lots of plug-ins, I think I might be very tempted by the Sony VAIO! I should also mention that the Sony VAIO costs more than the equivalent PowerMac - about £3000 as opposed to about £2500 - so it is not a cheaper option to buy, but it would save money in the long run by offering more productivity and speed of operation.

Another matter worth mentioning is that I am hearing that lots of people are busy converting their entire sample libraries from whichever hardware format they have been using, such as Akai or Emu, into a software sampler format, such as the EXS24 format for Logic Audio. This is because of the sheer convenience of having all the samples available from within the sequencing environment. People are even converting GigaSampler libraries into EXS24 format so they can be used directly in Logic and on the Mac rather than on the PC. The hardware samplers are mostly sitting there unused, until, perhaps, people need the extra outputs. And even this is getting sorted with the latest software samplers. Changing times, changing times!

Pick of the plugs

With such a vast range of plug-ins and virtual instruments to choose from, which should you go for? Obviously this has to be a personal choice based on features, price, availability - and needs. I will share my personal choices with you here, to help you with your decision-making.

For my TDM system, I would add Sound Replacer and Reverb One, Focusrite D2 and D3, Line 6 Amp Farm and Echo Farm, LexiVerb and Drawmer Dynamics. I would anticipate using most of these on most sessions for basic processing and effects. Then I would add Soft SampleCell. This is a very useful sample-replay plug-in to have available as it uses the same file format for its audio files as Pro Tools and integrates neatly enough into any TDM system.

You are spoilt for choice when it comes to third-party TDM plug-ins, but my favourites would have to be the Sony Oxford EQs, Serato's Pitch 'n Time, Antares Auto-Tune and Microphone Modeler, Metric Halo's SpectraFoo for metering, and GRM Tools for effects. The Waves Renaissance and Pro-FX bundles would be my choice for music production - but you can't really go wrong with any of the Waves plug-ins; they are all first rate. If I had to pick just one VST plug-in, being a guitarist, I would go for the excellent Hughes & Kettner Warp amplifier and speaker simulation. And if I were using Digital Performer or Pro Tools I would use Pluggo or VST Wrapper, or maybe the Spark FX Machine to plug this into my host software. Talking about Digital Performer, I would buy this software just to have access to the amazing Audio Ease plug-ins. Altiverb is one of the best concert hall reverbs I have come across, and the Nautilus and Rocket Science Bundles contain some of the most interesting and creative effects available at the present time.

Now as far as virtual instruments are concerned, the Waldorf PPG Wave 2.V and Native Instruments Pro-52, B4, FM7, Absynth and Battery would be my choices. For an easy to configure and fast to use sampler, I would go for SampleTank, but for more serious sample playback and synthesis, I would go for Unity Session. If I was working on 'dance' tracks (which I am not at the moment) I would definitely be working with Reason, ReCycle and ReBirth - the powerful team from Propellerheads. I would probably want to have Alkali and Melodyne around as well - especially for working on remixes. And as a keen guitar player, I could not live without AmpliTube!

Now when it comes to choosing a processing card to run plug-ins, I find this a tough choice. I like the idea of having the Pultec and LA2A compressors but I also like the idea of running some of the TC plug-ins on hardware, and being able to run various third-party plug-ins on this card, such as the Sony Oxford EQ. Or maybe I would settle for a dual 1.25 MHz PowerMac instead and trust that this could run plenty of native plug-ins without too much trouble.

As far as 'rolling your own' is concerned, I like the idea of being able to do this more than the actuality. I reckon that in practice I would buy DSPider and SynthSpider and mostly use the presets. I know – that sounds kind of sad, doesn't it! The thing is, life is just too short for me – I have another book to write and more music to record. But don't let me discourage you in any way! There's a world of plug-ins out there – go get some!

Appendix

This section lists all the websites for the manufacturers mentioned in the book.

Antares

ww.antares.com

Kantos
Auto-Tune
Mic Modeler
MDT
JVP
Infinity

Apogee

www.apogee.com

MasterTools

Arboretum

www.arboretum.com

HyperPrism
Harmony
Raygun
Ionizer
Realizer Pro

Audio Ease

www.audioease.com

Altiverb
Nautilus Bundle
Rocket Science Bundle
VST Wrapper

Audio Genetics

www.audio-genetics.com

Alkali

BIAS

www.bias.com

BIAS Peak
BIAS Deck

BitHeadz

www.bitheadz.com

BitHeadz Unity Session
BitHeadz Unity Discrete Drums
BitHeadz Retro AS-1
BitHeadz Unity DS-1
BitHeadz Phrazer

Celemony

www.celemony.com

Melodyne

Cool Stuff Labs

www.coolstufflabs.com

Generator X

Cycling '74

www.cycling74.com

Essential Instruments Pluggo Collection
Max/MSP

Digidesign

www.digidesign.com

Reverb One
Bruno/Reso
Maxim
Sound Replacer
DPP-1
DINR
D-Verb
D-Fi
D-Fx
Focusrite D2
Focusrite D3
Drawmer Dynamics
Line 6 Amp Farm
Line 6 Echo Farm
Aphex Aural Exciter
Big Bottom
Lexicon LexiVerb
Orange Vocoder

Soft SampleCell
Access Virus

Dsound

www.dsound1.com

Dsound V12 Multichannel Valve Interface Plug-in for PowerCore

DUY

www.duy.com

MAX
Shape
Wide
DaD Valve
DaD Tape
Z-Room
ReDSPider
DSPider
SynthSpider

Emagic

www.emagic.de

Emagic EVP88
Emagic EXS-24
ESB TDM
Sound Diver
Logic Audio

IK Multimedia

www.ikmultimedia.com
ikm@ikmultimedia.com
www.sampletank.com

T-Racks
SampleTank
Amplitube

INA-GRM

www.ina-grm.com

GRM Tools

Koblo

www.koblo.com
koblo@koblo.com

Koblo 9000

MacDSP

www.macdsp.com

Compressor Bank
Filter Bank
MC2000

Metric Halo Labs

www.metrichalo.com

SpectraFoo
Channel Strip

MOTU

www.motu.com

Digital Performer

Native Instruments

www.native.com
www.native-instruments.net

Native Instruments Reaktor
Native Instruments Dynamo
Native Instruments Pro 52
Native Instruments B-4
Native Instruments FM7
Native Instruments Battery
Native Instruments Absynth
Native Instruments Spektral Delay

TC|Works Modular
TC|Works Mercury 1

TC|Works Spark

TC PowerCore DSP card & Plug-ins

Universal Audio

www.uaudio.com

Universal Audio Powered Plug/ins/UAD-1 DSP card

Waldorf

www.waldorf-music.de
www.waldorf.com
Waldorf Wave PPG 2.v
www.waldorf-gmbh.de.com

Waldorf Wave PPG 2.v
Waldorf Attack

Wave Mechanics

www.wavemechanics.com

UltraTools
Speed

Waves

www.waves.com

Waves Bundle

C1 compressor/Gate
L1 UltraMaximizer
Q10 ParaGraphic Equalizer
S1 Stereo Imager
PS22 StereoMaker

PAZ Analyser
True Verb
DeEsser
MaxxBass
C4 Multiband Parametric Processor

Waves Renaissance Suite

Renaissance DeEsser
Renaissance Vox
Renaissance Bass
Renaissance Compressor
Renaissance Equalizer
Renaissance Reverberator

Waves Pro-FX

SuperTap
MondoMod
MetaFlanger
Doppler
Enigma
UltraPitch

Waves Restoration bundle

X-Noise
X-Click
X-Crackle
X-Hum

Waves Native Masters

Linear Phase EQ
Linear Phase Multiband
L2 UltraMaximizer

Index

www.focalpress.com

Join Focal Press on-line

As a member you will enjoy the following benefits:

- an email bulletin with **information on new books**
- a regular **Focal Press Newsletter**:
    - featuring a selection of new titles
    - keeps you informed of **special offers, discounts and freebies**
    - alerts you to **Focal Press news and events** such as author signings and seminars
- complete access to **free content** and reference material on the focalpress site, such as the focalXtra articles and commentary from our authors
- a **Sneak Preview** of selected titles (sample chapters) *before* they publish
- a chance to have your say on our **discussion boards** and **review books** for other Focal readers

Focal Club Members are invited to give us feedback on our products and services.
Email: worldmarketing@focalpress.com – we want to hear your views!

Membership is **FREE**. To join, visit our website and register. If you require any further information regarding the on-line club please contact:

Lucy Lomas-Walker
Email: l.lomas@elsevier.com
Tel: +44 (0) 1865 314438
Fax: +44 (0)1865 314572
Address: Focal Press, Linacre House,
Jordan Hill, Oxford, UK, OX2 8DP

Catalogue

For information on all Focal Press titles, our full catalogue is available online at www.focalpress.com and all titles can be purchased here via secure online ordering, or contact us for a free printed version:

USA
Email: christine.degon@bhusa.com
Tel: +1 781 904 2607 T

Europe and rest of world
Email: j.blackford@elsevier.com
Tel: +44 (0)1865 314220

Potential authors

If you have an idea for a book, please get in touch:

USA
editors@focalpress.com

Europe and rest of world
focal.press@elsevier.com